Dimensional Scaling in Chemical Physics

edited by

DUDLEY R. HERSCHBACH

Department of Chemistry,
Harvard University,
Cambridge, MA, U.S.A.

JOHN AVERY

H.C. Ørsted Institute,
University of Copenhagen, Denmark

and

OSVALDO GOSCINSKI

Department of Quantum Chemistry,
Uppsala University, Sweden

KLUWER ACADEMIC PUBLISHERS
DORDRECHT / BOSTON / LONDON

Library of Congress Cataloging-in-Publication Data

```
Dimensional scaling in chemical physics / edited by Dudley R.
   Herschbach, John Avery, and Osvaldo Goscinski.
        p.   cm.
     "August 27, 1992."
     Includes index.
     ISBN 0-7923-2036-0 (alk. paper)
     1. Quantum chemistry.  2. Scaling (Statistical physics)
   I. Herschbach, Dudley R.  II. Avery, John.  III. Goscinski, Osvaldo.
   QD462.6.S25D56  1993
   541.2'8--dc20                                        92-35070
```

ISBN 0-7923-2036-0 (HB)
ISBN 0-7923-2072-7 (PB)

Published by Kluwer Academic Publishers,
P.O. Box 17, 3300 AA Dordrecht, The Netherlands.

Kluwer Academic Publishers incorporates
the publishing programmes of
D. Reidel, Martinus Nijhoff, Dr W. Junk and MTP Press.

Sold and distributed in the U.S.A. and Canada
by Kluwer Academic Publishers,
101 Philip Drive, Norwell, MA 02061, U.S.A.

In all other countries, sold and distributed
by Kluwer Academic Publishers Group,
P.O. Box 322, 3300 AH Dordrecht, The Netherlands.

Printed on acid-free paper

DIMENSIONAL SCALING IN CHEMICAL PHYSICS

Contents

Preface

Dynamical correlations among strongly interacting particles pose formidable, recalcitrant problems for quantitative theory in atomic and molecular physics. The nonseparable, many-body character of such problems makes conventional methods of quantum mechanics extremely arduous or intractable. Over the past 50 years, great efforts have been devoted to electronic structure calculations, but even with the mighty computing power now available or in prospect, the difficulty of accurately evaluating many-body correlations remains daunting.

Dimensional scaling offers a new approach to quantum dynamical correlations. Although in statistical mechanics dimension has long served as a key diagnostic parameter, in quantum theory dimension was until recently considered only a crude tool of very limited scope. However, as applied to prototype electronic structure problems during the past decade, rather simple dimensional interpolation and extrapolation techniques have yielded remarkably good results for many-body effects. A crucial point has emerged: With dimensional scaling, the magnitude and number of strong dynamical interactions becomes far less troublesome than with conventional methods; it is the dimension dependence that matters. Many-body effects, when expressed in units of one- or few-body problems, tend to have only mild dependence on dimensionality. These exploratory studies have mapped out basic computational strategies and heuristic perspectives that now seem ripe for wider application.

This is the first book dealing with dimensional scaling methods in the quantum theory of atoms and molecules. Appropriately, it is a multiauthor production, derived chiefly from papers presented at a workshop held in June 1991 at the Ørsted Institute in Copenhagen. This week-long meeting created lively interactions and vigorous correlations among 20 exceptionally dynamic chemists and physicists. Although focused on dimensional scaling, the workshop included also several discussions of other unorthodox methods for treating nonseparable dynamical problems and electronic correlation.

In shaping the book, we have aimed to serve three needs: an introductory tutorial for this still fledgling field; a guide to the literature; and an inventory of current research results and prospects.

4

Part I treats basic aspects of dimensional scaling. Addressed to readers entirely unfamiliar with the subject, it provides both a qualitative overview, in a chapter with few formulae, and a tour of elementary quantum mechanics in D-dimensions, in a chapter treating simple examples in explicit detail. Part II surveys the research frontier. The eight chapters exemplify current techniques and outline results, serving as a bridge to otherwise less accessible and more technical journal articles. Major topics include atomic structure in the large-D and low-D limits; prototype hyperspherical and hypercylindrical systems; general computational strategies; two-electron excited states; correlation energy of metalic hydrogen; and dimensional interpolation of virial coefficients. Part III presents other methods. Two chapters discuss nonseparable dynamics, especially as treated by the hyperspherical method, and electron correlation in pseudomolecular excited states of atoms. Although procrustean conformity was not imposed, unifying and complementary themes are emphasized throughout the book.

With great pleasure we renew earnest thanks to our evangelical authors; to the Ørsted Institute for multidimensional hospitality, including particularly hyperdelicious dinners prepared by Chef Alex Nielsen; and to the Swedish and Danish Research Councils and the Carlsberg Foundation for grants which made the Copenhagen workshop possible.

The Editors

Part I
BASIC ASPECTS

Chapter 1

INTRODUCTION

Dudley R. Herschbach
Department of Chemistry
Harvard University
12 Oxford Street
Cambridge, MA 02138, USA

Abstract

Dimensional scaling methods are illustrated for some simple proto-type problems that exemplify characteristic features. For two-electron atoms, a chief testing ground, accurate electronic energies have been obtained by combining results from two very simple, exactly solvable limits in which the spatial dimension $D \to \infty$ or $D \to 1$. Between these limits, the scaled correlation energy is nearly a linear function of $1/D$, and interpolation yields results for $D = 3$ comparable to the best conventional configuration interaction calculations. Appealing heuristic concepts have also emerged. Among these are semiclassical properties at large-D akin to the prequantum valence models of Lewis and Langmuir and to the resonance theme of Pauling. An annotated guide to the D-scaling literature provides background for the current research presented in this book and for anticipated developments.

D. R. Herschbach et al. (eds.), Dimensional Scaling in Chemical Physics, 7–59.

The Basic Approach

This chapter has a threefold mission: to outline dimensional scaling concepts and methods, in qualitative fashion; to survey exploratory applications thus far made to archetype problems of atomic and molecular physics, especially electronic structure; and to highlight some opportunities and challenges. Curious visitors or prospective recruits to this young field will find roots and branches that extend to many other fields, some far distant from chemical physics. At present, however, the directly pertinent literature consists of only about 80 papers. These are listed here in an annotated bibliography [1-84] which also supplies some historical perspective. A complementary tour of elementary quantum mechanics in D-dimensions is provided in the next chapter.

Taking the dimension of space as a variable has become a customary expedient in statistical mechanics, in field theory, and in quantum optics [12,17,18,85-87]. Typically a problem is solved analytically for some "unphysical" dimension $D \neq 3$ where the physics becomes much simpler, and perturbation theory is employed to obtain an approximate result for $D = 3$. Most often the analytic solution is obtained in the $D \to \infty$ limit, and $1/D$ is used as the perturbation parameter. In quantum mechanics, this method has been extensively applied to problems with one degree of freedom, as reviewed by Chatterjee [60], but such problems are readily treated by other methods. Much more recalcitrant are problems involving two or more nonseparable, strongly- coupled degrees of freedom, the chief focus of the methods presented in this book.

My own pursuit of dimensional scaling stratagems was prompted by Witten's excellent tutorial article [12] in *Physics Today* (1980) about quarks, gluons, and "impossible problems" of quantum chromodynamics. Witten illustrated the utility of the large-D limit with a rough calculation for helium. His result for the ground state energy at $D = 3$ was off by 40%, pitiful compared with the 5% accuracy of conventional first-order perturbation theory or the 1.5% accuracy of the Hartree-Fock approximation. However, since I was teaching a quantum mechanics course and on the lookout for provocative problems, I tried setting up the helium example as a homework exercise. By merely recasting the large-D limit to factor out the hydrogenic portion, I found that a very simple calculation gave 1% accuracy. This encouraged me to try to make use of another unphysical limit, $D \to 1$, which had a known solution [5,7]. In order to interpolate, I assumed a geometric series in powers of $1/D$, fixing the parameters by means of the simple, exactly calculable $D \to \infty$ and $D \to 1$ limits. Setting $D = 3$ in the resulting series gave the correct energy within 0.002% [27]. The question whether this was a portent or a fluke provoked much further work, in collaboration with enterprising students and colleagues, which led us to other intriguing surprises.

The various dimensional scaling methods now available typically involve four steps:

(1) Generalize the problem to D-dimensions.

(2) Transform to a suitably scaled space to remove the major, generic D-dependence of the quantity to be determined.

(3) Evaluate the scaled quantity at one or more special D-values, such as $D \to \infty$, where the computation is relatively "easy."

(4) Obtain an approximation for $D = 3$ by relating it to the special D-values, usually by some interpolation or extrapolation procedure.

To illustrate these procedures in the simplest way, we examine two elementary examples, the *random walk* and the *hydrogenic atom*, before surveying other applications. In all cases we consider, the $D \to \infty$ limit will serve as the starting point. For electronic structure this limit proves to be beguilingly simple and exactly calculable for any atom or molecule. Nonseparable, many-body effects such as electronic correlation, so vexing for conventional methods, are fully included in the large-D limit. This is a great advantage, although the dimension dependence of these effects must be established in order to derive accurate results for $D = 3$. How well this can be accomplished for many-body systems remains an open question. The exemplary results we shall review for few-body systems suggest promising computational approaches and also bring out heuristic perspectives quite different from the customary orbital descriptions.

The Random Walk

Much use is made of random walks on periodic lattices to model many phenomena in polymer science and other domains. Here we consider in heuristic fashion just the large-D limit for a unconstained walk of N-steps on a hypercubic lattice [45]. The end-vector distribution function for such walks satisfies a diffusion equation closely related to the Schrödinger equation for a free particle [67]. For $D = 3$, to evaluate the distribution function or its moments requires large-scale computations, since for walks with thousands of steps the number of possible configurations becomes enormous. In contrast, the $D \to \infty$ limit is delightfully trivial. Suppose the first step is in the direction of a particular Cartesian axis; no later steps will take that direction since there are infinitely many other axis directions to choose among. In this limit, the walk thus is unique; it consists of one step in each direction. The coordinates of successive positions are simply:

$$(0, 0, 0, \ldots); (1, 0, 0, \ldots); (1, 1, 0, \ldots); \ldots$$

and all properties of the walk in the $D \to \infty$ limit can be readily evaluated in analytic form. The leading correction term, proportional to $1/D$, can also be computed explicitly, by taking D very large but finite so zthat there are no more than two steps in each direction. For example, in a walk of N steps the square of the radius of gyration about the nth principal axis direction, given by the mean square distance of the walker from that axis, is

$$< R_n^2 > = \frac{N}{\pi^2 n^2} \left(1 + \frac{3}{4D} + \ldots \right), \tag{1}$$

where $n = 1, 2, 3, \ldots D$ and higher terms are of order $1/D^2$ [45]. The n^{-2} dependence (curiously, like that for the hydrogen atom energy levels!) predicts that the mean squares of the radii about the first three principal axes are in the ratio $< R_1^2 >:< R_2^2 >:< R_3^2 > = 1 : 1/4 : 1/9$. The distribution functions for the individual squared radii can also be evaluated [45]; aside from normalization, these have the form

$$P_N(R_n^2) \propto x^{\frac{D}{2}-1} exp \left(-\frac{1}{2} D x \right) \tag{2}$$

where $x \equiv R_n^2 / < R_n^2 >_\infty$ and the subscript ∞ indicates the leading term of Eq.(1). The peak occurs at

$$x_{max} = 1 - (2/D) \tag{3}$$

and the standard deviation is given by

$$\Delta x \equiv [< x^2 > - < x >^2]^{\frac{1}{2}} = \left(\frac{2}{D}\right)^{\frac{1}{2}}. \tag{4}$$

Thus, for large-D, the width shrinks as $D^{-\frac{1}{2}}$ and the distribution ultimately becomes a delta-function centered at $x = 1$.

These simple limiting results compare well with those for $D = 3$ found from computer simulations. The ratios of the mean squared principal radii are 1: 0.228 : 0.0847 and the form of the distribution functions closely resembles Eq.(2), with the peak position a few percent higher and the width a few percent narrower.

The "shape" of the random walk [45] is characterized by an average asphericity parameter, given in the large-D limit by

$$< A_D >= \frac{2}{5} - \frac{12}{175D} + \ldots \tag{5}$$

The computer simulations yield $< A_3 >= 0.39 \pm 0.004$, a value only about 3% larger than the estimate of 0.377 obtained from the two leading terms of Eq.(5). Such accuracy is adequate for many applications, and could be improved by systematic evaluation of higher order contributions.

The large-D limit even yields exact results [45,68] for some properties involving the sum of squares of the principal radii of gyration, $R^2 = R_1^2 + R_2^2 + \ldots + R_D^2$. This provides a measure of the "size" or end-to-end extent of the random walk. For long, unrestricted open chain walks, both the mean of R^2 and its distribution $P(R^2)$ can be evaluated analytically. The approximate large-D results of Eqs. (1) and (2) actually agree exactly with the analytic results; in particular, the mean $< R^2 >= N/6$, independent of the dimensionality. The size of these simple walks hence varies as $N^{\frac{1}{2}}$ for any D. For other types such as self-avoiding walks [45], the critical exponent can differ markedly from 1/2 for low D and depend strongly on dimensionality.

The Hydrogenic Atom

The four typical steps of dimensional scaling are readily carried out for a hydrogenic atom, but here we describe these qualitatively, postponing to Chapter 2 any derivations. A key feature is that the D-dimensional case can be cast into the same form as that for $D = 3$, so the hydrogenic atom can be solved exactly in any dimension.

Step (1): Generalize to D-dimensions. The Schrödinger equation may be generalized to D-dimensions in various ways. The procedure used throughout this book is simply to endow all vectors with D Cartesian components, as illustrated in Fig. 1 for spherical polar coordinates. Thereby the Laplacian operator in the kinetic energy ∇_D^2 and the Jacobian volume element J_D are modified, but the potential energy retains the same form as for $D = 3$. Table 1 exhibits simplifying reductions. For any central force potential, V(r), dependent only on the radial distance between a pair of particles, the squared angular momentum operator L_{D-1}^2 is a constant of the motion; its eigenfunctions Y_{D-1} are hyperspherical harmonics [52], discussed by John Avery in Chapter 5. The angular dependence hence can be separated just as for $D = 3$. Furthermore, the D-dependence can be removed from the Laplacian and the Jacobian by setting $\Psi = J_D^{-\frac{1}{2}}\Phi$ and solving, not for the wavefunction Ψ, but for Φ, the square root of the probability distribution function, $|\Phi|^2 = J_D|\Psi|^2$. This transformation gives rise to a centrifugal potential that contains all the explicit D-dependence. The radial wave equation for the Jacobian-weighted probability amplitude Φ is identical to the familiar equation for $D = 3$ except that the orbital angular momentum quantum number l in the centrifugal potential is replaced by $\Lambda = l + \Delta$, with the sole dimension dependence contained in $\Delta = \frac{1}{2}(D - 3)$.

Accordingly, for a hydrogenic atom in D-dimensions the energy levels are given by

$$E_{n,D} = -\frac{1}{2}\left[\frac{Z}{(n+\Delta)}\right]^2, \tag{6}$$

in hartree atomic units, with Z the nuclear charge, $n \equiv p + l + 1$ the familiar principal quantum number ($n = 1, 2, 3,...$) and p the number of radial nodes ($p = 0, 1, 2, ...$). Since the dimension dependence enters via $l + \Delta$, for any central force problem an isomorphism exists between angular momentum and dimensionality such that each half-unit increment in l is equivalent to a unit increment in D, as perhaps first noted by Van Vleck [6]. For instance, among hydrogen atom states with no radial nodes ($p = 0$), the $1s$ ground states ($l = 0$) for $D = 5, 7, 9,...$, respectively, have the same values for $\Lambda = l + \Delta$ and therefore the same energies and radial wavefunctions as the excited $2p, 3d, 4f, \ldots$ states ($l = 1, 2, 3,...$) of the $D = 3$ atom. This *interdimensional degeneracy* permits the ladder of excited states for any given number p of radial nodes and given D to be generated from the lowest state with that p by the transcription $D \rightarrow D + 2l$.

Step (2): Transform to D-scaled Space. The motivation for this transformation is evident from the form of the hydrogenic ground-state energy,

$$E_D = -\frac{1}{2}\left(\frac{Z}{\kappa}\right)^2 \tag{7}$$

and the corresponding radial probability distribution,

$$|\Phi_D|^2 = \left(\frac{Zr}{\kappa}\right)^{2\kappa} exp(-2Zr/\kappa) \tag{8}$$

(unnormalized) where $\kappa = \frac{1}{2}(D - 1)$. The energy vanishes as $D \rightarrow \infty$ and becomes singular as $D \rightarrow 1$. This strong variation with D can be eliminated by introducing suitable D-dependent units for the distance and energy scales [7,27], chosen to reduce to unity at $D = 3$. As indicated in Table 1, an appropriate scaling for large-D inflates the radial distance by a factor proportional to κ^2/Z, which is the radius of the maximum in the probability distribution of Eq.(8). This scaling, quadratic in D, offsets the quadratic D-dependence of the centrifugal energy. For low-D, an analysis [7,25,39] reviewed by David Goodson and Mario López-Cabrera in Chapter 4 shows that the appropriate scaling inflates distance by a factor proportional to κ/Z. Both these distance scalings correspond to compressing the energy scale by Z^2/κ^2, so the scaled hydrogenic ground-state energy becomes -1/2, independent of D.

If we use the large-D scaling for r and normalize the radial probability distribution at its maximum, located at $r = 1$ in the scaled

Table 1. Central force problem in D-dimensions.

Quantity	*Equation*
Schrödinger equation:	$[-\frac{1}{2}\nabla_D^2 + V(r)]\Psi_D = E_D\Psi_D$
Laplacian:	$\nabla_D^2 = \frac{1}{r^{D-1}}\frac{\partial}{\partial r}(r^{D-1}\frac{\partial}{\partial r}) - \frac{L_{D-1}^2}{r^2}$
Squared angular momentum :	$L_{D-1}^2 Y_{D-1} = \ell(\ell + D - 2)Y_{D-1}$
Jacobian factor:	$J_D = r^{D-1}(\sin\theta)^{D-2}$
Probability density:	$\Phi_D = J_D^{\frac{1}{2}}\Psi_D$
Radial equation:	$[-\frac{1}{2}\frac{d^2}{dr^2} + \frac{\Lambda(\Lambda+1)}{2r^2} + V(r)]\Phi_D = E_D\Phi_D$
Centrifugal potential:	$\frac{\Lambda(\Lambda+1)}{2r^2}$, with $\Lambda = \ell + \frac{1}{2}(D-3)$
Scaled equation, via $r \to \kappa^2 r$:	$[-\frac{1}{2\kappa^2}\frac{d^2}{dr^2} + W(r)]\Phi_D = \kappa^2 E_D\Phi_D$
Effective potential:	$W_D = \frac{f_D}{2r^2} + V(r)$
	$f_D = \frac{\Lambda(\Lambda+1)}{\kappa^2} \to 1$ for $\kappa \to \infty$

units, the distribution becomes

$$|\Phi_D|^2 = r^{D-1} exp[-(D-1)(r-1)] \qquad (9)$$

Figure 2 shows how the form of this distribution changes as the dimension varies between 2 and 100. The scaled standard deviation is given by

$$\Delta r \equiv [<r^2> - <r>^2]^{\frac{1}{2}} = \frac{D^{\frac{1}{2}}}{D-1}, \qquad (10)$$

so for large-D the distribution shrinks in width as $D^{-\frac{1}{2}}$ and ultimately becomes a delta-function centered at $r = 1$. This sharply peaked density for $D \rightarrow \infty$ corresponds to a *pseudoclassical* limit in which the electron is located at a fixed radial distance. In contrast, for $D \rightarrow 1$ the density becomes extremely broad, so this may be regarded as a *hyperquantum* limit.

If in analogous fashion we introduce a scaled size parameter, $r \equiv x/x_{max}$, for the random walk problem and normalize the squared radii distribution $P_N(R_n^2)$ of Eq.(2) at its maximum, the result takes the same form as Eq.(9) for the hydrogenic radial probability distribution, except that D is replaced by $\frac{D}{2}$. With this change, Figure 2 applies as well to the random walk problem. This illustrates the generic character of dimensional scaling.

Step (3): Evaluate Dimensional Limits. When dimension-scaled units are introduced, the scaled equations for the pseudoclassical large-D and hyperquantum low-D domains take markedly different forms. As seen in Table 1, in the large-D regime the radial kinetic energy is quenched because the electron acquires an effective mass proportional to $\kappa^2 \sim D^2$. However, the centrifugal kinetic energy survives because this effective mass factor cancels against the quadratic D-dependence introduced in Step (1) into the orbital angular momentum. Thus the effective potential $W(r)$, the sum of the scaled centrifugal and Coulombic terms, has the dominant role. In the pseudoclassical $D \rightarrow \infty$ limit, this potential is simply

$$W(r) = \frac{1}{2r^2} - \frac{1}{r}, \qquad (11)$$

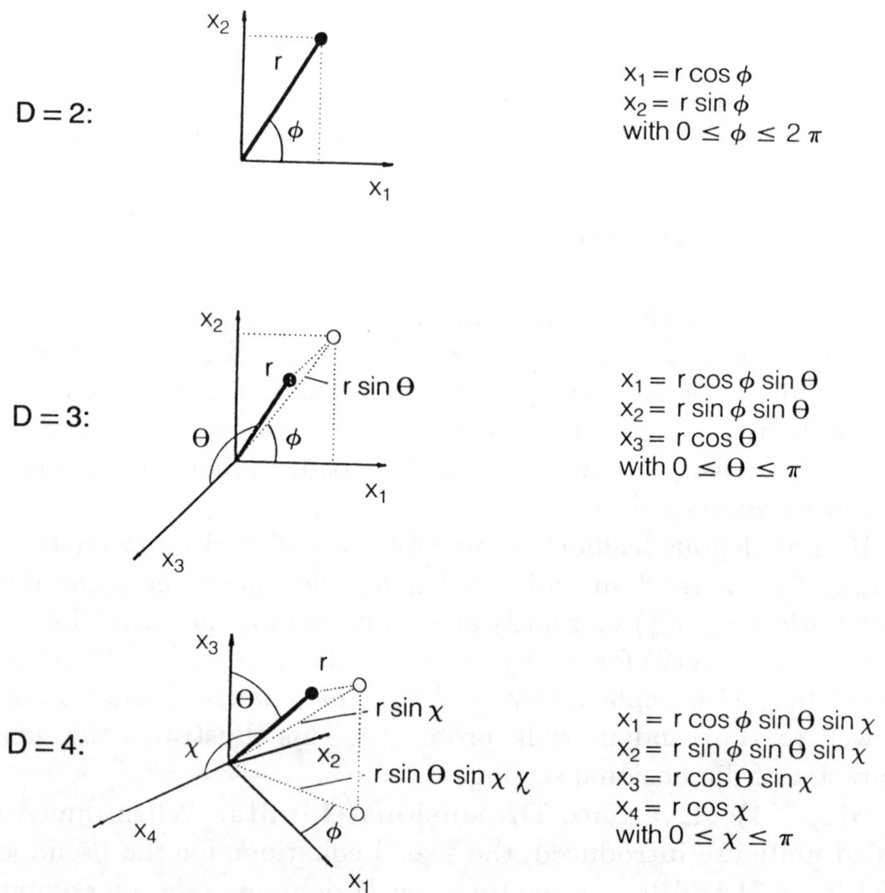

Figure 1. Transformations defining D-dimensional hyperspherical polar coordinates in terms of Cartesian coordinates, illustrated for $D = 2, 3, 4, \ldots$ On going from the D to $D+1$ case, a further Cartesian axis x_{D+1} is added; the radius vector $\mathbf{r}$ is then projected on this axis via the cosine of the new polar angle θ_{D+1} and projected on the D-dimensional subspace via the sine of that angle.

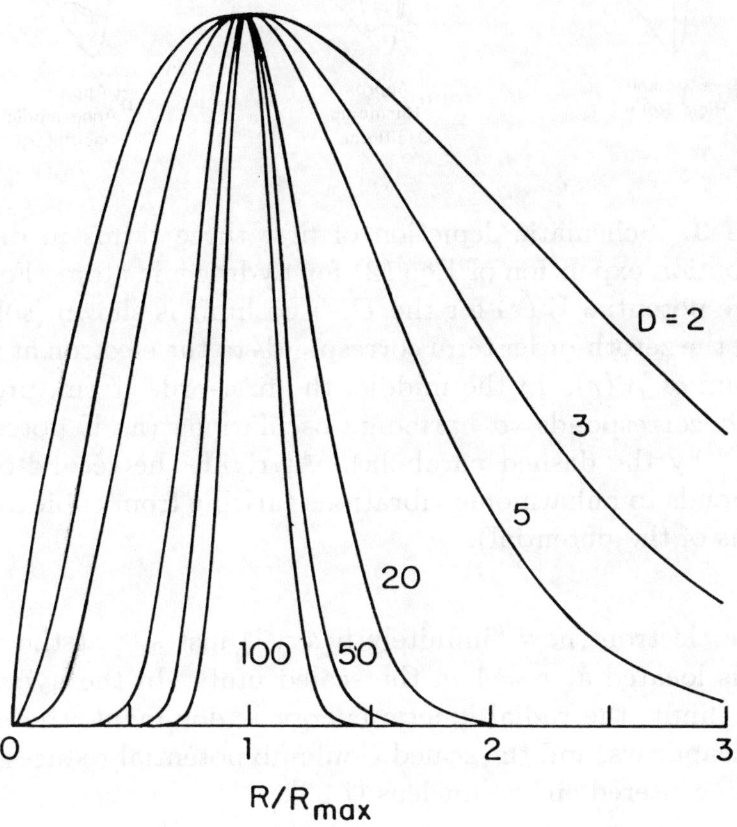

Figure 2. Radial probability distribution for a ground-state hydrogenic atom in D-dimensional space. The curves (labeled with values of D) are normalized at the maximum, which occurs at $r_m = \frac{\kappa^2}{Z}$, in units of bohr radii, with $\kappa = \frac{(D-1)}{2}$.

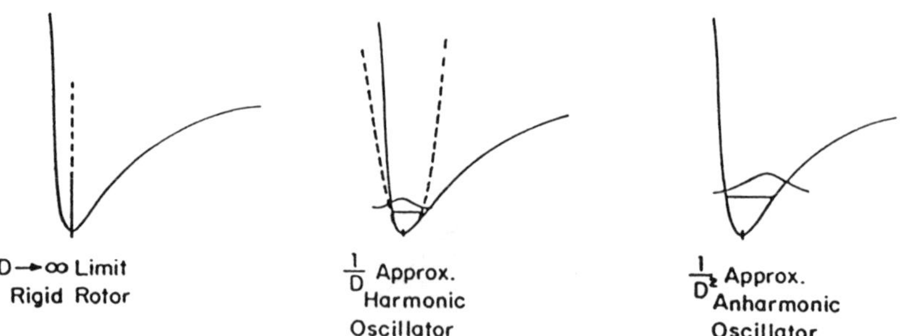

Figure 3. Schematic depiction of first three terms in dimensional perturbation expansion of Eq.(12) for hydrogenic atom. For each the effective potential $W(r)$ for the $D \to \infty$ limit is shown (solid curve). At left, the zeroth-order term corresponds to the electron at rest at the minimum of $W(r)$. In the middle, the first-order term, proportional to $1/D$, corresponds to harmonic oscillations (as if potential were replaced by the dashed parabola). At right, the second-order term corresponds to anharmonic vibrations (arising from cubic and quartic portions of the potential).

and the electron (now "infinitely heavy") just sits at the minimum, which is located at $r = 1$ in the scaled units. In the hyperquantum $D \to 1$ limit, the radial kinetic energy is dominant, the centrifugal term disappears, and the scaled Coulomb potential reduces to a delta function centered on the nucleus [7,39].

Step (4): Relate $D = 3$ to Limits. Since the exact dimension dependence of the energy is known for the hydrogenic atom, as given by Eqs. (6) or (7), our scaling scheme has exploited this to render the scaled energy independent of D. In this ideal case, evaluating the energy at any D determines it for all D; no interpolation or perturbation expansion is needed. To illustrate the general procedure, however, we note some aspects of a dimensional expansion about the $D \to \infty$ limit.

As depicted in Fig. 3, expanding the hydrogenic energy in powers of $1/D$ about this limit is tantamount to treating the atom like a diatomic molecule and evaluating contributions arising from various powers of the displacement from the minimum of the $W(r)$ potential.

For the ground-state, the result has the form

$$E_D = E_\infty \left[1 + \frac{2}{D} + \frac{3}{D^2} + \ldots\right] \tag{12}$$

The leading term corresponds to a rigid rotor approximation, with the electron fixed at the potential minimum and $E_\infty = -\frac{1}{2}(\frac{2Z}{D})^2$. The term in $1/D$ corresponds to the zero-point energy of a harmonic oscillator with a quadratic force constant determined by the curvature of the potential at its minimum. The terms involving higher powers of $1/D$ correspond to anharmonic vibrations governed by cubic, quartic, and higher derivatives of the potential. For D finite but very large, the effective mass of the electron will be large enough to confine it to small amplitude oscillations and to ensure that this motion is decidedly in the semiclassical regime. As D decreases further, the oscillations will become increasingly anharmonic and increasingly semiquantal. Eventually, for low D the electron will undergo the wild excursions characteristic of a strongly quantal regime.

Although the perturbation expansion of Eq.(12) is convergent at $D = 3$, the convergence is quite slow; e.g., to attain an accuracy of 1% requires 6 terms, and 0.005% requires 11 terms [27]. The slow convergence of the $1/D$ expansion about the large-D limit is due to the singular behavior of the energy at low-D, arising from the $(D-1)^{-2}$ dependence of the exact result of Eq.(7). For the hydrogenic atom, this singular factor accounts for all the terms in the $1/D$ expansion. Other systems are less simple, yet the hydrogenic case exemplifies the virtue of incorporating any known dimensional singularities into the scaling scheme.

Overview of D-Scaling for Electronic Structure

Before surveying instructive aspects of results obtained for two- and many-electron atoms, and few-electron molecules, we outline some general features. Again, these pertain to the typical four steps of dimensional scaling.

Step (1): Many-Body Systems. Regardless of the number of particles or their masses and charges, the Schrödinger equation in D-dimensions can be set up by generalizing the hydrogenic case in the usual way. For states with zero total angular momentum, the result may be cast into the same form as that for $D = 3$, with an additional term that contains all the explicit D-dependence [69]. This term is a scalar function of coordinates, not involving any derivative operators; it represents a centrifugal potential and depends quadratically on D. The evaluation of the centrifugal potential reduces to a purely geometrical exercise, as described in Chapter 2 and in Chapter 6 by Agnes Tan and John Loeser. For most states with nonzero total angular momentum, the formulation is much less simple. A general procedure for such states [80] is given in Chapter 8 by Martin Dunn and Deborah Watson.

Step (2): D-scaling, Hydrogenic or Uniform. In treating the energy levels of atoms, we can use the same D-scaled units for distance and energy that we employed for the hydrogenic atom. This will not remove all of the D-dependence but will ensure that the scaled dimensional limits are finite. The *hydrogenic scaling* makes the unit of distance became proportional to the *square* of $\kappa = \frac{1}{2}(D - 1)$ at large-D whereas it is *linear* in κ at low-D. This change does not matter when computing the energy (inversely proportional to κ^2 in both regimes), since averaging over the electronic coordinates causes the scaling factors for distance to cancel out.

In treating properties of atoms or molecules that involve a length scale that is not integrated out, we need a different scaling scheme. For example, consider the simplest molecule, H_2^+. In the Born-Oppenheimer approximation, the electronic energy depends parametrically on the internuclear distance, R. Use of hydrogenic scaling, which scales R differently in the $D \to \infty$ and $D \to 1$ limits, makes the electronic energy curves $E_D(R)$ incommensurate in these limits [48]. A simple cure is provided by a *uniform scaling* procedure, discussed in Chapter 6 by Tan and Loeser and in Chapter 7 by David Goodson and Mario López-Cabrera. The net effect is to make the distance unit proportional to $\frac{D(D-1)}{6}$ for all D. This scale factor interpolates smoothly between the hydrogenic dependence, proportional to D^2 at large-D and to $(D - 1)$ at low-D, and reduces to unity at $D = 3$. As well as reconciling the dimensional limits, the uniform scaling procedure brings a major part of the chemical bonding contributions into both limits [48].

Step (3): Dimensional Limits. For any atom or molecule, all terms in the Coulombic potential will scale as r^{-1}, whereas all terms in the kinetic energy (both those involving derivative operators and those in the scalar centrifugal potential) scale as r^{-2}. Accordingly, introducing D-scaled units yields limits qualitatively like those illustrated in Figs. 2 and 3 for the hydrogenic atom. At large D, the dimension dependence disappears from both the scaled centrifugal term U and the Coulombic terms V, whereas the kinetic energy terms involving derivatives all become proportional to $1/D^2$. In effect, the factor $\hbar^2/m_e$ involving Planck's constant and the electronic mass, which occurs in the unscaled kinetic energy, is replaced by $1/D^2$ in the scaled version, whereas this factor cancels from the scaled centrifugal potential. The limit $D \to \infty$ thus is tantamount to $\hbar \to 0$ and/or $m_e \to \infty$ in the kinetic energy. In the D-scaled space, the electrons assume fixed positions relative to the nuclei and each other, a *unique* geometrical configuration (as in the random walk problem). This corresponds to the minimum of the effective potential $W = U + V$, the sum of the scaled centrifugal and Coulombic terms. We call this the *Lewis structure*; it can be calculated exactly and provides a rigorous version of the qualitative electron-dot formulas, introduced 75 years ago [88] and still widely used.

The pseudoclassical large D limit is not the same as the conventional classical limit obtained by $\hbar \to 0$ for a fixed dimension [17,18]. Since the unscaled centrifugal potential is proportional to $\hbar^2$, it does not contribute to the conventional limit. With dimensional scaling, however, the centrifugal terms introduce barriers which prevent the electrons from falling into the nucleus or colliding with each other. In short, adding "extra" angular momentum as represented by D cures some major ills of the old quantum theory and thus invites use of modern semiclassical methods.

In the $D \to 1$ limit, the kinetic terms involving derivatives remain, the centrifugal terms drop out, and the scaled Coulombic potentials metamorphize into delta functions. This *hyperquantum* limit is tantamount to $\hbar \to \infty$ in the unscaled wave equation. For electronic structure, the low-D limit is generally less useful than the large-D limit, because only the ground state of a delta-function potential [2,5,9] is bound and that can accommodate only two electrons. However, for any atom or molecule, the dominant D-dependence of the ground-state energy, incorporated in the scaled units, comes from a second-order pole at $D = 1$. The coefficient of this term is the residue of the pole singularity and is given by the eigenvalue of the corresponding delta-function potential. Hyperquantum singularities at $D = 1$ are a characteristic feature of Coulombic systems which result from the divergence of the expectation values of the potential terms at particle coalescences [32,39].

Step (4): Interpolation or Perturbation Calculations. Exploiting the results obtained for $D \to \infty$ and/or $D \to 1$ or other special D-values to construct a good approximation to the "real-world" result for $D = 3$ is the crucial step. It is greatly facilitated by using scaled units (Step 2) to render the remaining D-dependence of the scaled quantities smooth and mild. Thus far, two-electron atoms have served as the chief testing ground. As outlined below, both interpolation between the large-D and low-D limits and perturbation expansions in powers of $1/D$ have yielded quite good results.

Since the relevant dimensional parameter is $1/D$, the pseudoclassical large-D limit is closer to $D = 3$ than is the hyperquantum low-D limit. As in Fig. 3, for D finite but very large, equivalent to a very heavy electronic mass, the electrons are confined to harmonic oscillations about the fixed positions attained in the $D \to \infty$ limit. We call these motions *Langmuir vibrations*, to acknowledge his prescient suggestion 70 years ago [89] that "the electrons could...rotate, revolve, or oscillate about definite positions in the atom." In a dimensional perturbation expansion the first-order term, proportional to $1/D$, corresponds to these harmonic vibrations, whereas higher-order terms correspond to anharmonic contributions. Standard methods for analysis of molecular vibrations [90] thus become directly applicable to electronic structure. These methods are semiclassical in form and far simpler, both conceptually and computationally, than the conventional orbital formulation.

Two-Electron Atom Ground States

Dimensional scaling for two-electron atoms has been examined at several levels of approximation. The most comprehensive treatment was provided by Loeser [30,31]. He generalized to arbitrary D both the Hylleras-Pekeris and Hartree-Fock variational algorithms, and thereby obtained very accurate ground-state total energies and correlation energies for $Z = 1$ to 6 and for $D = \infty$ down to below $D = 1$. As well as enabling tests of approximations, these essentially exact results revealed some striking regularities. Figure 4 exhibits the smooth variation with $1/Z$ and $1/D$ of the dimension-scaled ground-state energy.

Dimensional Limits. In the $D \to \infty$ limit, the dimension-scaled effective potential for a two-electron atom [27], obtained in the way outlined for the hydrogenic atom, is simply

$$W(r_1, r_2, \theta) = \frac{1}{2}\left(\frac{1}{r_1^2} + \frac{1}{r_2^2}\right)\frac{1}{\sin^2\theta} + V(r_1, r_2, \theta), \qquad (13)$$

where θ is the angle between the electron-nucleus radii r_1 and r_2, and V denotes the familiar Coulombic terms for electron-nucleus attraction and electron-electron repulsion. The energy and geometry of the Lewis structure found from the minimum of the W-potential are given by

$$\epsilon_\infty = -\frac{1}{r_m^2 \sin^2\theta_m}$$

$$r_{1m} = r_{2m} = \frac{1}{(1 + \cos\theta_m)^2} \qquad (14)$$

$$\cos\theta_m = -\frac{\lambda}{64}[\lambda + (128 + \lambda^2)^{\frac{1}{2}}],$$

with $\lambda = 1/Z$; the units for energy are Z^2/κ^2 and for distance κ^2/Z. These results pertain to $Z > 1.2334$, a critical value [46] below which the potential minimum becomes unsymmetrical ($r_{1m} \neq r_{2m}$). The Lewis structures are strongly *bent* rather than linear because θ_m, the angle between the electron radii at the potential minimum, reflects the competition between centrifugal repulsion (minimal for $\theta = 90°$) and interelectron repulsion (minimal for $\theta = 180°$). For the helium atom, the Lewis structure has $r_m = 1.214a_o$ and $\theta_m = 95.3°$; these values are, respectively, about 0.4% larger and 4% smaller than the most probable radius and angle found for $D = 3$ in high quality variational calculations [91].

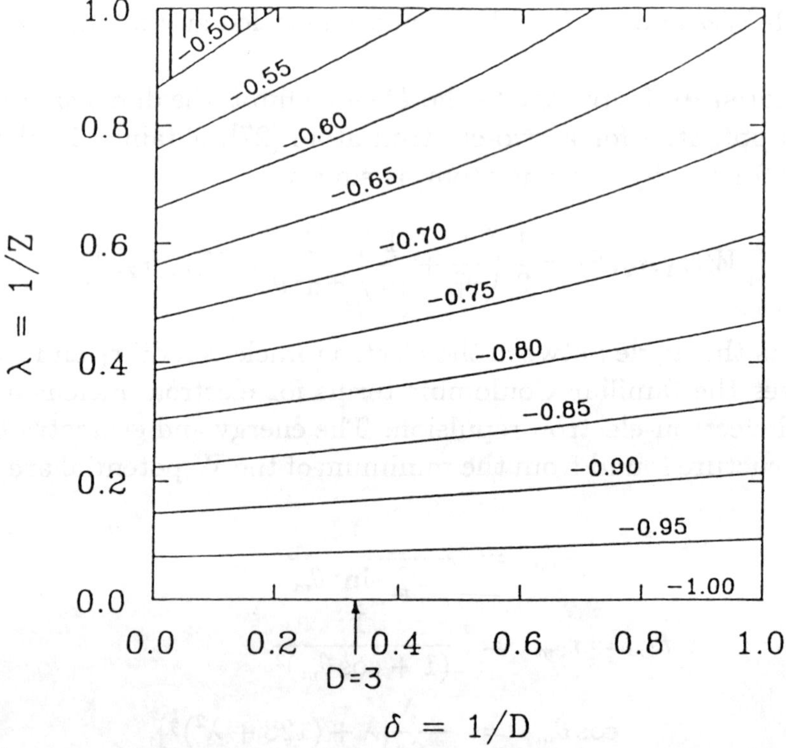

Figure 4. Contour map of dimension-scaled energy (in units of $\frac{Z^2}{\kappa^2}$ hartrees) for the ground state of the two-electron atom as a function of the inverse spatial dimension $\delta = 1/D$ and inverse nuclear charge $\lambda = 1/Z$. Hatching indicates the region beyond the first ionization continuum edge, where no stable bound state exists.

In the $D \rightarrow 1$ limit, the centrifugal term disappears and the dimension-scaled Coulombic potential is replaced by a corresponding delta-function potential [7,39]. This permits an exact solution by numerical quadrature [5].

First-Order Interpolation. The dimension dependence is dominated by singularities at $D = 1$, arising from a second-order pole (like the hydrogenic atom but with a different residue) and a confluent first-order pole. Deducting the readily calculable contributions from these poles markedly improves the efficacy of dimensional interpolation. The simplest approximation of this kind yields

$$E_D(1s^2) = -\left(\frac{2Z}{(D-1)}\right)^2 [\epsilon_1 + (1-\delta)\epsilon_1' + (1-\delta)^2(\tilde{\epsilon}_\infty + \delta\tilde{\epsilon}_\infty')] \quad (15)$$

where $\delta = 1/D$ and the coefficients are functions of Z. This approximation was constructed [25,39,46] by combining two first-order perturbation calculations, one performed at the $D \rightarrow 1$ limit, the other at the $D \rightarrow \infty$ limit. The coefficients ϵ_1 and ϵ_1' are the residues of the second- and first-order poles at $D = 1$, whereas $\tilde{\epsilon}_\infty = \epsilon_\infty - \epsilon_1 - \epsilon_1'$ and $\tilde{\epsilon}_\infty' = 2\epsilon_\infty + \epsilon_\infty' - 2\epsilon_1 - \epsilon_1'$ are obtained from the minimum and curvature of the effective potential at $D \rightarrow \infty$, corrected for contributions from the $D = 1$ singularities. For helium, Eq.(15) with $Z = 2$ and $D = 3$ is in error by only 0.14%. That is about ten-fold smaller than the error of the Hartree-Fock approximation, so for helium this quite simple first-order dimensional interpolation yields about 90% of the correlation energy.

Hydride Ion and Symmetry Breaking. The hydride ion provides an instructive example, since the restricted Hartree-Fock theory fails to predict the stability of the ground state. This occurs because even though the nominal configuration is $1s^2$, on average one electron is much further from the nucleus than the other. As noted under Eq.(14), the large-D limit predicts that for low Z such symmetry breaking will occur because the effective potential no longer has a single minimum. Figure 5 shows how the Lewis structures and the corresponding potentials change in the symmetry breaking region. For $Z < Z_o = 1.228$, the symmetric configuration of Eq. (14) becomes a saddle point rather than a minimum, and there are two equivalent unsymmetrical minima that differ by interchange of r_{1m}

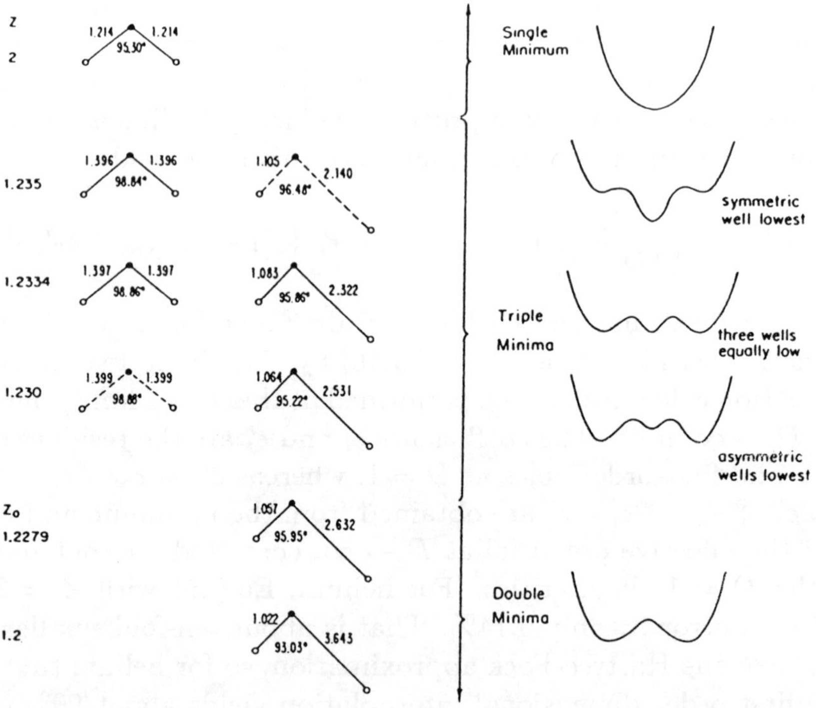

Figure 5. Variation of Lewis structures for two-electron atoms in symmetry breaking region (at left). Corresponding values of Z, θ_m, and r_m are shown, with distance units relative to the same-Z hydrogenic atom. Solid lines indicate structures that pertain to global mimima of effective potentials, dashed lines those that pertain to less stable local minima as pictured schematically (at right).

and r_{2m}. Actually, the symmetry-breaking transition begins slightly above Z_o, where both the single symmetrical minimum and the pair of unsymmetrical minima coexist [46].

For $Z = 1$ and $D = 3$, only the ground state is bound, while for an infinitesimally larger nuclear charge, there are infinitely many bound states. For $Z = 1$ and $D = \infty$ the symmetry breaking is drastic: one of the electrons escapes into the ionization continuum. Doren found that the very simple first-order dimensional expansion of Eq. (15), determined just from the hydrogenic contribution at $D \to \infty$ and the hyperquantum pole terms at $D \to 1$, gives both the ground state and the doubly excited $2p^2\ ^3P^e$ state for the hydride ion at $D = 3$ with an accuracy of a few tenths of a percent [46]. Both states are obtained in a single stroke, by virtue of an exact interdimensional degeneracy [7,8] between the excited state at $D = 3$ and the ground state for $D = 5$. In this approximation, the poles at $D = 1$ are solely responsible for the stability of the atom, so much of the effect of short-range electron repulsion evidently is contained in the residues of those poles. This rudimentary dimensional scaling calculation is far simpler than the Hartree-Fock method but again accounts for much of the correlation energy.

Large-Order $1/D$ Perturbation Expansion. Highly accurate energies have been computed by extending the perturbation expansion in powers of $1/D$ to large order [41,77,82]. This has also elucidated the singularity structure in the $D \to \infty$ limit, which exhibits aspects of both an essential singularity and a square-root branch point. After the first few terms the expansion diverges strongly, as the perturbation coefficients grow factorially. However, summation techniques incorporating the singularity structure nevertheless yield excellent results. For the ground state of helium, 30 terms give 9 significant figures. The attainable accuracy appears to be limited only by accumulation of roundoff error in the expansion coefficients. The perturbation methods and results are described in Chapter 7 by Goodson and López-Cabrera.

Correlation Energy. The dimension dependence of the correlation energy proved to be remarkably simple [26,37]. The correlation energy, defined as the difference between the exact nonrelativistic energy and the Hartree-Fock approximation, $\Delta E_D = E_D - E_D^{HF}$, can be calculated exactly in both the $D \to 1$ and $D \to \infty$ limits. As seen in Fig. 6, we find that the dimension-scaled ΔE_D is nearly a linear function of $1/D$ between these limits. Linear interpolation between the exactly known limits yields the correlation energy for $D = 3$ and $Z \geq 2$ within about 0.3%; adding this to the Hartree-Fock energy then gives the total ground-state energy within 0.005% or better. This accuracy is comparable to the best available configuration interaction calculations.

Also displayed in Fig. 6 is the variation of the correlation energy with nuclear charge. For a given D, the quantity $Z^2 \Delta E_D$ (in D-scaled units) is approximately linear in $1/Z$ and varies only by about 25% over the full range of nuclear charge. The ratio $\Delta E_D/\Delta E_\infty$ is quite insensitive to Z; for $D \geq 2$ and $Z \geq 2$, it varies by less than 5%. For $D = 3$ and $Z \geq 2$, the ratio is given by $1.49 + (0.134/Z)$ within 0.5% or better.

These results indicate that the correlation energy for $D = 3$ is closely related to that for the $D \to \infty$ limit. The possibility of exploiting this for many-electron systems is very inviting. The computation of ΔE_∞ only requires finding the minimum of the effective potential for the full problem and its value at a nearby point for the Hartree-Fock approximation.

Hartree-Fock Version. As with the analogous mean field approximation in statistical mechanics, the error in the Hartree-Fock approximation is expected to diminish as D increases. This is because fluctuations decrease in proportion to $D^{-\frac{1}{2}}$, as illustrated in Fig. 2 for the hydrogenic atom. However, whereas the mean field approximation for critical exponents of phase transitions becomes exact [86] for sufficiently large D, for the Hartree-Fock approximation the correlation energy remains nonzero and relatively large even for $D \to \infty$. As a function of the total energy, ΔE_D for the helium atom varies from 2.3% at the $D \to 1$ limit to 1.5% for $D = 3$ to 0.99% at the $D \to \infty$ limit.

The origin of this residual Hartree-Fock error at the large$-D$ limit is readily identified [40]. The Hartree-Fock wavefunction, constructed as a product of one-electron orbitals, lacks any explicit dependence on the angle θ between the electron radii. Hence this angle enters only in the Jacobian volume element, which contains $(\sin \theta)^{D-2}$; therefore

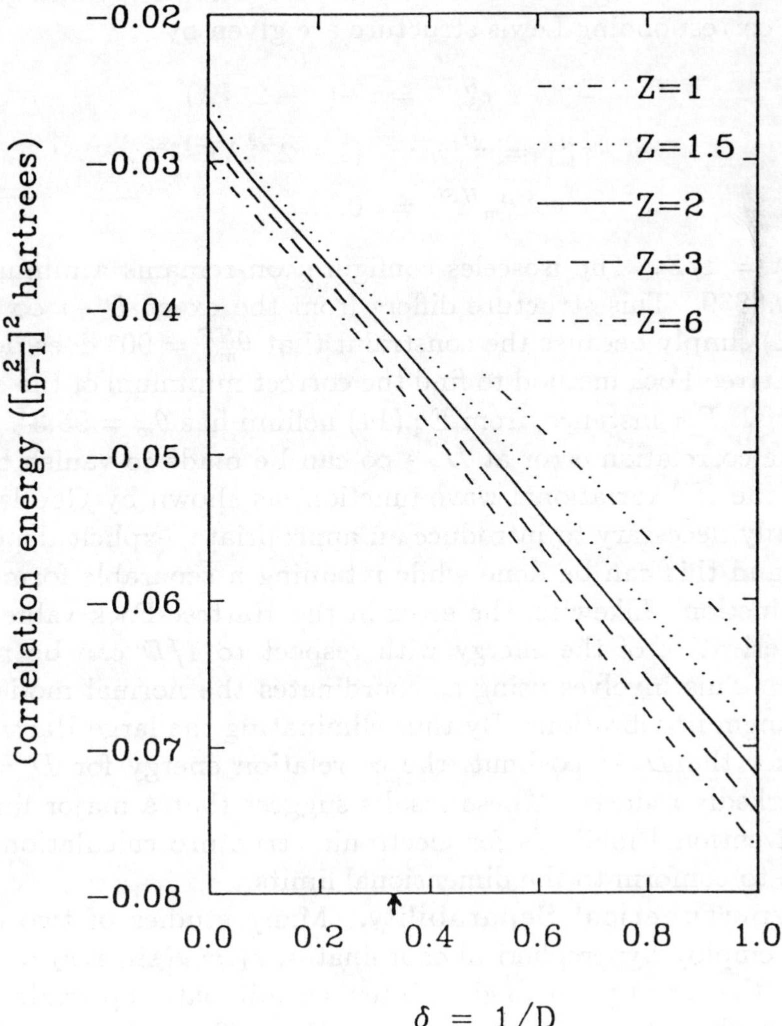

Figure 6. Correlation energy for ground states of two-electron atoms as a function of $\delta = 1/D$. Values for real atoms (He, Li^+, C^{4+}) may be read off at $\delta = 1/3$ (indicated by arrow). For nuclear charge $Z \geq 2$, the maximum deviations from linearity are about 1% . (The curve for $Z = 1$ terminates at $\delta = 1/2$ because the hydride ion is unbound in the Hartree-Fock approximation, and correlation energies therefore are not well-defined for $D > 2$.)

as $D \to \infty$, the angle becomes fixed at 90°. The energy and geometry of the corresponding Lewis structure are given by

$$\begin{aligned}
\epsilon_\infty^{HF} &= -(1 - 2^{-\frac{3}{2}}\lambda)^2 \\
r_{1m}^{HF} = r_{2m}^{HF} &= (1 - 2^{-\frac{3}{2}}\lambda)^{-1} \\
\cos\theta_m{}^{HF} &= 0,
\end{aligned} \tag{16}$$

with $\lambda = 1/Z$. The isosceles configuration remains a minimum for $Z > 0.8839$. This structure differs from the exact $D \to \infty$ limit of Eq.(14) simply because the constraint that $\theta_m^{HF} = 90°$ does not allow the Hartree-Fock method to find the correct minimum of the effective potential. For instance, from Eq.(14) helium has $\theta_m = 95.3°$.

The correlation error at $D \to \infty$ can be made to vanish by modifying the HF variational wave function, as shown by Goodson [40]. It is only necessary to introduce an appropriate, explicit dependence on θ, and this can be done while retaining a separable form for the wavefunction. Likewise, the error in the Hartree-Fock value for the first derivative of the energy with respect to $1/D$ can be made to vanish. This involves using as coordinates the normal modes q_i for the Langmuir vibrations. By thus eliminating the large Hartree-Fock error for the $D \to \infty$ limit, the correlation energy for $D = 3$ will be markedly reduced. These results suggest that a major limitation of conventional methods for electronic structure calculations is the failure to conform to the dimensional limits.

Hyperspherical Separability. Many studies of two electron atoms employ hyperspherical coordinates, $r_1 = R\sin\alpha, r_2 = R\cos\alpha$, and θ, the usual polar angle. Often an adiabatic approximation is invoked which treats the hyperradius R as effectively separable. Although computations have provided evidence for the approximate separability, the physical reason for this remained unclear. However, it is readily shown [40,65,69] that for two-electron atoms not subject to symmetry breaking, the hyperradius separability is exact for $D \to \infty$ and nearly exact to order D^{-2}. For the symmetric Lewis structure, the Langmuir vibrational mode corresponding to symmetric stretching of r_1 and r_2 is approximately separable. At large D this mode becomes proportional to the hyperradius. Thus the large$-D$ limit reveals that the symmetry of the Lewis structure fosters the separability of the hyperradius.

Charge Renormalization. Another means to exploit the beguiling simplicity of the $D \to \infty$ limit is suggested by a heuristic analogy with renormalization group theory [87]. In view of the smooth variation of the electronic energy with D and Z, displayed in Fig. 4, we expect that an effective value of the nuclear charge can be found such that for that value, denoted Z_∞, the scaled energy at $D \to \infty$ is the same as for $D = 3$ with the actual nuclear charge. This requires

$$\epsilon_3(Z) = \epsilon_\infty(Z_\infty). \tag{17}$$

The requisite value of Z_∞ is readily obtained by inverting Eq.(14); *e.g.*, for helium we find 2.343. The increase above the actual charge, $\Delta Z = Z_\infty - Z$, offsets the enhanced centrifugal repulsion of the $D \to \infty$ limit.

Such an evaluation of Z_∞ is only of descriptive interest, since it requires knowledge of the $D = 3$ energy. However, we can try to evaluate Z_∞ or ΔZ in some other way and thereby predict the $D = 3$ result. Figure 7 illustrates a scheme that makes use of the Hartree-Fock energies for $D = 3$ and $D \to \infty$. By inverting Eq.(16) we can find the charge Z_∞^{HF} which makes the Hartree-Fock energy for $D \to \infty$ become the same as that for $D = 3$. For helium this charge is 2.293. The increment in the effective charge, $\Delta Z^{HF} = Z_\infty^{HF} - Z$, provides only a lower bound for ΔZ because the Hartree-Fock results omit the correlation energy. To make an approximate allowance for this, we can determine the increment ΔZ^{CE} needed to make the Hartree-Fock energy coincide with the exact energy for $D \to \infty$ and thus reduce the correlation energy to zero in that limit. For helium, this increment is 0.047. Then we take the sum, $\Delta Z^{HF} + \Delta Z^{CE}$, as an estimate for ΔZ.

Table 2 gives results obtained with this extremely simple charge renormalization scheme. It affords good estimates for the $D = 3$ ground-state energy, even for low Z. The strategy of renormalizing the $D \to \infty$ limit to simulate $D = 3$ results could be applied as well to variational improvements on the Hartree-Fock approximation that include some of the correlation energy.

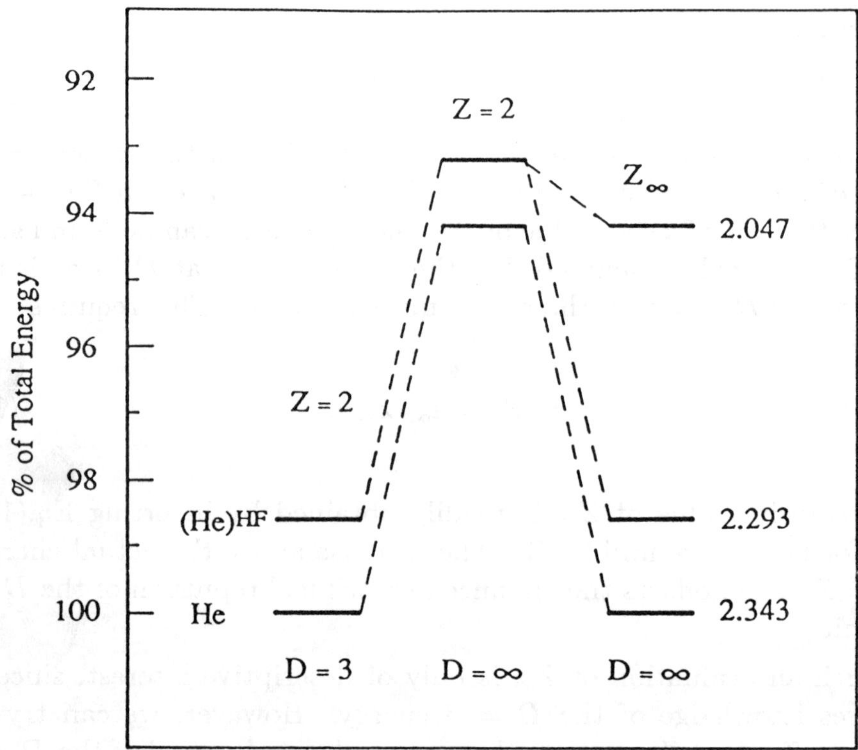

Figure 7. Charge renormalization procedure for ground state of helium. At left are shown exact and Hartree-Fock energies for $D = 3$. In middle are corresponding energies for $D \to \infty$ limit. At right are energies derived from Eq.(17) by adjusting nuclear charge to find value Z_∞ for which the $D \to \infty$ limit will coincide with the $D = 3$ energy.

Table 2. Increments $\Delta Z = Z_\infty - Z$ in nuclear charge required to match exact and Hartree-Fock energies for $D \to \infty$ to those for $D = 3$.

Z	ΔZ	ΔZ^{HF}	ΔZ^{CE}	$\Delta Z^{HF} + \Delta Z^{CE}$	$\% Error$
1.2	0.255	0.195	0.049	0.244	0.45
1.5	0.285	0.230	0.048	0.278	0.28
2	0.343	0.293	0.047	0.340	0.04
3	0.468	0.422	0.046	0.468	-
6	0.858	0.814	0.045	0.859	-

Two-Electron Atom Excited States

Dimensional scaling for excited states of two-electron atoms has thus far explored only a few aspects. Here we briefly discuss exact and approximate coincidences between the energy levels of certain excited states for $D = 3$ and ground states of higher dimension, as well as the pseudomolecular level patterns of intrashell states that result from strong electron correlation. Dimensional perturbation theory and higher angular momentum states are treated in Chapter 8 by Goodson, Dunn, and Watson.

Exact and Near Interdimensional Degeneracies. As noted under Eq.(6), the spectrum of the one-electron atom, or any central force potential, exhibits extensive interdimensional degeneracies; all eigenstates with a given value of $D + 2l$ have the same energy. For the two-electron atom most of these interdimensional degeneracies are destroyed by electron correlation. However, some persist as approximate degeneracies and certain of those between $D = 3$ and $D = 5$ remain exact [8,35]. Thus the energies of $^3P^e, ^1P^e, ^3D^o$, and $^1D^o$ states for $D = 3$ are respectively identical to those for $^1S^e, ^3S^e, ^1P^o$, and $^3P^o$ states at $D = 5$.

These exact interdimensional degeneracies have been exploited [68] to obtain accurate energies for doubly excited $2pnpP^e$ states of helium at $D = 3$, with $n = 2 - 6$, by calculating energy eigenvalues for the singly excited $1s(n-1)sS^e$ states at $D = 5$. The lowest of these states, with $n = 2$, lie above the first ionization threshold for $D = 3$; yet the excited P^e and D^o states that are below the second ionization threshold are nonetheless bound states embedded in the $He^+(n = 1) + e^-$ continuum. The symmetry of these states prevents them from autoionizing, since they would have to decay to He^+ in a $1p$ state. This is the physical reason why such states are isomorphous with bound S^e and P^o states for $D = 5$. Higher P^e and D^o states, above the second ionization threshold for $D = 3$, do autoionize; those states are degenerate with autoionizing S^e and P^o states for $D = 5$.

Near degeneracies between the ground states for $D \geq 5$ and certain $D = 3$ excited intrashell states are particularly striking [47,79]. In analogy to the hydrogenic case, we find that $D \to D + 2L$ identifies non-S excited states that lie close to a ground $1s^2\ ^1S^e$ state of higher dimension. This amounts to "extrapolating" the exact inter-dimensional degeneracy of the $D = 5$ ground state with $2p^2 \; {}^3P^e$ for $D = 3$, so that the ground states for $D = 7, 9, 11, \ldots$ correspond approximately to the excited $3d^2 \; {}^1D^e$, $4f^2 \; {}^3F^e$, $5g^2 \; {}^1G^e$,... states for $D = 3$. The energies of these odd$-D$ states, $E_D(1s^2 \; {}^1S^e)$, are well approximated by Eq.(15), our simple first-order formula, which becomes more accurate as D is increased. Furthermore, as illustrated in Fig. 8, other excited intrashell supermultiplet states [92] are nearly degenerate with these ground states of higher D. Many analogous sequences exist that sprout from other ${}^1S^e$ "generator states" of higher D; for example, that arising from $E_D(2p^2 \; {}^1S^e)$ is included in Fig. 8.

The pattern of approximate interdimensional degeneracies of two-electron atoms has been nicely elucidated [79] by means of the adiabatic molecular orbital description, presented in Chapter 12 by Jan-Michael Rost and John Briggs. The MO description and $D-$dimensional generalization are naturally complementary, since both emphasize angular momentum. For the H_2^+ problem, the MO prototype, there are myriad exact interdimensional degeneracies [62] because $D \to D + 2$ is equivalent to $m \to m + 1$, increasing by unity the projection of the electronic angular momentum on the internuclear axis. When the MO's are transcribed to treat two-electron motion, many approximate or vestigial degeneracies persist.

A perturbation treatment, using exact or near interdimensional degeneracies as a zeroth-order solution, is an attractive prospect. Especially for autoionizing states, this might substantially facilitate computing resonance widths, as well as energies. Increasing D markedly enhances semiclassical methods and treating S-states is much easier than high-L states.

Pseudomolecular Energy Level Patterns.   In intrashell states, the two electrons have the same principal quantum number so are nominally at the same distance from the nucleus. The electrons hence exhibit strongly correlated, collective motions and the energy levels and wavefunctions resemble those for molecular rovibational states. Kellman and Herrick [92] discovered these patterns and showed that many features could be interpreted in terms of an empirical rovibrator model, e-core-e, analogous to a linear triatomic molecule. This model has since been examined and refined by anal-

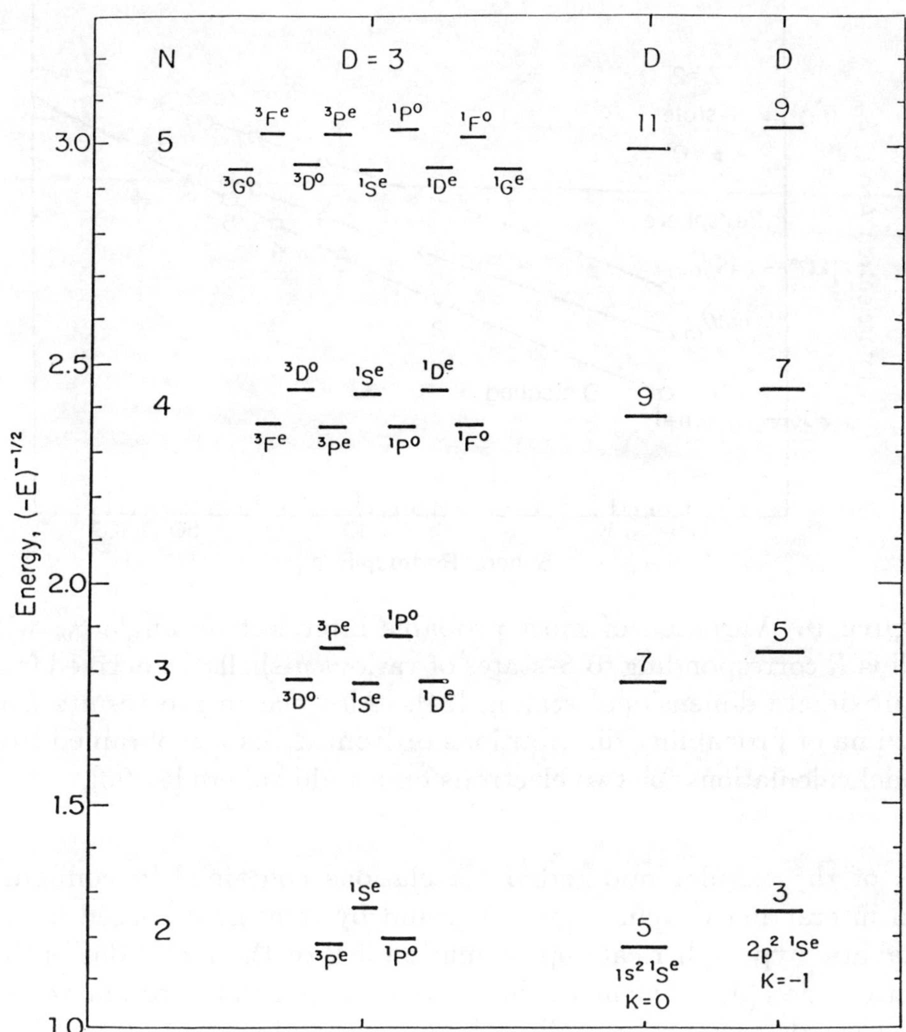

Figure 8. Correlation of quasidegenerate excited energy levels of $D = 3$ helium (at left) with corresponding $^1S^e$ "generator states" for $D = 3, 5, 7, 9, 11$ (at right). Ordinate scale plots $(-E)^{-\frac{1}{2}}$, where E (in hartree atomic units) is energy below the double-ionization limit $(He^{2+} + 2e^-)$.

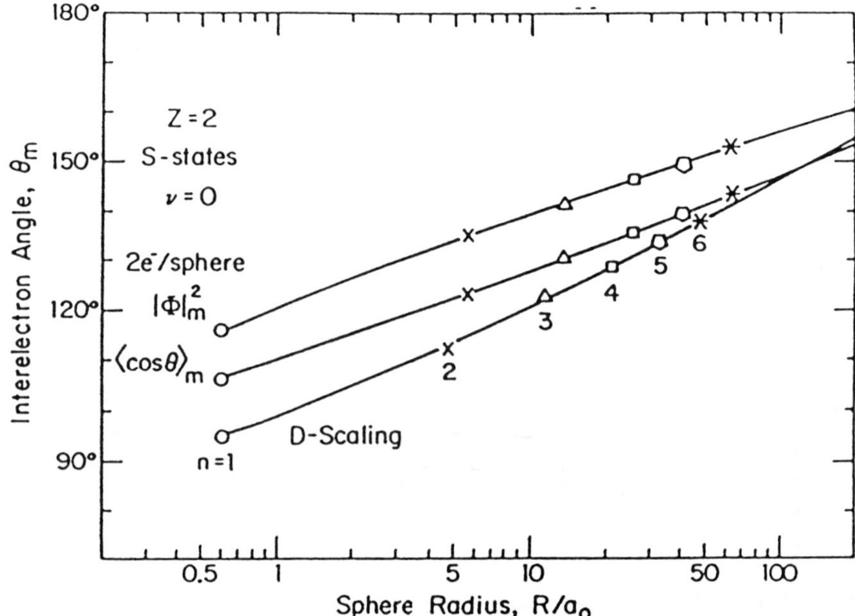

Figure 9. Variation of most probable interelectron angle θ_m with radius R corresponding to S-states of various n-shells, as derived from approximate dimensional scaling [24]. Also shown are results from maxima of probability distributions or from $< \cos\theta >$ obtained from model calculations for two electrons on a rigid sphere [47,96].

ysis of the angular and radial correlations contained in configuration interaction wavefunctions [93], and by treatments based on the adiabatic hyperspherical approximation [94] or the molecular orbital transcription [95]. The model has also been applied to ordinary quasi-two-electron systems, including the ground and excited states of the alkaline earth atoms and alkali negative ions [93]. Since dimensional scaling renders all atoms pseudomolecular at large-D, it offers another approach to the rovibrator model for intrashell states [47], particularly to the electronic geometry of these states.

For the ground state of helium the Lewis structure is symmetrical but strongly bent, as seen in Eq. (14) and Fig. 5. Likewise, D-scaling predicts that the excited intrashell states are *bent* rather than linear. Figure 9 shows the electron geometry estimated from a rough approximation examined by van der Merwe [24]. By setting $r_1 = r_2$

and freezing θ, he obtained for S-states an integrable, quasihydrogenic problem and thus determined the variation with n of the Lewis structure. The apparent conflict with the linear rovibrator model has been resolved by taking account of the angle dependence of the Jacobian volume element [47]. Again, the bent geometry predicted by D-scaling agrees well with Jacobian-weighted variational results. Figure 9 includes for comparison curves derived from model calculations for two electrons on a rigid sphere [96]; although previously regarded as consistent with linear geometry, this model also gives bent structures when the Jacobian is included.

The full two-electron Hamiltonian, including angular momentum [97], indeed corresponds to an *asymmetric rotor*, floppy but bent. This provides simpler and more explicit interpretations of the intrashell features than obtained from the linear model. The intrashell atomic terms for any given L can be readily transcribed from asymmetric rotor states [47]. These are specified by L and its projections K_a and K_c, respectively, on the principal axes of least and largest moment of inertia. The corresponding atomic term is a singlet if K_a and K_c are both even or both odd, and a triplet otherwise. The parity of the atomic state is the same as that of the K_c quantum number. The excited intrashell states that are nearly degenerate with ground state higher$-D$ states are all L_{L0} rotor states.

Many-Electron Atoms

A first-order dimensional perturbation treatment treatment of ground-state N-electron atoms has been devised by Loeser [38]. He greatly simplified the analysis by postulating that in the Lewis structure, the electrons are equidistant from one another and equidistant from the nucleus. Then the minimization of the effective potential can be carried out analytically. It involves only two free parameters, a single distance r_m and a single angle θ_m, and the energy and geometry for

$D \to \infty$ are obtained from

$$
\begin{aligned}
\epsilon_\infty &= -\frac{N}{2r_m^2}(1-x)(1+\frac{x}{1-Nx}) \\
r_m &= [(1-x)/(1-Nx)]^2 \\
\cos\theta_m &= -x/(1-x),
\end{aligned}
\tag{18}
$$

where x is the smallest positive root of a quartic equation,

$$
8Z^2x^2(2-Nx)^2 = (1-x)^3.
\tag{19}
$$

The minimum puts the electrons at the corners of a regular N-point simplex, a "hypertetrahedron," while the nucleus lies along an axis that passes perpendicularly through the centroid. This is quite literally a generalization of the cubical atom postulated by Lewis in the prequantum era [88].

The high symmetry of the $D \to \infty$ structure also simplifies evaluating the electronic vibrational modes, and to order $1/D$ this can also be carried out analytically. All atoms have just five distinct Langmuir modes, except for the smallest atoms, which have even fewer (one for H, three for He, four for Li). To take account of the Pauli Principle, Loeser assumed that the spin and symmetry rules are unaltered in D-space and related the various Langmuir vibrational modes to the familiar electron configurations $|n_1l_1 \ldots n_Nl_N >$ in the double limit $D, Z \to \infty$, where both representations become exact. Also, to avoid or reduce contributions from symmetry breaking transitions, he deleted terms beyond the lowest nonvanishing order in $1/Z$ from the Langmuir vibrations.

In this way, Loeser obtained total energies with maximum errors of only about 1% , for atoms with up to $N \sim 100$ electrons. This is comparable in accuracy to single-ζ Hartree-Fock results. Computing the Lewis + Langmuir terms requires far less effort; the time required increases only as $N^{\frac{4}{3}}$ with the number of electrons. Loeser also found that the dependence of neutral atom energies on the nuclear charge is roughly $Z^{\frac{12}{5}}$ for values of $Z \leq 150$, as observed for real atoms, and roughly $Z^{\frac{7}{3}}$ for very large Z, in accord with the asymptotic result given by the Thomas-Fermi theory.

Of most interest is the correlation energy, which is easily evaluated. The Hartree-Fock version of the $D \to \infty$ limit yields simply

$$\begin{aligned}
\epsilon_\infty^{HF} &= -\frac{N}{2}[(1 - 2^{-\frac{3}{2}}(N-1)\lambda]^2 \\
r_m^{HF} &= [(1 - 2^{-\frac{3}{2}}(N-1)\lambda]^{-1} \\
\cos\theta_m{}^{HF} &= 0,
\end{aligned} \tag{20}$$

and there are only two distinct Langmuir vibrational frequencies. At the same first-order of approximation, Loeser obtained correlation energies that agree within the estimated uncertainties with most known values. Thus, although the first-order dimensional perturbation treatment cannot give good ionization energies or term values, it does approximate the correlation effects well. It is remarkable that such a simple approximation, which treats electrons as if they were classical particles vibrating harmonically about fixed sites, appears to provide a better starting point for determining correlation energies than does the rather elaborate quantum treatment based on an orbital description.

The H_2^+ and H_2 Molecules

In applying dimensional scaling to molecules, we encounter some new issues, thus far examined throughly only for H_2^+ and partially for H_2. The effective potential for the $D \to \infty$ limit undergoes symmetry breaking as the nuclear geometry is varied, thereby acquiring multiple minima that correspond to different Lewis structures. Tunneling among these minima represents Pauling resonance among the Lewis structures. Also, as noted already, the scaling scheme used for atoms needs to be modified for molecules. These features, briefly exemplified here, are treated in Chapter 6. A high-order dimensional perturbation treatment of H_2^+ is described in Chapter 7 by Goodson and López-Cabrera.

For the H_2^+ molecule-ion, the Schrödinger equation for fixed nuclei can be solved exactly in any dimension, since it is separable in spheroidal coordinates. By virtue of the cylindrical symmetry, an exact interdimensional degeneracy links the D-dependence to the orbital angular momentum projection m on the internuclear axis: $D \to D+2$ corresponds to $m \to m+1$. Consequently, for any odd D the energies and wavefunctions can be derived by suitably scaling excited states of the $D = 3$ molecule. The degeneracies are independent of the number of ellipsoidal and hyperboloidal nodal surfaces, given in united atom notation by the dimension independent quantum numbers k and $l - |m|$, respectively. Frantz [53,62,63] has computed eigenparameters accurate to 10 to 15 digits for several states with k and $l - |m|$ up to three for a wide range of internuclear distance and for D up to 100, corresponding to $|m|$ up to fifty for $D = 3$. He also determined [48] the effective potentials for the $D \to \infty$ limit for both H_2^+ and H_2.

Symmetry Breaking and Tunneling. When the separation R between the nuclei is small, the Lewis structure for H_2 has both electrons in the plane bisecting the molecular axis, but when R becomes large enough the effective potential acquires two pairs of double minima [48]. One pair corresponds to localizing each electron on a different nucleus; the other pair, much less favorable energetically, has both electrons on one or the other nucleus. Figure 10 illustrates these "electronic isomers" by showing the dependence on R of the dihedral angle ϕ between two planes hinged on the molecular axis, each containing one of the electrons. When viewed along the molecular axis, for small R the Lewis structure resembles that for the helium atom ($R = 0$) and ϕ is near 95°. As the critical internuclear distance is approached, the optimum dihedral angle opens up by a few degrees. At larger R, the dihedral angle decreases rapidly towards 90° for the most favorable structure (peroxide-like) but opens up further for the less favorable one (amino-like).

Thus, in the $D \to \infty$ limit the effective potentials change form as R varies, exhibiting minima that correspond to a different Lewis structure for each distinct valence bond configuration. Figure 11 shows another plot of the H_2 effective potential which brings out this aspect. As D is decreased, equivalent to lowering the electronic mass, tunneling among these various minima in the effective potential becomes increasingly important; this corresponds to resonance among the different valence bond structures.

Accordingly, a key requisite for treating molecules is the ability to evaluate tunneling in many degrees of freedom. The instanton method

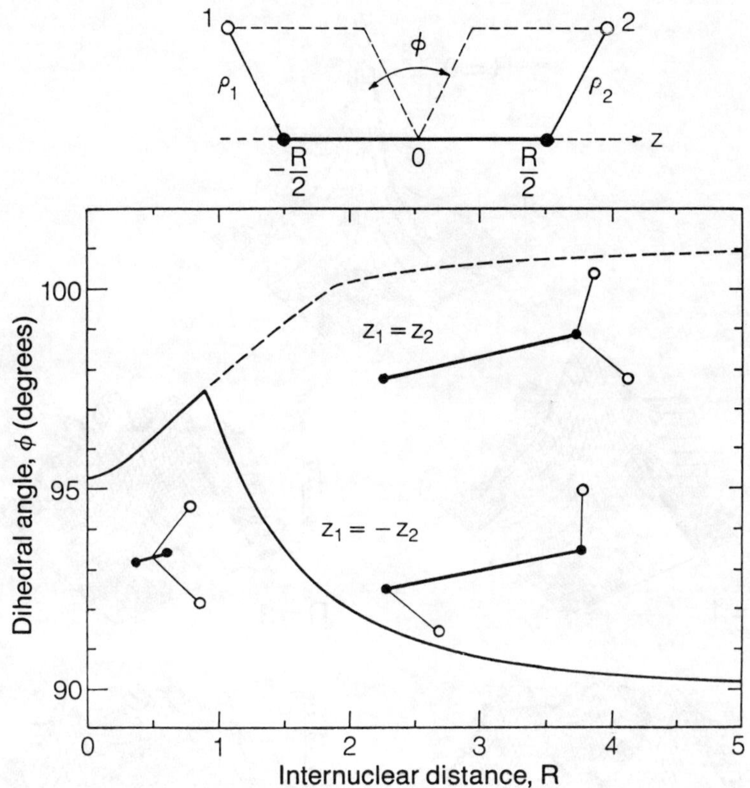

Figure 10. Variation of dihedral angle φ for Lewis structures of hydrogen molecule with internuclear distance (Bohr units). Throughout, radial distance of electrons from molecular axis ($r_1 = r_2 = r_m$) corresponds to the minimum of effective potential for $D \to \infty$. For small R, electrons lie in the plane bisecting the molecular axis and the dihedral angle is close to the value 95.3° pertaining to the united atom limit. Symmetry breaking occurs at two critical points: $R_c \approx 0.9111$, where $\varphi_m = 97.51°$ and $r_m = 0.9195$; and $R_c \approx 1.9137$, where $\varphi_m = 100.14°$ and $r_m = 1.3532$. Stick figures show typical structures at smaller and larger R.

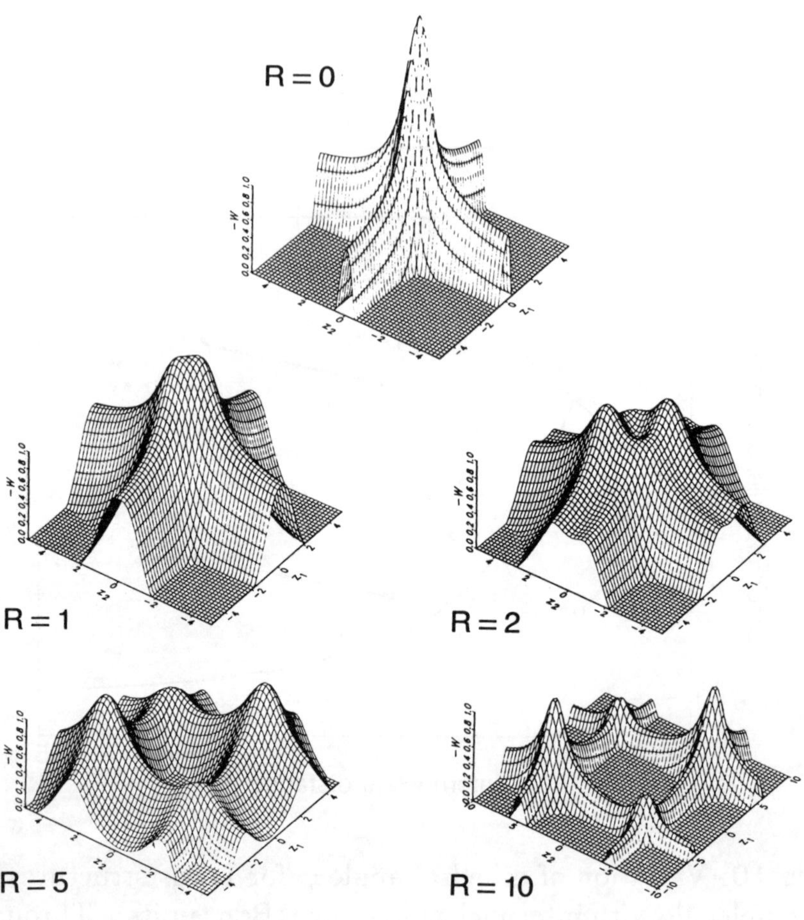

Figure 11. Dependence on internuclear distance R of the H_2 effective potential surface as function of the electronic coordinates z_1 and z_2 at constant r and φ (cf. Fig. 10). Negative energies are plotted to better exhibit changes in the minima as R varies. The surface is symmetric in z until the first splitting occurs at $R_c \approx 0.9111$, which forms a pair of global minima with $z_1 = -z_2$; the second splitting at $R_c \approx 1.9137$ forms another pair of minima with $z_1 = z_2$.

[98] appears promising in this regard, and it is also particularly congenial for the $D \to \infty$ limit. This method examines the evolution of the system in imaginary time, equivalent to motion in real time in an inverted potential like that of Fig. 11. The classically forbidden trajectories under the actual potential barrier separating minima then become allowed trajectories in the inverted potential. In Chapter 6, Sabre Kais describes instanton calculations for tunneling in H_2^+. By using both spheroidal and cylindrical coordinates, he examines tunneling in both one [76] and two degrees of freedom [72]. He finds the energy splittings due to tunneling, which correspond to Pauling resonance energies, are in good agreement with the numerical results of Frantz. Since the instanton calculations are semiclassical, the known mass dependence of the action integral and fluctuation factor provides a generic scaling law that links the $D \to \infty$ limit to $D = 3$. Use of the effective potential for the large$-D$ limit, which is exactly calculable from *classical* electrostatics, thereby yields quantitative results for electronic tunneling, an intrinsically *quantal* phenomenon.

Uniform Scaling. As with atoms, the effective potentials for molecules comprise rescaled centrifugal terms as well as the familiar Coulombic terms. In the customary Born-Oppenheimer fashion the portion involving electronic coordinates is evaluated for fixed R . This electronic energy, which does not contain the nuclear repulsion term(s), exhibits the minima that define the Lewis structures. In his study of H_2^+ and H_2 , Frantz made use of the same hydrogenic scaling previously employed for atoms. With this scaling, the united atom $(R \to 0)$ and separated atom $(R \to \infty)$ limits are both correct, and the electronic energy at the global minimum of the effective potential for $D \to \infty$ varies smoothly between these limits [48]. However, when the nuclear repulsion is added, the net interaction is repulsive. Thus, at least with hydrogenic-scaling, for these molecules the combined effect of centrifugal and nuclear repulsion precludes bonding in the $D \to \infty$ limit. The hydrogenic scaling also makes this limit incommensurate with the $D \to 1$ limit, as already discussed (p. 20).

Figure 12 illustrates how these drawbacks are cured by means of the uniform scaling, designed to accommodate both the high$-D$ and low$-D$ limits. This scheme uses the same scaled unit for energy, but changes the distance unit. The unscaled, physical internuclear

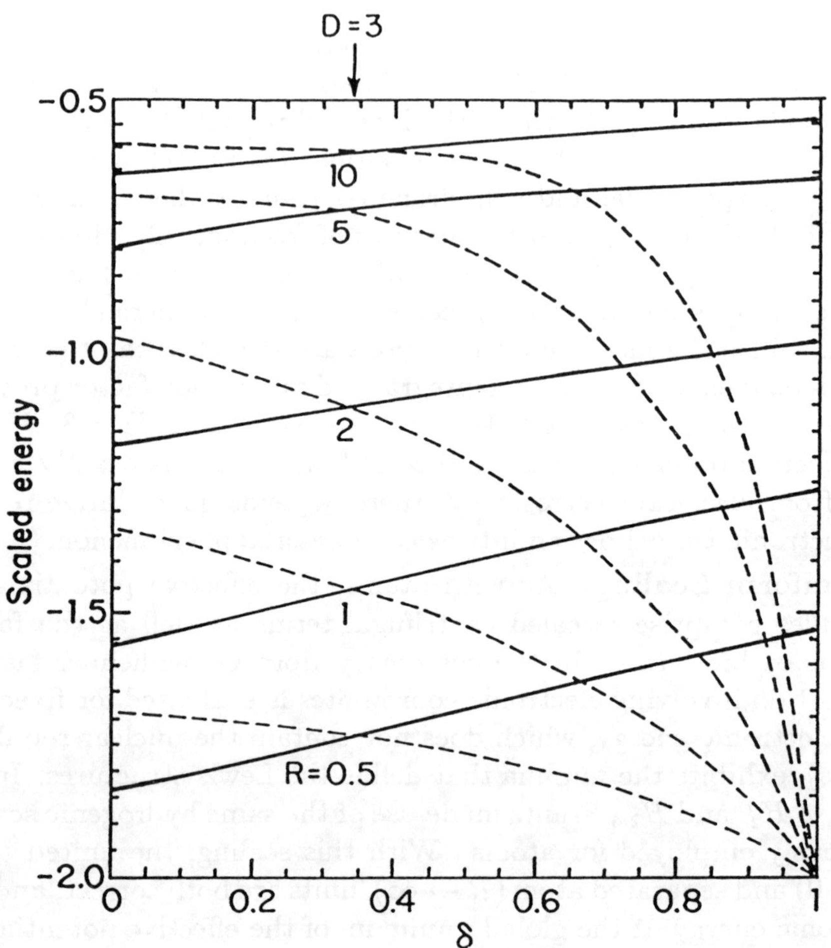

Figure 12. Electronic energy of ground-state H_2^+ (in units of κ^{-2} hartrees) as a function of $\delta = 1/D$ for fixed values of scaled internuclear distance, R_H (dashed curves) or R_U (solid curves), defined in Eq.(21). For $D = 3$, the scalings reduce to unity; thus curves are labeled simply by unscaled R (bohr units).

distance R is related to R_H and R_U, the high$-D$ hydrogenic and uniform scaled versions, by

$$R = \left(\frac{D-1}{2}\right)^2 R_H = \frac{D(D-1)}{6} R_U. \tag{21}$$

The energy E_D as a function of R_H (shown as dashed curves) varies rather strongly with $\delta = 1/D$. In particular, as $D \to 1$ the energy descends to the united atom limit $(R \to 0)$ for all values of R_H. This does not happen when the energy is expressed instead as a function of R_U (solid curves) because the uniform scaling incorporates the correct limiting behavior. With uniform scaling, the electronic energy also lies lower at the $D \to \infty$ limit. This offsets enough of the nuclear repulsion to enable the large$-D$ limit to provide a fairly good approximation to the molecular potential energy curve [48,55]. Also, $E_D(R_U)$ becomes approximately linear in $1/D$, thereby facilitating dimensional interpolation.

Although the uniform scaling was devised to take account of singularity structure in the dimensional limits, as described in Chapters 6 and 7, it also has a simple connection with the hydrogenic atom. The D-dependence of the scaling factors in Eq. (21) correspond to different expectation values [27] for the ground state atom, namely:

$$< r^{-1} >^{-1} \sim \left(\frac{D-1}{2}\right)^2 \quad \text{and} \quad < r > \sim \frac{D(D-1)}{4}. \tag{22}$$

This suggests that the uniform scaling procedure is likely to prove suitable for any molecule.

Assessments and Prospects

Dimensional scaling is still largely in an exploratory era. The emphasis has been on establishing basic features and techniques, as well as assessing strategies for applications to nonseparable dynamics. Most of the work described in this book thus predates or belongs to an "early two-electron age." This is prologue to the crucial test: accurately evaluating many-body correlation effects. Here we summarize some pertinent aspects, options, and obstacles. Table 3 provides a concise guide to the chief topics treated in previous studies and subsequent chapters of this book. Although the focus is on electronic structure, many results also apply to any problem involving dynamical correlations among strongly interacting particles.

General Aspects. The key general theme is the virtue of exploiting the easily calculable dimensional limits. The large$-D$ limit has received the most attention, and deservedly so. Since the significant variable is $1/D$, the pseudoclassical $D \rightarrow \infty$ limit is closer to the "real world" at $D = 3$ than is the less simple hyperquantum $D \rightarrow 1$ limit. For any system the $D \rightarrow \infty$ limit can be evaluated exactly, regardless of nonseparability, the strength of interactions, or the number of degrees of freedom. This limit thus provides a new, exact reference level for evaluating energy differences, the usual task in chemical physics. In any approximate treatment of a dynamical problem, it should be standard practice to compare the large$-D$ limit of the approximation with the exact limit. This can reveal unrecognized shortcomings and suggest means for improvement, as in the Hartree-Fock case [40].

A remarkable and apparently general property of the $D \rightarrow \infty$ limit is described in Chapters 5 and 6 by Vladimir Popov. In studies of systems with one degree of freedom, it was discovered that $1/n$ expansions (equivalent to $1/D$) could be constructed in the usual way even when the stationary point (at $r = r_0$) of the effective potential became a maximum rather than a minimum. For quasistationary states or resonances located near or above such barriers, at which r_0 and the energy become complex, this yields simple and surprisingly accurate means to evaluate both the resonance position and its width. The method is found to work very well for several cases, including the H atom in external fields [44,49,59] and H_2 rotational resonances [84]. Another important aspect of complex $D-$scaling has been developed recently by Rost [83] to evaluate bound state energies in the presence of symmetry breaking for three-body Coulombic systems such as H^- and H_2^+. He obtains an analytic first-order approximation that yields good results. Complex D-scaling should find wide application.

Since the dimensional limits include all electron correlation, the ability to approximate the far more difficult $D = 3$ solution by interpolation or perturbation expansions does not depend on the magnitude of the electronic interactions but only on the dimension de-

Table 3. Inventory of D-scaling studies of two- electron atoms (TEA), many-electron atoms (MEA), few- electron molecules (FEM) and other systems: citations to annotated bibliography and later chapters of this book.

The $D \to \infty$ limit: TEA[12,13,24,27,28,65,83]; MEA [38, Ch.3]; FEM[48,Ch.6.3];[34,83,Ch.2,Ch.5.1]

The $D \to 1$ limit: TEA[5,7,9,10,25,27,39]; FEM[2, Ch.4.1]

1st or 2nd order $1/D$ expansions: TEA[13,27,40,81,83,Ch.2]; MEA [38]; FEM[33,44,45,59,Ch.5.3,Ch.6.2]

Dimensional interpolation: TEA[27]; FEM[48,Ch.6.3]

Large-order $1/D$ expansions: TEA[36,41,82,Ch.7.1,Ch.8.1]; FEM[77,Ch.7.1];[15,19,20,22,29,49,50,57,58,64]

Variant $D-$scaling schemes: [48,Ch.6.3,Ch.7.1]

Complex D-scaling: TEA[83];[56,84,Ch.5.3,Ch.6.2]

Many-body $D-$dimensional Hamiltonian: [38,69,80,Ch.3,Ch.6]

Variational calculations: TEA[7,30]

Hartree-Fock approx.: TEA[10,31,40]; MEA[74,Ch.4.2]; FEM[Ch.6.3]

Correlation energy: TEA[26,37,40]; MEA [38,Ch.3];

Dimensional singularities: TEA[25,27,32,33,39, Ch.4.1,Ch.7.1,Ch.7.3]; FEM[77,Ch.7.1]; [32,33,66]

Symmetry breaking: TEA[23,24,46,Ch.2,Ch.7.3]; MEA [38,Ch.3]; FEM[48,62,Ch.6.3]

Tunneling, resonances, quasistationary states: FEM[72,76,Ch.6.4]

Relativistic aspects: [11,21,66]

Interdimensional degeneracies: TEA[7,8,35,47,68,79,Ch.8]; FEM[8,62]; [6,Ch.6.1]

Excited electronic states: TEA[30,47,68,79,Ch.8,Ch.12]; FEM[62,79]

Hyperspherical aspects: TEA[28,40,65]; MEA [69]; [75,Ch.5,Ch.11]

Two-dimensional systems: [1,16,23,43,61,74,Ch.4.2]

Other central potentials: [20,32,33,54,58,60,84,Ch.5.3]

Other many-body systems: [50,70,Ch.9,Ch.10]

Properties other than energy: [12,53,Ch.4.1,Ch.5,Ch.7.2]

Reviews of $1/D$ methods: [12,14,17,18,36,51,55,56,60,73]

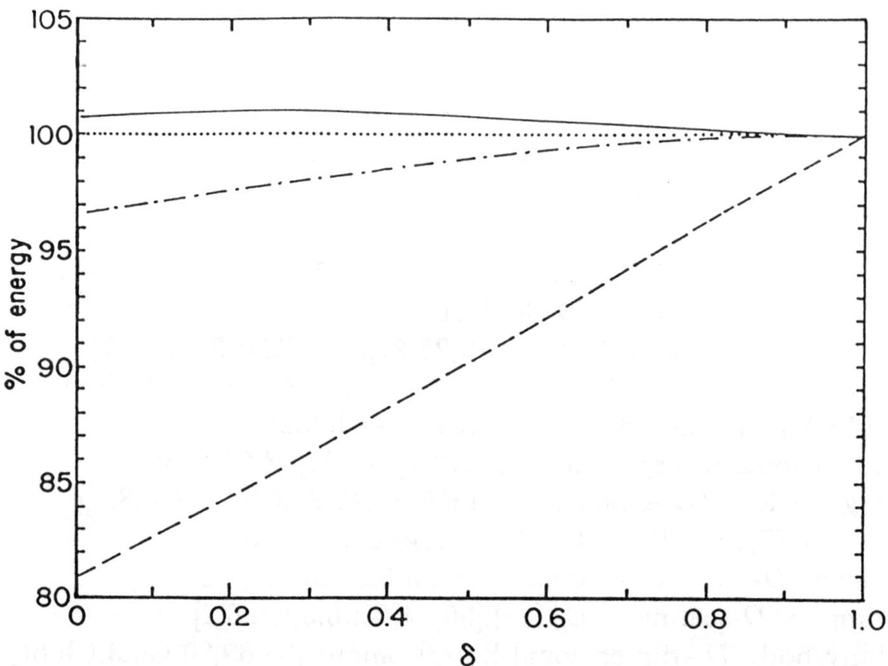

Figure 13. Cumulative contributions (% of total) to electronic energy of ground-state H_2^+ arising from dimensional singularities. For scaled distance $R_U = 1$. Lowest curve (dashed) shows contribution from second-order pole at $D = 1$; middle curve (dot-dashed) sum of first- and second-order poles at $D = 1$; uppermost curve (solid) adds contributions from singularties at $D \rightarrow \infty$ limit.

pendence. It is heartening that the D-dependence has proven to be simple for several prototype examples. The most striking of these are the correlation energy of two-electron atoms [26,37], shown in Fig. 3; the tunneling or resonance splittings in H_2^+ [72], shown in Fig. 2 of Chapter 6.4; and the cluster integrals for virial coefficients of a hard sphere fluid [70] , treated in Chapter 10.

For electronic energies, the major D-dependence is found to be governed by singularities at the dimensional limits. These appear to have generic forms (branch points at $D \rightarrow \infty$, coincident single and double poles at $D \rightarrow 1$). After these singularities are taken into account, the remaining D-dependence is quite mild. Figure 13 illustrates this for H_2^+ [77]. All but about 1% of the ground-state electronic energy comes from the dimensional singularities, and the remainder varies only slightly between $D \to \infty$ and $D = 3$. Such results suggest that eventually a general scaling scheme may be developed, with all significant singularities incorporated into the scaled distance and energy units, so that even low-order perturbation theory would often suffice to determine accurately the remaining weak D-dependence.

Computational Strategies. The essential requisites needed to extend dimensional scaling to many-body systems are generalized procedures for constructing the D-dimensional Hamiltonian and $1/D$ perturbation expansion. These are now at hand, and work is underway in our laboratory to implement a many-body computer code [99]. Our chief strategy is to use the Hartree-Fock (HF) method in the conventional way but employ a dimensional perturbation expansion to evaluate the correlation energy. In principle, this is straightforward, since the $1/D$ expansions for the complete problem and the HF approximation are computed in basically the same fashion. If high-order $1/D$ calculations are required, nontrivial complications enter the HF version. These arise both because the HF expansion is not about global the minimum of the effective potential and averages are required to uncouple the electronic coordinates. The perturbation expansion can also provide wavefunctions and quantities other than the energy. The computations differ radically from the familiar variational methods based on orbital descriptions. Rather than evaluating integrals and solving secular equations, dimensional perturbation chiefly requires computing derivatives of the centrifugal and Coulombic potentials and matrix multiplications. In practice, techniques for reducing round-off error and storage requirements have an important role and may limit the scope of the method. However, the $1/D$ expansion is well suited to parallel processing and thus can take full advantage of the new generation of massively parallel computers.

There are other inviting approaches. One form of the many- body D-dimensional Schrödinger equation [69] is well suited to variational calculations. It enables any existing computer code to be generalized to D dimensions simply by adding the matrix elements for the centrifugal potential. To exploit the opportunity to connect $D = 3$ results with the exactly solvable large$-D$ regime, the variational trial functions need to be capable of simulating well the $D-$dependence of the true wavefunction. For the most part, means to implement this criterion will have to be explored empirically. Scaling procedures for many-body systems also have to be pursued empirically and variational methods are well suited to that task. Also inviting is the use of D−scaling in conjunction with the density functional method [100]. Since the accuracy and scope of the density functional method is limited chiefly by the correlation energy functional, D−scaling may provide a significant improvement. The charge renormalization procedure illustrated in Fig. 7 likewise might be extended to many-body systems.

Anticipated Obstacles. Among the complicating factors, two seem likely to be the most challenging for many-body ground-state systems. (1) The topology of the effective potential for $D \to \infty$ may become awkward when symmetry breaking generates multiple minima and saddle points. (2) In their present form, neither the hydrogenic nor the uniform scaling schemes builds in shell structure in the $D \to \infty$ limit, although this emerges when the electronic vibrations are antisymmetrized [38]. Excited electronic states are subject to other lurking difficulties. (3) In the transition between the large−D regime and $D = 3$, the order of energy levels may change, inducing many curve-crossings. (4) Also, the usual scaling schemes cause the excited states to collapse [30] to the ground-state in the large−D limit.

Means to contend with these obstacles must be assessed by direct calculations, but results for few-body systems offer some guidance. Item (1) may not prove to be a major handicap for $1/D$ expansions. The complex scaling results noted above indicate that the localization attained in the large−D limit copes well with symmetry breaking of the kind illustrated in Figs. 5 and 11. Likewise, item (3) may not be prohibitive; in Chapter 8 Goodson and Watson obtain good results for an excited state of helium that is subject to an infinite number of curve crossings in the transition to $D = 3$. The drawbacks (2) and (4) of the current scaling schemes are easily handled for systems with one degree of freedom, as also described in Chapter 5 by Popov. Rather than developing an expansion in terms of a parameter such as $\frac{1}{2}(D-1)$ which becomes unity for $D = 3$, the appropriate scaling parameter to use is $n + \frac{1}{2}(D - 3)$, with n the principal quantum number. For two-electron atoms, van der Merwe [24] devised an analogous procedure to obtain the results of Fig. 9, using approximate semiclassical quantization to renormalize the scaling. This same effect is achieved for many-electron atoms by the subhamiltonian method presented in Chapter 3 by Zheng Zhen and John Loeser.

Extending D-scaling to many-body systems will surely bring out limitations not yet apparent. In view of the quality and character of the results found for few-body systems, however, it seems not unreasonable to hope that these methods will continue to benefit from what Wigner called the "unreasonable effectiveness of mathematics" in the natural sciences [101].

Heuristic Aspects. Beyond its promise as a means to enhance computational methods, dimensional scaling offers new heuristic perspectives. In particular, every atom or molecule acquires a new symmetry associated with the electronic geometry of its Lewis structure. Trends in these structures and the vibrational modes of the electrons, easily evaluated and visualized, may provide guidance in interpreting spectral properties, stereochemistry or reactivity. Tracing out such features may prove a useful complement to the traditional orbital pictures, particularly for analysis of electronic pathways in reactions, because the readily calculable Lewis structures and Langmuir vibrations include much of the electron correlation.

For heuristic analysis, the pseudoclassical character of the large$-D$ regime is a great advantage. It also addresses a persistent philosophical question of quantum theory: Might not "hidden" classical variables exist? At first blush, the fixed electronic geometry for Lewis structures and defined modes for Langmuir vibrations appear to violate the uncertainty principle. Computing electronic tunneling from classical trajectories on a classical electrostatic potential seems even more egregious. However, dimensional scaling of the coordinates implies that the conjugate momenta are scaled inversely, so the commutators and the uncertainty principle remain invariant [27]. In effect, we transform to a strange space which brings out reticent classical structure and *hides the quantum mechanics.* Our seemingly classical calculations at large D are still quantum mechanical. This is a fundamental reason why dimensional scaling gives surprisingly good results.

Acknowledgements

The work at Harvard reviewed here has come from happy collaboration with singular graduate students, postdoctoral fellows, and faculty visitors. In chronological sequence: John Loeser, Doug Doren, David Goodson, Don Frantz, Agnes Tan, Debbie Watson, Stella Sung, Raphy Levine, John Morgan, Sabre Kais, Mario López-Cabrera, John Avery, Jan-Michael Rost, and John Briggs. We are grateful for support received from the Venture Research Unit of BP International Limited and the Office of Naval Research.

Annotated Bibliography

This list is intended to include all papers (up to mid-1992) dealing with applications of dimensional scaling or equivalent methods to electronic structure. A few other pertinent studies of kindred problems are included; others appear in the references for later chapters. To afford some historical perspective, papers are sorted by vintage. Table 3 and the text provide a guide to the major topics and also supply some cross references not made apparent by the brief annotations.

Precursors and Harbingers: Prior to 1985

1. W. Kohn and J.M. Luttinger, Phys. Rev. **98**, 915 (1955). *H atom in $D = 2$.*
2. A.A. Frost, J. Chem. Phys. **25**, 1150 (1956). *Delta-function model ($D = 1$).*
3. G.A. Gallup, J. Mol. Spectrosc. **3**, 673 (1959). *Oscillator in D-dimensions.*
4. J. Louck, J. Mol. Spectrosc. **4**, 285, 298, 334 (1960). *Angular momentum in $D-dimensions$.*
5. C.M. Rosenthal, J. Chem. Phys. **55**, 2474 (1971). *$1/Z$ expansion for delta-function model ($D = 1$) of two-electron atom.*
6. J.H. Van Vleck, in *Wave Mechanics, the First Fifty Years*, edited by W.C. Price, et al. (Butterworths, London, 1973), pp. 26-37. *Interdimensional degeneracies for central force systems.*
7. D.R. Herrick and F.H. Stillinger, Phys. Rev. A **11**, 42 (1975). *Variational calculations for D-dimensional two-electron atoms.*

8. D.R. Herrick, J. Math. Phys. **16,** 281 (1975). *Interdimensional degeneracies.*

9. I.R. Lapidus, Am. J. Phys. **43,** 790 (1975). *Delta function (D = 1) model for two-electron systems.*

10. Y. Nogami, M. Vallieres, and W. van Dijk, Am. J. Phys. **44,** 886 (1976). *Analytic solution, Hartree-Fock for delta function (D = 1) helium.*

11. M.M. Nieto, Am. J. Phys. **47,** 1067 (1979). *D-dimensional Klein-Gordon equation.*

12. E. Witten, Phys. Today **33 (7),** 38 (1980). *Tutorial on 1/D expansions.*

13. L.D. Mlodinow and N. Papanicolaou, Ann. Phys.(N.Y.), **128,** 314 (1980); **131,** 1 (1981). *Algebraic operator method for large D; two-electron atom.*

14. L.D. Mlodinow, in *Progress in Particle and Nuclear Physics,*edited by D. Wilkinson (Pergamon, New York, 1982), Vol. 8, p. 387. *General aspects.*

15. C.M. Bender, L.D. Mlodinow, and N. Papanicolaou, Phys. Rev. A **25,** 1305 (1982). *1/D expansion for H atom in magnetic field.*

16. T. Ando, A.B. Fowler, and F. Stern, Rev. Mod. Phys. **54,** 437 (1982). *Electronic systems in two dimensions.*

17. L.G. Yaffe, Rev. Mod. Phys. **54,** 407 (1982). *Review, 1/D applications.*

18. L.G. Yaffe, Phys. Today **36 (8),** 50 (1983). *Tutorial on 1/D applications.*

19. J. Ader, Phys. Lett. **97A,** 178 (1983). *Moment method for 1/D expansion.*

20. U. Sukhatme and T. Imbo, Phys. Rev. D **28,** 418 (1983). *Shifted 1/D expansion.*

21. J.L. Miramonteo and C. Pajares, Nuovo Cimento **84,** 10 (1984). *D−dimensional relativistic equations.*

22. L.D. Mlodinow and M.P. Schatz, J. Math. Phys. **25,** 943 (1984). *1/D expansion for one degree of freedom.*

23. F.J. Asturias and S.R. Aragon, Am. J. Phys. **53,** 893 (1985). *H atom and periodic table in two dimensions.*

Focus on Electronic Structure: Since 1985

24. P. du T. van der Merwe, J. Chem. Phys. **81,** 5976 (1984); 82, 5293 (1985); Phys. Rev. A **34,** 3452 (1986). *Large−D limit for two-electron atoms.*

25. D.J. Doren and D.R. Herschbach, Chem. Phys. Lett. **118,** 115 (1985). *Use of $D = 1$ singularities.*

26. J.G. Loeser and D.R. Herschbach, J. Phys. Chem. **89,** 3444 (1985). *Two-electron correlation energy linear in $1/D$.*

27. D.R. Herschbach, J. Chem. Phys. **84,** 838 (1986). *D−interpolation.*

28. O. Goscinski and V. Mujica, Int. J. Quantum Chem. **29,** 897 (1986). *Large D, general aspects.*

29. V.S. Popov, V.M. Vainberg, and V.D. Mur, Yad. Fiz. **44,** 1103 (1986); Sov. J. Nucl. Phys. **44,** 714 (1986). *Summation methods, $1/n$ higher order.*

30. J.G. Loeser and D.R. Herschbach, J. Chem. Phys. **84,** 3882 (1986). *Hylleraas-Pekeris calculations for D-dimensional two-electron atoms.*

31. J.G. Loeser and D.R. Herschbach, J. Chem. Phys. **84,** 3893 (1986). *Hartree-Fock for D−dimensional two-electron atoms.*

32. D.J. Doren and D.R. Herschbach, Phys. Rev. A **34,** 2654 (1986). *Large−D general aspects, including singularity structure.*

33. D.J. Doren and D.R. Herschbach, Phys. Rev. A **34,** 2665 (1986). *Convergence properties of $1/D$ expansions.*

34. P. du T. van der Merwe, Phys. Rev. D **33,** 3383 (1986); Phys. Rev. A **36,** 3446 (1987); **38,** 1187 (1988). *Large−D limit for three-particle systems.*

35. D.J. Doren and D.R. Herschbach, J. Chem. Phys. **85,** 4557 (1986). *Exact and near interdimensional degeneracies.*

36. J.G. Loeser and D.R. Herschbach, J. Chem. Phys. **86,** 2114 (1987). *$1/D$ expansions for two-electron atoms.*

37. J.G. Loeser and D.R. Herschbach, J. Chem. Phys. **86,** 3512 (1987). *D−dependence of correlation energy for two-electron atoms.*

38. J.G. Loeser, J. Chem. Phys. **86,** 5635 (1987). *Large−D for many-electron atoms.*

39. D.J. Doren and D.R. Herschbach, J. Chem. Phys. **87,** 433 (1987). *Two-electron atoms near $D \to 1$ limit.*

40. D.Z. Goodson and D.R. Herschbach, J. Chem. Phys. **86,** 4997 (1987). *Electron correlation and Hartree-Fock at large-D.*

41. D.Z. Goodson and D.R. Herschbach, Phys. Rev. Lett. **58,** 1628 (1987). *Recursive $1/D$-expansion for two-electron atoms.*

42. D.R. Herschbach, J. Chem. Soc. Faraday Disc. **84,** 465 (1987). *Review.*

43. K. Tanaka, M. Kobashi, T. Shichiri, T. Yamabe, D.M. Silver, and H.J. Silverstone, Phys. Rev. B **35,** 2513 (1987). *H atom in electric field, $D = 2$.*

44. V.M. Vainberg, V.D. Mur, V.S. Popov, and A.V. Sergeev, Pis'ma Zh. Eksp. Teor. Fiz. **46,** 178 (1987); JETP Lett. **46,** 225 (1987). *$1/n$ expansion for H atom in electric field.*

45. J. Rudnick and G. Gaspari, Science **237,** 384 (1987). *Random walks; $1/D$.*

46. D.J. Doren and D.R. Herschbach, J. Phys. Chem. **92,** 1816 (1988). *Symmetry breaking, hydride ion.*

47. D.R. Herschbach, J.G. Loeser, and D.K. Watson, Z. Phys. D **10,** 195 (1988). *Pseudomolecular geometry of doubly-excited two-electron atoms.*

48. D.D. Frantz and D.R. Herschbach, Chem. Phys. **126,** 59 (1988). *Large$-D$ limit for H_2^+ and H_2 .*

49. V.M. Vainberg, V.D. Mur, V.S. Popov, A.V. Sergeev, and A.V. Shcheblykin, Teor. Mate. Fiz. **74,** 399 (1988); Theo. Math. Phys. **74,** 269 (1988). *Recursive $1/n$ expansion; H atom Stark effect.*

50. A.A. Belov and Yu.E. Lozovik, Zh. Eksp. Teor. Fiz. **94,** 38 (1988); Sov. Phys. JETP **87,** 2413 (1988). *N identical bodies, $1/D$ expansion.*

51. D.R. Herschbach, At. Phys. **11,** 63 (1989). *Review.*

52. J. Avery, *Hyperspherical Harmonics; Applications in Quantum Theory* (Kluwer, Dordrecht, 1989).

53. D.D. Frantz, D.R. Herschbach, and J.D. Morgan III, Phys. Rev. A **40,** 1175 (1989). *Accurate calculations for H_2^+ at large-D.*

54. S.Kais, D.R. Herschbach, and R.D. Levine, J. Chem. Phys. **91,** 7791 (1989). *$D-$scaling and symmetry operations.*

55. D.R. Herschbach, Proc. Welch Fd. Conf. Chem. Res. **32,** 95 (1989). *Review.*

56. V.D. Mur, V.S. Popov, and A.V. Sergeev, Zh. Eksp. Teo. Fiz. **97,** 32 (1990); Sov. Phys. JETP **70,** 16 (1990). *Bound and quasistationary states from $1/n$ expansions; two Coulomb centers; three-body problem.*

57. V.D. Mur and V.S. Popov, Zh. Eksp. Teor. Fiz. **97,** 1729 (1990); Sov. Phys. JETP **70,** 975 (1990). *The $1/n$ nethod; semiclassical character.*

58. A.A. Belov, Yu. E. Lozovik, and V.A. Mandel'shtam, Zh. Eksp. Teor. Fiz. **98,** 25 (1990); Sov. Phys. JETP **71,** 12 (1990). *$1/D$ expansion for van der Waals interaction coefficients of two H atoms.*

59. V.S. Popov, V.D. Mur, A.V. Sergeev, and V.M. Vainberg, Phys. Lett. A **149,** 418 (1990). *H atom in parallel magnetic and electric fields.*

60. A. Chatterjee, Phys. Repts. **186,** 249 (1990). *Review large-N expansions.*

61. R.K. Bhaduri, S. Das Gupta, and S.J. Lee, Am. J. Phys. **58,** 983 (1990). *Thomas-Fermi atoms in two dimensions.*

62. D.D. Frantz and D.R. Herschbach, J. Chem. Phys. **92,** 6668 (1990). *Symmetry breaking and interdimensional degeneracies in H_2^+.*

63. D.D. Frantz and D.R. Herschbach, Comp. Chem. **14,** 225 (1990). *Computer program for H_2^+ in arbitrary D.*

64. S.S. Stepanov and R.S. Tutik, Zh. Eksp. Teor. Fiz. **100,** 415 (1991); Sov. Phys. JETP **73,** 227 (1991). *Expansion in Planck's constant; excited states; relation to WKB, $1/n$, and $1/D$ methods.*

65. J. Avery, D.Z. Goodson and D.R. Herschbach, Int. J. Quantum Chem. **39,** 657 (1991). *Approximate separation of many-body hyperradius at large-D.*

66. D.Z. Goodson, J.D. Morgan III, and D.R. Herschbach, Phys. Rev. A **43,** 4617 (1991). *D-singularity analysis of relativistic H atom.*

67. E.P. Rapso, S.M. deOliveira, A.M. Nemirovsky, and M.D. Coutinho-Filho, Am. J. Phys. **59,** 633 (1991). *Random walks.*

68. D.Z. Goodson, D.K. Watson, J.G. Loeser, and D.R. Herschbach, Phys. Rev. A **44,** 97 (1991). *Energies of doubly-excited two-electron atoms from interdimensional degeneracies.*

69. J. Avery, D.Z. Goodson, and D.R. Herschbach, Theo. Chem. Acta **81,** 1 (1991). *Many-body, general D Schrödinger equation.*

70. J.G. Loeser, Z. Zhen, S. Kais, and D.R. Herschbach, J. Chem. Phys. **95,** 4525 (1991). *$D-$interpolation for hard-sphere virial coefficients.*

71. S.M. Sung and D.R. Herschbach, J. Chem. Phys. **95,** 7437 (1991). *Spheroidal eigenfunctions of H atom.*

72. S. Kais, J.D. Morgan III, and D.R. Herschbach, J. Chem. Phys. **95,** 9028 (1991). *Electronic tunneling in two degrees of freedom, H_2^+.*

73. A. Gonzalez, Few-Body Systems **10,** 43 (1991). *Qualitative properties, 3 and 4 bodies, $1/D$ expansion.*

74. P. Pyykkö and Y.-F. Zhao, Int. J. Quantum Chem. **40,** 527 (1991). *Hartree-Fock atoms in two dimensions.*

75. J. Avery and D.R. Herschbach, Int. J. Quantum Chem. **41,** 673 (1992). *Hyperspherical Sturmian basis functions.*

76. S. Kais, D.D. Frantz, and D.R. Herschbach, Chem. Phys. **161,** 393 (1992). *Large$-D$ electronic tunneling in H_2^+ .*

77. M. López-Cabrera, D.Z. Goodson, D.R. Herschbach, and J.D. Morgan III, Phys. Rev. Lett. **68,** 1992 (1992). *Large-order $1/D$ expansion for H_2^+; singularity structure.*

78. D.R. Herschbach, in *Chemical Bonding: Structure and Dynamics,* edited by A. Zewail (Academic Press, New York, 1992), pp. 175-222. *Review.*

Currently in Press

79. J.M. Rost, S.M. Sung, D.R. Herschbach, and J.S. Briggs, Phys. Rev. A **46,** (1992). *Molecular orbital description in D of doubly excited atoms.*

80. M. Dunn and D.K. Watson, Phys. Rev. A (1992). *Higher angular momentum states in $D-$dimensions.*

81. D.Z. Goodson and D.R. Herschbach, Phys. Rev. A (1992). *Approximate $1/D$ summation methods.*

82. D.Z. Goodson, M. López-Cabrera, D.R. Herschbach, and J.D. Morgan III, J. Chem. Phys. *Large-order $1/D$ expansion for two-electron atoms.*

83. J.M. Rost, J. Phys. Chem. (1993). *Complex scaling, weakly bound three particle Coulomb systems.*

84. S. Kais and D.R. Herschbach, J. Chem. Phys. (1993). *Complex scaling, quasistationary states.*

Other References Cited

85. T.H. Berlin and M. Kac, Phys. Rev. **86,** 821 (1952).

86. H.E. Stanley, Phys. Rev. **176,** 718 (1968).

87. K.G. Wilson, Revs. Mod. Phys. **55,** 583 (1983).

88. G.N. Lewis, J. Am. Chem. Soc. **38,** 762 (1916).

89. I. Langmuir, J. Am. Chem. Soc. **41,** 868 (1919). See also, J.H. Van Vleck, Pure Appl. Chem. **24,** 235 (1970).

90. E.B. Wilson, J.C. Decius, and P.C. Cross, *Molecular Vibrations* (McGraw-Hill, New York, 1955).

91. S. Chanderasekhar and G. Herzberg, Phys. Rev. **98,** 1050 (1955).

92. M.E. Kellman and D.R. Herrick, Phys. Rev. A **21,** 418 (1980); **22,** 1517, 1536 (1980); D.R. Herrick, Adv. Chem. Phys. **52,** 1 (1983).

93. R.S. Berry and J.L. Krause, Adv. Chem. Phys. **70,** 35 (1988) and work cited therein. See also Chapter 12 of this book.

94. S. Watanabe and C.D. Lin, Phys. Rev. A **34,** 823 (1986) and work cited therein. See also Chapter 11 of this book.

95. J.M. Rost and J.S. Briggs, J. Phys. B **24,** 4293 (1991) and work cited therein. See also Chapter 12 of this book.

96. P.C. Ojha and R.S. Berry, Phys. Rev. A **36,** 1575 (1987).

97. S.I. Nikitin and V.N. Ostrovsky, J. Phys. B **18,** 4349, 4371 (1985).

98. W.H. Miller, J. Chem. Phys. **58,** 1664 (1973).

99. Work in progress by Tim Germann and Carol Traynor; many-body $1/D$ expansion formulated by Martin Dunn, David Goodson, and John Morgan III.

100. R.G. Parr and W. Yang, *Density Functional Theory of Atoms and Molecules* (Oxford University Press, New York, 1989).

1.   59

101. E. Wigner, *Symmetries and Reflections* (Greenwood Press, Westport, Conn., 1978), p. 222.

Chapter 2

TUTORIAL

Dudley R. Herschbach
Department of Chemistry
Harvard University
12 Oxford Street
Cambridge, MA 02138, USA

Abstract

Generalization of quantum mechanics to D spatial dimensions is illustrated explicitly for a few elementary examples. For central force problems the sole effect is to augment the orbital angular momentum by $l \to l + \frac{1}{2}(D-3)$. As shown by Rost, this relation holds even for $D = 1$, for which $l = 0$ and $l = 1$ correspond to eigenstates of even and odd parity. A key theorem for S states of any N-body system is demonstrated for the $N = 3$ case: the D-dimensional Hamiltonian can be cast in the same form as $D = 3$, with the addition of a scalar centrifugal potential that contains the sole dependence on D as a quadratic polynomial. For two-electron atoms, the $D \to \infty$ limit and the first-order correction in $1/D$ are discussed for both the complete Hamiltonian and the Hartree-Fock approximation.

This chapter provides an introductory tour, limited to elementary examples but deriving "from scratch" features that underlie any meth-

D. R. Herschbach et al. (eds.), Dimensional Scaling in Chemical Physics, 61–80.
© 1993 *Kluwer Academic Publishers. Printed in the Netherlands.*

ods employing the spatial dimension D as an interpolation or scaling parameter. After some general remarks about D as a parameter and the $1/D$ expansion, we discuss simple central force problems, for which D is isomorphous with angular momentum. We treat particularly the hydrogenic atom and harmonic oscillator. We then discuss prototype noncentral systems, including two-electron atoms and the H_2^+ and H_2 molecules, to illustrate how to recast the Hamiltonians in a canonical form, introduce D-scaling, and evaluate the large-D limit.

Role of D as a Parameter. Throughout this book D denotes the number of Cartesian components of a vector. The kinetic energy operator and volume element accordingly change form as D is changed. However, for the potential energy we choose to retain the same form as for $D = 3$. Formally, we are free to do this, but it may seem strange. For the usual generalization of a Coulombic potential, obtained from the D-dimensional Laplace equation, the radial dependence is $1/r^{D-2}$. This is not useful for our purposes, since for $D > 3$ a hydrogenic atom with such a potential does not have stable bound states [1].

What we refer to as a "D-dimensional" system thus appears at first glance to be a hybrid, with the kinetic energy set up in D-dimensions but the potential energy in $D = 3$. A similarity transformation removes the dependence on D from the kinetic energy derivatives and the volume element, however. The full Hamiltonian then takes the same form as in $D = 3$, except for the addition of a scalar centrifugal potential that contains the only explicit D-dependence. This holds for any form of the potential or number of particles [2]. Accordingly, we can just as well regard D as a continuous parameter. Varying D enables us to generate a family of kindred Hamiltonians to link the desired energy levels and wavefunctions for $D = 3$ to those for some special or limiting D-values, where computations become easier.

Note that the whole procedure could almost be formulated with $D = 3$ fixed, by merely adding the centrifugal potential with an adjustable prefactor designed to vanish at $D = 3$. The caveat "almost" pertains to the form of the centrifugal potential, which indeed is determined by the transformation introduced to remove the D-dependence of the kinetic energy and the volume element [2-4].

The $1/D$ Perturbation Expansion. In many problems the

Hamiltonian does not contain any physical parameter suitable for a perturbation expansion. For instance, on introducing atomic units for distance and energy (bohr radii/Z and Z^2 hartree units), the Hamiltonian for a hydrogenic atom becomes independent of all the physical parameters (Planck's constant, reduced mass, electronic and nuclear charges). More typically, the Hamiltonian contains parameters, but zeroth-order solutions for special values of these are not tractable or good starting approximations. In such cases, resort is usually made to variational calculations. Another approach is to concoct an perturbation expansion by treating a quantity (usually taken to be fixed) as a free, variable parameter. D-scaling is such an approach. It is widely applicable because the large-D limit is usually relatively easy to solve. With a judicious choice of scaling, this limit often can provide a good zeroth approximation. Furthermore, its pseudoclassical character facilitates evaluating the coefficients of the $1/D$ perturbation expansion.

Central Force Problems

For a spherically symmetric potential, the Schrödinger equation according to our prescription is

$$\left[-\frac{1}{2}\nabla_D^2 + V(r) \right] \Psi_D = E_D \Psi_D \tag{1}$$

in appropriate reduced units. The D-dimensional Laplacian is defined by

$$\nabla_D^2 \equiv \sum_{k=1}^{D} \frac{\partial^2}{\partial x_k^2} \tag{2}$$

in terms of Cartesian coordinates $x_k (k = 1, 2, ..., D)$. Polar coordinates are defined by $x_k = r f_k(\Omega_{D-1})$, as illustrated in Fig. 1 of Chapter 1, where Ω_{D-1} denotes a set of $D - 1$ angles. The radial coordinate r is the radius of a D-dimensional sphere,

$$r \equiv \left[\sum_{k=1}^{D} x_k^2 \right]^{\frac{1}{2}} \tag{3}$$

The Laplacian in polar coordinates is

$$\nabla_D^2 = K_{D-1}(r) - \frac{L_{D-1}^2}{r^2} \tag{4}$$

where

$$K_{D-1}(r) \equiv \frac{1}{r^{D-1}} \frac{\partial}{\partial r}\left(r^{D-1} \frac{\partial}{\partial r}\right)$$

and L_{D-1} is a generalized orbital angular momentum operator, a function of the $D-1$ angles. For the moment, we defer the derivation in order to arrive quickly at the main result.

Angular Momentum Isomorphism. The eigenvalues of L_{D-1}^2 may be found by a tidy method due to Kramer. By virtue of the spherical symmetry, the radial and angular dependence of the wavefunction are separable; thus $\Psi_D(r, \Omega_{D-1}) = R(r)Y(\Omega_{D-1})$, where the angular factor is an angular momentum eigenfunction,

$$L_{D-1}^2 Y(\Omega_{D-1}) = CY(\Omega_{D-1}) \tag{5}$$

and C is the eigenvalue to be determined. When determining the angular factor $Y(\Omega_{D-1})$, we can fix r at any value and choose any convenient magnitude for $V(r)$. Hence by setting $V(r) = E_D$ we reduce Eq.(1) to the Laplace equation, $\nabla_D^2 \Psi_D = 0$, which must yield the same angular solutions. Since the desired solutions must be regular near the origin $r = 0$, they may be most simply constructed by writing the wavefunction as a homogeneous polynomial in the Cartesian coordinates. This has the form

$$\Psi_D = r^l Y(\Omega_{D-1}), \tag{6}$$

where $l = 0, 1, 2, \ldots$ is the order of the polynomial. Then the Laplace equation gives

$$\nabla_D^2 r^l Y(\Omega_{D-1}) = [l(l + D - 2) - C] r^{l-2} Y(\Omega_{D-1}) = 0, \tag{7}$$

and thus $C = l(l+D-2)$. The radial part of the Hamiltonian therefore is given by

$$H_D = -\frac{1}{2} K_{D-1}(r) + \frac{l(l + D - 2)}{2r^2} + V(r). \tag{8}$$

The corresponding radial Schrödinger equation, $H_D \Psi_D = E_D \Psi_D$, is simplified by $\Psi_D = r^{-\frac{(D-1)}{2}} \Phi_D$, which gives

$$[-\frac{1}{2}\frac{\partial^2}{\partial r^2} + \frac{\Lambda(\Lambda+1)}{2r^2} + V(r)]\Phi_D = E_D \Phi_D \tag{9}$$

where the sole dependence on dimension appears in

$$\Lambda = l + \frac{1}{2}(D - 3). \tag{10}$$

Therefore the radial equation for the probability amplitude Φ_D of any D-dimensional central force problem is the same as that for $D = 3$, but with the orbital angular momentum given by Eq.(10). This is the key result, establishing that D is isomorphic with the orbital angular momentum [5], such that $D \to D + 2$ is equivalent to $l \to l + 1$.

Note that the lowest dimension for which this derivation of Eq.(10) holds is $D = 2$; then $Y = Y(\phi)$ is a function of a single angle (cf. Fig. 1 of Chapter 1). The formulation can be extended without change to include the $D = 1$ case, however, as pointed out by Jan Rost. Since for $D = 1$ there is no angular variable, Y is constant and therefore $C = 0$ according to Eq.(5). However, in this case Eqs. (6) and (7) are still satisfied for $l = 0$ or 1. These solutions correspond to $\Psi_D \sim r^l$, which if $l = 0$ is even and if $l = 1$ is odd with respect to the parity operation, $r \to -r$. Hence Eq.(10) can include $D = 1$ if the allowed values of the "angular momentum" quantum number, $l = 0$ or 1, are construed to label the states of even and odd parity, respectively.

Table 1 summarizes the D-dependence. To conform with convention, we write m rather than l for the $D = 2$ case. With l or $m = 0, 1, 2, \ldots$ the solutions to problems with D even are among those for $D = 2$ and solutions to problems with D odd are among those with $D = 3$.

Laplacian Operator. To derive Eq.(4), we first note that

$$\frac{\partial}{\partial r} = \sum_k \frac{\partial x_k}{\partial r}\frac{\partial}{\partial x_k} = \sum_k f_k(\Omega)\frac{\partial}{\partial x_k}. \tag{11}$$

Hence

$$\frac{\partial}{\partial r}(r^{D-1}\frac{\partial}{\partial r}) = (D-1)r^{D-2}\sum_k f_k\frac{\partial}{\partial x_k} + r^{D-1}\sum_k f_k\sum_j \frac{\partial x_j}{\partial r}\frac{\partial^2}{\partial x_j \partial x_k}. \tag{12}$$

Table 1. Angular Momentum Eigenvalues in D-dimensions

D	L_{D-1}^2	Λ	$\Lambda(\Lambda+1)$
1	0	-1	0
1	1	0	0
2	m^2	$m-\frac{1}{2}$	$m^2-\frac{1}{4}$
3	$l(l+1)$	l	$l(l+1)$
4	$l(l+2)$	$l+\frac{1}{2}$	$(l+\frac{1}{2})(l+\frac{3}{2})$
5	$l(l+3)$	$l+1$	$(l+1)(l+2)$
D	$l(l+D-2)$	$l+\frac{D-3}{2}$	$(l+\frac{D-3}{2})(l+\frac{D-1}{2.})$

Then we find

$$K_{D-1}(r) = \frac{(D-1)}{r^2}\sum_k x_k\frac{\partial}{\partial x_k} + \frac{1}{r^2}\sum_k\sum_j x_k x_j\frac{\partial^2}{\partial x_k\partial x_j}. \tag{13}$$

The squared angular momentum operator [6] is defined by

$$L_{D-1}^2 \equiv \sum_{j=2}^{D}\sum_{k=1}^{j-1}(x_k p_j - p_k x_j)^2 \tag{14}$$

with $p_k \equiv -i\partial/\partial x_k$. Therefore,

$$L_{D-1}^2 = \sum_{j=2}^{D}\sum_{k=1}^{j-1}\left[-x_k^2\frac{\partial^2}{\partial x_j^2} - x_j^2\frac{\partial^2}{\partial x_k^2} + x_k\frac{\partial}{\partial x_k} + x_j\frac{\partial}{\partial x_j} + 2x_k x_j\frac{\partial}{\partial x_k}\frac{\partial}{\partial x_j}\right] \tag{15}$$

When the double sums are written out explicitly, the first two terms are seen to give

$$-r^2\sum_{k=1}^{D}\frac{\partial^2}{\partial x_k^2} + \sum_{k=1}^{D}x_k^2\frac{\partial^2}{\partial x_k^2}$$

and the third and fourth terms give

$$(D-1)\sum_{k=1}^{D}x_k\frac{\partial}{\partial x_k}.$$

Thus we obtain

$$\frac{L_{D-1}^2}{r^2} = -\sum_k \frac{\partial^2}{\partial x_k^2} + \frac{(D-1)}{r^2} \sum_k x_k \frac{\partial}{\partial x_k} + \frac{1}{r^2} \sum_k \sum_j x_k x_j \frac{\partial^2}{\partial x_k \partial x_j}. \quad (16)$$

On rearrangement this yields Eq.(4).

The dependence of the Laplacian on angles is implicit in Eq.(16). This may be made explicit, as worked out in detail by Louck [7], by expressing L_{D-1}^2 in terms of a sequence of generalized orbital angular momentum operators:

$$
\begin{aligned}
L_1^2 &= -\frac{\partial^2}{\partial \theta_1^2} \\[2ex]
L_2^2 &= -\frac{1}{\sin \theta_2} \frac{\partial}{\partial \theta_2} \left(\sin \theta_2 \frac{\partial}{\partial \theta_2} \right) + \frac{L_1^2}{\sin^2 \theta_2} \\[1ex]
&\text{and} \\[1ex]
L_k^2 &= -\frac{1}{\sin^{k-1} \theta_k} \frac{\partial}{\partial \theta_k} \left(\sin^{k-1} \theta_k \frac{\partial}{\partial \theta_k} \right) + \frac{L_{k-1}^2}{\sin^2 \theta_k}
\end{aligned}
\quad (17)
$$

for $k = 2, 3, \ldots D - 1$. Here the angles are defined in a nested fashion [7] as in Fig. 1 of Chapter 1, with the ranges: $0 \le \theta_1 \le 2\pi; 0 \le \theta_k \le \pi$. In terms of the notation customary for $D = 2$ and 3, we have: $\theta_1 \equiv \phi$ and $L_1 \equiv L_z$; and $\theta_2 \equiv \theta, L_2 \equiv L$. Note that these operators are independent of the dimensionality, in the sense that the same expression holds for L_k^2 regardless of whether it is calculated in a space of dimension $(k + 1)$, or $(k + 2)$, or $(k + 3), \ldots$ etc.

Transformation to Unit Jacobian. The radial part of the Jacobian volume element is $J_D = r^{(D-1)}$; thus, the transformation used to obtain Eq.(9) reduces the Jacobian factor to unity for the radial probability distribution, given by $|\Phi_D|^2 = J_D |\Psi_D|^2$. With this transformation, $\Psi_D = r^{-\kappa} \Phi_D$ and $\kappa = \frac{1}{2}(D - 1)$, the first term of

Eq.(8) becomes

$$K_{D-1}(r) \;=\; \frac{1}{r^{2\kappa}}\frac{\partial}{\partial r}r^{2\kappa}[-\kappa r^{-\kappa-1}\Phi + r^{-\kappa}\frac{\partial\Phi}{\partial r}]$$

$$=\; \frac{1}{r^{2\kappa}}\frac{\partial}{\partial r}[-\kappa r^{\kappa-1}\Phi + r^{\kappa}\frac{\partial\Phi}{\partial r}]$$

$$=\; \frac{1}{r^{2\kappa}}[-\kappa(\kappa-1)r^{\kappa-2}\Phi - \kappa r^{\kappa-1}\frac{\partial\Phi}{\partial r} + \kappa r^{\kappa-1}\frac{\partial\Phi}{\partial r} + r^{\kappa}\frac{\partial^2\Phi}{\partial r^2}]$$

$$=\; r^{-\kappa}[\frac{\partial^2\Phi}{\partial r^2} - (\frac{D-1}{2})(\frac{D-3}{2})\frac{\Phi}{r^2}].$$

$$(18)$$

On combining this with the second term of Eq.(8) and noting that

$$l(l+D-2) + (\frac{D-3}{2})(\frac{D-1}{2}) = (l+\frac{D-3}{2})(l+\frac{D-1}{2}) \quad (19)$$

we obtain Eqs.(9) and (10), with $\Lambda = l + \frac{1}{2}(D-3)$.

Hydrogenic Atom

In Chapter 5, John Avery and Jens Peder Dahl give elegant treatments of the D-dimensional hydrogenic atom, both in direct and in momentum space. Here we consider only a few elementary properties. With $V(r) = 1/r$ in Eq.(9), these properties are readily derived from $D = 3$ results by exploiting the isomorphism of Eq.(10). This gives the energy levels as

$$E_{n,D} = -\frac{1}{2}\frac{Z^2}{(n+\Delta)^2}, \qquad \text{with} \quad n = 1,2,\ldots \quad (20)$$

in hartree atomic units, where $n \equiv p+l+1$ is the principal quantum number, $p = 0,1,2,\ldots$ the number of radial nodes, and $\Delta \equiv \frac{1}{2}(D-3)$. The energy level formula holds even when D is not an integer [8]. For brevity, we consider only $D \geq 2$, to avoid subtle aspects that enter for $D = 1$ because of the singularity at $r = 0$ of the Coulombic potential [9]. Likewise, we consider only the D-dimensional ground state, with no radial nodes ($p = 0$), in order to illustrate most simply the $1/D$ expansion and some properties of the radial probability distribution

[10]. The closely related $1/n$ expansion for central force problems, including excited states, is described in Chapter 5 by Popov.

Dimensional perturbation expansion. Since as $D \to \infty$ the centrifugal term blows up in Eq.(9), in the large-D regime a natural choice of scaling is $r \to \Lambda(\Lambda + 1)R$. The radial Schrödinger equation then becomes

$$\frac{1}{\Lambda(\Lambda + 1)} \left[-\frac{1}{2\Lambda(\Lambda + 1)} \frac{d^2}{dR^2} + W(R) \right] \Phi_D = E_D \Phi_D \tag{21}$$

where the effective potential is

$$W(R) = \frac{1}{2R^2} - \frac{1}{R}. \tag{22}$$

The factor of $1/\Lambda(\Lambda + 1)$ outside the brackets goes into the unit of energy; that inside has the role of an effective mass and quenches the radial derivative. In the $D \to \infty$ limit, the system merely sits at the minimum of the effective potential, at $R = R_m$, so the scaled distance becomes fixed. The scaled energy remains finite and is given by

$$\Lambda(\Lambda + 1)E_D \to \epsilon_\infty = Z^2 W(R_m), \tag{23}$$

At the minimum, $R_m = 1$ and $W(R_m) = -\frac{1}{2}$. The perturbation expansion about this limit is most simply obtained by expanding $W(R)$ in powers of the displacement, $\xi = (R - R_m)/R_m = (R - 1)$. This gives

$$\begin{aligned} W(R) - W(R_m) &= \frac{1}{2} \frac{\xi^2}{(1 + \xi)^2} \\ &= \frac{1}{2}\xi^2(1 - 2\xi + 3\xi^2 - 4\xi^3 + 5\xi^4 - \dots) \end{aligned} \tag{24}$$

The leading term describes harmonic oscillations about the minimum of $W(R)$; since the effective mass is $\Lambda(\Lambda + 1)$, the zero-point energy contributed by this term is given by $[\Lambda(\Lambda + 1)]^{-\frac{1}{2}}$. The contributions from the cubic, quartic,...terms can be evaluated by ordinary perturbation theory. (*Cf.* Fig. 3 of Chapter 1.) For the ground state, we thus obtain

$$\epsilon_\infty = -\frac{1}{2}Z^2 \left[1 - \frac{1}{[\Lambda(\Lambda + 1)]^{\frac{1}{2}}} + \frac{1}{2\Lambda(\Lambda + 1)} + \cdots \right]. \tag{25}$$

One step remains. The factor $\Lambda(\Lambda + 1)$ was the natural choice to scale the distance and energy because this eliminated all dimension dependence from the effective potential. It is not suitable for linking the large-D regime to $D = 3$, however. For the ground-state ($l = 0$), we have $\Lambda = \frac{1}{2}(D - 3)$, which vanishes as $D \to 3$, so the expansion in the form of Eq.(25) becomes singular. We need to recast the expansion in terms of a parameter which remains finite at $D = 3$. This can be done by expanding $\Lambda(\Lambda + 1)$ wherever it appears in powers of $1/D$, using

$$\Lambda(\Lambda + 1) \equiv (\frac{D - 3}{2})(\frac{D - 1}{2}) = (\frac{D}{2})^2(1 - \frac{3}{D})(1 - \frac{1}{D}). \qquad (26)$$

The result for the unscaled energy is then

$$E_D = -\frac{1}{2}Z^2(\frac{2}{D})^2 \left[1 + \frac{2}{D} + \frac{3}{D^2} + \frac{4}{D^3} + \ldots\right] \qquad (27)$$

In this form, the expansion is nonsingular and convergent for $D > 1$. Indeed, it is readily summed to give the exact ground state energy, in agreement with Eq.(20) for $n = 1$, which is proportional to κ^{-2}, where $\kappa \equiv \frac{1}{2}(D - 1)$. The convergence of Eq.(27) is quite slow, however. The optimum procedure is to reexpand Eq.(25) in terms of a parameter which eliminates as much of the energy dependence as possible. For the hydrogenic ground state that parameter clearly is κ; if we reexpand Eq.(25) via $\Lambda(\Lambda + 1) = \kappa^2(1 - \frac{1}{\kappa})$, the result collapses to a single term, $-\frac{1}{2}Z^2/\kappa^2$, the correct ground state energy.

Other Properties.   The isomorphism of Eq.(10) permits all quantities pertaining to the ground state of a D-dimensional hydrogenic atom to be obtained from those for the $D = 3$ states with no radial nodes, by the transcription $n = l + 1 = \kappa = (D - 1)$. This is how Eqs.(8-10) and Fig. 2 of Chapter 1 were obtained [10]. As another example, here we evaluate the standard deviation of the radial momentum. For a $D = 3$ atom, this is given by $\Delta p = n^{-1}(2l + 1)^{-\frac{1}{2}}$. Transcription to the D-dimensional atom and scaling (by κ^{-2}, since $p = -i\partial/\partial r$) gives $\Delta p = (D - 1)/(D - 2)^{\frac{1}{2}}$; thus the scaled momentum spread becomes singular as $D^{\frac{1}{2}}$ in the large-D limit. Combining this with Eq.(10) of Chapter 1, we obtain the uncertainty product,

$$\Delta r \Delta p = \frac{1}{2}\left[\frac{D}{D - 2}\right]^{\frac{1}{2}}. \qquad (28)$$

This is independent of the scaling and remains finite as $D \to \infty$.

In the scaling procedure illustrated in Eq.(21), we blithely dropped the radial derivative, $\frac{\partial^2}{\partial r^2}$, because it is multiplied by the reciprocal of the effective mass, which vanishes as D^{-2} in the limit $D \to \infty$. Since the probability amplitude becomes a delta function in that limit [10], we need to check whether its curvature might become singular more rapidly than D^2 and thereby invalidate the quenching of the radial derivative. In fact we find that $<p^2>=(\Delta p)^2$, so the scaled expectation value of the radial derivative becomes singular only as the first power of D and can legitimately be dropped in the large-D limit.

Harmonic Oscillator

For a D-dimensional isotropic harmonic oscillator, with $V(r) = \frac{1}{2}\omega^2 r^2$ in Eq.(9), the isomorphic link gives the energy levels as

$$E_D = \left(2v + \Lambda + \frac{3}{2}\right)\hbar\omega = \left(2v + l + \frac{1}{2}D\right)\hbar\omega \qquad (29)$$

with $v = 0, 1, 2, \ldots$; again this formula holds even if D is not an integer [8]. Provided that $D \geq 2$, the spectrum as expected corresponds to superposing D oscillators along the Cartesian axes; the zero-point energy is $\frac{1}{2}D\hbar\omega$. However, once again the $D = 1$ case is less straightforward. According to Eq.(29),

$$E_{D=1} = \left(2v + \frac{1}{2}\right)\hbar\omega, \qquad \text{with} \quad v = 0, 1, 2, \ldots; l = 0. \qquad (30)$$

This does not coincide with the correct result, proportional to $(v+\frac{1}{2})$. The situation is resolved by the parity criterion of Rost noted under Eq.(10). Since Eq.(30) corresponds to $l = 0$, it pertains to the *even* parity solutions with an even number $2v$ of nodes. The *odd* parity components, obtained from Eq.(29) with $l = 1$, are

$$E_{D=1} = \left(2v + 1 + \frac{1}{2}\right)\hbar\omega, \qquad \text{with} \quad v = 0, 1, 2, \ldots; l = 1. \qquad (31)$$

These solutions have an odd number $2v + 1$ of nodes. Together the even levels ($\frac{E}{\hbar\omega} = \frac{1}{2}, \frac{5}{2}, \ldots$) and the odd levels ($\frac{E}{\hbar\omega} = \frac{3}{2}, \frac{7}{2}, \ldots$) correctly fill out the $D = 1$ ladder.

In the same fashion illustrated above for the hydrogenic atom, all quantities pertaining to the ground state of a D-dimensional isotropic oscillator can be obtained from those for the $D = 3$ states with no radial nodes, by the transcription $v = l = \kappa - 1 = \frac{1}{2}(D - 3)$. Although the properties of the atom and oscillator differ markedly for $D = 3$, merely scaling the probability distribution to its maximum, located at $r = \kappa^2$ for the atom and at $r = \kappa^{\frac{1}{2}}$ for the oscillator, brings out resemblances. It is an instructive exercise, recommended to the reader, to carry out the D-transcriptions for various expectation values and (with ample use of the Stirling approximation for gamma functions) to examine their behavior at large-D.

Two-Electron Atoms

We now sketch the treatment of S-states of two-electron atoms in a fashion analogous to that given for the hydrogenic atom. In setting up the D-dimensional Hamiltonian, we demonstrate that it can likewise be cast in the same form as that for $D = 3$, with the addition of a scalar centrifugal potential which contains the sole dependence on D as a quadratic polynomial. This is a key theorem, valid for S states of any N-body system [2]. We then introduce D-scaling, find the minimum of the effective potential for $D \to \infty$, and examine the harmonic force constants for the electronic vibrations, which determine the first-order term of the $1/D$ expansion. This shows how electron repulsion induces symmetry breaking when the nuclear charge Z decreases below a critical value, $Z_c \approx 1.237$ [11]. We also briefly discuss the Hartree-Fock approximation [12].

Hamiltonian in D-dimensions. For S-states, the wavefunction depends only on the electron-nucleus radii r_1 and r_2 and the angle θ between these radii. In its customary form , the Schrödinger equation is

$$\left[-\frac{1}{2}(\nabla_1^2 + \nabla_2^2) + V(r_1, r_2, \theta) \right] \Psi_D = E_D \Psi_D \qquad (32)$$

The D-dimensional Laplacian in these coordinates, worked out long before [13] the advent of quantum mechanics, is

$$\nabla_1^2 + \nabla_2^2 = K_{D-1}(r_1) + K_{D-1}(r_2) - \left(\frac{1}{r_1^2} + \frac{1}{r_2^2} \right) L_{D-1}^2 \qquad (33)$$

in terms of the operators defined in Eqs.(4) and (17). Among the $D-1$ angles on which L^2_{D-1} depends, only θ is nonseparable, since it is the only angle that appears in the kinetic and potential energy. Hence, L^2_{D-2} will be a constant of the motion, and we use

$$L^2_{D-1} = -\frac{1}{\sin^{D-2}\theta}\frac{\partial}{\partial\theta}\left(\sin^{D-2}\theta\right)\frac{\partial}{\partial\theta} + \frac{L^2_{D-2}}{\sin^2\theta} \tag{34}$$

according to Eq.(17), where L^2_{D-2} can be replaced by its eigenvalue $L(L+D-3)$, which vanishes for S-states. The Jacobian factor [7, 13] is given by

$$J_D = (r_1 r_2)^{D-1}\sin^{D-2}\theta \tag{35}$$

If we now introduce $\Psi_D = J_D^{-1/2}\Phi_D$ and carry out the differentiations as in Eq.(18), we find the radial terms $K_{D-1}(r)\Psi_D$ are the same as for the hydrogenic atom and the angular term $L^2_{D-1}\Psi_D$ is

$$-(\sin\theta)^{\frac{1}{2}(D-2)}\left\{\frac{\partial^2\Phi}{\partial\theta^2} + \left(\frac{D-2}{2}\right)\left[\left(\frac{D-2}{2}\right) - \left(\frac{D-4}{2}\right)\frac{1}{\sin^2\theta}\right]\Phi\right\} \tag{36}$$

Assembling these terms gives the Schrödinger equation for the probability amplitude as

$$(T + U + V)\Phi_D = E_D\Phi_D \tag{37}$$

where derivatives appear only in

$$T = -\frac{1}{2}\left[\frac{\partial^2}{\partial r_1^2} + \frac{\partial^2}{\partial r_2^2} + \left(\frac{1}{r_1^2} + \frac{1}{r_2^2}\right)\frac{\partial^2}{\partial\theta^2}\right] \tag{38}$$

The dimension dependence appears only in the centrifugal term

$$U = \frac{1}{2}\left(\frac{1}{r_1^2} + \frac{2}{r_2^2}\right)\left[-\frac{1}{4} + \left(\frac{D-2}{2}\right)\left(\frac{D-4}{2}\right)\frac{1}{\sin^2\theta}\right] \tag{39}$$

and V is the Coulombic potential. This form, although unconventional for electronic structure studies, is customary in molecular spectroscopy (with $U+V$ replaced by a vibrational potential). In keeping with the quasicartesian form of the T terms, the Jacobian for the $|\Phi_D|^2$ function is unity.

Another convenient form for the Schrödinger equation can be obtained [2] by factoring the Jacobian as $J_D = J_3 J_{D-3}$, so that

$$J_3 = r_1^2 r_2^2 \sin\theta \qquad \text{and} \qquad J_{D-3} = (r_1 r_2 \sin\theta)^{D-3} \qquad (40)$$

Then the transformation $\Psi_D = J_{D-3}^{-1/2} \Phi_D$ yields again Eq.(37), but the derivative terms become

$$T = -\frac{1}{2}\left[K_2(r_1) + K_2(r_2) + \left(\frac{1}{r_1^2} + \frac{1}{r_2^2}\right) \frac{1}{\sin\theta} \frac{\partial}{\partial\theta}\left(\sin\theta \frac{\partial}{\partial\theta}\right)\right] \qquad (41)$$

and the centrifugal term becomes

$$U = \frac{1}{2}\left(\frac{1}{r_1^2} + \frac{1}{r_2^2}\right)\left(\frac{D-3}{2}\right)^2 \frac{1}{\sin^2\theta} \qquad (42)$$

The Schrödinger equation thus is the same as for $D = 3$, except for the addition of the D-dependent centrifugal potential, and now the Jacobian for the $|\Phi_D|^2$ function is J_3. This is the preferred form for calculations employing conventional variational methods. It allows any existing computer code to be extended to D dimensions simply by adding the matrix elements for the centrifugal potential of Eq.(42). These results are readily generalized to the N-body problem, for any mass distribution of the particles or form of the interaction potential [2].

Large-D Regime. For $D \to \infty$, any scheme that scales the radial distances as D^2 in such a way as to cancel the D-dependence of the centrifugal potential will yield the same effective potential in the scaled coordinates,

$$W = U + V = \frac{1}{2}\left(\frac{1}{r_1^2} + \frac{1}{r_2^2}\right)\frac{1}{\sin^2\theta} - \frac{1}{r_1} - \frac{1}{r_2} + \frac{\lambda}{(r_1^2 + r_2^2 - 2r_1 r_2 \cos\theta)^{1/2}} \qquad (43)$$

with $\lambda = 1/Z$. Differences in the precise choice of the scaling factor or the form of the Schrödinger equation only enter into the effective mass. When the $1/D$ expansion is recast [3, 10] in a way analogous to Eq.(26), such differences are taken into account.

If the effective potential has a single, symmetric minimum ($Z > Z_c, r_{1m} = r_{2m}$), the conditions $(\partial W/\partial r)_m = 0$ and $(\partial W/\partial\theta)_m = 0$

yield analytic formulas for the location and scaled energy at the minimum, $\varepsilon_\infty = W(r_{1m}, r_{2m}, \theta)$. These are given in Eq.(14) of Chapter 1. In this case, analytic formulas can also be obtained for the second derivatives that govern the harmonic vibrations [10]. Table 2 lists the vibrational force constants and normal mode frequencies determined [10, 12] using Wilson's FG matrix treatment, the procedure standard for molecular vibrations [14]. The radial displacements from the minimum are combined in the usual way to form symmetry coordinates,

$$A = 2^{-1/2}(\Delta r_1 - \Delta r_2) \qquad \text{and} \qquad S = 2^{-1/2}(\Delta r_1 + \Delta r_2) \qquad (44)$$

corresponding to asymmetric and symmetric stretching motions. The angular displacement $\Delta\theta$ is itself a symmetry coordinate. The potential of Eq.(43) in the harmonic approximation thus becomes

$$W = \varepsilon_\infty + \frac{1}{2}F_{AA}A^2 + \frac{1}{2}F_{SS}S^2 + F_{S\theta}S\Delta\theta + \frac{1}{2}F_{\theta\theta}(\Delta\theta)^2 \qquad (45)$$

Since the asymmetric stretching mode is separable in the harmonic regime, its frequency is simply given by $\omega_A = (F_{AA}/\mu)^{1/2}$, with μ the effective mass. The symmetric stretching and bending frequencies, ω_S and ω_θ, must be obtained from a 2 x 2 secular equation that includes $F_{S\theta}$, the coupling force constant [14]. For the ground electronic state, the zero-point vibrational energy, given by the sum of the frequencies, determines the coefficient of $1/D$ in the first-order perturbation expansion. Excited S-states correspond to one or more vibrational quanta in these modes.

The force constants and frequencies (listed for $\mu = 1$) are given in terms of $\eta = \cos\theta_m$ and ε_∞, both of which are functions only of Z. As discussed in Chapter 1, the deviation of $\cos\theta_m$ from zero is a measure of electron correlation. Since $F_{S\theta}$ is simply proportional to $\cos\theta_m$, the coupling between the symmetric stretching and bending modes quite directly reflects the electron correlation. This coupling is in fact weak, especially at large Z. For the stretching modes, $F_{AA} < F_{SS}$ over the full range of Z; at least in the harmonic regime, it is always easier to move an electron away from the nucleus if the other electron moves closer. This effect increases markedly as Z decreases, and F_{AA} becomes negative for $Z < 1.2279$. The symmetric configuration r_{1m}

Table 2. Harmonic vibrations of two-electron atoms

Constants	*Normal Mode Frequencies*

$$F_{AA} = |\epsilon_\infty|^2(1 + 6\eta - 3\eta^2) \qquad \omega_A = |\epsilon_\infty|(1 + 6\eta - 3\eta^2)^{\frac{1}{2}}$$

$$F_{SS} = |\epsilon_\infty|^2(1 - \eta^2) \qquad \omega_S = 2^{-\frac{1}{2}}|\epsilon_\infty|(5 - 6\eta + 5\eta^2 - X)^{\frac{1}{2}}$$

$$F_{\theta\theta} = \tfrac{1}{8}|\epsilon_\infty|\frac{(1 - \frac{3}{2}\eta + \frac{3}{2}\eta^2)}{1 - \eta^2} \qquad \omega_\theta = 2^{-\frac{1}{2}}|\epsilon_\infty|(5 - 6\eta + 5\eta^2 + X)^{\frac{1}{2}}$$

$$F_{S\theta} = \tfrac{\sqrt{2}}{4}|\epsilon_\infty|^{\frac{3}{2}}\eta \qquad X = (9 - 36\eta + 94\eta^2 - 84\eta^3 + 33\eta^4)^{\frac{1}{2}}$$

$$\eta = \cos\theta_m = -\tfrac{\lambda}{64}[\lambda + (128 + \lambda^2)^{\frac{1}{2}}]$$

$$\tilde{F}_{AA} = (1 - 4\hat{\lambda})(1 - \hat{\lambda})^3 \qquad \tilde{\omega}_A = (1 - 4\hat{\lambda})^{\frac{1}{2}}(1 - \hat{\lambda})^{\frac{3}{2}}$$

$$\tilde{F}_{SS} = (1 - \hat{\lambda})^4 \qquad \tilde{\omega}_S = (1 - \hat{\lambda})^2$$

$$\tilde{F}_{rr} = (1 - \tfrac{5}{2}\hat{\lambda})(1 - \hat{\lambda})^3 \qquad \tilde{\omega}_r = (1 - \tfrac{5}{2}\hat{\lambda})^{\frac{1}{2}}(1 - \hat{\lambda})^{\frac{3}{2}}$$

$$\hat{\lambda} = 2^{-\frac{3}{2}}\lambda \quad \text{and} \quad \lambda = 1/Z$$

$= r_{2m}$ then becomes a saddle point rather than a minimum and the potential has two equivalent unsymmetrical minima that differ by interchange of r_{1m} and r_{2m}. In fact, the symmetry breaking sets in at a slightly higher nuclear charge, $Z_c \approx 1.237$, where the pair of unsymmetric minima began to grow in [11].

Another instructive aspect is noteworthy. Numerical analysis of the dependence of the ground state energy on Z and D has shown that for a special nuclear charge Z^*, which varies somewhat with D, the first three terms of the $1/D$ expansion give the exact energy [10]. This value is roughly estimated as $Z^* \approx 12.3$ and does not vary much for $1 < D < 5$. Truncating the $1/D$ expansion at three terms corresponds to retaining in the effective potential only quadratic, cubic, and quartic terms for electron vibrations about the minimum. The variation of the vibrational constants with Z and the effective mass parameter, specified by D, is such that for Z^* the quartic oscillator model has the same lowest eigenvalue as does the full, exact atomic potential. This curious situation deserves study.

Hartree-Fock Version. For $D \to \infty$, the scaled energy of the Hartree-Fock model [12] is found from the effective potential function for the full problem, Eq.(43), but with the constraint that $\cos \theta_m^{HF} = 0$. The minimum consistent with this constraint defines r_m^{HF} and $\varepsilon_\infty = W(r_{1m}^{HF}, r_{2m}^{HF}, 90°)$ This yields the formulas given in Eq.(16) of Chapter 1.

Developing a HF version of the $1/D$ expansion is actually more involved than for the full problem, however. The HF approximation for the vibrational modes cannot be obtained directly by specializing results for the full case. To represent a wavefunction independent of θ, we must preaverage over the angular dependence of the exact Hamiltonian. This introduces two major changes: (1) the interelectron repulsion term, λ/r_{12}, is replaced by its averaged power series expansion about $\theta = 90°$; and (2) the coefficient of the centrifugal potential changes in a way tantamount to replacing D by $D + 1$ in the unscaled Hamiltonian. The latter change comes from reducing the number of degrees of freedom from three to two.

Table 2 includes the force constants and harmonic frequencies obtained in this way [12]. The tildes serve to emphasize that the decorated quantities pertain to the HF structure for $D \to \infty$, not that for the full problem. Since the radial dependence of the HF wavefunction is also separable, coupling terms involving products of Δr_1 and Δr_2 in the vibrational potential must also be averaged out. Accordingly, the HF zero-point energy for electronic vibrations comes solely from a doubly-degenerate stretching mode, denoted here by $\tilde{\omega}_r$. The corresponding force constant, $\tilde{F}_{rr}$, is roughly the average of those for the symmetric and asymmetric stretching modes, $\tilde{F}_{SS}$ and $\tilde{F}_{AA}$. The magnitude and Z-dependence of the latter resemble closely that for the constants for the full problem, F_{SS} and F_{AA}, but it is $\tilde{F}_{rr}$ that governs the HF zero-point energy. As Z decreases, $\tilde{F}_{rr}$ eventually becomes negative, but not until $Z < 0.884$. The HF averages that suppress electron correlation thus also delay the onset of symmetry breaking. This exemplifies another unrealistic consequence of the HF approximation.

The H_2^+ and H_2 Molecules

Here we consider only the first step in D-scaling for molecules, the generalization of the electronic Hamiltonian [15]. Other aspects are reviewed in Chapter 1 and current work is described in Chapters 4 and 7 by Goodson and López-Cabrera and in Chapter 5 by Tan and Loeser. In our discussion, the main point is to show how the kinship of dimension and angular momentum is modified for these linear molecules as compared with their united atom cousins.

For simplicity, we consider D-dimensional cylindrical coordinates. These comprise a linear coordinate z orthogonal to a $(D-1)$ subspace. The subspace is specified by spherical coordinates: ρ, the radius of the $(D-1)$ hypersphere, and Ω_{D-2}, a set of $(D-2)$ angles. Accordingly, we take $z = x_D$ and define ρ by Eq.(3) with $D \to D-1$. For $D = 3$, in addition to z and ρ there is a single azimuthal angle of rotation. As pictured in Fig. 10 of Chapter 1, the nuclei are located on the z-axis at $-R/2$ and $+R/2$, respectively. In H_2^+ the Coulombic interaction depends only on R and (ρ, z), the pair of coordinates that locate the electron. In H_2, the interaction involves five electronic coordinates: (ρ_1, z_1) and (ρ_2, z_2) and the dihedral angle ϕ between the pair of planes that contain the electrons and the molecular axis.

Table 3 summarizes the D-dependent quantities for H_2^+ and H_2, in

Table 3. Dimension dependent terms for H_2^+ and H_2.

	H_2^+ *Molecule*	H_2 *Molecule*
∇_D^2	$K_{D-2}(\rho) + \frac{\partial^2}{\partial z^2} - \frac{L_{D-2}^2}{\rho^2}$	$\sum_i K_{D-2}(\rho_i) + \frac{\partial^2}{\partial z_i^2} - (\frac{1}{\rho_1^2} + \frac{1}{\rho_2^2})L_{D-2}^2$
L_{D-2}^2	$\lvert m\rvert(\lvert m\rvert + D - 3)$	$\frac{1}{\sin^{D-2}\phi}\frac{\partial}{\partial\phi}(\sin^{D-2}\phi\frac{\partial}{\partial\phi}) - \frac{L_{D-3}^2}{\sin^2\phi}$
J_D	$\rho^{D-2}(\sin\phi)^{D-3}$	$(\rho_1\rho_2)^{D-2}(\sin\phi)^{D-3}$
U	$\frac{\Lambda(\Lambda+1)}{2\rho^2}$	$\frac{1}{2}(\frac{1}{\rho_1^2} + \frac{1}{\rho_2^2})[-\frac{1}{4} + \frac{\Lambda(\Lambda+1)}{\sin^2\phi}]$
Λ	$\lvert m\rvert + \frac{D-4}{2}$	$\lvert m\rvert + \frac{D-5}{2}$

a form which enables easy comparison with the hydrogenic atom and two-electron atom. Since the $(D-1)$ dimensional subspace is spherical, the Laplacians for these molecules can be transcribed from their united atoms. In fact, ∇_D^2 for these molecules is simply given by setting $D \to D-1$, $r_i \to \rho_i$, and adding the Cartesian second derivative term(s), $\partial^2/\partial z_i^2$. The angular momentum operator L_{D-2}^2 can be replaced by its eigenvalue in the case of H_2^+, since all of the $D-2$ angles are separable, including the azimuthal angle which exists for $D = 3$. The corresponding quantum number, denoted by $\lvert m\rvert = 0, 1, 2, \ldots$ is the magnitude of the projection of the orbital angular momentum on the linear axis z. The L_{D-2}^2 operator is not a constant of the motion in the case of H_2, since the dihedral angle ϕ is nonseparable. The other angles are all separable, however. As in Eq.(34) we need only step down by $D \to D-1$ to reach an operator that is constant, namely L_{D-3}^2, with eigenvalue $\lvert m\rvert(\lvert m\rvert + D - 4)$. Included in Table 3 are the expressions for the centrifugal potentials obtained as usual by incorporating the square root of the Jacobian into the probability amplitude. The correspondence with the united atoms is evident.

Acknowledgements

This chapter, chiefly derived from notes compiled a decade ago to
introduce students to the rudiments of dimensional scaling, incorporates suggestions made by its readers over the years. I particularly
thank David Goodson, John Loeser, Jan-Michael Rost, Stella Sung,
and Tony Tanner for their comments and contributions.

References

1. See, *e.g.*, W. Büchel and I.M. Freeman, Am. J. Phys. **37**, 1222
   (1969).
2. J. Avery, D.Z. Goodson, and D.R. Herschbach, Theo. Chem.
   Acta **81**, 1 (1991).
3. J.G. Loeser, J. Chem. Phys. **86**, 5635 (1987); see also Chaps.
   3 and 6.
4. E. Witten, Phys. Today **33 (7)**, 38 (1980).
5. D.R. Herrick, J. Math Phys. **16**, 281 (1975).
6. B.R. Judd, *Angular Momentum Theory for Diatomic Molecules*
   (Academic Press, New York, 1975), pp. 31 and 53.
7. J.D. Louck, J. Mol. Spect. **4**, 285, 298, 334 (1960).
8. See, *e.g.*, L.D. Landau and E.M. Lifshitz, *Quantum Mechanics*
   (Addison-Wesley, Reading, MA, 1958), Sec. 36.
9. See H.N. Núñez-Yépez, C.A. Vargas, and A.L. Salas-Brito, Phys.
   Rev. A **39**, 4306 (1989) and work cited therein.
10. D.R. Herschbach, J. Chem. Phys. **84**, 838 (1986).
11. D.J. Doren and D.R. Herschbach, J. Phys. Chem. **92**, 1816
    (1988).
12. D.Z. Goodson and D.R. Herschbach, J. Chem. Phys. **86**, 4997
    (1987).
13. M.J.M. Hill, Trans. Cambridge Philos. Soc. **13**, 36 (1883).
14. E.B. Wilson, J.C. Decius, and P.C. Cross, *Molecular Vibrations*
    (Dover Publ., New York, 1980); pp. 69-74; 113-114.
15. D.D. Frantz and D.R. Herschbach, Chem. Phys. **126**, 59 (1988).

Part II

THE RESEARCH FRONTIER

Chapter 3

LARGE-D LIMIT FOR N-ELECTRON ATOMS

Zheng Zhen [1] and John Loeser
Department of Chemistry, Oregon State University
Corvallis, OR 97331-4003

Abstract

The large-D limit for a many-electron atom is one in which the electrons assume positions which are fixed relative to one another and to the nucleus. Results at $D=3$ can be obtained from this classical limit through use of a 1/D expansion. It is generally preferable, however, to modify the large-D hamiltonian so that it reflects the dominant finite-D effects. We describe a simple procedure for performing such a modification without destroying the classical character or analytic simplicity of the original large-D limit. The resulting "subhamiltonians" incorporate approximations to the kinetic terms analogous to those which the Hartree-Fock approximation invokes in the potential terms. Because of their complementary characters, the subhamiltonian and Hartree-Fock methods can be used to rectify each others' deficiencies. In particular, subhamiltonians can be used to study correlation energies and effects. Total correlation energies for all neutral atoms with

---

[1]current address: Department of Radiation Oncology, Division of Medical Physics, University of Pittsburgh, Pittsburgh, PA 15213

D. R. Herschbach et al. (eds.), Dimensional Scaling in Chemical Physics, 83–114.
© 1993 *Kluwer Academic Publishers. Printed in the Netherlands.*

$Z \leq 118$ have been obtained; there is good overall agreement with known values, though the breakdown of the energies indicates the presence of significant systematic errors. Dynamical effects of electron correlation can also be studied through analysis of the correlation-induced changes in the subhamiltonian interelectron distances.

Introduction

In this chapter we apply dimensional scaling techniques to the problem of electronic structure in many-electron atoms. As usual in the dimensional scaling approach, the motivating idea is to generalize the problem to spaces of arbitrary dimensionality D, treat it at one or more values of D where it's particularly easy to do so, and finally relate the results obtained back to $D = 3$. The $D \to \infty$ limit turns out to be the easiest place to treat the many-electron atom. In fact, one can obtain [1] analytic solutions for this limit, as well as for the first-order corrections at finite D. With some work these results can be used to calculate approximate solutions at $D = 3$. However, the raw $D \to \infty$ solutions do not correspond very well with our common notions of what an atom looks like.

The primary purpose of this chapter is to show how the $D \to \infty$ limit of the hamiltonian for a many-electron atom can be altered so that it accounts approximately for the effects of finite D. The resulting hamiltonians, which we will call subhamiltonians, are almost as simple as the $D \to \infty$ limit hamiltonians from which they are constructed. They are therefore in many respects quite crude. On the other hand, they incorporate no approximations which would destroy many-body effects, and so turn out to be quite useful for studying electron correlation.

Comparisons between dimensional scaling and more familiar techniques for treating atomic electronic structure, specifically orbital-based methods and density functional methods, will be drawn throughout this chapter. Such comparisons will be used to clarify the nature of the approximations arising during dimensional scaling treatments and to motivate the construction of subhamiltonians. The most important features of the overall solutions obtained for many-electron atoms are described, as well as results pertaining to electron correlation. Finally, avenues for further development, again motivated by comparisons with familiar techniques, will be offered in the last section.

Three limits and their associated models

The Hamiltonian for a many-electron atom,

$$\mathcal{H} = -\tfrac{1}{2} \sum_{i=1}^{N} \nabla_i^{\,2} - \sum_{i=1}^{N} \frac{Z}{r_i} + \sum_{\substack{i=1 \\ i<j}}^{N} \sum_{j=1}^{N} \frac{1}{r_{ij}}, \tag{1}$$

has at least three natural limiting forms which lead to simple yet useful perturbation expansions. These are the large-atom or Thomas-Fermi limit ($N\to\infty$ for fixed Z/N), the limit of infinite nuclear charge ($Z\to\infty$ for fixed N), and the limit of large spatial dimensionality ($D\to\infty$ for fixed Z and N). The first two limits are familiar and well-characterized, while the third is not.

The goal of this chapter is to use the $D\to\infty$ limit, but it is useful to draw some analogies and contrasts with the other limits. First, we note that all three limits appear on a somewhat equal footing if the hamiltonian is written out with explicit sums over the Cartesian axes and over the protons in the nucleus:

$$\begin{aligned}
\mathcal{H} = & -\tfrac{1}{2} \sum_{i=1}^{N} \sum_{\mu=1}^{D} \frac{\partial^2}{\partial x_{i\mu}^{\,2}} - \sum_{i=1}^{N} \sum_{r=1}^{Z} \left[\sum_{\mu=1}^{D} x_{i\mu}^{\,2} \right]^{-1/2} \\
& + \sum_{\substack{i=1 \\ i<j}}^{N} \sum_{j=1}^{N} \left[\sum_{\mu=1}^{D} (x_{i\mu} - x_{j\mu})^2 \right]^{-1/2} .
\end{aligned} \tag{2}$$

Table 1 outlines the basic characteristics of the three limits, including the asymptotic factors needed to finitize the length and energy scales, and the natural perturbation parameters for expansions about the limits. Finally, a graphical portrayal of the three limits is given in Figure 1, which shows for the argon atom the manner in which each limit "evolves" into the solution for the real atom as the relevant parameter is finitized.

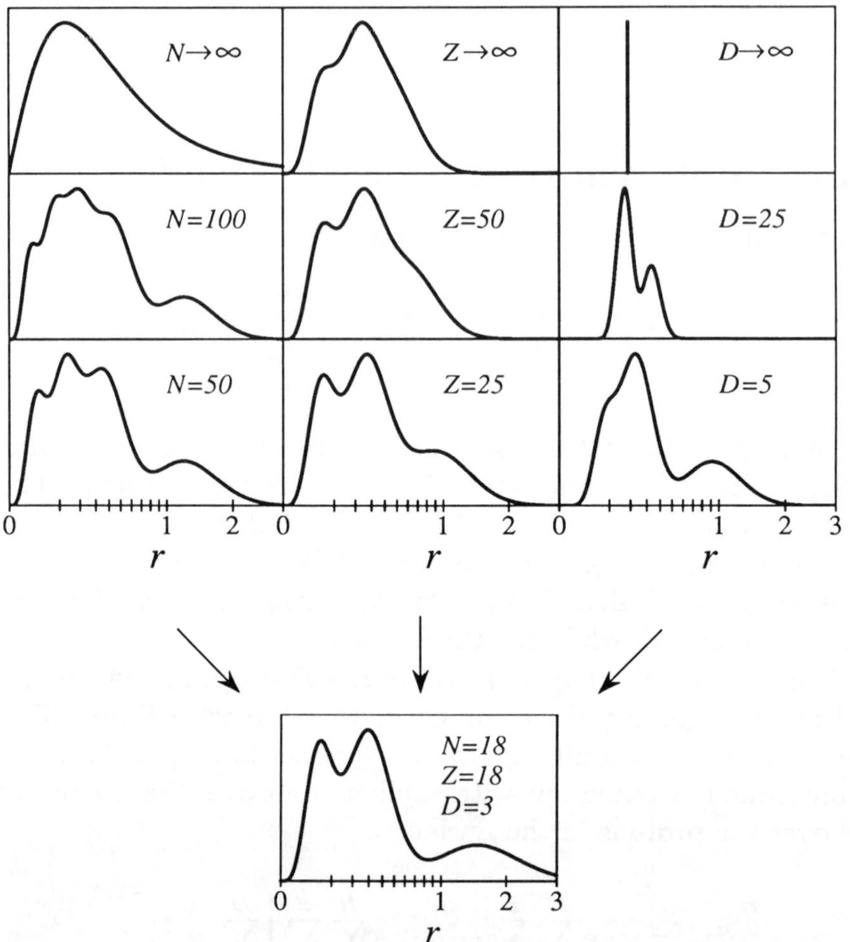

Figure 1. Evolution of the radial electron density for argon from each of the three limits described in Table 1. The limits (top row) are those given by the Thomas-Fermi approximation, the non-interacting electron model, and the large-dimension limit. The abscissae are linear in $\sqrt{r}$, where r is given in units of $(18/N)^{1/3}\,a_o$ for the first column, $(18/Z)\,a_o$ for the second, and $(D/3)^2\,a_o$ for the third. All curves except the Thomas-Fermi limit were obtained by adding Slater distributions whose exponents were determined from the subhamiltonian minima described in Sec. 4.

Table 1. Comparison of three limits for many-electron atoms [1,2,3].

	$N\to\infty$	$Z\to\infty$	$D\to\infty$
theory of the limit	Thomas- Fermi approximation	non-interacting electron model	large-dimension limit
nature of the limit: electrons behave	statistically	independently	classically
asymptotic behavior of energies	$N^{7/3}$	Z^2	D^{-2}
asymptotic behavior of distances	$N^{-1/3}$	Z^{-1}	D^2
perturbation expansion parameter	$N^{-1/3}$	Z^{-1}	D^{-1}
effective theory for finite N, Z, D:	density functionals	Hartree-Fock approximation	sub-hamiltonians

Below we will make the most use of the $Z\to\infty$ and $D\to\infty$ limits, and it will be convenient to have the Schrödinger equation in a form which can be analyzed easily in both limits. To obtain such a form, it is first necessary to use reduced units, so that expectation values for distances and energies will remain finite in the two limits; appropriate units (Table 1) are Ω/Z Bohr radii for distances and Z^2/Ω hartrees for energies, where $\Omega = \beta(\beta-N)$ and $\beta = \frac{D-1}{2}$. It is also advantageous to rewrite the Schrödinger equation as one for the probability amplitude Φ (related to the wave function Ψ by $|\Phi|^2 = J|\Psi|^2$, where J is the Jacobian which appears in the volume element). If one chooses the internal coordinates to be the electron-nucleus distances ρ_i and the cosines for the interelectron angles at the nucleus γ_{ij}, then the resulting S-state Schrödinger equation is [1]

$$(T' + T^\circ + V^\circ + V')\,\Phi = \epsilon\Phi, \tag{3}$$

where

$$\Omega\,T' = -\tfrac{1}{2}\sum_{i=1}^{N}\left(\frac{\partial^2}{\partial\rho_i^2} + \sum_{j\neq i}\sum_{k\neq i}\frac{\gamma_{jk}-\gamma_{ji}\gamma_{ik}}{\rho_i^2}\frac{\partial^2}{\partial\gamma_{ji}\partial\gamma_{ik}}\right)$$

$$T^\circ = \sum_{i=1}^{N}\frac{1}{2\rho_i^2}\frac{\Gamma^{(i)}}{\Gamma}$$

$$V^\circ = -\sum_{i=1}^{N}\frac{1}{\rho_i}$$

$$Z\,V' = \sum_{\substack{i=1\\i<j}}^{N}\sum_{j=1}^{N}\frac{1}{\sqrt{\rho_i^2 + \rho_j^2 - 2\rho_i\rho_j\gamma_{ij}}}.$$

Here Γ is the Gramian determinant $|\gamma_{jk}|_{1\leq j\leq N}^{1\leq k\leq N}$ for all N electrons, and $\Gamma^{(i)}$ is the Gramian determinant for all electrons except the ith. The definitions and properties of Gramian determinants are reviewed in the appendix to this chapter. Here we merely note that the effect of the generalized centrifugal potential T° is to work against those configurations which have small measure in the space of internal coordinates. Roughly speaking, this means that it favors those configurations in which electrons are far from the nucleus, and orthogonal to each other.

Eq. (3) shows that the $Z \to \infty$ and $D \to \infty$ limits are complementary to one another: the $D \to \infty$ limit is to the kinetic terms what the $Z \to \infty$ limit is to the potential terms. More precisely, the $D \to \infty$ limit amounts to ignoring the harder set of kinetic terms (the second derivatives T'), while the $Z \to \infty$ amounts to ignoring the harder set of potential terms (the interelectron repulsions V'). Moreover, one can systematically develop perturbation expansions ($1/D$ and $1/Z$ expansions) about each of these limits. It will be shown below that this complementarity is very useful, since it makes it possible to use one approach to correct the shortcomings of the other.

We now review the $D \to \infty$ solution for the many-electron atom problem. Since $\Omega \sim \frac{1}{4}D^2$, the limit is obtained by simply dropping the second derivative terms from Eq. (4),

$$\mathcal{H}_\infty = T^\circ + V^\circ + V'. \tag{4}$$

The solution is given by the minimum of this non-differential hamiltonian. Numerically, one finds that for a wide range of atoms and ions (including all neutral atoms with $Z < 14$ and all positive ions with $N/Z \lesssim 0.936$) the global minimum of $\mathcal{H}_\infty$ has maximal symmetry, meaning that all electrons are equidistant from the nucleus and equiangular with respect to one another. The solution is then given by [1]

$$\rho_\infty = \left(\frac{1 - \xi/N}{1 - \xi} \right)^2$$

$$\theta_\infty = \cos^{-1}\left(\frac{\xi}{\xi - N} \right) \tag{5}$$

$$\epsilon_\infty = -\frac{1}{2}\left(\frac{1 - \xi}{1 - \xi/N} \right)^3 (N - N\xi + \xi),$$

where ξ is the smallest positive root of $8NZ^2\xi^2(2 - \xi)^2 = (N - \xi)^3$. One can think of this solution in the following way: First place the nucleus at the origin and each electron along a distinct Cartesian axis at distance ρ_∞ from the nucleus; then open up the structure symmetrically (like a high-dimensional umbrella) until the interelectron angles all equal θ_∞. For neutral atoms and positive ions, ρ_∞ lies in the range $1.0 - 1.6$, and θ_∞ in the range $90° - 96°$.

The $D \to \infty$ solution is very simple, but it is clearly quite distant from the $D = 3$ solution. In particular, the equality of all electron-nucleus distances means that there is no shell structure. (This only appears at order $1/D$, when T' first comes into play; it may be noted that for larger atoms, where the global minimum of $\mathcal{H}_\infty$ no longer has maximal symmetry, the solutions do display a kind of meta-shell structure, but one which is apparently unrelated to the normal shell structure.) At first sight, the angular features of the solution also appear to be troublesome, since a set of angles all in the vicinity of $90°$ cannot be represented by a three-dimensional figure. However, this aspect turns out to be quite appropriate and desirable for any classical model of the electronic structure of an atom. This is simply because one would expect each interparticle distance in a classical model to correspond to some kind of average value in the true wavefunction, and any set of such representative values will in general define a higher-dimensional figure. For example, the set of expectation values for the 10 interparticle distances in beryllium, as obtained from a high-accuracy calculation, defines a four-dimensional figure.

Improved models for finite N, D, and Z

The three "raw" limits described in the preceeding section are valuable insofar as they are conceptually and analytically simple. However, each lies rather far from the domain of real atoms, and as a consequence is not especially useful as a starting point for quantitative treatments. In order to obtain a more useful starting point, it is necessary to insert at least some of the higher-order effects (of finite N, Z, or D) into the zeroth- order picture.

Several common electronic structure methods, as well as the sub-hamiltonian approximation we are about to describe, can be viewed in this way. First, various density functional approximations may be viewed as improvements upon Thomas-Fermi ($N \to \infty$) theory obtained by reintroducing some aspects of discreteness (finite N). In particular, it is usually desirable to revert to a discrete sum for the kinetic part of the hamiltonian, and only treat the potential terms using density functionals, as in the $X\alpha$ method [4]. Similarly, the Hartree and Hartree-Fock approximations may be viewed as improvements upon the non-interacting electron model ($Z \to \infty$) in which interelectron repulsions have been reintroduced in an approximate way [5]. Finally, a subhamiltonian constitutes an improvement upon the large-dimension limit model in which the effects of the derivative terms have been reintroduced.

The motivating idea for subhamiltonians can be seen in the following sequence of reductions for a one-particle Hamiltonian (for the probability amplitude Φ):

$$\mathcal{H} \quad = \quad \tfrac{1}{2}p_{\text{rad}}^2 + \tfrac{1}{2}p_{\text{ang}}^2 + V \tag{6}$$

$$\Big\downarrow \quad \text{valid for central } V$$

$$\mathcal{H}_{\text{rad}} \quad = \quad \tfrac{1}{2}p_{\text{rad}}^2 + \frac{\ell(\ell+1)}{2\rho^2} + V \tag{7}$$

$$\Big\downarrow \quad \text{valid for Coulombic } V$$

$$\mathcal{H}_{\text{sub}} \quad = \quad \frac{n^2 - \ell(\ell+1)}{2\rho^2} + \frac{\ell(\ell+1)}{2\rho^2} + V. \tag{8}$$

This reduction is amplified in Table 2. Each step represents a simplification achieved at the cost of restricted applicability and mode of representation [6]. The first step, which is valid for any central potential, depends upon the angular problem having been solved elsewhere, and results in a new hamiltonian (the radial hamiltonian) which yields only radial eigenfunctions. However, the simplified hamiltonian has the same spectrum as the original one. Similarly, the second step, which is valid only for Coulombic potentials, depends upon the radial problem having been solved as well, and results in a new hamiltonian (the subhamiltonian) which yields neither radial nor angular eigenfunctions. However, the subhamiltonian is still a nontrivial representation, and yields (by minimization) the correct eigenspectrum. For example, the subhamiltonian for a hydrogen atom is just

$$\mathcal{H}_{\text{sub}} = \frac{n^2}{2\rho^2} - \frac{1}{\rho}; \tag{9}$$

Table 2. Stepwise reduction of a one-particle hamiltonian.

hamiltonian	general	polar	radial	sub
symmetry $SO(\cdots)$	1	2	3	4
potential	$V(\rho,\theta,\varphi)$	$V(\rho,\theta)$	$V(\rho)$	$1/\rho$
kinetic operator in reduced form	$-\frac{1}{2}\nabla^2(\rho,\theta,\varphi)$	$-\frac{1}{2}\nabla^2(\rho,\theta)$ $+$ $m^2/2\rho^2\sin^2\theta$	$-\frac{1}{2}\nabla^2(\rho)$ $+$ $\ell(\ell+1)/2\rho^2$	$n^2/2\rho^2$
eigenfunctions $\begin{cases} \text{from } \mathcal{H} \rightarrow \\ \text{assumed} \rightarrow \end{cases}$	$\Phi(\rho,\theta,\varphi)$	$\Phi(\rho,\theta)$ $\times$ $\Omega_m(\varphi)$	$\Phi(\rho)$ $\times$ $\Omega_{\ell m}(\theta,\varphi)$	$\Omega_{n\ell m}(\rho,\theta,\varphi)$

minimization gives $\rho_n = n^2$, which is the radius (in Bohr radii) of a circular Bohr orbit for quantum number n, and $E_n = -1/(2n^2)$, which is the correct energy (in hartrees) for quantum number n.

The subhamiltonian for a one-electron atom has just been introduced through analogy with the familiar construction of radial hamiltonians. There are several other routes to the same result. Of particular relevance here is the method of shifted $1/D$ expansions [7], wherein one treats a problem as if it had extra degrees of freedom beyond D in order to achieve better convergence in the perturbation expansion. (Note that one does not alter the hamiltonian, which is still D-dimensional.) For a hydrogen atom with principle quantum number $n = m+1$, it can be shown that an expansion in powers of $\beta+m$ (where again $\beta = \frac{D-1}{2}$) will give the exact energy at zeroth order, and identically zero at all higher orders. The effective potential for the shifted $1/D$ expansion is just

$$\mathcal{H}_{\text{sub}} = \frac{(1+m/\beta)^2}{2\rho^2} - \frac{1}{\rho}, \tag{10}$$

which is the dimensionally generalized version of Eq. (9).

Many-electron atom subhamiltonians

Subhamiltonians for many-electron atoms may be constructed in a similar way. Roughly speaking, the procedure involves boosting the centrifugal potentials of the $D \to \infty$ limit hamiltonian in order to simulate the effects of the omitted second derivatives.

Consider an N-electron atom of nuclear charge Z, and let $n_i = m_i + 1$ denote the principle quantum number of the ith electron. A simple subhamiltonian may be constructed by boosting the centrifugal potential for each electron in Eq. (4) by the same factor used for a one-electron atom:

$$\mathcal{H}_{\text{simple}} = \sum_{i=1}^{N} \frac{(1+m_i/\beta)^2}{2\rho_i^2} \frac{\Gamma^{(i)}}{\Gamma} - \sum_{i=1}^{N} \frac{1}{\rho_i} + \frac{1}{Z} \sum_{\substack{i=1 \\ i<j}}^{N} \sum_{j=1}^{N} \frac{1}{\sqrt{\rho_i^2 + \rho_j^2 - 2\rho_i\rho_j\gamma_{ij}}}.$$

(11)

Note that for $D=3$ we have $\beta \equiv \frac{D-1}{2} = 1$, so $1+m_i/\beta = n_i$.

Minimization of this subhamiltonian with respect to the $N(N+1)/2$ internal coordinates ρ_i and γ_{ij} yields a classical model for the electronic structure of a many-electron atom. It turns out that the solutions generally appear to possess the full symmetry of the subhamiltonian, meaning that all electrons within a shell will be equivalent. If one just assumes this, then the subhamiltonian becomes especially simple. For example, for a $D = 3$ atom or ion with configuration $K^2 L^{N-2}$ (where K and L denote the first two shells — recall that electrons are differentiated on the basis of principle quantum number only), one has

$$\mathcal{H}_{\text{simple}} = 2 \left(\frac{1}{2\rho_{\text{K}}^2} \frac{\Gamma^{(\text{K})}}{\Gamma} - \frac{1}{\rho_{\text{K}}} \right) + (N-2) \left(\frac{4}{2\rho_{\text{L}}^2} \frac{\Gamma^{(\text{L})}}{\Gamma} - \frac{1}{\rho_{\text{L}}} \right)$$
$$+ \frac{1}{Z} \left(\frac{1}{\sqrt{2\rho_{\text{K}}^2(1-\gamma_{\text{KK}})}} + \frac{\frac{1}{2}(N-2)(N-3)}{\sqrt{2\rho_{\text{L}}^2(1-\gamma_{\text{LL}})}} + \frac{2(N-2)}{\sqrt{\rho_{\text{K}}^2 + \rho_{\text{L}}^2 - 2\rho_{\text{K}}\rho_{\text{L}}\gamma_{\text{KL}}}} \right)$$

(12)

where

$$\frac{\Gamma^{(\text{K})}}{\Gamma} = \frac{1 + (N-3)\gamma_{\text{LL}} - (N-2)\gamma_{\text{KL}}^2}{\left(1-\gamma_{\text{KK}}\right)\left([1+\gamma_{\text{KK}}][1+(N-3)\gamma_{\text{LL}}] - 2(N-2)\gamma_{\text{KL}}^2\right)}$$

$$\frac{\Gamma^{(L)}}{\Gamma} = \frac{[1+\gamma_{KK}][1+(N-4)\gamma_{LL}]-2(N-3)\gamma_{KL}^2}{\left(1-\gamma_{LL}\right)\left([1+\gamma_{KK}][1+(N-3)\gamma_{LL}]-2(N-2)\gamma_{KL}^2\right)}.$$

$$(13)$$

Minimization with respect to the five variables ρ_K, ρ_L, γ_{KK}, γ_{KL}, and γ_{LL} gives the electronic structure and energy.

For an explicit example of a solution, we consider the three-shell problem of argon ($K^2L^8M^8$). The minimum of $\mathcal{H}_{simple}$, now expressed in unreduced units (Bohr radii for distances, hartrees for energy), is at

$$r_K = 0.0568 \quad r_L = 0.2946 \quad r_M = 1.5063$$

$$\theta_{KK} = 90.57° \quad \theta_{KL} = 90.09° \quad \theta_{KM} = 90.00°$$
$$\theta_{LL} = 90.65° \quad \theta_{LM} = 90.10°$$
$$\theta_{MM} = 91.29°$$

$$(14)$$

$$E_{simple} = -509.904$$

This solution describes a (high-dimensional) configuration of electrons which are localized in three concentric shells. The geometry corresponds roughly to a set of expectation values or probability maxima for the real atom. (For comparison, the shell radii computed from Slater's rules [8], using $r_n^{eff} = n^2/Z_n^{eff}$, are $r_K = 0.0565$, $r_L = 0.2888$, and $r_M = 1.3333$, while the true nonrelativistic ground state energy [9] is -527.5 a.u. No angular expectation values are currently available for comparison, though one can anticipate that all values would be somewhere in the vicinity of 90°.)

Before considering other atoms, we note a problem with the above subhamiltonian, and make a correction for it. The problem is that we are seeking to incorporate into the centrifugal potential the effects of both the radial and angular second derivatives, but have paid attention to only the radial part of the problem in writing down Eq. (11). To make this clear, we note that the simple subhamiltonian could also be written

$$\mathcal{H}_{simple} = \hat{T}^\circ + V^\circ + V',$$

$$(15)$$

where $\hat{\,}$ indicates that variables in Eq. (4) have been scaled according to

$$\hat{} : \quad \begin{cases} \rho_i & \to \quad \dfrac{\beta}{\beta+m_i}\,\rho_i \\[2em] \gamma_{ij} & \to \quad \gamma_{ij}. \end{cases} \tag{16}$$

In order to correct for this deficiency, we observe that angular motion between any electron pair is generally associated with the sums of the quantum numbers of the individual electrons [1]. This suggests the following alternative form for the subhamiltonian:

$$\mathcal{H}_{\text{sub}} = \tilde{T} + V^\circ + V', \tag{17}$$

where $\tilde{}$ means that the following scalings have been performed:

$$\tilde{} : \quad \begin{cases} \rho_i & \to \quad \dfrac{\beta}{\beta+m_i}\,\rho_i \\[2em] \gamma_{jk}^{(i)} & \to \quad \dfrac{\beta+m_i}{\beta+m_j+m_k}\,\gamma_{jk}^{(i)}. \end{cases} \tag{18}$$

Here the superscript in $\gamma_{jk}^{(i)}$ indicates that this variable occurs in the expansion for $\Gamma^{(i)}/\Gamma$. It is important to note that there is really insufficient justification for Eq. (18) at this time. It represents a first attempt to construct a many-electron atom subhamiltonian, and should be considered only as a model or a rough approximation. For now, however, we will work with the subhamiltonian given by Eqs. (17) and (18). (In Sec. 6 we will describe a procedure by which one can probably construct many-electron atom subhamiltonians in a less *ad hoc* manner.)

The subhamiltonians given by Eqs. (17) and (18) can be minimized without difficulty. Again it is found that the minima generally possess full symmetry, so that all electrons within a shell are equivalent. For an atom or ion with configuration $K^2 L^{N-2}$, the subhamiltonian takes the form

$$\mathcal{H}_{\text{sub}} = 2\left(\frac{1}{2\rho_K^2}\frac{\Gamma^{(K)}}{\Gamma} - \frac{1}{\rho_K}\right) + (N-2)\left(\frac{4}{2\rho_L^2}\frac{\Gamma^{(L)}}{\Gamma} - \frac{1}{\rho_L}\right)$$

$$+\frac{1}{Z}\left(\frac{1}{\sqrt{2\rho_K^2(1-\gamma_{KK})}}+\frac{\frac{1}{2}(N-2)(N-3)}{\sqrt{2\rho_L^2(1-\gamma_{LL})}}+\frac{2(N-2)}{\sqrt{\rho_K^2+\rho_L^2-2\rho_K\rho_L\gamma_{KL}}}\right)$$

(19)

where

$$\frac{\Gamma^{(K)}}{\Gamma}=\frac{1+(N-3)\frac{1}{3}\gamma_{LL}-(N-2)(\frac{1}{2}\gamma_{KL})^2}{\left(1-\gamma_{KK}\right)\left([1+\gamma_{KK}][1+(N-3)\frac{1}{3}\gamma_{LL}]-2(N-2)(\frac{1}{2}\gamma_{KL})^2\right)}$$

$$\frac{\Gamma^{(L)}}{\Gamma}=\frac{[1+2\gamma_{KK}][1+(N-4)\frac{2}{3}\gamma_{LL}]-2(N-3)\gamma_{KL}^2}{\left(1-\frac{2}{3}\gamma_{LL}\right)\left([1+2\gamma_{KK}][1+(N-3)\frac{2}{3}\gamma_{LL}]-2(N-2)\gamma_{KL}^2\right)}.$$

(20)

As an example of a solution, we again consider argon (and again revert to unscaled units of Bohr radii and hartrees):

$$r_K=0.0569 \quad r_L=0.2946 \quad r_M=1.5013$$

$$\begin{aligned}
\theta_{KK}&=90.56° & \theta_{KL}&=90.25° & \theta_{KM}&=90.02° \\
& & \theta_{LL}&=91.37° & \theta_{LM}&=90.33° \\
& & & & \theta_{MM}&=93.04°
\end{aligned}$$

(21)

$$E_{\text{sub}}=-510.275$$

Overall, this is not too different from the result obtained from the simple subhamiltonian. The primary difference is that the angles are a little larger, and that the energy is slightly lower.

Under the assumption that its global minimum will have maximal symmetry, the subhamiltonian for any atom is readily minimized. The results for the energy are given in Fig. 2, while numbers for the rare gases are given in Table 3. It can be seen that the binding energy is always underestimated, though the error decreases from about 5% for small atoms to about 1% for very large atoms. (The asymptotic limit is 0.87%.) The electronic geometries are also generally speaking representative of expectation values except for the valence shell(s), whose radii tend to be significantly too large.

The deficiencies in the solutions derived from subhamiltonians can be at least partly attributed to the classical nature of the model. For

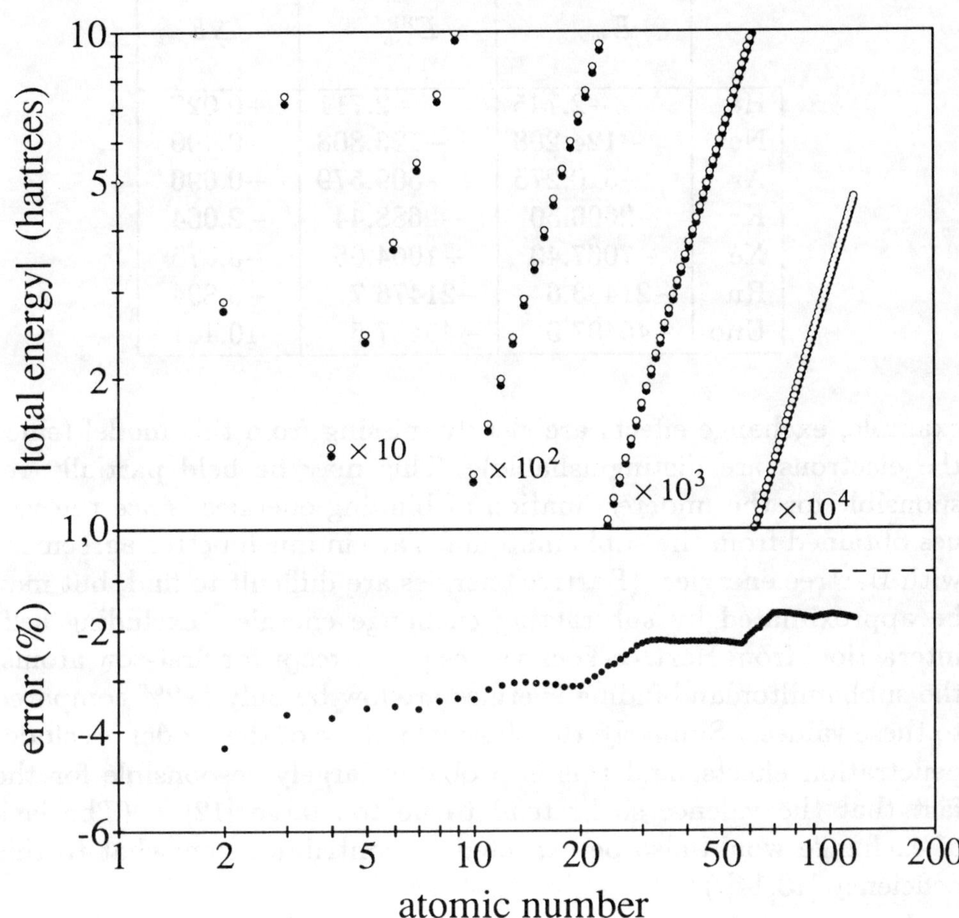

Figure 2. <u>Upper panel</u>: Log-log plot of total nonrelativistic atomic energies ($1 \leq N = Z \leq 118$) given by subhamiltonian minima ($\bullet$) and by accurate conventional calculations [10] ($\circ$). <u>Lower panel</u>: Percent error in the energies determined from subhamiltonians. Dashed line indicates the error as $N = Z \to \infty$.

Table 3. Energies computed for rare gases (in hartrees).

	E_{sub}	$E_{\text{sub}}^{\text{HF}}$	ΔE_{sub}
He	-2.745	-2.711	-0.027
Ne	-124.208	-123.808	-0.400
Ar	-510.275	-509.579	-0.696
Kr	-2690.50	-2688.44	-2.064
Xe	-7067.43	-7064.05	-3.375
Rn	-21483.6	-21476.7	-6.894
Uuo	-45497.9	-45487.5	-10.461

example, exchange effects are clearly missing from this model (since the electrons are distinguishable). This may be held partially responsible for the underestimation of binding energies, since the values obtained from the subhamiltonians are in much better agreement with Hartree energies. (Hartree energies are difficult to find, but may be approximated by subtracting exchange energies, excluding self-interaction, from Hartree-Fock values [11]; except for first-row atoms, the subhamiltonian binding energies are low by only 1–2% compared to these values.) Similarly, the classical nature of the model precludes penetration effects, and this is probably largely responsible for the fact that the valence shells tend to be too large [12]. (The lack of exchange would also be expected to contribute somewhat to this deficiency [13,14].)

In spite of their rough appearance, the solutions given by subhamiltonians are in one sense quite sophisticated, for they are fully correlated. This is because, in contrast to the other electronic structure methods discussed in Secs. 2 and 3, the subhamiltonian approximation does not invoke any averaging or uniformization of the potential terms of the hamiltonian. Thus, many-body effects are still present. These may be extracted by constructing and minimizing an uncorrelated (Hartree-Fock) version of the subhamiltonian, and comparing the results with those obtained above. (Because exchange is missing, we could equally well designate the uncorrelated version a Hartree subhamiltonian, but in order to agree with standard usage we will apply the label Hartree-Fock, or HF.)

The HF subhamiltonian is exceptionally simple. Since HF averaging removes any explicit dependence upon the interelectron angles from the potential terms of the Hamiltonian, these angles will be determined solely by the minima of the Gramian determinant ratios $\Gamma^{(i)}/\Gamma$. The minima are achieved when all interelectron angles are 90°. (A simple geometric proof follows from the observation, elaborated in the appendix, that Γ is the volume of the skewed N-dimensional cube defined by unit vectors pointing to the N electrons, while $\Gamma^{(i)}$ is defined similarly for all but the ith electron; thus each Gramian ratio is just the height of one unit vector relative to the base defined by the others.) The HF minimum can therefore be obtained by setting all $\gamma_{jk} = 0$ in the subhamiltonians considered above (either $\mathcal{H}_{\text{simple}}$ or $\mathcal{H}_{\text{sub}}$), which gives

$$\mathcal{H}^{\text{HF}}_{\text{sub}} = \sum_{i=1}^{N} \frac{(1+m_i/\beta)^2}{2\rho_i^2} - \sum_{i=1}^{N} \frac{1}{\rho_i} + \frac{1}{Z} \sum_{i=1}^{N} \sum_{\substack{j=1 \\ i<j}}^{N} \frac{1}{\sqrt{\rho_i^2 + \rho_j^2}}. \tag{22}$$

Again, $\beta = 1$ for $D = 3$.

The Hartree-Fock subhamiltonian is readily minimized, especially if one makes the assumption of maximal symmetry (which again appears in general to be justified). For example, for $K^2 L^{N-2}$ atoms or ions the symmetrized HF subhamiltonian is just

$$\begin{aligned}
\mathcal{H}^{\text{HF}}_{\text{sub}} = {} & 2 \left(\frac{1}{2\rho_K^2} - \frac{1}{\rho_K} \right) + (N-2) \left(\frac{4}{2\rho_L^2} - \frac{1}{\rho_L} \right) \\
& + \frac{1}{Z} \left(\frac{1}{\sqrt{2\rho_K^2}} + \frac{\frac{1}{2}(N-2)(N-3)}{\sqrt{2\rho_L^2}} + \frac{2(N-2)}{\sqrt{\rho_K^2 + \rho_L^2}} \right),
\end{aligned} \tag{23}$$

and for argon, the minimum is now

$$r_K = 0.0568 \quad r_L = 0.2946 \quad r_M = 1.5085 \tag{24}$$

$$E^{\text{HF}}_{\text{sub}} = -509.579$$

Overall, the solutions are quite similar to those obtained from the correlated subhamiltonian, except of course that the angles are now all 90°. In particular, one again finds that the radii are representative of the true radii except in the valence shells (which are too diffuse), and that energies are systematically low relative to the true Hartree-Fock energies (by 1–5%, with the largest errors for small atoms).

We note that the global minimization approach to the Hartree-Fock approximation just described is equivalent to the minimization of a set of coupled one-electron subhamiltonians. That is, one could instead minimize (by iteration to self-consistency) the set of one-electron subhamiltonians

$$\mathcal{H}^{\text{HF}}_{\text{sub},i} = \underbrace{\frac{(1+m_i/\beta)^2}{2\rho_i{}^2} - \frac{1}{\rho_i}}_{I_i} + \sum_{j\neq i}\underbrace{\frac{1/Z}{\sqrt{\rho_i{}^2 + \rho_j{}^2}}}_{J_{ij}}. \tag{25}$$

At self-consistency, the minimum of $\mathcal{H}^{\text{HF}}_{\text{sub},i}$ defines an orbital energy ϵ_i. Since the sum of orbital energies double-counts the interelectron repulsions, the total energy is given by either of the expressions

$$\begin{aligned}
E^{\text{HF}}_{\text{sub}} &= \sum_{i=1}^{N}\epsilon_i - \sum_{\substack{i=1 \\ i<j}}^{N}\sum_{j=1}^{N}J_{ij} \\
&= \sum_{i=1}^{N}I_i + \sum_{\substack{i=1 \\ i<j}}^{N}\sum_{j=1}^{N}J_{ij}.
\end{aligned} \tag{26}$$

It isn't hard to check that this procedure, which parallels the usual approach to the Hartree-Fock approximation, is equivalent to the global minimization.

Electron correlation

We will model electron correlation as the difference between the exact and Hartree-Fock subhamiltonian minima. In particular, we will approximate the correlation energy by

$$\begin{aligned}
\Delta E &\equiv E - E^{\text{HF}} \\
&\approx E_{\text{sub}} - E^{\text{HF}}_{\text{sub}},
\end{aligned} \tag{27}$$

and we will view the geometric changes in the subhamiltonian minima as modeling the changes in expectation values in the true electronic distribution induced by electron correlation.

Since the minima of subhamiltonians are only extremely rough models for real-world electronic structures, it is clear that their use for the modeling of correlation effects is only justified if there is a high degree of cancellation of errors. That is, the errors associated with the use of the subhamiltonian model in the correlated and un-correlated versions of the problem must be almost equal if the small difference associated with correlation is to be found. To see why such cancellation may be expected, we consider schematically the nature of the subhamiltonian and Hartree-Fock approximations. Recall that one could write the exact many-electron atom hamiltonian (for the probability amplitude) as $\mathcal{H}=T'+T^\circ+V^\circ+V'$, where the four sets of terms were respectively second derivatives, centrifugal potentials, electron-nucleus attractions, and electron-electron repulsions. Both the subhamiltonian and Hartree-Fock approximations may be viewed as adulterations to one set of terms in $\mathcal{H}$: the subhamiltonian approximation amounted to replacement of the second derivatives T' by additional centrifugal potentials (which may conveniently be denoted $\langle T'\rangle^\circ$), while the Hartree-Fock approximation involved simulation of the interelectron terms V' by means of additional central potentials (which may be denoted $\langle V'\rangle^\circ$). Thus, the energies in Eq. (27) are the eigenvalues of four hamiltonians connected according to the scheme

$$
\begin{array}{ccc}
\mathcal{H} & \xrightarrow{\ \langle\ \rangle^\circ \text{ on } V'\ } & \mathcal{H}^{\mathrm{HF}} \\[2pt]
\Big\downarrow{\scriptstyle\langle\ \rangle^\circ \text{ on } T'} & & \Big\downarrow{\scriptstyle\langle\ \rangle^\circ \text{ on } T'} \\[2pt]
\mathcal{H}_{\mathrm{sub}} & \xrightarrow{\ \langle\ \rangle^\circ \text{ on } V'\ } & \mathcal{H}^{\mathrm{HF}}_{\mathrm{sub}}
\end{array}
\qquad (28)
$$

Now the justification using Eq. (27) is basically the idea that the horizontal and vertical approximations operate at mutually exclusive sites in the hamiltonian, so that one can expect the approximations to be orthogonal to (or decouple from, or commute with) one another. Of course, the subhamiltonians themselves make it clear that all terms in the hamiltonian come into play in determining correlation effects, so the decoupling is only approximate.

The correlation energies obtained from Eq. (27) are more accurate than one might expect. This can be seen in Fig. 3, which shows the calculated correlation energies of neutral atoms through $Z = 118$ along with available reference values. The reference values plotted are "experimental" values [15,16] (total ionization energies minus accurate Hartree-Fock energies, with corrections for relativistic effects); second-order Møller-Plesset calculations [17,18]; and semi-empirical density functional calculations [19]. The subhamiltonian correlation energies are in very good overall agreement with the known values.

In spite of the good agreement with the known values, there are a couple of reasons why the subhamiltonian correlation energies for larger atoms should be regarded with some caution at this time. First, the subhamiltonians given by Eqs. (17) and (18) are still somewhat *ad hoc*. In fact, part of the justification for keeping this form was the fact that it worked so well. (In Sec. 6 we will discuss an alternative approach to constructing subhamiltonians which should remove the arbitrariness of the present form.) Second, some of the details of the calculations to be discussed below suggest that cancellation of errors may contribute to the agreement with known correlation energies.

With this caution in mind, we consider the correlation energies obtained from the subhamiltonians as preliminary estimates for the actual atomic correlation energies. The numbers show that the total correlation energies continue their faster-than-linear increase through the rest of the periodic table. Correlation energies of larger atoms are sometimes estimated by assuming a contribution of -0.04 a.u. per electron [20]. The basis for this rule was the approximate constancy of the correlation energy per electron for the second row atoms (which until fairly recently were the largest atoms for which total correlation energies were available). According to Fig. 3, this rule (the dotted line) leads to underestimation of the correlation energy in larger atoms by a factor of about two. The asymptotic behavior of the subhamiltonian correlation energies has not yet been analyzed, but it is not inconsistent with that obtained from large-D limit [1], namely $-0.0136\,Z^{4/3}$.

It can also be seen in Fig. 3 that the increase in the correlation energy is particularly rapid when inner shells are being filled, for example

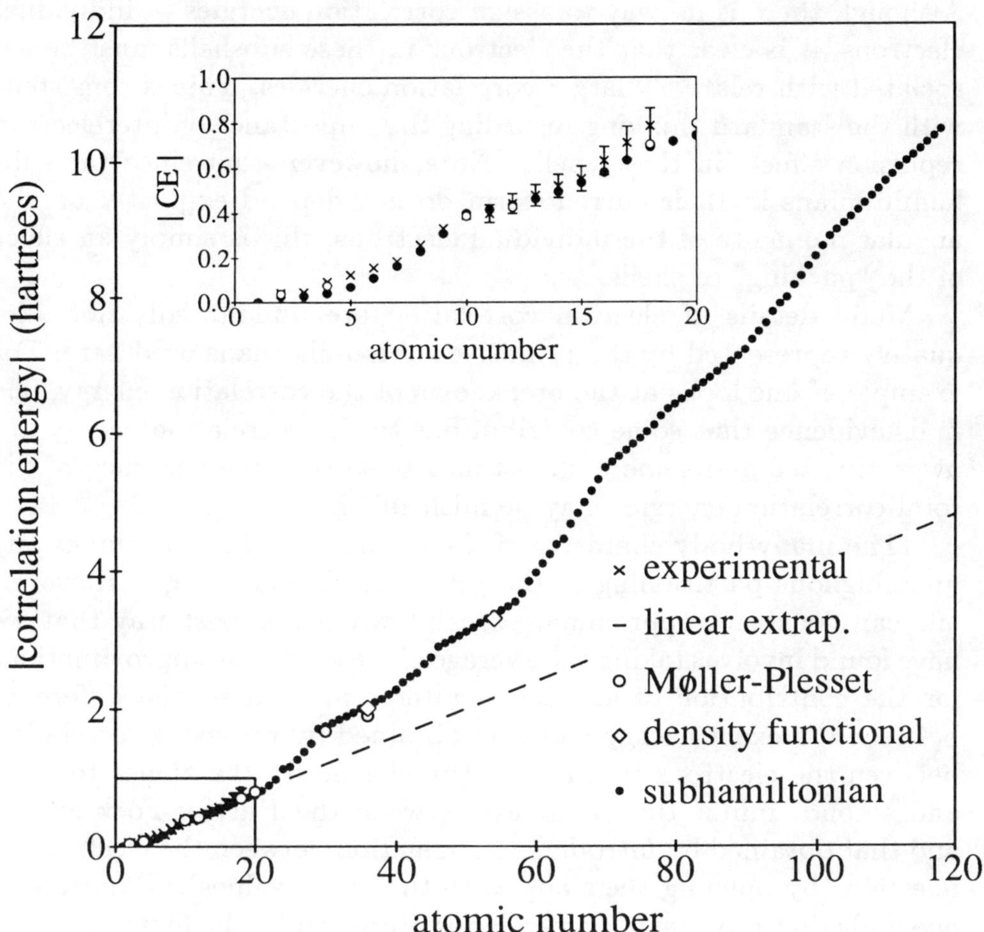

Figure 3. Comparison of total atomic correlation energies from several sources: difference between full and HF subhamiltonians (•); difference between total ionization energies and accurate HF calculations with relativistic corrections [15,16] (×); linear extrapolation of the latter values [20] (– – –); second-order Møller-Plesset calculations [17,18] (○); and semi-empirical density functional calculations [19] (◇).

for $N = 21\text{–}30$ (the $3d$ subshell) and $N = 57\text{–}71$ (the $4f$ subshell). Although there is no way to assign correlation energies to individual electrons, it is clear that the electrons in these subshells must be associated with relatively larger correlation energies. This is consistent with the standard thinking regarding the importance of interelectron repulsion effects in these shells. Note, however, that since the subhamiltonians in their current form do not depend explicitly on the angular momenta of the individual electrons, this is simply an effect of the "packing" of shells.

Many details of electron correlation are undoubtedly not adequately represented by the primitive subhamiltonians used here. For example, if one looks at the breakdown of the correlation energy, one finds evidence that some contributions to the correlation energy are overestimated and some underestimated, so that the accuracy of the total correlation energies may be misleading.

[The many-body character of the subhamiltonians precludes any unambiguous partitioning of the total correlation energy. However, one can construct approximate breakdowns. The best way that we have found involves taking the average of two different approximations for the contribution of any set of interactions: first, the difference between the exact energy and that obtained by *removing* correlation between the electrons in question (by closing up the angles to 90°); and second, minus the difference between the Hartree-Fock energy and that obtained by *introducing* correlation between the electrons in question (by opening their angles to the exact values). Empirically, one finds that the sum of pair energies produced by the first procedure underestimates the total correlation energy, while that given by the second procedure overestimates the total. However, the average of the two procedures almost reproduces the total correlation energy. We therefore consider the pair energies given by this average as a partitioning of the total.]

As an example, for argon one finds the breakdown (in hartrees)

$$\Delta E_{\text{KK}} = -0.031 \quad \Delta E_{\text{KL}} = -0.023 \quad \Delta E_{\text{KM}} = -0.000$$
$$\Delta E_{\text{LL}} = -0.422 \quad \Delta E_{\text{LM}} = -0.027 \tag{29}$$
$$\Delta E_{\text{MM}} = -0.193$$

$$\Delta E = -0.696$$

For comparison, one can consider the breakdown of the correlation energy given by second-order Møller-Plesset calculations [18]. These are only approximate, but they are the best calculations currently available, and moreover they can be partitioned into a sum of pair energies [21]. (True $D = 3$ correlation energies, like those obtained from the subhamiltonians, cannot be partitioned unambiguously.) The breakdown for argon is

$$\Delta E_{\mathrm{KK}} = -0.038 \quad \Delta E_{\mathrm{KL}} = -0.040 \quad \Delta E_{\mathrm{KM}} = -0.003$$
$$\Delta E_{\mathrm{LL}} = -0.291 \quad \Delta E_{\mathrm{LM}} = -0.081 \tag{30}$$
$$\Delta E_{\mathrm{MM}} = -0.253$$

$$\Delta E \;=\; -0.706$$

These numbers, together with those for other atoms for which comparison values are available [18], suggest that in general the subhamiltonian model will overestimate intrashell contributions to the correlation energy, except for the innermost and outermost shells, and underestimate intershell contributions. It might be argued that the process of forcing the electrons into shells at well-defined radii with respect to the nucleus, as in the subhamiltonian model, would be expected to increase intrashell correlation and decrease intershell correlation, except perhaps for the innermost and outermost shells. It might even be argued that there should be some degree of cancellation, so that the total correlation energy would remain fairly constant in spite of the shifting breakdown. However, other factors may also contribute, such as an inability of the subhamiltonian approximation to account for degeneracy effects in the valence shell. In any case, the inadequate modeling of the breakdown of the correlation energy must be viewed as a defect of the subhamiltonian model described above.

Of course, the lack of any reference to spin or orbital angular momentum quantum numbers within the subhamiltonians also means that any attempt at further dissection of the correlation energies will only lead to results which are uniformized over shells. This deficiency might be addressed by the program for improvement of the subhamiltonian model described in the next section.

Finally, it should be noted that the lack of exchange would be expected to lead to some overestimation of correlation effects, since exchange "pre- correlates" parallel-spin pairs of electrons and thereby decrcases the effect of true correlation. In larger atoms, however, this effect is not very important for the bulk of the electrons, since the Fermi hole between parallel-spin electrons (whose radius is of the order of the interelectron distance r_s) is small compared to the Coulomb screening distance (which is of order $r_s^{1/2}$) [22].

Discussion

The electronic structures generated by subhamiltonians are similar to those envisioned by G. N. Lewis and Irving Langmuir a decade before the development of quantum mechanics [23]. In the Lewis-Langmuir atomic models, point-like electrons were arranged symmetrically in shells about the nucleus. An octet, for example, had one electron at each corner of a cube. The subhamiltonian minima also describe symmetric configurations of point-like electrons. However, the structures are generally not three-dimensional. For this reason we refer to them as "hyper-Lewis" structures.

In retrospect, one can see why any attempt to represent the wavefunction of a many-electron system by a system of point particles *should* give a geometry which can't be visualized in three-space. For example, any set of representative (as opposed to instantaneous) interparticle distances, such as a set of expectation values, will generally define a structure which is higher than three-dimensional. In fact, the set of expectation values for an N-electron atom can be expected to define an N-dimensional figure, just like the subhamiltonian minimum.

Clearly the subhamiltonians described in this chapter represent only a first approximation, and they could undoubtedly be improved upon. For example, one might through more careful consideration of nodal structure [24] or expectation values [25] introduce distinctions based on the angular momentum quantum numbers ℓ_i, thereby generating subshell structure. Also, one might introduce quantum defects [26], which would have the effect of drawing in the presently overextended valence shells. There is, however, a more promising and less arbitrary way to improve the upon the subhamiltonians described in this chapter.

The route to improvement may once again be viewed through analogy with the more familiar Hartree-Fock method (Table 4). The idea is simply to view both the subhamiltonian and Hartree-Fock approximations as being the zeroth-order terms in perturbation expansions. Thus, for the Hartree-Fock approach, one writes

$$\mathcal{H} = \underbrace{T' + T^\circ + V^\circ + \langle V'\rangle^\circ}_{\mathcal{H}^{\text{HF}}} + \mu\left(V' - \langle V'\rangle^\circ\right), \qquad (31)$$

and systematically develops the solutions to the exact Schrödinger equation as perturbation expansions in μ (the Møller-Plesset expansion). Similarly, for the subhamiltonian approach one can write

$$\mathcal{H} = \underbrace{\langle T'\rangle^\circ + T^\circ + V^\circ + V'}_{\mathcal{H}_{\text{sub}}} + \nu\left(T' - \langle T'\rangle^\circ\right), \qquad (32)$$

and in principle one could try to improve upon the subhamiltonian approximation through a perturbation expansion in ν. Of course, the subhamiltonian model discussed in this chapter is still far from being the kinetic equivalent of the Hartree-Fock approximation. (Rather, it might be considered to be the analog of, say, an approximation using a single-zeta Hartree wavefunction.) This raises the question, however, of what the kinetic equivalent of the Hartree- Fock approximation might be.

In order to address this question, we first make note of the physical content of the perturbation expansion in ν. This is similar to that already encountered in the straightforward $1/D$ expansion [1]. First, the unperturbed hamiltonian is just the subhamiltonian, so the zeroth-order solution is given simply by the subhamiltonian minimum. For higher orders, we need to consider the effect of the perturbation, which is just the difference between the derivatives T' and the subhamiltonian approximation to the effect of those derivatives, $\langle T'\rangle^\circ$. At first order, T' will give rise to harmonic vibrations in all normal modes of the zeroth-order structure, with quantum numbers corresponding to the electron configuration; this is a positive contribution to the first-order energy. On the other hand, $-\langle T'\rangle^\circ$ clearly gives a negative contribution.

Table 4. Comparison of origins, characteristics, and extensions for Hartree-Fock and subhamiltonian models.

Method	$X' = V'$ $\left(\begin{smallmatrix}\text{interelectron}\\\text{repulsions}\end{smallmatrix}\right)$	$X' = T'$ $\left(\begin{smallmatrix}\text{second}\\\text{derivatives}\end{smallmatrix}\right)$
primitive theory (X' neglected)	non-interacting electron model	large-dimension limit approximation
perturbation expansion (in X')	$\dfrac{1}{Z}$ expansion	$\dfrac{1}{D}$ expansion
effective theory for finite Z or D ($X' \to \langle X'\rangle^\circ$)	Hartree-Fock approximation	subhamiltonian approximation
perturbation expansion (in $X' - \langle X'\rangle^\circ$)	Møller-Plesset expansion	(unnamed)
represented at zeroth order	exchange	correlation

The analogy between the new (ν) and Møller-Plesset (μ) expansions now suggests the natural way to improve the subhamiltonians. Just as the true Hartree-Fock solution (the Hartree-Fock limit) is characterized by the lack of any first-order corrections in the Møller-Plesset expansion (Brillouin's Theorem), so might one *define* the optimal subhamiltonian by the condition that all first-order effects in the resulting perturbation expansion vanish. The natural way to do this would probably be to demand that the positive and negative contributions cancel within each normal mode. This may be viewed as a many-particle generalization of the idea of shifted $1/D$ expansions [7]. With this definition, the subhamiltonian would probably need to be constructed by iteration to self-consistency, just like the Hartree-Fock hamiltonian. We note that for a hydrogen atom, the procedure just outlined leads right back to the hydrogenic subhamiltonian described in Sec. 3.

It appears that subhamiltonians constructed according to this new prescription would at least partially rectify several key deficiencies of the approximate forms discussed in this chapter. First, some simple experimentation within the Hartree-Fock approximation reveals that the adjustments will be modest in the inner shells, but will decrease the extent of the centrifugal boost in the valence shells. Thus, the procedure will have essentially the same effect as introducing quantum defects. This means that the valence shells will not be as overextended as they are in the current model. Second, the current complete symmetry among the electrons of a shell would probably be broken once the parameters of the centrifugal boost are allowed to adjust themselves in response the the vibrational state. In other words, the self-consistent subhamiltonian would probably have subshell structure. Finally, since symmetries with respect to particle interchange are reflected in the vibrational state, and therefore in the self-consistent subhamiltonian, there is even room for exchange effects to enter.

The ν perturbation expansion could in principle be carried to higher order, allowing one to obtain better $D = 3$ solutions from it. However, it would require quite a bit of effort to develop the expansion, and there is no reason to believe that it would be competitive with more familiar approaches, such as the μ expansion. (In fact, one might expect convergence problems to render it less useful). For now, it seems best to continue to regard the subhamiltonians as an auxiliary to conventional electronic structure methods. The combination of numerical accuracy and conceptual simplicity evidenced by the calculations described in this chapter suggest that subhamiltonians could turn out to be a useful alternative to other post-Hartree-Fock methods.

Appendix: Gramian determinants and centrifugal potentials

Gramian determinants and their ratios play a key role in the dimensional scaling treatment of many-electron atoms (and other many-body problems), so we summarize in this appendix several properties of these objects.

Consider a set of N unit vectors $\hat{\mathbf{r}}_i$ rooted at a common point, which can be taken to be the origin. Their Gramian determinant is defined as [27]

$$\Gamma = \begin{vmatrix} \hat{\mathbf{r}}_1 \cdot \hat{\mathbf{r}}_1 & \hat{\mathbf{r}}_1 \cdot \hat{\mathbf{r}}_2 & \cdots & \hat{\mathbf{r}}_1 \cdot \hat{\mathbf{r}}_N \\ \hat{\mathbf{r}}_2 \cdot \hat{\mathbf{r}}_1 & \hat{\mathbf{r}}_2 \cdot \hat{\mathbf{r}}_2 & \cdots & \hat{\mathbf{r}}_2 \cdot \hat{\mathbf{r}}_N \\ \vdots & \vdots & \ddots & \vdots \\ \hat{\mathbf{r}}_N \cdot \hat{\mathbf{r}}_1 & \hat{\mathbf{r}}_N \cdot \hat{\mathbf{r}}_2 & \cdots & \hat{\mathbf{r}}_N \cdot \hat{\mathbf{r}}_N \end{vmatrix} = \begin{vmatrix} 1 & \gamma_{12} & \cdots & \gamma_{1N} \\ \gamma_{21} & 1 & \cdots & \gamma_{2N} \\ \vdots & \vdots & \ddots & \vdots \\ \gamma_{N1} & \gamma_{N2} & \cdots & 1 \end{vmatrix} \tag{33}$$

where $\gamma_{jk} = \gamma_{kj} = \cos\theta_{jk}$. Numerically, Γ is equal to the square of the N- dimensional volume of the skewed hypercube defined by the $\hat{\mathbf{r}}_i$. Thus, it vanishes if the N unit vectors are linearly dependent, and Γ is essentially a quantitative measure of their independence.

Expansion of the determinant gives

$$\Gamma = 1 - \sum_{i \neq j}\sum \gamma_{ij}^2 + \sum_{i \neq j \neq k}\sum\sum 2\gamma_{ij}\gamma_{jk}\gamma_{ki}$$

$$- \sum_{i \neq j \neq k \neq \ell}\sum\sum\sum \left(2\gamma_{ij}\gamma_{jk}\gamma_{k\ell}\gamma_{\ell i} - \gamma_{ij}^2\gamma_{k\ell}^2 \right) + \dots . \tag{34}$$

where each sum is over all products that remain inequivalent after recognizing that $\gamma_{jk} = \gamma_{kj}$. The n-fold sum in this expansion consists of all contributions of degree n, and describes the non-orthogonality contributions of order n. In electronic structure applications, the vectors which describe the electronic positions are more or less orthogonal, and this is a fairly convergent expansion. Thus the leading-order contributions are the pairwise interactions.

Now consider N vectors $\mathbf{r}_i$ of arbitrary lengths $r_i = |\mathbf{r}_i|$. Although Gramian determinants are sometimes defined for non-unit vectors, it is more convenient for electronic structure applications to continue to keep the radial and angular aspects of the problem separated. Thus, we define the Gramian determinant for these vectors as above, with $\hat{\mathbf{r}}_i = \mathbf{r}_i/r_i$. The N-dimensional volume of the parallelotope (*i.e.*, generalized parallelepiped) defined by the $\mathbf{r}_i$ is clearly just $r_1 r_2 \cdots r_N \sqrt{\Gamma}$.

Similarly, if $\Gamma^{(i)}$ denotes the Gramian determinant for all but the ith vector (the above determinant with row i and column i deleted), then $r_1 \cdots r_{i-1} r_{i+1} \cdots r_N \sqrt{\Gamma^{(i)}}$ is the $(N-1)$-dimensional volume define by all but the ith vector. Therefore the ratio of these quantities, $r_i \sqrt{\Gamma/\Gamma^{(i)}}$, is the height of $\mathbf{r}_i$ with respect to the hyperplane define by the other $N-1$ vectors. For $N=3$, for example, $r_1 r_2 r_3 \sqrt{\Gamma}$ is the volume of the parallelepiped, $r_2 r_3 \sqrt{\Gamma^{(1)}}$ is the area of one face, and $r_1 \sqrt{\Gamma/\Gamma^{(1)}}$ is the height of $\mathbf{r}_1$ with respect to that face.

The generalized centrifugal potential T° defined by Eq. (3) can now be given a concrete geometric interpretation. The centrifugal potential of electron i is given simply by $T_i^\circ = 1/(2h_i^2)$, where $h_i = r_i \sqrt{\Gamma/\Gamma^{(i)}}$ is the distance of electron i from the hyperplane defined by the other electrons. (Thus, it might be more appropriate, albeit awkward, to call the T° a "cohyperplanifugal" potential, since its effect is to work against any configuration in which particles become cohyperplanar.) Loosely speaking, the effect of T° is to work against those configurations of particles which have small measure in the total configuration space.

The Gramian determinant ratios $\Gamma^{(i)}/\Gamma$ occurring in the centrifugal potential can also be expanded out in analogy with Eq. (34),

$$\frac{\Gamma^{(i)}}{\Gamma} = 1 + \sum_{\substack{j \\ (j \neq i)}} \gamma_{ij}^2 - \sum_{\substack{j \neq k \\ (j,k \neq i)}} \sum 2\gamma_{ij}\gamma_{jk}\gamma_{ki} + \ldots, \tag{35}$$

and again the series is fairly rapidly convergent since the cosines γ_{ij} are generally small. It isn't hard to see that truncation of this series after the quadratic terms results in a complete decoupling of the angular correlations between different electron pairs (since such truncation leaves no term in the hamiltonian with more than one angle γ_{ij}). It may be that different approximations to this series (exponential, geometric, etc.) will correspond with some of the approximations commonly used in treating electron correlation.

In the treatment of larger atoms, the evaluation of the Gramian determinants is the most time-consuming step. In the absence of assumptions regarding symmetry, it would undoubtedly be best to use elimination methods (requiring $\sim n^3$ operations) rather than explicit expansions ($\sim n!$ operations) for the evaluations. However, when the matrices are highly structured, explicit expansions become more useful. This is the case, for example, if one assumes that all electrons within a shell are equivalent. Suppose that the atom has Σ shells, with N_σ electrons in shell σ. Then the Gramian determinant has the form

$$\Gamma = \begin{vmatrix} \Gamma_{11} & \Gamma_{12} & \cdots & \Gamma_{1\Sigma} \\ \Gamma_{21} & \Gamma_{22} & \cdots & \Gamma_{2\Sigma} \\ \vdots & \vdots & \ddots & \vdots \\ \Gamma_{\Sigma 1} & \Gamma_{\Sigma 2} & \cdots & \Gamma_{\Sigma\Sigma} \end{vmatrix}, \tag{36}$$

where $\Gamma_{\sigma\sigma'}$ is an $N_\sigma \times N_{\sigma'}$ block, and the diagonal and off-diagonal blocks are of the forms

$$\Gamma_{\sigma\sigma} = \begin{pmatrix} 1 & \gamma_{\sigma\sigma} & \cdots & \gamma_{\sigma\sigma} \\ \gamma_{\sigma\sigma} & 1 & \cdots & \gamma_{\sigma\sigma} \\ \vdots & \vdots & \ddots & \vdots \\ \gamma_{\sigma\sigma} & \gamma_{\sigma\sigma} & \cdots & 1 \end{pmatrix} \qquad \Gamma_{\sigma\sigma'} = \begin{pmatrix} \gamma_{\sigma\sigma'} & \gamma_{\sigma\sigma'} & \cdots & \gamma_{\sigma\sigma'} \\ \gamma_{\sigma\sigma'} & \gamma_{\sigma\sigma'} & \cdots & \gamma_{\sigma\sigma'} \\ \vdots & \vdots & \ddots & \vdots \\ \gamma_{\sigma\sigma'} & \gamma_{\sigma\sigma'} & \cdots & \gamma_{\sigma\sigma'} \end{pmatrix}. \tag{37}$$

Expansion of such a determinant leads to an expression which factors to

$$\Gamma = (1 - \gamma_{11})^{N_1 - 1}(1 - \gamma_{22})^{N_2 - 1} \cdots (1 - \gamma_{\Sigma\Sigma})^{N_\Sigma - 1}\, \Upsilon. \tag{38}$$

The unfactorizable part Υ is a multinomial in the $\gamma_{\sigma\sigma'}$, at most quadratic in any one of these, with coefficients depending on the various N_σ. The number of terms in Υ depends on the number of shells as shown:

shells:	1	2	3	4	5	6	7
terms:	2	5	15	54	234	1218	7530

We don't know how to write down a general expression for Υ (and therefore Γ). However, the following algorithm allows one to determine a general expression for any atom with Σ shells from the expression for an atom with just 2 electrons in each shell: First, evaluate the $2\Sigma \times 2\Sigma$ determinant for the two-per-shell atom symbolically, and factor it. The resulting expression will be in the form of Eq. (38) with all $N_\sigma = 2$. Next, raise each factor $(1 - \gamma_{\sigma\sigma})$ to the appropriate power, namely $N_\sigma - 1$. Finally, in Υ replace each occurrence of $\gamma_{\sigma\sigma}$ by $(N_\sigma - 1)\gamma_{\sigma\sigma}$, and each occurrence of $\gamma_{\sigma\sigma'}$ by $\frac{1}{2}\sqrt{N_\sigma N_{\sigma'}}\gamma_{\sigma\sigma'}$. This yields an expression for Γ valid for any atom with Σ shells. Note also that the determinant $\Gamma^{(\sigma)}$ can be obtained by simply decrementing N_σ by 1 in this general expression.

Acknowledgments

This material is based upon work supported by the National Science Foundation under Grant. No. CHE-9007620 and by the Donors of The Petroleum Research Fund administered by the American Chemical Society. We thank John Summerfield for his valuable comments and questions.

References

1. J. G. Loeser, *J. Chem. Phys.* **86**, 5635 (1987).
2. W. Thirring, *A Course in Mathematical Physics*, Vol. 3 (Springer, NY, 1979), pp. 3 and 260.
3. B.-G. Englert, *Semiclassical Theory of Atoms* (Springer, NY, 1988), pp. 1–26.
4. T. Ziegler, *Chem. Rev.* **91**, 651 (1991).
5. S. M. Blinder, *Am. J. Phys.* **33**, 431 (1965).
6. S. M. Omohundro, *Geometric Perturbation Theory in Physics* (World Scientific, Singapore, 1986), p. 29.
7. A. Chatterjee, *Phys. Rep.* **186**, 249 (1990).
8. W. Kauzmann, *Quantum Chemistry* (Academic, NY, 1957), p.332.
9. J. C. Slater, *Quantum Theory of Matter*, 2nd ed. (McGraw-Hill, NY, 1968), p. 362.

10. J. P. Desclaux, *At. Data Nucl. Data Tables* **12**, 311 (1973).

11. J. C. Slater and J. H. Wood, *Int. J. Quant. Chem.* **4**, 3 (1971).

12. G. Plato, *Ann. Phys. (Leipzig) 5. Folge* **21**, 745 (1935).

13. D. R. Hartree and W. Hartree, *Proc. Roy. Soc.* **A166**, 450 (1938).

14. P. Gombás, *Theorie und Lösungsmethoden des Mehrteilchenproblems der Wellenmechanik* (Birkhäuser, Basel, 1950), p. 238.

15. A. Veillard and E. Clementi, *J. Chem. Phys.* **49**, 2415 (1968).

16. D. Feller, C. M. Boyle, and E. R. Davidson, *J. Chem. Phys.* **86**, 3424 (1987).

17. K. Jankowski, P. Malinowski, and M. Polasik, *J. Phys. B* **12**, 3157 (1979); *J. Chem. Phys.* **76**, 448 (1982).

18. V. Termath, W. Klopper, and W. Kutzelnigg, *J. Chem. Phys.* **94**, 2002 (1991).

19. J. P. Perdew, *Phys. Rev. B* **33**, 8822 (1986).

20. R. D. Cowan, *The Theory of Atomic Structure and Spectra* (Univ. of California, Berkeley, 1981), p. 203.

21. A. Szabo and N. S. Ostlund, *Modern Quantum Chemistry* (McGraw-Hill, NY, 1989) p. 352.

22. J. P. Perdew, in P. Vashishta, R. K. Kalia, and R. F. Bishop (eds.), *Condensed Matter Theories*, Vol. 2 (Plenum, NY, 1987), p. 89

23. W. B. Jensen, *J. Chem. Ed.* **61**, 191 (1984).

24. J. G. Leopold and I. C. Percival, *J. Phys. B: Atom. Molec. Phys.* **13**, 1037 (1980).

25. P. Gombás, *Pseudopotentiale* (Springer, NY, 1967).

26. M. J. Seaton, *Rep. Prog. Phys.* **46**, 167 (1983).

27. F. R. Gantmacher, *The Theory of Matrices*, Vol. 1 (Chelsea, NY, 1977), p. 247.

Chapter 4

LOW D REGIME

4.1   The One-Dimension Limit

David Z. Goodson and Mario López-Cabrera
Department of Chemistry
Harvard University
12 Oxford Street
Cambridge, MA 02138, USA

Abstract

With appropriate dimensional scalings, the $D \to 1$ limit of Schrödinger equations for coulombic systems provides important information about the dimension dependence of energy eigenvalues. The energy typically has a second-order pole at $D = 1$, the residue of which can often be exactly determined. We demonstrate this with some simple examples and then review a systematic procedure for characterizing a class of dimensional singularities found in coulombic problems.

Introduction

Dimensional continuation of the Schrödinger equation has yielded valuable insights into the properties of atoms and molecules. If we consider the energy eigenvalues to be a function $E(D)$ of the

D. R. Herschbach et al. (eds.), Dimensional Scaling in Chemical Physics, 115–130.
© 1993 *Kluwer Academic Publishers. Printed in the Netherlands.*

dimensionality D of Cartesian coordinate space, then the physical solution $E(3)$ is a special case of a more general problem and thus can be appreciated within a broader context. The approach to $D = 3$ from above has recently received much attention [1]. In E. A. Abbott's *Flatland* [2] the visitor from a three-dimensional space is able to understand much that seems mysterious to the two-dimensional flatlanders; similarly, various properties of the three-dimensional solutions of Schrödinger's equation are easier to understand from the perspective of higher dimensions. In the case of the helium atom, for example, the large-dimension limit sheds light on such mysteries as the apparent molecule-like motions of the electrons [3], the approximate separability of the wavefunction in terms of certain types of collective coordinates [4-6], and the rates of convergence of variational calculations [7-9]. Analysis of the Schrödinger equation at $D = 5$ has yielded information about three-dimensional excited states, explaining some striking similarities between seemingly unrelated eigenstates [9].

The approach to three-dimensions from below can also be quite instructive. In fact, the $D \to 1$ limit offers a particularly useful vantage point. This might seem surprising, as a one-dimensional atom behaves rather oddly. Its ground-state energy, for example, is infinite. One can introduce a dimensional scaling [10,11] to make this quantity finite, but then only the $1s$ electrons remain bound, and they tend to locate themselves at the cusps of the wavefunction, far from their most probable positions in three dimensions. Nevertheless, Loeser and Herschbach [11,12] discovered several years ago that by introducing some simple dimensional scalings and then interpolating between $E(1)$ and $E(\infty)$ they could obtain remarkably accurate estimates for $E(3)$. This result prompted an extensive study of the $D \to 1$ limit by Doren and Herschbach [13-17], which showed that some important physical effects are modeled at one dimension that are difficult to see from the large-dimension limit. The limits $D \to 1$ and $D \to \infty$ are complementary. They are complementary in the sense of the numerical interpolations of Loeser and Herschbach, but also in the sense that Bohr used the word. The large-dimension limit is a classical limit, with the effects of quantum mechanics approximated by a simple additional term in the potential function; the one-dimension limit has been referred to as a *hyperquantum limit* [11], because it emphasizes, in fact exaggerates, certain quantum effects.

It is clear from the success of the dimensional interpolation schemes [11,12] and from more recent work [15,18-23] that in order to obtain accurate numerical results from expansions about the large-dimension limit it is important to include information from one dimension. In this Chapter we describe how to define and solve one-dimensional Schrödinger equations for coulombic systems and then summarize the essential results concerning dimensional singularity analysis at $D = 1$. Section II contains a survey of some simple one-dimensional atoms and molecules. In Section III we review in some detail a method of dimensional singularity analysis developed by Doren and Herschbach [14,16]. The one-dimension energy eigenvalues turn out to be poles in the function $E(D)$, which result from the divergence at particle coalescences of the expectation values of coulomb potentials. Removing these poles is an important first step in the calculation of $E(3)$ from a $1/D$ expansion. We also briefly discuss the behavior of the energy at $D = -1, -3, -5, \ldots$. Excited states have poles at these negative dimensionalities.

Schrödinger's Equation in One Dimension

We begin with the hydrogen atom. The first step is to define what we mean by a one-dimensional Schrödinger equation. Perhaps the most straightforward approach would be simply to choose

$$\left(-\frac{1}{2}\frac{d^2}{dx^2} - \frac{1}{|x|} \right) \Psi = E\Psi, \tag{1}$$

by analogy with the three-dimensional case. However, our interest here is not in one-dimensional physics as such, but rather in the one-dimensional *limit* of the physically correct three-dimensional Schrödinger equation. Therefore, we begin with the standard D-dimensional continuation of the radial equation [10]

$$\left[-\frac{1}{2}\frac{d^2}{dr^2} - \frac{(D-1)}{2r}\frac{d}{dr} - \frac{1}{r} \right] \Psi = E\Psi, \tag{2}$$

and consider the behavior of its solutions in the limit $D \to 1$. Eq. (2) for arbitrary D can be solved [10] in the same manner as in the usual three-dimensional case. The ground-state eigenvalue is

$$E(D) = \frac{-2}{(D-1)^2} \tag{3}$$

and the expectation value of the radial coordinate is [11]

$$\langle r \rangle = \left(\frac{D-1}{2}\right)\left(\frac{D}{2}\right). \tag{4}$$

In the limit $D \to 1$ the electron becomes coincident with the nucleus, resulting in an infinite binding energy.

One might therefore conclude that a one-dimensional universe would be so foreign as to be of little relevance to a three-dimensional physicist. Suppose, however, that we introduce the change of variable

$$r = \left(\frac{D-1}{2}\right)\tilde{r}. \tag{5}$$

(The factor of 2 is arbitrary; we have included it simply as a matter of convenience, so that $\tilde{r} = r$ when $D = 3$. In general any scale factor that is proportional to $D - 1$ in the limit $D \to 1$ would work as well.) If we substitute Eq. (5) into Eq. (2), then we find that

$$\left[-\frac{1}{2}\frac{d^2}{d\tilde{r}^2} - \frac{(D-1)}{2\tilde{r}}\frac{d}{d\tilde{r}} - \frac{(D-1)}{2\tilde{r}}\right]\Psi = \tilde{E}\Psi, \tag{6}$$

with the new eigenvalue

$$\tilde{E} = \left(\frac{D-1}{2}\right)^2 E. \tag{7}$$

Eqs. (5) and (7) can be thought of as changes in the distance and energy units, respectively. In these new, dimension-scaled, units the radial expectation value, $\langle \tilde{r} \rangle = \frac{1}{2}D$, has a much tamer dimension dependence, and the new ground-state energy, $\tilde{E} = -\frac{1}{2}$, is completely independent of D. Thus, the one-dimension limit of the scaled Schrödinger equation, Eq. (6), is not such a bad model for the physical, three-dimensional, problem.

But what are we to make of the new dimension-scaled potential energy function, $\tilde{V}(\tilde{r}) = (D-1)/2\tilde{r}$, when $D = 1$? Herrick and Stillinger [10] have observed that $(D-1)/2r = \delta(x)$ in the one-dimensional limit, where x is the cartesian coordinate, $-\infty < x < \infty$, and $\delta(x)$ is the Dirac delta function. This surprising result is a consequence of the dimension dependence of the Jacobian volume element, $r^{D-1}dr$, which leads to the identity [14]

$$\int_0^\infty \Psi^* \frac{1}{r}\Psi r^{D-1}dr = \frac{1}{D-1}\int_0^\infty \Psi^*\delta(r)\Psi dr$$

$$+ \int_0^1 [\Psi^*(r)\Psi(r) - \Psi^*(0)\Psi(0)]\frac{1}{r}r^{D-1}dr$$

$$+ \int_1^\infty \Psi^*\frac{1}{r}\Psi r^{D-1}dr. \tag{8}$$

The second and third terms on the right-hand side are finite at $D = 1$, so they become insignificant compared to the first term in the one-dimension limit. Note that $\delta(r) = 2\delta(x)$, since the former is normalized on the domain $0 \le r < \infty$ while the latter is normalized on a domain that is twice as large, $-\infty < x < \infty$. Thus,

$$\int_0^\infty \Psi^* \left(\frac{D-1}{2\tilde{r}}\right) \Psi\tilde{r}^{D-1}d\tilde{r} \to \int_{-\infty}^\infty \Psi^*\delta(\tilde{x})\Psi d\tilde{x}. \tag{9}$$

Hence, our prescription for constructing the one-dimension limit of a Schrödinger equation is to replace the coulomb potential $1/r$ with $\delta(\tilde{x})$.

Let us return to the hydrogen atom. The Laplacian operator in one dimension is simply d^2/dx^2, so our one-dimensional Schrödinger equation is

$$\left[-\frac{1}{2}\frac{d^2}{dx^2} - \delta(x)\right] \Psi = \tilde{E}\Psi, \tag{10}$$

instead of Eq. (1). (We write x, instead of $\tilde{x}$, with the understanding that from now on we will be using the scaled distance units.) Integration of Eq. (10) over the real axis gives the following condition on the first derivatives of the wavefunction:

$$\frac{d}{dx}\Psi(x)\bigg|_{x\to 0_-} - \frac{d}{dx}\Psi(x)\bigg|_{x\to 0_+} = 2\Psi(0). \tag{11}$$

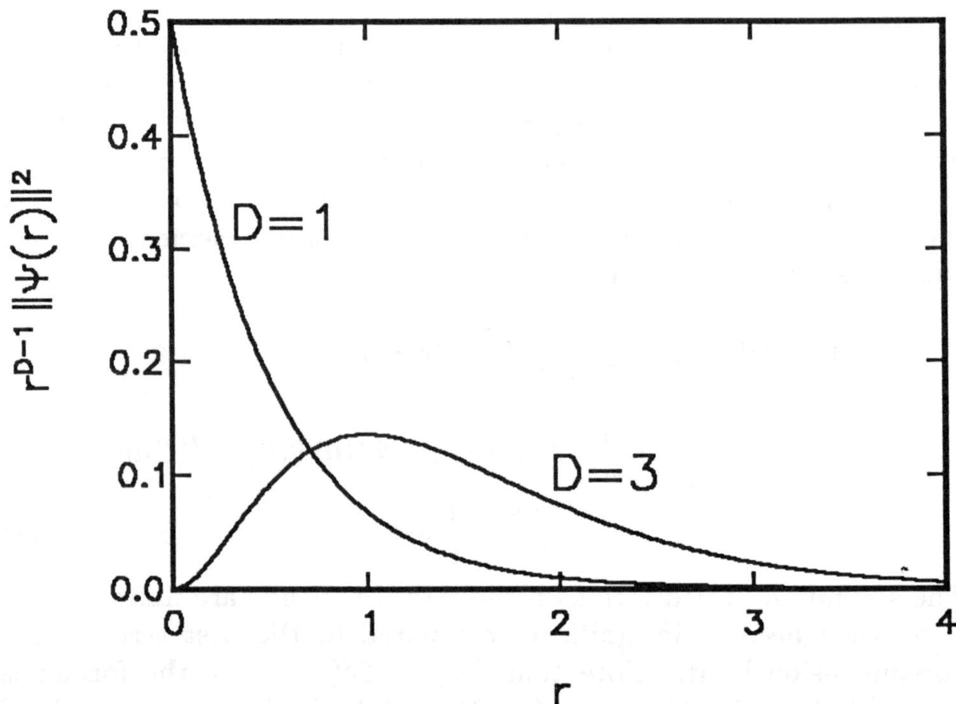

Figure 1. Radial probability distribution for the dimension-scaled hydrogen atom, for $D = 1$ and $D = 3$.

We find that the *only* bound-state eigenfunction is

$$\Psi(x) = e^{-|x|}. \tag{12}$$

Substituting this into Eq. (10), we find the eigenvalue $\tilde{E} = -\frac{1}{2}$, as we noted above. Now suppose that we undo the energy scaling of Eq. (7), expressing the energy in simple atomic units. The result, then, for the unscaled one-dimensional eigenvalue is identical to Eq. (3), which is the exact solution for arbitrary D. Thus, we have determined the full solution $E(D)$ by solving only the particularly easy case $D = 1$.

Fig. 1 compares the radial probability distributions, $r^{D-1}|\Psi(r)|^2$, at $D = 3$ and $D = 1$ in our dimension-scaled coordinate system. At $D = 3$ the most probable radial distance is $r = 1$; although the wavefunction is finite at $r = 0$, the D-dimensional volume element r^{D-1} insures that the probability is zero for the electron to be coincident with the nucleus. At $D = 1$ the volume element is unity and $r = 0$ is the most probable configuration. This seemingly unphysical result in fact contains some important physical information that is absent from the semiclassical $D \to \infty$ limit. With appropriate dimensional scalings [11], the large-D limit of the radial probability distribution becomes sharply peaked at $r = 1$. Since $r = 1$ is also the most probable location at $D = 3$, the large-D limit can be an excellent qualitative model for the physical system. However, for an accurate treatment of three-dimensional problems with coulomb potentials it is important to also include the behavior of the wavefunction at the cusps, where particles become coincident [24]. The one-dimension limit strongly emphasizes the behavior at these cusps. Thus, an extrapolation between $D \to 1$ and $D \to \infty$ [11,12] in effect builds into the large-D limit the proper cusp conditions. We will make this argument more precise in the next Section.

We can compute other properties of the one-dimensional hydrogen atom as well. Consider, for example, the behavior of the atom when subjected to an external field. Interaction with a uniform electric field $\mathcal{E}$ introduces a perturbing potential $H_\mathcal{E} = -\mathcal{E} \cdot \mathbf{r}$. Within first-order perturbation theory this term has no effect, since the ground state of hydrogen does not have a permanent dipole moment. Within second-order perturbation theory, however, there is a shift in the energy due to an induced dipole moment; the shift is proportional to the square of the field strength, $\Delta E = -\frac{1}{2}\alpha_p \mathcal{E}^2$, and the constant α_p is called the *polarizability*. Suppose that we define the one-dimension limit of the perturbing potential in the most straightforward way, $H_\mathcal{E} \to -\mathcal{E}x$. The first-order correction to the wavefunction, Ψ_1 is determined by the equation

$$(H_0 - \tilde{E}_0)\Psi_1(x) = \left(\frac{D-1}{2}\right)^3 \mathcal{E}x\Psi_0(x), \tag{13}$$

with the factor $(D-1)^3$ due to the dimensional scaling of the distance and energy units. Thus, in the limit $D \to 1$ the field does not cause an energy shift.

In the case of the unperturbed hydrogen atom we found that a direct solution at $D = 1$ gave an unphysical result, but once we introduced appropriate dimensional scalings we found that the energy at $D = 1$ became equivalent to the solution at $D = 3$. Let us attempt to improve our polarizability calculation in a similar fashion. Suppose we express the field strength in the form

$$\mathcal{E} = \left(\frac{2}{D-1}\right)^3 \tilde{\mathcal{E}}, \tag{14}$$

with the scaled field strength $\tilde{\mathcal{E}}$ assumed to be independent of dimension. Then the right-hand side of Eq. (13) remains nonzero in the one-dimension limit. The solution for Ψ_1 is

$$\Psi_1(x) = -\tfrac{1}{2}\tilde{\mathcal{E}}\exp[-x(x^2 + |x|)]. \tag{15}$$

Substituting this into Eq. (13) yields $\Delta\tilde{E} = -\tfrac{5}{8}\tilde{\mathcal{E}}^2$. Undoing the scalings of E and $\mathcal{E}$, we find that

$$\alpha_p = \frac{5}{64}(D-1)^4. \tag{16}$$

The exact solution for arbitrary D is [22]

$$\alpha_p(D) = \frac{1}{64}(D-1)^4(D+1)(D+\tfrac{3}{2}). \tag{17}$$

Thus, the one-dimension limit for the polarizability does not give the full solution; however, it differs only by a quadratic polynomial in D.

We consider next the simplest molecule, H_2^+. Let R be the internuclear distance, within the Born-Oppenheimer approximation. Expressing distance and energy in the dimension-scaled units, the one-dimensional Schrödinger equation becomes

$$\left[-\frac{1}{2}\frac{d^2}{dx^2} - \delta(x + \tfrac{1}{2}\tilde{R}) - \delta(x - \tfrac{1}{2}\tilde{R})\right]\Psi = \tilde{E}\Psi, \tag{18}$$

with the scaled internuclear distance $\tilde{R}$

$$R = \left(\frac{D-1}{2}\right)\tilde{R}. \tag{19}$$

(This dimensional scaling of the internuclear distance is an important subtlety, which we will discuss further in Chapter 7.1.) Eq. (18) was studied by Frost [25], who suggested it as an approximate model for three-dimensional H_2^+. The unnormalized solution for the symmetric bound-state wavefunction is

$$\Psi(x) = \begin{cases} (1 + e^{k\widetilde{R}})e^{kx} & \text{for } x < -\frac{\widetilde{R}}{2}, \\ e^{kz} + e^{-kz} & \text{for } -\frac{\widetilde{R}}{2} \le x \le \frac{\widetilde{R}}{2}, \\ (1 + e^{k\widetilde{R}})e^{-kx} & \text{for } x > \frac{\widetilde{R}}{2}, \end{cases} \quad (20)$$

where k is determined by the transcendental equation

$$k - 1 - e^{-k\widetilde{R}} = 0. \quad (21)$$

The energy eigenvalue is $\widetilde{E} = -\frac{1}{2}k^2$. The one-dimensional probability distribution is peaked at the nuclei while the three-dimensional solution is peaked between the nuclei and displaced somewhat from the internuclear axis.

Schrödinger's equation for S states of the one-dimensional helium atom is

$$\left[-\frac{1}{2}\left(\frac{d^2}{dx_1^2} + \frac{d^2}{dx_2^2} \right) - 2\delta(x_1) - 2\delta(x_2) + \delta(x_1 - x_2) \right] \Psi = \widetilde{E}\Psi. \quad (22)$$

This equation was studied by Rosenthal [26] as an approximate model for three-dimensional helium. He was able to reduce the problem to the solution of a one-variable integral equation, which can be solved numerically for $\widetilde{E}$ to arbitrary precision. Only electrons in $1s$ orbitals are bound. Thus, the ground state binds both electrons but all excited states ionize.

Coulombic Poles

The most straightforward way to connect the large-D limit to the physical three-dimensional solution is to use an asymptotic expansion, which for coulombic problems has the form [1,18]

$$D^2 E(\delta) = \sum_{i=0}^{\infty} E_i \delta^{-i}, \quad (23)$$

where $\delta \equiv 1/D$. The factor D^2 is a dimensional scaling of the energy units that is needed to insure that the limit $D \to \infty$ be well-defined.

Unfortunately, one finds in practice that there are values of δ for which the summation in Eq. (23) does not converge, and even when it does converge the rate of convergence tends to be rather slow. The reason for these convergence problems is the presence of singularities in the function $D^2 E(\delta)$. If the summation in Eq. (23) is truncated at any finite order, then this function is approximated as a polynomial in δ. A polynomial is an analytic function, so if there are values of δ such that the exact solution is singular then Eq. (23) will not be able to provide an accurate model near the singular points.

A characteristic feature of ground state energies of coulombic systems is the presence of a second-order pole at $\delta = 1$. The origin of these poles has been explained by Doren and Herschbach [14,16] in terms of an analysis of the Schrödinger equation at particle coalescences. Their method of analysis allows one to predict the locations and types of a certain class of dimensional singularities without actually having to solve for the function $E(\delta)$. We will illustrate this first for central-potential problems and then for many-particle systems.

The D-dimensional matrix element of an operator f is defined as

$$\langle f \rangle \equiv \int_0^\infty \Psi^* f \Psi r^{D-1} dr \Big/ \int_0^\infty \Psi^* \Psi r^{D-1} dr. \tag{24}$$

Our principal interest is in coulombic potentials, so let us assume that $V(r) \sim r^{-1}$ for small r. Now suppose $\Psi(r)$ is nonzero at $r = 0$, and consider the term $\langle V \rangle \sim \langle r^{-1} \rangle$ in the expectation value of the Hamiltonian. Then the integrand of the numerator in Eq. (24), with $f = V$, is proportional to r^{D-2} when r is small. Thus, the integral will be divergent for $D = 1$.

This argument identifies the location of a singularity, but it does not elucidate its type, for example, whether it is a pole or a branch point. Since the divergence occurs at the limit $r \to 0$, the singularities of E can be characterized by determining the behavior of Ψ at small r. In the neighborhood of the origin, $\Psi(r)$ formally has the expansion

$$\Psi(r) \sim c_0 + c_1 r + c_2 r^2 + \dots, \tag{25}$$

where the c_i are functions of D. Substituting Eq. (25) into the Schrödinger equation and expanding the potential in the form

$$V(r) \sim -r^{-1}(V_0 + V_1 r + V_2 r^2 + \cdots), \tag{26}$$

we obtain a recursive solution for the c_i,

$$c_i = -\frac{2}{i(D+i-2)}\left(Ec_{i-2} + \sum_{j=0}^{i-1} V_j c_{i-1-j}\right), \qquad (27)$$

with $c_{-1} \equiv 0$. Suppose that $c_0 \neq 0$. Then $c_1 = -2V_0 c_0 (D-1)^{-1}$, which is singular at $D = 1$.

To see how this singularity in the wavefunction translates into a singularity in E, we introduce the dimensional scaling of the distance units by replacing r with $(\frac{D-1}{2})\tilde{r}$ according to Eq. (5). Then the Schrödinger equation takes the form

$$\left[-\frac{1}{2}\frac{d^2}{d\tilde{r}^2} - \frac{(D-1)}{2\tilde{r}}\frac{d}{d\tilde{r}} - \left(\frac{D-1}{2}\right)V(\tilde{r})\right]\Psi = \tilde{E}\Psi, \qquad (28)$$

where $\tilde{E} \equiv (\frac{D-1}{2})^2 E$. In place of Eq. (27) we now have

$$c_i = -\frac{2\tilde{E}}{i(D+i-2)}c_{i-2} - \frac{1}{i}\left(\frac{D-1}{D+i-2}\right)\sum_{j=0}^{i-1} V_j c_{i-1-j}\left(\frac{D-1}{2}\right)^j, \qquad (29)$$

which is regular at $D = 1$. Thus, the dimensional scaling that removes the singularity from the wavefunction introduces a second-order pole into the energy, $E = 4\tilde{E}(D-1)^{-2}$. The residue $4\tilde{E}$ is determined from the eigenvalue of the one- dimensional Schrödinger equation constructed from the prescription given in the previous Section.

Eq. (27) implies that the wavefunction will have dimensional singularities at those values of D such that $(D+i-2)$ can be zero; that is, $D=0, -1, -2,\ldots$, in addition to the case $D = 1$ just discussed. However, when $D < 1$ these singularities will appear in both the numerator and the denominator of Eq. (24) for $\langle H \rangle$; they cancel out, leaving E regular. As long as $c_0 \neq 0$, the only such dimensional singularity in E will be at $D = 1$.

Our assumption that $c_0(D)$ be nonzero at $D = 1$ is not generally valid. In fact, one can prove [14] that the assumption will hold only for the ground state; $c_0(1) = 0$ for any excited state, which then implies that E is regular at $D = 1$. We can locate the excited-state singularities by determining the value of D at which the matrix element $\langle V \rangle$ diverges, using Eq. (27) as before. Suppose that $c_0 = 0$

but $c_1 \neq 0$. Then the integrand at small r will behave as r^D, which implies that $\langle V \rangle$ will diverge at $D = -1$. Similarly, if the first nonzero coefficient of the expansion of the wavefunction is c_j, then $\langle V \rangle$ will diverge at $D = 1 - 2j$. Therefore, we need to determine for each excited state the value of j corresponding to the first nonzero coefficient at the potentially singular values of $D = -1, -3, -5, \ldots$. A detailed analysis [14] of the scaled recursion relation, Eq. (29), reveals that $j = n_r$, where n_r is the number of radial nodes in Ψ. The singularity occurs at $D = 1 - 2n_r$ and the scaling $r \to (D - 1 + 2n_r)\tilde{r}$ moves the singularity from the wavefunction to the eigenvalue, with the result $E \sim (D + 2n_r - 1)^{-2}\tilde{E}$.

The analysis thus far only applies to states with $l = 0$. For any central-potential problem there exists [27] an isomorphism between angular momentum and dimensionality such that a unit increment in l is equivalent to an increment of two units in D. This *interdimensional degeneracy* allows us to generate the energy for nonzero l from the $l = 0$ result through the transcription $D \to D + 2l$. The general form then for the energy of the one-electron atom in the neighborhood of this dimensional singularity is

$$E \sim \tilde{E}(D + 2n - 3)^{-2}, \tag{30}$$

where $n = n_r + l + 1$ is, by definition, the principal quantum number. Thus, the dimension-scaled eigenvalue $\tilde{E}$ at $D = 3 - 2n$ gives the residue of a second-order pole of the unscaled energy function $E(D)$.

This singularity analysis is applicable to any Schrödinger equation for a central potential, and the resulting singularity structure depends only on the behavior of the potential in the limit $r \to 0$. Any system whose potential at small r is proportional r^{-1} will have a second-order pole according to Eq. (30). However, it is important to note that this second-order pole is only the most singular behavior at $D = 3 - 2n$; in general, one can also expect a coincident first-order pole,

$$E \sim \frac{a_{-2}}{(D + 2n - 3)^2} + \frac{a_{-1}}{D(D + 2n - 3)}. \tag{31}$$

In principle there could also be other types of singularities at $D = 3 - 2n$ as long as they do not diverge as quickly as a second-order pole. In the case of the one-electron atom the residue a_{-1} of the first-order pole is equal to zero, but for other systems, the Yukawa potential for example [14], it is nonzero.

The analysis is also applicable to relativistic problems. Consider the Klein-Gordon equation , which can be written in the form [20]

$$\left[-\frac{1}{2}\frac{d^2}{dr^2} - \frac{(D-1)}{2r}\frac{d}{dr} + V(r) - \left(\frac{\gamma^2}{2}\right)\frac{1}{r^2}\right]\Psi = \epsilon\Psi, \qquad (32)$$

with $\gamma = Z\alpha$, where α is the fine structure constant. This equation describes a spinless particle in a central potential. The eigenvalue ϵ is not the energy of the system, but is related to the energy according to

$$\epsilon = \frac{(1 + \frac{1}{2}\gamma^2 E)}{(1 + \gamma^2 E)^2}\, E. \qquad (33)$$

Eq. (32) has the same form as the Schrödinger equation except that the potential energy has added to it an additional term proportional to r^{-2}. Again we assume that $V(r)$ behaves as r^{-1} near the origin. It is straightforward to derive the recursion relation for the coefficients of the expansion of the wavefunction, corresponding to Eq. (27), and then introduce a dimensional scaling of r that transfers the singularity from the wavefunction to the energy, with the result [20]

$$\epsilon \sim \tilde{\epsilon}(\bar{D} + 2n - 3)^{-2}, \qquad (34)$$

where

$$\bar{D} = \left[(D - 2 + 2l)^2 - 4\gamma^2\right]^{1/2} + 2 - 2l. \qquad (35)$$

$\bar{D} - D$ is of order γ^2, so the pole in the eigenvalue is only slightly displaced from its location in the nonrelativistic case. Solving Eq. (34) for E in terms of $\tilde{\epsilon}$ gives

$$E \sim -\gamma^{-2} + \gamma^{-2}\left(1 - 2\tilde{\epsilon}\gamma^2\left\{D^\dagger - D^* + [(D - D^\dagger)^2 - 4\gamma^2]^{1/2}\right\}^{-2}\right)^{-1/2}, \qquad (36)$$

with

$$D^* \equiv 3 - 2n, \qquad D^\dagger \equiv 2 - 2l. \qquad (37)$$

$\tilde{\epsilon}$ is a parameter that has not been determined by this analysis, but otherwise the form Eq. (36) is in complete agreement with the exact solution. The exact result corresponds to $\tilde{\epsilon} = 2$.

The functional dependence on δ of the relativistic energy is clearly considerably more complicated than that of the nonrelativistic solution, Eq. (3). To better elucidate the nonrelativistic limit, we can expand Eq. (36) in powers of γ^2. If $D^\dagger \neq D^*$, then

$$E = -2\xi^2 + \gamma^2 \left[4\beta^3(\xi - \eta) + 4\beta^2\xi^2 + 4\beta\xi^3 + 3\xi^4 \right] + \cdots, \qquad (38)$$

where $\xi \equiv (D - D^*)^{-1}$, $\eta \equiv (D - D^\dagger)^{-1}$, and $\beta \equiv (D^\dagger - D^*)^{-1}$. Higher order terms in γ^2 contain higher powers of ξ and η. If $D^\dagger = D^*$, then we have

$$E = -2\xi^2 - 2\gamma^2\xi^4 - 4\gamma^4\xi^6 - \cdots. \qquad (39)$$

The zeroth-order term is of course simply the nonrelativistic coulombic pole at $D = D^*$. Higher orders in the expansion introduce higher-order poles at D^* as well as additional poles at $D = D^\dagger$.

For many-particle systems one must consider matrix elements $\langle r_{ij}^{-1} \rangle$ where i and j represent any two-particles. Each of these integrals diverge at $r_{ij} = 0$ in the one-dimension limit introducing a second-order pole at $D = 1$ in the energy. Thus, the singularity structure at $D = 1$ is essentially the same as for the central potentials. However, in contrast to the one-electron atom, in the case of molecules and many-electron atoms there will be additional singularities at other values of D. Consider the following argument, due to Doren and Herschbach [16]. The wavefunction can always be expressed as an expansion in terms of hydrogenic eigenfunctions, but in order to obtain the exact wavefunction it is necessary to include an infinite number of terms, corresponding to increasingly higher excitations. We showed above that the coulombic pole for a state of the one-electron atom occurs at $D = 3 - 2n$, where n is the principal quantum number. Thus, the hydrogenic expansion would seem to imply that any many-particle system will have an infinite sequence of coulombic poles at $D = -1, -3, -5, \ldots$, in addition to the poles at $D = 1$. The convergence of the $1/D$ expansion depends on the singularity structure on the energy as a function of the variable $\delta \equiv 1/D$. In terms of δ, the hydrogenic expansion implies a sequence of poles between -1 and 0, clustering at 0, which would result in a nonisolated singularity at $\delta = 0$. This argument is not mathematically rigorous [20], but its conclusion is basically correct. In Chapter 7.1 we show that there is a rather complicated singularity at $\delta = 0$ that has a strong effect on the convergence of the $1/D$ expansion.

References

1. D. R. Herschbach, At. Phys. **11**, 63 (1989); A. Chatterjee, Phys. Rep. **186**, 249 (1990).
2. E. A. Abbott, *Flatland: A Romance of Many Dimensions* (Roberts Brothers, Boston, 1891).
3. D. R. Herschbach, J. G. Loeser and D. K. Watson, Z. Phys. D **10**, 195 (1988).
4. O. Goscinski and V. Mujica, Int. J. Quantum Chem. **29**, 897 (1986).
5. D. Z. Goodson and D. R. Herschbach, J. Chem. Phys. **86**, 4997 (1987).
6. J. Avery, D. Z. Goodson and D. R. Herschbach, Int. J. Quantum Chem. **39**, 657 (1991).
7. J. G. Loeser and D. R. Herschbach, J. Chem. Phys. **84**, 3882 (1986).
8. J. D. Morgan III, to be published; see also Chapter 7.3.
9. D. Z. Goodson, D. K. Watson, J. G. Loeser and D. R. Herschbach, Phys. Rev. A **44**, 97 (1991).
10. D. R. Herrick and F. H. Stillinger, Phys. Rev. A **11**, 42 (1975).
11. D. R. Herschbach, J. Chem. Phys. **84**, 838 (1986).
12. J. G. Loeser and D. R. Herschbach, J. Phys. Chem. **89**, 3444 (1985).
13. D. J. Doren and D. R. Herschbach, Chem. Phys. Lett. **118**, 115 (1985).
14. D. J. Doren and D. R. Herschbach, Phys. Rev. A **34**, 2654 (1986).
15. D. J. Doren and D. R. Herschbach, Phys. Rev. A **34**, 2665 (1986).
16. D. J. Doren and D. R. Herschbach, J. Chem. Phys. **87**, 433 (1987).
17. D. J. Doren and D. R. Herschbach, J. Phys. Chem. **92**, 1816 (1988).

18. D. Z. Goodson and D. R. Herschbach, Phys. Rev. Lett. **58**, 1628 (1987).

19. D. D. Frantz and D. R. Herschbach, Chem. Phys. **126**, 59 (1988).

20. D. Z. Goodson, J. D. Morgan III, and D. R. Herschbach, Phys. Rev. A **43**, 4617 (1991).

21. D. Z. Goodson and D. R. Herschbach, submitted to Phys. Rev. A.

22. M. López-Cabrera, Ph.D. thesis, Dept. of Physics, The Univ. of Michigan, Ann Arbor (1991).

23. M. López-Cabrera, D. Z. Goodson, D. R. Herschbach, and J. D. Morgan III, to be published; see also Chapters 7.1 and 8.1.

24. J. D. Morgan III, in *Numerical Determination of the Electronic Structure of Atoms, Diatomic and Polyatomic Molecules*, Vol. 271 of *NATO Advanced Study Institute, Series C: Mathematical and Physical Sciences*, edited by M. Defranceschi and J. Delhalle (Kluwer, Dordrecht, 1989), pp. 49-84.

25. A. A. Frost, J. Chem. Phys. **25**, 1150 (1956).

26. C. M. Rosenthal, J. Chem. Phys. **55**, 2474 (1971).

27. J. H. Van Vleck, in *Wave Mechanics, The First Fifty Years*, edited by W. C. Price *et al.* (Butterworths, London, 1973), pp. 26-37.

4.2 Flatland: The Periodic System for $D = 2$

Pekka Pyykkö
**Department of Chemistry, University of Helsinki,
Et. Hesperiankatu 4, SF-00100 Helsinki, Finland
and
Yongfang Zhao
Institute of Atomic and Molecular Physics, Jilin University,
130023 Changchun, PRC**

Abstract

The Hartree-Fock problem for $D = 2$ has been solved for $1 \leq Z \leq 24$ using a Gaussian basis and assuming $1/r$ Coulomb interactions. The order of occupation of the one-electron states is $1s \ll 2s < 2p < 3s < 3p < 4s \leq 3d < 4p$, like in the $D = 3$ case. The 1s shell is particularly small and strongly bound, making the 2D hydrogen a "superhalogen" or, possibly, a "molecular noble gas" and the 2D helium a "superinert gas". In contrast to 3D, the electron configurations $4s^1 3d^2$ and $4s^2 3d^3$ are preferred for the 2D "Sc" and "Cu", respectively. The six first 2D atoms have stronger and the later ones weaker valence bonding energies than do their 3D analogs. Previously unpublished orbital energies and $\langle r \rangle$ expectation values are given.

Introduction

The serious scientific reason for the present work [1] was the idea that the dimensionality, D, could be regarded as a variable around its usual physical value, D=3, and that this could yield deeper insight to the structure of the usual Periodic System. Alternatively, in systems like very anisotropic semiconductors, the real physics could be approximately of the D=2 type.

On the more jocular side, the starting point was a problem [2] in the 20th International Chemistry Olympiad, a teen-ager contest, where the participants were asked to construct the Periodic System for D=2, see Fig. 1-2. This supposedly correct answer was based on

131

D. R. Herschbach et al. (eds.), Dimensional Scaling in Chemical Physics, 131–138.

analogies, only, and as the Chairman of the International Jury one of us felt that a Hartree-Fock-level study of the problem would be motivated. We learned only later that the 2D Periodic System with the lnr electron-electron interaction had been considered by Asturias and Aragon [3]. The corresponding Thomas-Fermi problem has now been exactly solved [4], using the correct 2D physical laws [5]. Other references missed from [1] include the papers on one-electron [6-8] or two-electron atoms [9] or the one on the hydrogen molecule and molecule ion [10], showing that a localized Lewis pair emerges for large D. Solid-state problems, eventually with magnetic fields, have also been discussed (see e.g.[11-12]).

Roald Hoffmann has pointed out to us that the Bohr atom, a 2D model applied on the 3D world, works because the missing zero-point energy does cancel the wrong energy law of $-(Z/(n-\frac{1}{2}))^2$, compared to $-(Z/n)^2$, for 2D and 3D, respectively. Note that the energy only depends on the principal quantum number, $n = 1, 2, 3, ...$ in both cases.

In a broad sense, the science and technology of a two-dimensional universe, dubbed "planiverse", has been discussed by Dewdney and others [13-15].

Method

We consider the eigenstates of the Hamiltonian

$$H = \sum h_i + \sum g_{ij}$$

$$h_i = -(1/2)\nabla_i^2 - Z/r_i$$

$$g_{ij} = 1/r_{ij} \tag{1}$$

$$\nabla_i^2 = r^{-1}\frac{\partial}{\partial r}(\frac{r\partial}{\partial r}) + r^{-2}\frac{\partial^2}{\partial \phi^2}$$

To keep the parallelism to the 3D case, we thus assume r^{-1} electron-electron and electron-nucleus interactions, despite the fact that the Poisson equation in the 2D case would require $ln(r)$ interactions. The details of the solution are worked out in [1]. A single-determinant solution is sought for the closed-shell case. The open-shell solutions are

SOLUTIONS

Theoretical task 1

a) The Flatlandian periodic table:

1 $1s^1$							2 $1s^2$
3 $|2s^1$	4 $|2s^2$			5 $|2s^2 2p^1$	6 $|2s^2 2p^2$	7 $|2s^2 2p^3$	8 $|2s^2 2p^4$
9 $|3s^1$	10 $|3s^2$			11 $|3s^2 3p^1$	12 $|3s^2 3p^2$	13 $|3s^2 3p^3$	14 $|3s^2 3p^4$
15 $|4s^1$	16 $|4s^2$	17 $|4s^2 3d^1$	18 $|4s^2 3d^2$	19 $|4s^2 3d^3$	20 $|4s^2 3d^4$	21 $|4s^2 3d^4 4p^1$	22 $|4s^2 3d^4 4p^2$	23 $|4s^2 3d^4 4p^3$	24 $|4s^2 3d^4 4p^4$

b)

The element of life: 5

$C_2H_6 \longleftrightarrow$

$C_2H_4 \longleftrightarrow$

$C_6H_{12} \longleftrightarrow$

There are no aromatic ring compounds.

c) Sextet rule

10 electron rule

Figure 1. The "correct solution" [2].

d) The ionization energies and the trends in electronegativity:

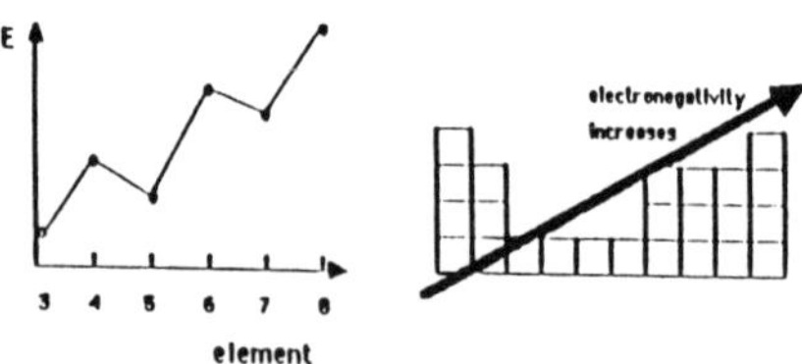

e) The molecular orbital diagram of the homonuclear X_2 molecules:

| The energies of atomic orbitals of free atoms | The energies of the molecular orbitals of homonuclear diatomic molecules | The energies of atomic orbitals of free atoms |

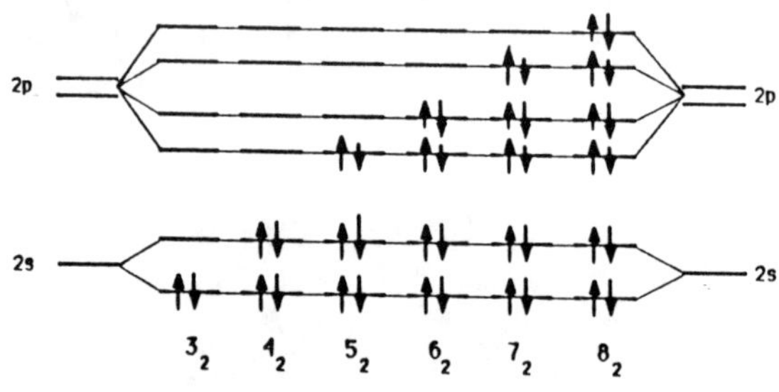

stable unstable stable stable stable unstable

f) The Lewis structures and geometries:

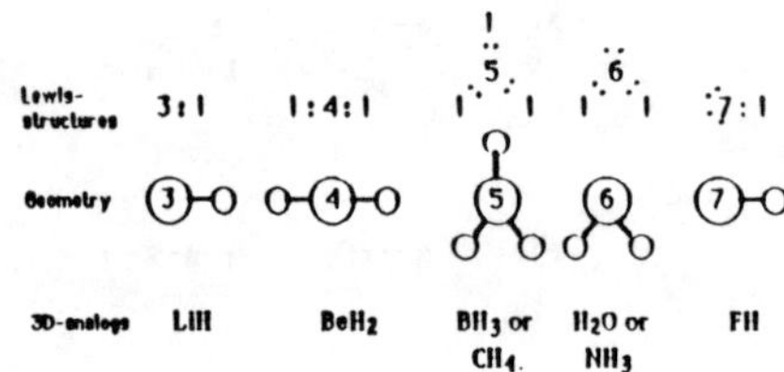

g) The three-dimensional analogs and the states of Flatlandian elements:

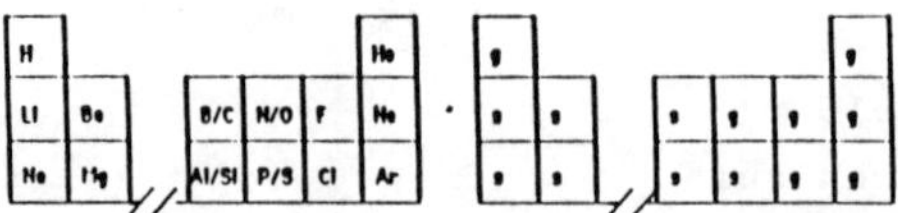

Figure 2. The "correct solution" [2](cont.).

Table 1. The 2D Periodic System according to the present Hartree-Fock calculations.

H	He								
$1s$	$1s^2$								
Li	Be					B	N	F	Ne
$2s$	$2s^2$					$2p$	$2p^2$	$2p^3$	$2p^4$
Na	Mg					Al	P	Cl	Ar
$3s$	$3s^2$					$3p$	$3p^2$	$3p^3$	$3p^4$
K	Ca	Sc	Mn	Cu	Zn	Ga	As	Br	Kr
$4s$	$4s^2$	$4s3d^2$	$4s^23d^2$	$4s^23d^3$	$4s^23d^4$	$4p$	$4p^2$	$4p^3$	$4p^4$

either single- or double-determinant ones. "Even-tempered" Gaussian basis functions with orbital exponents

$$\alpha_i = \alpha_0 \beta^n$$

were used. The numerical results were obtained with $\beta = 2$. The numbers of basis functions are given in ref [1]. The program used was a modified version of Pitzer's 3D program [16]. As the eigenvalue problem is solved iteratively in it, care had to be taken to avoid converging to excited states.

Results

The names of the 2D elements are a matter of convention. The choice of the usual group 1, 2 and 13 names for the 2D s^1, s^2 and p^1 elements is fairly obvious. For the 2D p^2 ones, the 3D group 15 names were chosen, both having half-filled p shells. The 2D p^3 group is the obvious analog of the 3D group 17, both having one hole in their p shells. Analogously, the 2D d elements were identified with Sc, Mn, Cu and Zn, respectively.

The total energies for the various ground-state candidates are given in ref. [1]. The guesses in Fig. 2 turned out to be correct,

Table 2. Orbital energies (a.u.) of the 2D atomic ground states. In cases like $_7$F, $[(2p_{-1})^2(2p_{+1})^1]$, a weighted average is taken. Note that of the transition metals Sc is $4s^1$ and the other three $4s^2$.

Z		1s	2s	3s	4s	2p	3p	4p	3d
1	H	1.99999							
2	He	3.91049							
3	Li	11.0148	0.32322						
4	Be	21.6194	0.42749						
5	B	35.6873	0.69427			0.48050			
6	N	53.1406	0.98436			0.71321			
7	F	74.2047	1.42416			0.68176			
8	Ne	98.6530	1.88687			0.78109			
9	Na	127.397	3.10571	0.17626		1.73057			
10	Mg	159.967	4.60634	0.20038		2.94424			
11	Al	196.356	6.38642	0.30806		4.43285	0.23583		
12	P	236.526	8.40733	0.41777		6.15949	0.33974		
13	Cl	280.569	10.7571	0.58991		8.21131	0.30005		
14	Ar	328.384	13.3401	0.76061		10.4953	0.32882		
15	K	380.318	16.4959	1.22543	0.10586	13.3478	0.69421		
16	Ca	436.162	20.0153	1.76872	0.11193	16.5609	1.13125		
17	Sc	495.157	23.5677	2.05068	0.17241	19.8185	1.33510		0.14841
18	Mn	558.407	27.4508	2.43492	0.11680	23.4031	1.64763		0.37008
19	Cu	625.453	31.5975	2.81618	0.11955	27.2513	1.95434		0.32789
20	Zn	696.271	35.9744	3.18197	0.12155	31.3315	2.25019		0.33154
21	Ga	771.067	40.8147	3.78014	0.17647	35.8730	2.76333	0.15829	0.61108
22	As	849.747	46.0249	4.48417	0.23109	40.7825	3.37420	0.21943	0.97028
23	Br	932.355	51.6517	5.33165	0.32753	46.1063	4.12276	0.18016	1.45810
24	Kr	1020.69	57.6621	6.25013	0.41926	51.8030	4.93649	0.18865	2.00210

except for the two transition metals, [17]Sc and [19]Cu which are, at Hartree-Fock level, $4s3d^2$ and $4s^23d^3$, respectively. This means that in the 2D model, there is no clear-cut $3d^1$ group 3 rare earth and no clear-cut $4s^1$ coinage metal. Rather the 2D Sc is a mixture of a rare earth and "Cr", with a half-filled d shell and the 2D Cu a mixture of a coinage metal and 3D Ni ($3d^84s^2$). The rest of the electron configurations are analogous to their 3D counterparts. The reason is the hydrogen-like degeneracy of the n=3 levels and the comparable screening effects for 2D and 3D.

The biggest difference is the very small radius and large binding energy of the 2D 1s shell, making the 2D hydrogen a "superhalogen" or, alternatively, a "diatomic inert gas", an unknown concept in 3D. Should that occur, the 2D water would be spontaneously reduced to hydrogen and oxygen. Similarly the 2D He is a "superinert gas". Given the tendency of the 1s shell to become small and strongly bound

Table 3. Orbital radii,$\langle r \rangle$ (in a.u.).

Z		1s	2s	3s	4s	2p	3p	4p	3d
1	H	0.50000							
2	He	0.30940							
3	Li	0.19089	2.43300						
4	Be	0.13809	1.73120						
5	B	0.10821	1.26927			1.46603			
6	N	0.08898	1.01589			1.11062			
7	F	0.07555	0.83137			0.99317			
8	Ne	0.06565	0.70973			0.85589			
9	Na	0.05804	0.59275	4.27908		0.64833			
10	Mg	0.05201	0.50692	3.29925		0.52731			
11	Al	0.04712	0.44272	2.51582		0.44575	3.04754		
12	P	0.04307	0.39301	2.07317		0.38667	2.38959		
13	Cl	0.03966	0.35335	1.74098		0.34179	2.17300		
14	Ar	0.03675	0.32103	1.51713		0.30640	1.90151		
15	K	0.03423	0.29414	1.29326	6.92447	0.27782	1.49996		
16	Ca	0.03204	0.27141	1.12640	5.44222	0.25418	1.25821		
17	Sc	0.03013	0.25196	1.01669	4.58141	0.23425	1.12379		3.05540
18	Mn	0.02843	0.23506	0.93702	5.07027	0.21724	1.02445		1.56692
19	Cu	0.02690	0.22030	0.86543	4.88340	0.20256	0.93796		1.48786
20	Zn	0.02553	0.20730	0.80859	4.76119	0.18976	0.86853		1.36687
21	Ga	0.02430	0.19577	0.75114	3.77683	0.17852	0.80032	4.37223	1.07212
22	As	0.02317	0.18548	0.70045	3.18585	0.16855	0.73855	3.54968	0.89787
23	Br	0.02215	0.17622	0.65484	2.70978	0.15965	0.68427	3.33274	0.78035
24	Kr	0.02122	0.16784	0.61412	2.39230	0.15164	0.63660	2.97412	0.69428

when D decreases, the small size of the actual, 3D H and He could be seen as a tail of this trend for D=3.

The orbital energies and average radii, $\langle r \rangle$, of the 2D atomic ground states are shown in Tables 2-3 below.

References

1. P. Pyykkö and Y.-F. Zhao, *The elements of Flatland: Hartree-Fock atomic ground states in two dimensions for Z = 1-24*, Int. J. Quantum Chem. 39 (1991) 000-000.
2. R. Laitinen and T.A. Pakkanen, in *XX International Chemistry Olympiad Finland 1988*, Report (1988) pp. 52-53.
3. F.J. Asturias and S.R. Aragon, *The hydrogenic atom and the periodic table of the elements in two spatial dimensions*, Am. J. Phys. 53 (1985) 893-899.
4. R.K. Bhaduri, S. Das Gupta and S.J. Lee, *The exactly solvable Thomas- Fermi atom in two dimensions*, Am. J. Phys. 58 (1990) 983-986.

5. I.R. Lapidus, *Fundamental units and dimensionless constants in a universe with one, two, and four space dimensions*, Am. J. Phys. 49 (1981) 890-891.

6. K. Andrew and J. Supplee, *A hydrogenic atom in d-dimensions*, Am. J. Phys. 58 (1990) 1177-1183.

7. M.M. Nieto, *Hydrogen atom and relativistic pi-mesic atom in N-space dimensions*, Am. J. Phys. 47 (1979) 1067-1072.

8. S.G. Kamath, *Classical solutions of Maxwell's equations in (2+1)-dimensional electrodynamics with the Chern-Simons term*, Phys. Rev. A 40 (1989) 6791-6799.

9. D.R. Herrick and F.H. Stillinger, *Variable dimensionality in atoms and its effect on the ground state of the helium isoelectronic sequence*, Phys. Rev. A 11 (1975) 42-53.

10. D.D. Frantz and D.R. Herschbach, *Lewis electronic structures as the large- dimension limit for H_2^+ and H_2 molecules*, Chem. Phys. 126 (1988) 59-71.

11. S.K. Ghosh and A.K. Dhara, *Density-functional theory of two-dimensional electron gas in a magnetic field*, Phys. Rev. A 40 (1989) 6103-6106.

12. R. Ferrari, *Two-dimensional electrons in a strong magnetic field: A basis for single-particle states*, Phys. Rev. B (1990).

13. M. Gardner, *The pleasures of doing science and technology in the planiverse*, Sci. Am. 243 (1) (1980) 14-20.

14. A.K. Dewdney (Ed.), *Symposium on Two-Dimensional Science and Technology*, Univ. of Western Ontario, London, Ont. (1981).

15. A.K. Dewdney and I.R. Lapidus (Eds.), *The Second Symposium on Two- Dimensional Science and Technology*, Turing Omnibus, London, Ont. (1986), as quoted in [3, 5].

16. R.M. Pitzer, QCPE 587 (1990).

Chapter 5

HYPERSPHERICAL SYMMETRY

5.1 D-Dimensional Hydrogenlike Orbitals

John Avery
H.C. Ørsted Institute
University of Copenhagen
DK-2100 Copenhagen, Denmark

Abstract

The Schrödinger equation for the D-dimensional analogue of hydrogen (equation (88)) can be solved exactly, both in direct space and in reciprocal space; and in both cases the solutions involve hyperspherical harmonics. In this section we shall discuss the close relationship between hyperspherical harmonics, harmonic polynomials, and exact D-dimensional hydrogenlike wave functions. We shall also discuss the importance of these functions in dimensional scaling and in the hyperspherical method.

D. R. Herschbach et al. (eds.), Dimensional Scaling in Chemical Physics, 139–164.

Introduction

Every student of atomic physics or quantum chemistry is familiar with the central role played by the hydrogen atom. Its importance derives from the fact that the Schrödinger equation for hydrogen can be solved exactly, coupled with the fact that approximate wave functions for many other systems can be built up from superpositions of hydrogenlike wave functions. For similar reasons, the D-dimensional analogue of hydrogen (equation (88)) plays a central role both in dimensional scaling [1-6] and in the hyperspherical method [7-26] (which is discussed in a later chapter of this book by Professor Fano). It is therefore worthwhile to look closely at the exact wave functions which can be obtained for the D-dimensional analogue of hydrogen, both in direct space and in momentum space.

The familiar 3-dimensional hydrogenlike wave functions in direct space can be expressed as confluent hypergeometric functions multiplied by spherical harmonics. We shall see that the D-dimensional hydrogenlike wave functions can also be expressed as confluent hypergeometric functions, but in this case they are multiplied by *hyperspherical harmonics.*

Hyperspherical harmonics are the D-dimensional generalizations of spherical harmonics; and all of the familiar theorems regarding spherical harmonics have D-dimensional generalizations. For example, the familiar sum rule for spherical harmonics can be generalized to a sum rule for hyperspherical harmonics, with Gegenbauer polynomials playing a role analogous to that played by Legendre polynomials in 3-dimensional space. In fact, Legendre polynomials are a special case of Gegenbauer polynomials; and just as the Green's function for the Laplace operator can be expanded in terms of Legendre polynomials, so also the Green's function of the generalized Laplace operator in a D-dimensional space can be expanded in terms of Gegenbauer polynomials.

In 1935, V.A. Fock [27,28] solved the Schrödinger equation for hydrogen in momentum space by a remarkable and beautiful method: He was able to show that when momentum space is mapped onto the surface of a 4-dimensional hypersphere by a suitable transformation, the hydrogen wave functions are proportional to 4-dimensional hyperspherical harmonics. Fock's momentum-space wave functions were the Fourier transforms of familiar 3-dimensional hydrogenlike orbitals; but he obtained them directly, by solving the integral form for the Schrödinger equation in reciprocal space, rather than by performing the Fourier transformation. A byproduct of Fock's analysis was the light which it cast on the high degeneracy of the energy levels, E_{nlm}, of hydrogen. From the spherical symmetry of the problem, one might expect the degeneracy to be $2l + 1$; but because the electron is moving in a Coulomb potential, the degeneracy is n^2. In direct space, this "unexpectedly" high degeneracy can be interpreted in terms of dynamical symmetry by invoking the Runge-Lenz vector. Fock's momentum-space analysis illuminated the problem in a different way by showing that the n^2 degeneracy corresponds exactly to the degeneracy of the 4-dimensional hyperspherical harmonics.

S.P. Alliluev [29,30] was able to obtain exact D-dimensional hydrogenlike wave functions in momentum space by a generalization of Fock's method. In Alliluev's treatment, Fock's transformation was generalized in such a way as to project D-dimensional momentum space onto the surface of a $(D+1)$-dimensional hypersphere [24]. The momentum-space hydrogenlike wave functions could then be shown to be proportional to $(D+1)$-dimensional hyperspherical harmonics.

From the above discussion, we can see that, both in direct space and in reciprocal space, the D-dimensional hydrogenlike wave functions involve hyperspherical harmonics; and we shall therefore devote a little space to discussing these functions. Hyperspherical harmonics are closely related to harmonic polynomials [24]. In fact, hyperspherical harmonics are nothing but harmonic polynomials, orthonormalized in an appropriate way, and divided by appropriate powers of the hyperradius. Let us therefore begin by looking briefly at the theory of harmonic polynomials.

Harmonic Polynomials

Let $x_1, x_2, ..., x_D$ be the Cartesian coordinates of a D-dimensional space, and let

$$f_n \equiv \prod_{j=1}^{D} x_j^{n_j} \tag{1}$$

where the n_j's are positive integers or zero and

$$n_1 + n_2 + ... + n_D = n \tag{2}$$

Then

$$\sum_{j=1}^{D} x_j \frac{\partial f_n}{\partial x_j} = n f_n \tag{3}$$

From (3) it follows that if ∇^2 is the generalized Laplacian operator

$$\nabla^2 \equiv \sum_{j=1}^{D} \frac{\partial^2}{\partial x_j^2} \tag{4}$$

and if R is the hyperradius [24]:

$$R^2 \equiv \sum_{j=1}^{D} x_j^2 \tag{5}$$

then

$$\nabla^2 \left(R^\beta f_\alpha \right) = \beta(\beta + D + 2\alpha - 2)R^{\beta-2} f_\alpha + R^\beta \nabla^2 f_\alpha \tag{6}$$

Let h_α be a homogeneous polynomial of order α satisfying

$$\nabla^2 h_\alpha = 0 \tag{7}$$

Such a homogeneous polynomial is said to be *harmonic*. We would like to resolve f_n into a series of harmonic polynomials of the form

$$f_n = h_n + R^2 h_{n-2} + R^4 h_{n-4} + ... \tag{8}$$

Since h_α is a linear combination of terms of the form shown in equation (1), it follows from (6) that

$$\nabla^2 \left(R^\beta h_\alpha \right) = \beta(\beta + D + 2\alpha - 2)R^{\beta-2} h_\alpha \tag{9}$$

Applying ∇^2 repeatedly to both sides of (8) we obtain

$$\nabla^2 f_n = 2(D + 2n - 4)h_{n-2} + 4(D + 2n - 6)R^2 h_{n-4} + ...$$

$$(\nabla^2)^2 f_n = 8(D + 2n - 6)(D + 2n - 8)h_{n-4} + ...$$

$$\tag{10}$$

and in general

$$(\nabla^2)^\nu f_n = \sum_{k=\nu}^{[\frac{1}{2}n]} \frac{(2k)!!}{(2k-2\nu)!!} \frac{(D+2n-2k-2)!!}{(D+2n-2k-2\nu-2)!!} R^{2k-2\nu} h_{n-2k} \tag{11}$$

For example, when n is even and $\nu = \frac{n}{2}$, we obtain

$$(\nabla^2)^{n/2} f_n = \frac{n!!(D+n-2)!!}{(D-2)!!} h_0 \tag{12}$$

or

$$h_0 = \frac{(D-2)!!}{n!!(D+n-2)!!} (\nabla^2)^{n/2} f_n \tag{13}$$

Equations (10) or (11) constitute a set of simultaneous equations which can be solved for h_α. For $n = 2$, we obtain:

$$\begin{aligned}
f_2 &= h_2 + R^2 h_0 \\
h_0 &= \frac{1}{2D} \nabla^2 f_2 \\
h_2 &= f_2 - \frac{R^2}{2D} \nabla^2 f_2
\end{aligned} \tag{14}$$

while for $n = 3$ we have:

$$\begin{aligned}
f_3 &= h_3 + R^2 h_1 \\
h_1 &= \frac{1}{2(D+2)} \nabla^2 f_3 \\
h_3 &= f_3 - \frac{R^2}{2(D+2)} \nabla^2 f_3
\end{aligned} \tag{15}$$

Thus, for example, we might wish to resolve the monomial $f_3 = x_1^2 x_2$ into a series of harmonic polynomials. To do this, we apply the generalized Laplace operator of equation (4) and we find that $\nabla^2 f_3 = 2x_2$. Then equation (15) tells us that the monomial f_3 can be resolved into an harmonic polynomial of order 3 and an harmonic polynomial of order 1 (i.e., $f_3 = h_3 + R^2 h_1$), where $h_3 = x_1^2 x_2 - R^2 x_2/(D+2)$ and $h_1 = x_2/(D+2)$.

Generalized Angular Momentum

The operator

$$\Lambda^2 \equiv -\sum_{i>j}^{D} \left(x_i \frac{\partial}{\partial x_j} - x_j \frac{\partial}{\partial x_i} \right)^2 \tag{16}$$

is called the generalized angular momentum operator or alternatively the Casimir operator. From equations (16), and (4), we can obtain the relation:

$$\Lambda^2 = -R^2 \nabla^2 + \sum_{i,j=1}^{D} x_i x_j \frac{\partial^2}{\partial x_i \partial x_j} + (D-1) \sum_{i=1}^{D} x_i \frac{\partial}{\partial x_i} \tag{17}$$

Let f_n be the function shown in equation (1). As we noticed in the previous section,

$$\sum_{i=1}^{D} x_i \frac{\partial f_n}{\partial x_i} = n f_n \tag{18}$$

Similarly, one can show that

$$\sum_{i,j=1}^{D} x_i x_j \frac{\partial^2 f_n}{\partial x_i \partial x_j} = n(n-1) f_n \tag{19}$$

Combining equations (17), (18) and (19), we obtain:

$$\Lambda^2 f_n = -R^2 \nabla^2 f_n + n(n + D - 2) f_n \tag{20}$$

Let h_λ be an harmonic polynomial of order λ. Such a polynomial is a linear combination of functions of the form f_n with $n = \lambda$; and in addition, it satisfies $\nabla^2 h_\lambda = 0$. Therefore, from equation (20), we have:

$$\Lambda^2 h_\lambda = \lambda(\lambda + D - 2) h_\lambda \tag{21}$$

Thus every harmonic polynomial of order λ in a D-dimensional space is an eigenfunction of generalized angular momentum corresponding to the eigenvalue $\lambda(\lambda + D - 2)$. When $D = 3$, Λ^2 reduces to the familiar angular momentum operator L^2, and the eigenvalue becomes $l(l+1)$. In the previous section, we saw that any homogeneous polynomial f_n can be resolved into a series of harmonic polynomials multiplied by powers of R. We now see that this is a resolution of f_n into eigenfunctions of generalized angular momentum. Thus, if O_λ is a projection operator corresponding to the λ'th eigenfunction of Λ^2,

$$O_\lambda [f_n] = R^{n-\lambda} h_\lambda \tag{22}$$

Angular Integrations

We now introduce a generalized solid angle element, $d\Omega$, defined by

$$dx_1 dx_2 ... dx_D = R^{D-1} dR \, d\Omega \tag{23}$$

The total solid angle can be evaluated by noticing that

$$\int_0^\infty dR \, R^{D-1} e^{-R^2} \int d\Omega = \prod_{j=1}^D \int_{-\infty}^\infty dx_j e^{-x_j^2} \tag{24}$$

But

$$\int_0^\infty dR \, R^{D-1} e^{-R^2} = \frac{1}{2}\Gamma(\frac{D}{2}) \tag{25}$$

and

$$\int_{-\infty}^\infty dx_j e^{-x_j^2} = \pi^{\frac{1}{2}} \tag{26}$$

Combining equations (24), (25) and (26), we obtain:

$$\int d\Omega = \frac{2\pi^{\frac{D}{2}}}{\Gamma\left(\frac{D}{2}\right)} \equiv S_D \tag{27}$$

When $D = 3$, the total solid angle in equation (27) reduces to the familiar value,

$$S_3 = \frac{2\pi^{\frac{3}{2}}}{\Gamma\left(\frac{3}{2}\right)} = 4\pi \tag{28}$$

We shall next show that if f_n is the function shown in equation (1), then if n is even,

$$\int d\Omega \, f_n = S_D R^n h_0 \tag{29}$$

where S_D is the total solid angle, and where h_0 is defined by equation (13). Since Λ^2 is Hermitian with respect to integration over the generalized solid angle, Ω, the eigenfunctions of Λ^2 corresponding to different eigenvalues are orthogonal:

$$\int d\Omega\, h_\lambda^* h_{\lambda'} = 0 \quad \text{if } \lambda \neq \lambda' \tag{30}$$

But h_0 is a constant, and therefore (30) implies that

$$\int d\Omega\, h_\lambda = 0 \quad \text{if } \lambda \neq 0 \tag{31}$$

Thus, if n is even,

$$\int d\Omega\, f_n = \int d\Omega\, (h_n + R^2 h_{n-2} + ... + R^n h_0) = S_D R^n h_0 \tag{32}$$

where $S_D \equiv \int d\Omega$. From (1) and (13), it follows that if all the n_j's are even,

$$h_0 = \frac{(D-2)!!}{(D+n-2)!!} \prod_{j=1}^{D} (n_j - 1)!! \tag{33}$$

while if any of the n_j's are uneven, h_0 vanishes. Equations (29) and (33) provide us with a powerful theorem for evaluating angular integrals in a D-dimensional space. Remembering that

$$f_n \equiv \prod_{j=1}^{D} x_j^{n_j} \tag{34}$$

we have for the case where all the $n_j's$ are even,

$$I(\mathbf{n}) \equiv R^{-n} \int d\Omega \prod_{j=1}^{D} x_j^{n_j} = \frac{(D-2)!! S_D}{(D+n-2)!!} \prod_{j=1}^{D} (n_j - 1)!! \tag{35}$$

while if any of the $n_j's$ are uneven, $I(\mathbf{n}) = 0$. When $D = 3$, and when all the $n_j's$ are even, we have:

$$\int d\Omega\, x_1^{n_1} x_2^{n_2} x_3^{n_3} = \frac{4\pi R^n}{(n+1)!!}(n_1 - 1)!!(n_2 - 1)!!(n_3 - 1)!! \tag{36}$$

where $n \equiv n_1 + n_2 + n_3$, while if any of the $n_j's$ are uneven, the integral vanishes.

The generalized Laplacian operator can be written in the form:

$$\nabla^2 = \sum_{j=1}^{D} \frac{\partial^2}{\partial x_j^2} = \frac{1}{R^{D-1}} \frac{\partial}{\partial R} R^{D-1} \frac{\partial}{\partial R} - \frac{\Lambda^2}{R^2} \qquad (37)$$

We can verify this relation by means of the following argument: The function f_n of equation (1) can be written in the form:

$$f_n = R^n \chi_{\mathbf{n}}(\Omega) \qquad (38)$$

where $\chi_{\mathbf{n}}(\Omega)$ is a pure function of the hyperangles. Therefore equations (18) and (19) imply that

$$\sum_{i=1}^{D} x_i \frac{\partial}{\partial x_i} = R \frac{\partial}{\partial R} \qquad (39)$$

and

$$\sum_{i,j=1}^{D} x_i x_j \frac{\partial^2}{\partial x_i \partial x_j} = R^2 \frac{\partial^2}{\partial R^2} \qquad (40)$$

Combining (17), (39) and (40), we obtain equation (37). When D=3, equation (37) reduces to the familiar expression for the Laplacian operator in terms of spherical polar coordinates:

$$\nabla^2 = \frac{1}{R^2} \frac{\partial}{\partial R} R^2 \frac{\partial}{\partial R} - \frac{L^2}{R^2} \qquad (41)$$

Hyperspherical Harmonics

Let us consider a set of harmonic polynomials, $h_{\lambda\mu}$, linearly independent, but all of order λ. The index μ is used here to distinguish between the different members of this set, all of which are eigenfunctions of Λ^2 belonging to the same eigenvalue:

$$\Lambda^2 h_{\lambda\mu} = \lambda(\lambda + D - 1) h_{\lambda\mu} \qquad \mu = 1, 2, 3, \ldots \qquad (42)$$

Each of the members of this set can be written in the form:

$$h_{\lambda\mu} = R^\lambda \mathcal{Y}_{\lambda\mu}(\Omega) \qquad (43)$$

where $\mathcal{Y}_{\lambda\mu}(\Omega)$ is a pure angular function. We can choose the harmonic polynomials, $h_{\lambda\mu}$, in such a way that they obey the orthonormality relations:

$$\int d\Omega \; h^*_{\lambda'\mu'} h_{\lambda\mu} = \delta_{\lambda'\lambda} \delta_{\mu'\mu} R^{\lambda+\lambda'} \tag{44}$$

Expressed in terms of the functions $\mathcal{Y}_{\lambda\mu}(\Omega)$, equation (44) becomes:

$$\int d\Omega \; \mathcal{Y}^*_{\lambda'\mu'} \mathcal{Y}_{\lambda\mu} = \delta_{\lambda'\lambda} \delta_{\mu'\mu} \tag{45}$$

The angular functions $\mathcal{Y}_{\lambda\mu}(\Omega)$ are called *hyperspherical harmonics.* When $D = 3$, they reduce to the familiar 3-dimensional spherical harmonics, $\mathcal{Y}_{lm}(\Omega)$. There are many different ways of choosing a set of hyperspherical harmonics (i.e., a set of orthonormal eigenfunctions of Λ^2). To illustrate this, we can consider the case where $D = 4$ and $\lambda = 1$. The set of functions

$$\mathcal{Y}_{1,\mu} = \frac{2^{\frac{1}{2}} x_\mu}{\pi R} \qquad \mu = 1, 2, 3, 4 \tag{46}$$

fulfil the orthonormality relations (45), as can be verified by means of equation (35). A more complete set of 4-dimensional hyperspherical harmonics of this type is shown in Table 1. Alternatively, the set of functions

$$\begin{aligned}
\mathcal{Y}_{1,0,0} &= -\frac{2^{\frac{1}{2}} x_4}{\pi R} \\
\mathcal{Y}_{1,1,1} &= -\frac{i(x_1 + ix_2)}{\pi R} \\
\mathcal{Y}_{1,1,0} &= -\frac{2^{\frac{1}{2}} x_3}{\pi R} \\
\mathcal{Y}_{1,1,-1} &= \frac{i(x_1 - ix_2)}{\pi R}
\end{aligned} \tag{47}$$

also fulfills the orthonormality relations. The hyperspherical harmonics shown in equation (47) are simultaneous eigenfunctions of $\Lambda^2_{(4)}$, $\Lambda^2_{(3)}$ and $\Lambda^2_{(2)}$, where

$$\Lambda^2_{(D)} \equiv -\sum_{i>j}^{D} \left(x_i \frac{\partial}{\partial x_j} - x_j \frac{\partial}{\partial x_i} \right)^2 \tag{48}$$

In other words, the indices labeling the hyperspherical harmonics shown in equation (47) are organized according to the chain of subgroups

$$SO(4) \supset SO(3) \supset SO(2) \qquad (49)$$

Gegenbauer Polynomials

Gegenbauer polynomials play a role in the theory of 3-dimensional spherical harmonics analogous to the role played by Legendre polynomials in the theory of 3-dimensional spherical harmonics; and in fact, Legendre polynomials are a special case of Gegenbauer polynomials. For $D = 3$, we can recall the familiar expansion:

$$\frac{1}{|\,\mathbf{x} - \mathbf{x}'\,|} = \frac{1}{R_>(1 + \epsilon^2 - 2\epsilon \mathbf{u} \cdot \mathbf{u}')^{1/2}} = \frac{1}{R_>} \sum_{l=0}^{\infty} \left(\frac{R_<}{R_>}\right)^l P_l(\mathbf{u} \cdot \mathbf{u}') \quad (50)$$

where

$$\epsilon \equiv \frac{R_<}{R_>} \qquad (51)$$

and where $\mathbf{u}$ and $\mathbf{u}'$ are unit vectors in the direction of $\mathbf{x}$ and $\mathbf{x}'$ respectively. The function $(1 + \epsilon^2 - 2\epsilon \mathbf{u} \cdot \mathbf{u}')^{-1/2}$ is the generating function for the Legendre polynomials, which are, by definition, the polynomials in $\mathbf{u} \cdot \mathbf{u}'$ found by collecting terms in ϵ^l in the Taylor series expansion of the generating function. The Gegenbauer polynomials are defined in a similar way, by means of the generating function:

$$\frac{1}{|\,\mathbf{x} - \mathbf{x}'\,|^{D-2}} = \frac{1}{R_>^{D-2}(1 + \epsilon^2 - 2\epsilon \mathbf{u} \cdot \mathbf{u}')^{\alpha}} = \frac{1}{R_>^{D-2}} \sum_{\lambda=0}^{\infty} \left(\frac{R_<}{R_>}\right)^{\lambda} C_{\lambda}^{\alpha}(\mathbf{u} \cdot \mathbf{u}')$$
$$(52)$$

where $\alpha \equiv (D - 2)/2$ and where $\mathbf{u}$ and $\mathbf{u}'$ are unit vectors in the directions of the D-dimensional vectors $\mathbf{x}$ and $\mathbf{x}'$:

$$\mathbf{u} \equiv \frac{\mathbf{x}}{R} = \frac{1}{R}(x_1, x_2, ..., x_D)$$

$$\mathbf{u}' \equiv \frac{\mathbf{x}'}{R} = \frac{1}{R}(x_1', x_2', ..., x_D') \qquad (53)$$

Like the Legendre polynomials, the Gegenbauer polynomials are found by expanding the generating function in a Taylor series, and collecting terms in powers of ϵ. It can easily be seen that (52) reduces to (50) for the case where $D = 3$. The first few Gegenbauer polynomials are shown in Table 2; and in general they are given by the series:

$$C_\lambda^\alpha(\mathbf{u} \cdot \mathbf{u}') = \sum_{t=0}^{[\lambda/2]} \frac{(-1)^t(\alpha + \lambda - t - 1)!(2\mathbf{u} \cdot \mathbf{u}')^{\lambda-2t}}{(\alpha - 1)!t!(\lambda - 2t)!} \tag{54}$$

If we choose the origin of our coordinate system in such a way that $\mathbf{x}' = 0$, then

$$\nabla^2 \frac{1}{|\mathbf{x} - \mathbf{x}'|^{D-2}} = \left(\frac{1}{R^{D-1}} \frac{\partial}{\partial R} R^{D-1} \frac{\partial}{\partial R} - \frac{\Lambda^2}{R^2}\right) R^{2-D} = 0 \quad R \neq 0 \tag{55}$$

(since Λ^2, acting on any function of R, gives zero). Combining (55) and (52), we obtain:

$$\nabla^2 \frac{1}{|\mathbf{x}' - \mathbf{x}|^{D-2}} = \sum_{\lambda=0}^{D} \frac{1}{R'^{\lambda+D-2}} \nabla^2 \left[R^\lambda C_\lambda^\alpha(\mathbf{u} \cdot \mathbf{u}')\right] = 0 \tag{56}$$

Since this relation must hold for many values of R', each term in the series must vanish separately; and thus,

$$\left(\frac{1}{R^{D-1}} \frac{\partial}{\partial R} R^{D-1} \frac{\partial}{\partial R} - \frac{\Lambda^2}{R^2}\right) R^\lambda C_\lambda^\alpha(\mathbf{u} \cdot \mathbf{u}') = 0 \tag{57}$$

But

$$\frac{1}{R^{D-1}} \frac{\partial}{\partial R} R^{D-1} \frac{\partial}{\partial R} R^\lambda = \lambda(\lambda + D - 2)R^{\lambda-2} \tag{58}$$

so that we obtain:

$$\left[\Lambda^2 - \lambda(\lambda + D - 2)\right] C_\lambda^\alpha(\mathbf{u} \cdot \mathbf{u}') = 0 \tag{59}$$

Since the Gegenbauer polynomials are eigenfunctions of Λ^2, it must be possible to express them as linear combinations of hyperspherical harmonics belonging to the same eigenvalue:

$$C_\lambda^\alpha(\mathbf{u} \cdot \mathbf{u}') = \sum_\mu a_{\lambda\mu} \mathcal{Y}_{\lambda\mu}(\Omega) \tag{60}$$

One can show [18] (using the fact that $C_\lambda^\alpha(\mathbf{u} \cdot \mathbf{u}')$ is invariant under rotations of the coordinate system) that

$$a_{\lambda\mu}(\Omega') = K_\lambda \mathcal{Y}_{\lambda\mu}^*(\Omega') \tag{61}$$

where K_λ is a constant. Thus the hyperspherical harmonics obey the sum rule:

$$C_\lambda^\alpha(\mathbf{u} \cdot \mathbf{u}') = K_\lambda \sum_\mu \mathcal{Y}_{\lambda\mu}^*(\Omega') \mathcal{Y}_{\lambda\mu}(\Omega) \tag{62}$$

which is the D-dimensional generalization of the familiar sum rule:

$$P_l(\mathbf{u} \cdot \mathbf{u}') = K_l \sum_m Y_{lm}^*(\Omega') Y_{lm}(\Omega) \tag{63}$$

The value of the constant K_l can be found by noticing that if

$$O_l[F(\Omega)] \equiv \frac{1}{K_l} \int d\Omega' P_l(\mathbf{u} \cdot \mathbf{u}') F(\Omega') = \sum_m Y_{lm}(\Omega) \int d\Omega' Y_{lm}^*(\Omega') F(\Omega') \tag{64}$$

then O_l is a projection operator corresponding to the lth eigenvalue of L^2. Thus if we apply O_l twice to any function, we must obtain the same result as when we apply it only once. This requirement can be used, in conjunction with our angular integration theorems, to show that

$$K_l = \frac{4\pi}{2l + 1} \tag{65}$$

and, in a similar way, we can show that in general,

$$K_\lambda = \frac{(D - 2)S_D}{2\lambda + D - 2} \tag{66}$$

where S_D is the total solid angle. If we set $\mathbf{u} = \mathbf{u}'$, so that $\mathbf{u} \cdot \mathbf{u}' = 1$, then (62) becomes:

$$C_\lambda^\alpha(1) = K_\lambda \sum_\mu \mathcal{Y}_{\lambda\mu}^*(\Omega) \mathcal{Y}_{\lambda\mu}(\Omega) \tag{67}$$

Integrating (67) over the solid angle, and making use of the orthonormality of the hyperspherical harmonics, we obtain:

$$g = \frac{1}{K_\lambda} C_\lambda^\alpha(1) S_D \tag{68}$$

where g is the number of linearly independent hyperspherical harmonics corresponding to a particular value of λ. From the Taylor series expansion of the generating function, one can show that $C_\lambda^\alpha(1)$ is given by the binomial coefficient,

$$C_\lambda^\alpha(1) = \frac{(\lambda + D - 3)!}{\lambda!(D - 3)!} \tag{69}$$

Combining (66), (68) and (69), we obtain the degeneracy of the hyperspherical harmonics:

$$g = \frac{(D + 2\lambda - 2)(\lambda + D - 3)!}{\lambda!(D - 2)!} \tag{70}$$

When $D = 3$ and $\lambda = l$, this becomes:

$$g = 2l + 1 \tag{71}$$

while when $D = 4$, we have:

$$g = (\lambda + 1)^2 \tag{72}$$

The Hydrogen Atom in Reciprocal Space

It is interesting to notice that when $D = 4$, the number of linearly independent hyperspherical harmonics belonging to a given value of λ is $(\lambda + 1)^2$, i.e., 1,4,9,16,.. and so on - exactly the same as the degeneracy of the solutions to the Schrödinger equation for a hydrogen atom. V. Fock was, in fact, able to show that the Fourier transforms of the hydrogen atom wave functions can be written in the form [27]:

$$\psi_{n,l,m}^t(\mathbf{k}) = M(k)\mathcal{Y}_{n-1,l,m}(\Omega) \tag{73}$$

where Ω is the solid angle in a 4-dimensional space defined by the unit vectors:

$$u_j = \frac{2k_0 k_j}{k_0^2 + k^2} \quad j = 1, 2, 3$$

$$\tag{74}$$

$$u_4 = \frac{k_0^2 - k^2}{k_0^2 + k^2}$$

and where $k_0^2 = -2E$. The function $M(k)$ is independent of the quantum numbers and is given by

$$M(k) = \frac{4k_0^{\frac{5}{2}}}{(k_0^2 + k^2)^2} \tag{75}$$

Fock's derivation of this result, expressed briefly, is as follows: The Fourier transformed Schrödinger equation for a hydrogen atom is an integral equation which can be written in the form [24,28],

$$(k_0^2 + k'^2)^2 \psi^t(\mathbf{k}') = \frac{Z}{2k_0\pi^2} \int d\Omega \frac{(k_0^2 + k^2)2\psi^t(\mathbf{k})}{|\mathbf{u} - \mathbf{u}'|^2} \tag{76}$$

where $\mathbf{u}$ and $\mathbf{u}'$ are unit vectors of the form shown in equation (74). The integral over $d\Omega$ is an integral over solid angle in a 4-dimensional space defined by these vectors. In other words, Fock's transformation, equation (74), maps the 3-dimensional $\mathbf{k}$-space onto the surface of a hypersphere in a 4-dimensional space. If we let

$$\psi^t(\mathbf{k}) = \frac{4k_0^{\frac{5}{2}}}{(k_0^2 + k^2)^2} F(\Omega) \tag{77}$$

then (76) takes on the simple form:

$$F(\Omega') = \frac{Z}{2\pi^2 k_0} \int d\Omega \, \frac{1}{|\mathbf{u} - \mathbf{u}'|^2} F(\Omega) \tag{78}$$

From equation (52) with $D = 4$ and $\alpha = 1$, we have

$$\frac{1}{|\mathbf{u} - \mathbf{u}'|^2} = \sum_{\lambda=0}^{\infty} C_\lambda^1(\mathbf{u} \cdot \mathbf{u}') \tag{79}$$

so that (78) becomes:

$$F(\Omega') = \frac{Z}{2\pi^2 k_0} \sum_{\lambda=0}^{\infty} \int d\Omega \, C_\lambda^1(\mathbf{u} \cdot \mathbf{u}') F(\Omega) \tag{80}$$

But from (62) we have

$$\int d\Omega \, C_\lambda^1(\mathbf{u} \cdot \mathbf{u}') F(\Omega) = K_\lambda O_\lambda \left[F(\Omega') \right] \tag{81}$$

where O_λ is a projection operator corresponding to the λth eigenvalue of Λ^2. When $D = 4$,

$$K_\lambda = \frac{2\pi^2}{\lambda + 1} \tag{82}$$

Thus we can rewrite (80) in the form:

$$F(\Omega) = \sum_{\lambda'=1}^{\infty} \frac{Z}{k_0(\lambda' + 1)} O_{\lambda'}[F(\Omega)] \tag{83}$$

If we let

$$F(\Omega) = \mathcal{Y}_{\lambda\mu}(\Omega) \tag{84}$$

then (83) becomes

$$\mathcal{Y}_{\lambda\mu}(\Omega) = \frac{Z}{k_0(\lambda + 1)} \mathcal{Y}_{\lambda\mu}(\Omega) \tag{85}$$

which will be satisfied if

$$\frac{Z}{k_0(\lambda + 1)} = 1 \tag{86}$$

Remembering that $k_0^2 = -2E$, and identifying $\lambda + 1$ with n, we have

$$E = -\frac{Z^2}{2n^2} \tag{87}$$

in agreement with the usual direct-space solution of the hydrogen atom problem.

Fock's result can easily be generalized to yield a reciprocal-space solution to the D-dimensional hydrogenlike wave equation:

$$\left(-\frac{1}{2}\nabla^2 - \frac{Z}{R}\right)\psi = E\psi \tag{88}$$

Here ∇^2 is the generalized Laplacian operator, Z is a constant, and R is the hyperradius. In the D-dimensional case, the Fock transformation

$$u_j = \frac{2k_0 k_j}{k_0^2 + k^2} \quad j = 1, 2, ..., D$$

$$\tag{89}$$

$$u_{D+1} = \frac{k_0^2 - k^2}{k_0^2 + k^2}$$

maps the D-dimensional **k**-space onto the surface of a $(D+1)$-dimensional hypersphere. Letting

$$\psi^t(\mathbf{k}) = \left[\frac{(2k_0)^{D+2}}{2(k_0^2 + k^2)^{D+1}}\right]^{\frac{1}{2}} F(\Omega) \tag{90}$$

we obtain an integral equation analogous to (78):

$$F(\Omega') = \frac{ZS_D(D-1)!!}{(2\pi)^D k_0} \int d\Omega_{D+1} \frac{F(\Omega)}{|\,\mathbf{u} - \mathbf{u}'\,|^{D-1}} \tag{91}$$

The integration over $d\Omega_{D+1}$ is an integration over solid angle in the $(D+1)$-dimensional space defined by the unit vectors shown in equation (89). The value of α appropriate to this space is $\alpha = (D-1)/2$, and, with this value of α, we obtain from (52):

$$\frac{1}{|\,\mathbf{u} - \mathbf{u}'\,|^{D-1}} = \sum_{\lambda=0}^{\infty} C_\lambda^\alpha(\mathbf{u} \cdot \mathbf{u}') \tag{92}$$

The equation analogous to (83) then becomes:

$$F(\Omega) = \sum_{\lambda=0}^{\infty} \frac{2Z}{k_0(2\lambda + D - 1)} O_\lambda^{D+1}\left[F(\Omega)\right] \tag{93}$$

which will be satisfied by

$$F(\Omega) = \mathcal{Y}_{n-1,\mu}(\Omega) \tag{94}$$

provided that

$$\frac{2Z}{k_0(2n + D - 3)} = 1 \tag{95}$$

Remembering that $k_0^2 = -2E$, we obtain the energy spectrum:

$$E = -\frac{2Z^2}{(2n + D - 3)^2} \tag{96}$$

The orthonormality properties of the hyperspherical harmonics can then be used to show that the Fourier-transformed D-dimensional hydrogenlike wave functions in equation (90) are properly normalized.

Direct-Space Solutions

In direct space, the D-dimensional analogue of the Schrödinger equation for hydrogen can be written in the form:

$$\left(-\frac{1}{2}\nabla^2 - \frac{Z}{R}\right)\psi = E\psi \tag{97}$$

Here ∇^2 is the generalized Laplace operator, defined by equation (37), while R is the hyperradius (equation (5)). In a later chapter of this book, Professor Fano will discuss the application of the hyperspherical method to nonseparable dynamical problems. Here we shall only note that if mass-weighted coordinates are used, the Schrödinger equation for any system interacting through Coulomb forces can be written in the form:

$$\left(-\frac{1}{2}\nabla^2 - \frac{Z(\Omega)}{R}\right)\psi = E\psi \tag{98}$$

where $Z(\Omega)$ is a pure function of the hyperangles, independent of R. the importance of equation (97) in the hyperspherical method derives from its close resemblance to (98).

Equation (97) can be solved in direct space by almost the same method which is used to solve the 3-dimensional Schrödinger equation for hydrogen: If $\mathcal{Y}_{l\mu}(\Omega)$ is a hyperspherical harmonic satisfying

$$\left[\Lambda^2 - l(l + D - 2)\right]\mathcal{Y}_{l\mu}(\Omega) = 0 \tag{99}$$

and if we let

$$s \equiv 2k_0 R \tag{100}$$

$$k_0 \equiv \frac{2Z}{2n + D - 3} \qquad n = 1, 2, 3, \ldots \tag{101}$$

$$E = -\frac{1}{2}k_0^2 = -\frac{2Z^2}{(2n + D - 3)^2} \tag{102}$$

and

$$N_{nl} = \frac{1}{(2l + D - 2)!}\left[\frac{(l + n + D - 3)!}{(n - 1 - l)!(2n + D - 3)}\right]^{\frac{1}{2}} \tag{103}$$

then the bound-state solutions to (97) can be written in the form:

$$\psi_{nl\mu} = R_{nl}(s)\mathcal{Y}_{l\mu}(\Omega) \tag{104}$$

where

$$R_{nl}(s) = (2k_0)^{D/2} N_{nl} s^l e^{-s/2} F(l+1-n \mid 2l+D-1 \mid s) \tag{105}$$

and where $F(a \mid b \mid s)$ is a confluent hypergeometric function defined by

$$F(a \mid b \mid s) \equiv 1 + \frac{a}{b}s + \frac{a(a+1)}{b(b+1)}\frac{s^2}{2!} + \dots \tag{106}$$

The first few wave functions are shown in Table 2. The reader may verify that when $D = 3$, these functions reduce to the familiar hydrogenlike bound-state wave functions. The continuum wave functions have almost the same form, the differences being that k_0 is imaginary [32] and that the normalization is different.

The constant, k_0, appears both in the direct-space and in the k-space wave functions. In both cases, k_0 is related to the energy by $k_0^2 = -2E$. For the reciprocal-space solutions, k_0 represents the radius of the hypersphere onto which momentum-space is mapped by the generalized Fock transformation (equation (89)). As we saw above, the momentum-space wave functions are proportional to hyperspherical harmonics on the surface of this hypersphere. The hyperspherical harmonics form a complete set, in the sense that any well-behaved function of the hyperangles can be expanded in terms of them. A set of hydrogenlike wave functions, all corresponding to the same value of k_0 (but with variable Z) is called a *Sturmian* basis [33-38,24]; and such a basis set has the degree of completeness just mentioned. However, if Z is held constant while k_0 is variable within the set, then the continuum functions are required for completeness.

Dimensional Scaling

In elementary textbooks, one often sees plots of the hydrogen radial distribution function, $R^2 \mid R_{nl}(R) \mid^2$. If we wish to find the expectation value of some function of R, we multiply that function by the radial distribution function and integrate over R. Notice that the *Jacobian weighting factor*, R^2, is included in the radial distribution function to take into account the R-dependence of the volume of a spherical shell of radius R and thickness dR. In a similar way, we can plot radial distribution functions of the form $s^{D-1} \mid R_{nl}(s) \mid^2$ for our D-dimensional solutions, and these can be used for finding the expectation value of any function of the hyperradius. (Here we use $s \equiv 2k_0 R$ as a variable, where k_0 is defined by equation (101)). The Jacobian weighting factor, s^{D-1}, is included in the distribution function to take into account the rate of increase of the volume of a thin hyperspherical shell as the hyperradius increases.

If we plot the radial distribution function for very large values of D, we can notice that it has an interesting property: As D becomes very large, the distribution function becomes more and more sharply localized. For the ground state, the maximum of the radial distribution function occurs when s has the value $s_m = D-1$. From equations (100) and (101), we can see that this corresponds to a value of the hyperradius given by

$$R_m \equiv \frac{D-1}{2k_0} = \frac{(D-1)^2}{4Z} \tag{107}$$

We can see the reason for the increasingly sharp localization of the radial distribution function with increasing D by noticing that it can be written in the form:

$$s^{D-1}|R_{1,0}(s)|^2 \sim s^{D-1}e^{-s} \sim \left(\frac{re^{-r}}{e^{-1}}\right)^{D-1} \tag{108}$$

where $r \equiv s/s_m$. The function $e|re^{-r}|$ has a single maximum at $r = 1$. When we raise this function to the power $D-1$ for higher and higher values of D, the resulting distribution function becomes vanishingly small except at its maximum. The half-width of this function in the high-D limit can be found by letting $r = 1 + \delta$ and solving

$$\left(\frac{re^{-r}}{e^{-1}}\right)^{D-1} \approx (1-\delta^2)^D = \frac{1}{2} \tag{109}$$

Taking the Dth root of this equation, we obtain:

$$1 - \delta^2 = \left(\frac{1}{2}\right)^{1/D} = e^{-ln2/D} \approx 1 - \frac{ln2}{D} \tag{110}$$

Thus the half-width becomes:

$$2\delta = 2\left(\frac{ln2}{D}\right)^{1/2} \tag{111}$$

In the case of the ground-state hydrogenlike wave function, ψ_0, we have:

$$\psi_0^t(\mathbf{k}) = \left(\frac{k_0}{S_{D+1}}\right)^{1/2}\left(\frac{2k_0}{k^2 + k_0^2}\right)^{(D+1)/2} \tag{112}$$

where

$$S_{D+1} \equiv \frac{2\pi^{(D+1)/2}}{\Gamma\left(\frac{D+1}{2}\right)} \tag{113}$$

is the total solid angle of a $D+1$-dimensional space. The momentum-space radial distribution function corresponding to ψ_0 is given by:

$$k^{D-1}\frac{k_0}{S_{D+1}}\left(\frac{2k_0}{k^2 + k_0^2}\right)^{D+1} \sim \left(\frac{2\tilde{k}}{\tilde{k}^2 + 1}\right)^{D+1} \tag{114}$$

where $\tilde{k} \equiv k/k_0$. This function of $\tilde{k}$ is shown in Figure 1 for various values of D. As D becomes very large, the k-space radial distribution function, plotted as a function of $\tilde{k}$, becomes sharply peaked at the momentum $k = k_0$ ($\tilde{k} = 1$). By an argument similar to that used above, we find that the half-width of this function is also $2\left(ln2/D\right)^{1/2}$. From the form of equations (108) and (114), it follows that, in the large-D regime, expectation values involving the radial distribution functions in both direct space and reciprocal space can conveniently be evaluated by the method of steepest descent [40].

It might seem, at first sight, that for the distribution functions both in direct space and in momentum space to become sharply localized in the high-D limit constitutes a violation of the Heisenberg uncertainty principle. However, we must remember that both r and $\tilde{k}$ are scaled coordinates, and that they are not canonically conjugate to each other. The reader may easily verify that if the distribution functions are plotted as functions of R and k, then as D becomes large, the volume of phase space in which both functions are appreciable is independent of D.

Table 1: D-dimensional hyperspherical harmonics

λ	$\mathcal{Y}_{\lambda\mu}(\Omega)$	ω
0	$\left[\frac{1}{S_D}\right]^{1/2}$	1
1	$\left[\frac{D}{S_D}\right]^{1/2} u_i \quad i = 1,...,D$	D
2	$\left[\frac{D(D+2)}{S_D}\right]^{1/2} u_i u_j \quad i \neq j$ $\left[\frac{D+2}{2(D-1)S_D}\right]^{1/2}(Du_i^2 - 1)$	$\frac{(D+2)(D-1)}{2}$

Table 2: D-dimensional hydrogenlike orbitals

Here $M(k) \equiv \left[\frac{2k_0}{k_0^2+k^2}\right]^{(D+1)/2} k_0^{1/2}$, while u_i and u_{D+1}, k_0, S_D, and S_{D+1} are defined respectively by equations (89), (101), (27), and (113).

n	l	$\psi^t(\mathbf{k})$	$\psi(\mathbf{x})$	ω
1	0	$M(k)\left[\frac{1}{S_{D+1}}\right]^{1/2}$	$\left[\frac{(2k_0)^D}{S_D(D-1)!}\right]^{1/2} e^{-k_0 R}$	1
2	0	$M(k)\left[\frac{D+1}{S_{D+1}}\right]^{1/2} u_{D+1}$	$\left[\frac{(2k_0)^D(D-1)}{S_D(D-2)!(D+1)}\right]^{1/2} e^{-k_0 R}\left(1 - \frac{2k_0 R}{D-1}\right)$	$D+1$
	1	$M(k)\left[\frac{D+1}{S_{D+1}}\right]^{1/2} u_i$	$\left[\frac{D(2k_0)^{D+2}}{S_D(D+1)!}\right]^{1/2} e^{-k_0 R} x_i$	

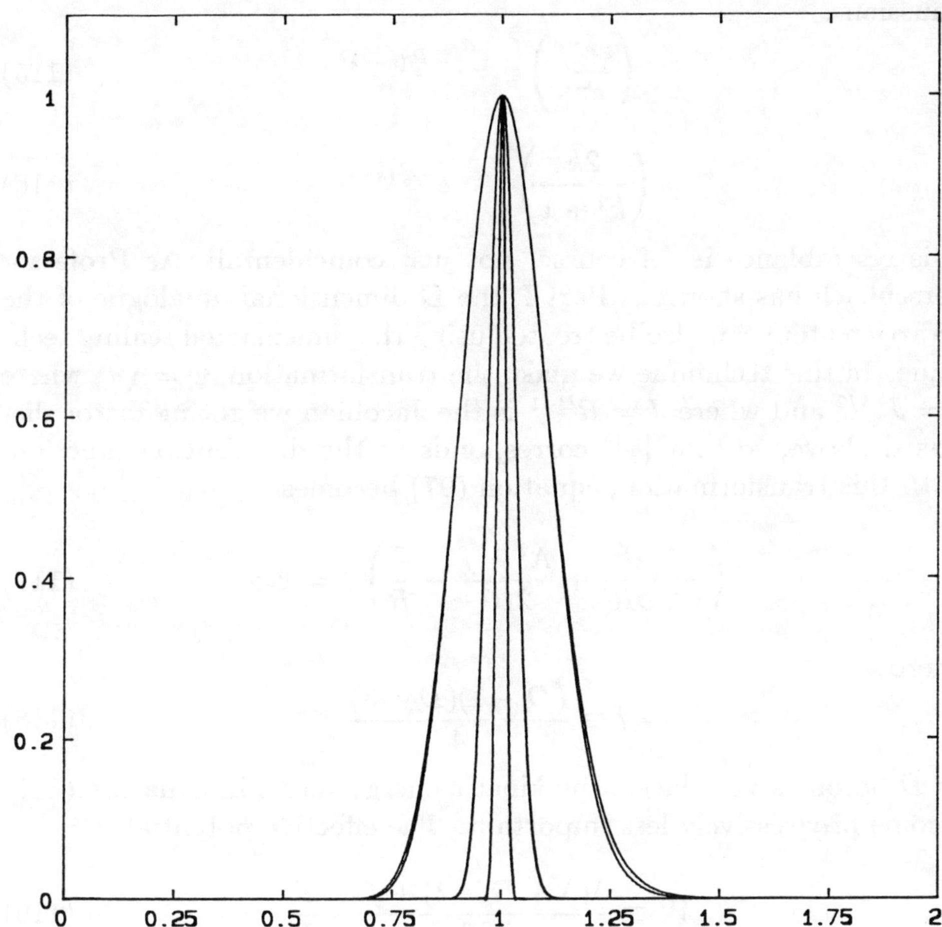

Figure 1. Radial distribution functions in direct space and in reciprocal space, for $D = 100$, $D = 1000$ and $D = 10000$. When plotted in terms of the scaled coordinates r and $\tilde{k}$, the distribution functions for high values of D are sharply peaked at $r = 1$ and $\tilde{k} = 1$; and they can be closely approximated by Gaussians (equations (115) and (116)). The direct- and reciprocal-space curves for $D = 100$ can be resolved, but for $D = 1000$ and $D = 10000$ they are indistinguishable.

We can see from Figure 1 that as D becomes large, the direct-space and momentum-space distribution functions begin to resemble Gaussians:

$$\left(\frac{re^{-r}}{e^{-1}}\right)^{D} \approx e^{-\frac{D}{2}(r-1)^2} \tag{115}$$

$$\left(\frac{2\tilde{k}}{\tilde{k}^2+1}\right)^{D} \approx e^{-\frac{D}{2}(\tilde{k}-1)^2} \tag{116}$$

This resemblance is, of course, not just coincidental! As Professor Herschbach has shown in Part I, the D-dimensional analogue of the hydrogen atom can also be treated using the dimensional scaling technique. In this technique we make the transformation, $\psi = \chi\phi$, where $\chi = J^{-1/2}$ and where $J = R^{D-1}$ is the Jacobian weighting factor discussed above, so that $|\phi|^2$ corresponds to the distribution function. With this transformation, equation (97) becomes

$$\left(-\frac{1}{2}\frac{\partial^2}{\partial R^2} + \frac{\Lambda^2+f}{2R^2} - \frac{Z}{R}\right)\phi = E\phi \tag{117}$$

where

$$f \equiv \frac{(D-1)(D-3)}{4} \tag{118}$$

As D becomes very large, the kinetic energy terms in equation (117) become progressively less important. The effective potential

$$W \equiv \frac{\lambda(\lambda + D - 2) + f}{2R^2} - \frac{Z}{R} \tag{119}$$

can be expanded in a Taylor series about its minimum, which occurs (when $\lambda = 0$) at the energy

$$E_m = -\frac{2Z^2}{(D-1)(D-3)} \tag{120}$$

In the high-D regime, the harmonic terms in the expansion are dominant; and to a close approximation, the ground-state solution for ϕ is just the ground state wave function of an harmonic oscillator. This function has a Gaussian form; its square is a Gaussian, and its Fourier transform is also a Gaussian. The agreement between the exact solutions and those found by the dimensional scaling technique can be made very precise by using the harmonic oscillator wave functions as a basis for expanding the solutions to equation (117). In the case of the D-dimensional analogue of hydrogen, solution by the dimensional scaling technique is unnecessary, since exact solutions are available. However, the problem serves as a prototype, from which we can learn much, and through which we can check the accuracy of the dimensional scaling method.

References

1. D.R. Herschbach, J. Chem. Phys. **84** 838 (1986).
2. J.G. Loeser and D.R. Herschbach, J. Phys. Chem. **89**, 3444-3447 (1985).
3. J.G. Loeser and D.R. Herschbach, J. Chem. Phys. **84**, 3882-3892, 3893-3900 (1986).
4. D.J. Doren and D.R. Herschbach, J. Chem. Phys., **85**, 4557-4562 (1986).
5. J.G. Loeser and D.R. Herschbach, J. Chem. Phys., **86**, 2114-2122, 3512-3521 (1987).
6. D.Z. Goodson and D.R. Herschbach, J. Chem. Phys. **86**, 4997-5008 (1987). 14. J.G. Loeser, J. Chem. Phys. **86**, 5635-5646 (1987).
7. J. Macek, J. Phys. B **1**, 831 (1968).
8. D.R. Herrick, J. Math. Phys. **16**, 1046 (1975).
9. J. Macek, Phys. Rev. A **31**, 2162 (1985).
10. J. Macek and K.A. Jerjian, Phys. Rev. A **33** 233 (1986).
11. U. Fano, Rep. Prog. Phys. **46**, 97 (1983).
12. U. Fano, Phys. Rev. A **22**, 260 (1980); A **24**, 2402 (1981).
13. U. Fano and A.R.P. Rao, *Atomic Collisions and Spectra*, Academic Press, (1986).
14. M.I. Haftel and V.B. Mandelzweig, Phys. Letters A **120**, 232 (1987).
15. A. Kupperman and P.G. Hypes, J. Chem. Phys. **84**, 5962 (1986).
16. M.E. Kellman and D.R. Herrick, Phys. Rev. A **22**, 1536, (1980).

17. D.R. Herrick, Adv. Chem. Phys. **52**, 1 (1983).

18. H. Klar, J. Phys. A **18**, 1561 (1985).

19. H. Klar and M. Klar, J. Phys. B **13**, 1057 (1980).

20. D.L. Knirk, J. Chem. Phys. **60**, 1 (1974).

21. T. Koga and T. Matsuhashi, J. Chem. Phys. **89**, 983 (1988).

22. C.D. Lin, Phys. Rev. A **23**, 1585 (1981).

23. J. Linderberg and Y. Öhrn, Int. J. Quant. Chem. **27**, 273 (1985).

24. J. Avery, *Hyperspherical Harmonics; Applications in Quantum Theory*, Kluwer Academic Publishers, Dordrecht, Netherlands, (1989).

25. J. Avery and F. Antonsen, Int. J. Quant. Chem., Symposium **23**, 159 (1989).

26. J. Avery, D.Z. Goodson and D. Herschbach, International J. Quantum Chem. **39** 657, (1991).

27. V.A. Fock, Kgl. Norske Videnskab. Forh. **31**, 138 (1958).

28. B.R. Judd, *Angular Momentum Theory for Diatomic Molecules*, Academic Press, (1975).

29. S.P. Alliluev, Sov. Phys. JETP **6**, 156 (1958).

30. M. Bander and C. Itzykson, Rev. Mod. Phys. **38**, 330, 346 (1966).

31. K. Andrew and J. Supplee, Am. J. Phys. **58**, 1177 (1990)

32. R.K. Peterkop, *Theory of Ionization of Atoms by Electron Impact*, translated by D.G. Hummer and E. Aronsen, Colorado Associated University Press, Boulder, Colorado, (1977).

33. H. Schull and P.-O. Löwdin, J. Chem. Phys. **30**, 617 (1959).

34. O. Goscinski, Preliminary Research Report No. 217, Quantum Chemistry Group, Uppsala University, (1968).

35. M. Rotenberg, Adv. At. Mol. Phys. **6**, 233 (1970).

36. E.J. Weniger, J. Math. Phys. **26**, 276 (1989).

37. J. Avery and D.R. Herschbach, Int. J. Quantum Chem. **41**, 673 (1992).

38. J. Avery, D.Z. Goodson and D. Herschbach, Theoretica Chemica Acta **91**, 1, (1991).

39. P.M. Morse and H. Feshbach, *Methods of Theoretical Physics*, McGraw-Hill, New York (1958), p. 437.

5.2 Ground-State Wigner Function for the D-dimensional Hydrogen Atom

Jens Peder Dahl
Technical University of Denmark
Chemical Physics, Chem. Dep. B, DTH 301
DK-2800 Lyngby, Denmark

Abstract

The Wigner function for the ground state of the D-dimensional hydrogen atom cannot be evaluated in an analytically closed form. It may, however, be evaluated analytically in a representation in which the wavefunction is written as a linear combination of gaussians. Using such a representation, we have determined the Wigner function for a number of D-values. The results are displayed through a series of contour maps.

Introduction

In previous papers, we have studied the ground-state Wigner function for the hydrogen atom [1] and some polyelectron atoms [2]. John Avery has suggested that it would be interesting to extend these studies to the hydrogen atom in D dimensions. Accordingly, we present here the ground-state Wigner function of the D-dimensional hydrogen atom for a selected series of D-values. These include $D = 2, 3, 10, 25, 100$.

To specify our notation, we begin with a definition of the spherical polar coordinates that we use, and write down the position and momentum ground-state wavefunction for the D-dimensional hydrogen atom. The Wigner function is then introduced in the succeeding section. Unfortunately, the form of the exact wavefunction does not allow an analytical evaluation of the Wigner function, but we show in a following section that an analytical evaluation is possible if the exact function is approximated by a linear combination of D-dimensional gaussians. An appropriate linear combination is accordingly determined in the next section, for each of the D-values under study. Fi-

165

D. R. Herschbach et al. (eds.), Dimensional Scaling in Chemical Physics, 165–178.
© 1993 *Kluwer Academic Publishers. Printed in the Netherlands.*

nally, the Wigner functions are evaluated and presented through a series of contour maps. A balanced treatment of the position and momentum coordinates is achieved by applying a dimensional scaling factor of $\left(\frac{D-1}{2}\right)^{\frac{3}{2}}$.

We use atomic units throughout.

Preliminaries

Let $\mathbf{x}$ and $\mathbf{k}$, respectively, be the position and momentum vectors in D-dimensional space, and let r and k be the corresponding hyperradii:

$$\mathbf{x} = (x_1, x_2, \ldots, x_D), \quad r^2 = x_1^2 + x_2^2 + \cdots + x_D^2 \tag{1}$$

and

$$\mathbf{k} = (k_1, k_2, \ldots, k_D), \quad k^2 = k_1^2 + k_2^2 + \cdots + k_D^2 \tag{2}$$

We may then introduce spherical polar coordinates of $\mathbf{x}$ by the definitions:

$$
\begin{aligned}
x_1 &= r\sin\alpha_D \sin\alpha_{D-1} \cdots \sin\alpha_3 \sin\alpha_2 & 0 \le r \le \infty \\
x_2 &= r\sin\alpha_D \sin\alpha_{D-1} \cdots \sin\alpha_3 \cos\alpha_2 & 0 \le \alpha_2 \le 2\pi \\
x_3 &= r\sin\alpha_D \sin\alpha_{D-1} \cdots \cos\alpha_3 & 0 \le \alpha_3 \le \pi \\
&\;\;\vdots \\
x_{D-1} &= r\sin\alpha_D \cos\alpha_{D-1} & 0 \le \alpha_{D-1} \le \pi \\
x_D &= r\cos\alpha_D & 0 \le \alpha_D \le \pi
\end{aligned}
\tag{4}
$$

A similar representation exists, of course, for the vector $\mathbf{k}$.

For the volume element, dx, we have the expression:

$$dx \equiv dx_1 dx_2 \ldots dx_D = r^{D-1} dr d\Omega \tag{4}$$

where the solid-angle element $d\Omega$ is given by:

$$d\Omega = (\sin\alpha_D)^{D-2}(\sin\alpha_{D-1})^{D-3}\cdots\sin\alpha_3 d\alpha_D \ldots d\alpha_3 \tag{5}$$

Integrating over all angles gives the total solid angle:

$$S_D \equiv \int d\Omega = \frac{2\pi^{D/2}}{\Gamma(D/2)} \tag{6}$$

With the above notation at hand, we may now specify the form of the position-space wavefunction $\psi_0(\mathbf{x})$ for the ground state of the D-dimensional hydrogen atom, [3]:

$$\psi_0(\mathbf{x}) = \sqrt{\frac{1}{S_D}} R_0(r) \tag{7}$$

The radial function $R_0(r)$ is:

$$R_0(r) = \left\{ \frac{1}{(n_0/2)^{2n_0+1}(2n_0)!} \right\}^{\frac{1}{2}} e^{-r/n_0} \tag{8}$$

where

$$n_0 \equiv \frac{D-1}{2} \tag{9}$$

It determines the ground-state energy through the relation

$$E_0 = -\frac{1}{2n_0^2} \tag{10}$$

The momentum-space wavefunction is the D-dimensional Fourier transform of $\psi_0(\mathbf{x})$. Its form is, [3]:

$$\phi_0(\mathbf{k}) = \left(\frac{k_0}{S_{D+1}} \right)^{\frac{1}{2}} \left(\frac{2k_0}{k^2 + k_0^2} \right)^{\frac{D+1}{2}} \tag{11}$$

where k_0 is the reciprocal of n_0:

$$k_0 = \frac{1}{n_0} \tag{12}$$

Both $\psi_0(\mathbf{x})$ and $\phi_0(\mathbf{k})$ are normalized to unity, i.e.:

$$\int \psi_0(\mathbf{x})^2 dx = \int \psi_0(\mathbf{x})^2 r^{D-1} dr d\Omega = \int_0^\infty R_0(R)^2 R^{D-1} dR = 1 \tag{13}$$

and

$$\int \phi_0(\mathbf{k})^2 dk = \int \phi_0(\mathbf{k})^2 k^{D-1} dk d\Omega = 1 \tag{14}$$

The Wigner Function

Being the Fourier transforms of each other, the two functions $\psi_0(\mathbf{x})$ and $\phi_0(\mathbf{k})$ contain the same physical information , but the form of one is evidently more oriented towards position-space, the form of the other more oriented towards momentum space. The so-called Wigner function [4] bears, on the other hand, a symmetric relation to the two spaces. For a general state, described by the wavefunction $\psi(\mathbf{x})$, it is defined as

$$f(\mathbf{x},\mathbf{k}) = \frac{1}{\pi^D} \int \psi(\mathbf{x} - \mathbf{x}')^* \psi(\mathbf{x} + \mathbf{x}') e^{-2i\mathbf{k}\cdot\mathbf{x}'} dx' \qquad (15)$$

Like a wavefunction, it contains all physical information about the state in question. We have, in particular, that

$$\int f(\mathbf{x},\mathbf{k}) dk = \psi(\mathbf{x})^* \psi(\mathbf{x}) \qquad (16)$$

and

$$\int f(\mathbf{x},\mathbf{k}) dx = \phi(\mathbf{k})^* \phi(\mathbf{k}) \qquad (17)$$

Thus, the Wigner function has the position and momentum densities as marginal densities. Partly for this reason, it is called a phase-space distribution function. We note, however, that although $f(\mathbf{x},\mathbf{k})$ is always real-valued, it will in general take negative as well as positive values. Hence, it cannot be interpreted as a true probability density in phase space. Its value at a given phase-space point is, nevertheless, a measure of the way the point supports the given quantum state. For a more precise meaning of this statement, and for general references to the literature, we refer to our detailed discussion of the three-dimensional case [1].

By inserting $\psi_0(\mathbf{x})$ from Eq. (7) into the defining relation (15), we obtain the ground-state Wigner function $f_0(\mathbf{x},\mathbf{k})$ for the D-dimensional hydrogen atom. The integral involved can, however, not be evaluated in a closed analytical form, and we shall therefore proceed in a similar way as in [1]: $\psi_0(\mathbf{x})$ is approximated by a linear combination of gaussians which is then inserted in (15). The resulting integrals can now be evaluated analytically. We shall develop the pertinent expressions in the following section.

Evaluation of the Wigner Function

By a D-dimensional gaussian we understand a normalized function of the form

$$\psi(\mathbf{x}; \alpha) = \left(\frac{2\alpha}{\pi}\right)^{\frac{D}{4}} e^{-\alpha r^2} \tag{18}$$

Since $r^2 = x_1^2 + x_2^2 + \cdots + x_D^2$, this is in fact a product of D one-dimensional Gaussians:

$$\psi(\mathbf{x}; \alpha) = \prod_{i=1}^{D} \left(\frac{2\alpha}{\pi}\right)^{\frac{1}{4}} e^{-\alpha x_i^2} \tag{19}$$

The overlap integral between two gaussians is therefore the Dth power of the overlap between two one-dimensional gaussians. That is:

$$\Delta_{ij} \equiv \langle \psi(\mathbf{x}; \alpha_i) | \psi(\mathbf{x}; \alpha_j) \rangle = \left(\frac{4\alpha_i \alpha_j}{(\alpha_i + \alpha_j)^2}\right)^{\frac{D}{4}} \tag{20}$$

Let us now approximate $\psi_0(\mathbf{x})$ by a linear combination of M gaussians with orbital exponents $\alpha_1, \alpha_2, \ldots, \alpha_M$, and also denote the approximating function by $\psi_0(\mathbf{x})$ for simplicity. We have then:

$$\psi_0(\mathbf{x}) = \sum_{i=1}^{M} c_i \psi(\mathbf{x}; \alpha_i) \tag{21}$$

with the normalization condition

$$\langle \psi_0(\mathbf{x}) | \psi_0(\mathbf{x}) \rangle \equiv \sum_{i=1}^{M} c_i^2 + 2 \sum_{i>j=1}^{M} c_i c_j \Delta_{ij} = 1 \tag{22}$$

Inserting the expansion (21) into the expression (15) for the Wigner function gives:

$$f_0(\mathbf{x}, \mathbf{k}) = \sum_{i=1}^{M} c_i^2 P_{ii}(\mathbf{x}, \mathbf{k}) + \sum_{i>j=1}^{M} c_i c_j \left(P_{ij}(\mathbf{x}, \mathbf{k}) + P_{ji}(\mathbf{x}, \mathbf{k})\right) \tag{23}$$

where we have used the definition

$$P_{ij}(\mathbf{x}, \mathbf{k}) = \frac{1}{\pi^D} \int \psi(\mathbf{x} - \mathbf{x}'; \alpha_i)^* \psi(\mathbf{x} + \mathbf{x}'; \alpha_j) e^{-2i\mathbf{k}\cdot\mathbf{x}'} d\mathbf{x}' \tag{24}$$

Straightforward integration gives:

$$P_{ij}(\mathbf{x}, \mathbf{k}) = \frac{1}{\pi^D} \Delta_{ij} e^{-\gamma_{ij} r^2} e^{-\frac{k^2}{\alpha_i + \alpha_j}} e^{-2i\tau_{ij}\mathbf{k}\cdot\mathbf{x}} \tag{25}$$

with

$$\gamma_{ij} = \frac{4\alpha_i \alpha_j}{\alpha_i + \alpha_j} \tag{26}$$

and

$$\tau_{ij} = \frac{\alpha_i - \alpha_j}{\alpha_i + \alpha_j} \tag{27}$$

Hence we get that

$$\begin{aligned}
f_0(\mathbf{x}, \mathbf{k}) &= \frac{1}{\pi^D} \sum_{i=1}^{M} c_i^2 e^{-2\alpha_i r^2} e^{-\frac{k^2}{2\alpha_i}} \\
&+ \frac{2}{\pi^D} \sum_{i>j=1}^{M} c_i c_j \Delta_{ij} e^{-\gamma_{ij} r^2} e^{-\frac{k^2}{\alpha_i + \alpha_j}} \cos(2\tau_{ij}\mathbf{k}\cdot\mathbf{x})
\end{aligned} \tag{28}$$

We note that the Wigner function for the spherically symmetric ground state only depends upon r, k and u, where u is the angle between the vectors $\mathbf{k}$ and $\mathbf{x}$:

$$\mathbf{k}\cdot\mathbf{x} = kr \cos u \tag{29}$$

In [1] we made a detailed study of the Wigner function's dependence on r, k and u for the three-dimensional hydrogen atom. We mentioned, in particular, that for several purposes it is sufficient to know the function obtained from (28) by integrating over u, and this was in fact the only kind of function that we considered for the poly-electron case. Hence, we shall also limit ourselves to this function in the present work. Actually, we shall also integrate over the $2D - 3$ angles that define the $\mathbf{x}, \mathbf{k}$ plane, but this merely introduces a constant factor. In addition, we multiply by $r^{D-1} k^{D-1}$ so that the result becomes a radial Wigner function $F_0(r, k)$, normalized such that

$$\int_0^\infty \int_0^\infty F_0(r, k)\, dr\, dk = 1 \tag{30}$$

The marginal densities of $F_0(r, k)$ are the radial densities $S_D r^{D-1} \psi_0(\mathbf{x})^2$ and $S_D k^{D-1} \phi_0(\mathbf{k})^2$.

When performing the mentioned integrations it is no restriction to assume that the vector $\mathbf{k}$ is directed along the Dth axis of the coordinate system, i.e., we may put $\mathbf{k} = (0, 0, ..., k)$. This makes the angle u identical with the angle α_D in (3). The remaining $2D - 3$ angles may then be taken to be $\alpha_2, \alpha_3, ..., \alpha_{D-1}$ and the $D - 1$ angular coordinates of $\mathbf{k}$. Integration over the angular coordinates of $\mathbf{k}$ introduces a factor S_D, and integration over $\alpha_2, \alpha_3, ..., \alpha_{D-1}$ introduces a factor S_{D-1}. Hence we get:

$$
\begin{aligned}
F_0(r, k) &= \frac{S_D^2}{\pi^D} \sum_{i=1}^{M} c_i^2 e^{-2\alpha_i r^2} e^{-\frac{k^2}{2\alpha_i}} \\
&+ \frac{2S_D S_{D-1}}{\pi^D} \sum_{i>j=1}^{M} c_i c_j \Delta_{ij} e^{-\gamma_{ij} r^2} e^{-\frac{k^2}{\alpha_i + \alpha_j}} G_{ij}(r, k)
\end{aligned}
\tag{31}
$$

where

$$
G_{ij}(r, k) = \int_0^{\pi} \cos(2\tau_{ij} k r \cos u)(\sin u)^{D-2} du
\tag{32}
$$

To evaluate the integral over u we note that a definition of the Bessel function $J_\nu(z)$ (for $\Re(\nu) > -\frac{1}{2}$) is, [5]:

$$
J_\nu(z) = \frac{(\frac{z}{2})^\nu}{\pi^{\frac{1}{2}} \Gamma(\nu + \frac{1}{2})} \int_0^{\pi} \cos(z \cos u)(\sin u)^{2\nu} du
\tag{33}
$$

By substituting this result together with the expression (6) for S_D we get finally:

$$
\begin{aligned}
F_0(r, k) &= \frac{4}{\Gamma(\frac{D}{2})^2} \sum_{i=1}^{M} c_i^2 e^{-2\alpha_i r^2} e^{-\frac{k^2}{2\alpha_i}} \\
&+ \frac{8}{\Gamma(\frac{D}{2})} \sum_{i>j=1}^{M} c_i c_j \Delta_{ij} e^{-\gamma_{ij} r^2} e^{-\frac{k^2}{\alpha_i + \alpha_j}} \left(\frac{1}{\tau_{ij} k r}\right)^{\frac{D-2}{2}} J_{\frac{D-2}{2}}(2\tau_{ij} k r)
\end{aligned}
\tag{34}
$$

To proceed, we need the values of the orbital exponents α_i and the expansion coefficients c_i. They are determined in the following section.

The Gaussian Expansions

To facilitate the calculations, we have chosen to let the M gaussians in the expansion (21) be an even-tempered basis set, i.e., we have assumed that the orbital exponents $\alpha_1, \alpha_2, \ldots, \alpha_M$ form a geometric progression:

$$\alpha_i = \alpha_1 \beta^{i-1}, \qquad i = 1, 2, \ldots, M \tag{35}$$

This leaves us with only two non-linear parameters, viz. α_1 and β. It is a characteristic feature of even-tempered basis sets that the overlap integral between two neighboring gaussians, $\psi(\mathbf{x}; \alpha_i)$ and $\psi(\mathbf{x}; \alpha_{i+1})$, is independent of i. As it is easily seen from the expression (20), this overlap has the value

$$\Delta \equiv \Delta_{i,i+1} \equiv \langle \psi(\mathbf{x}; \alpha_i) | \psi(\mathbf{x}; \alpha_{i+1}) \rangle = \left(\frac{4\beta}{(1+\beta)^2} \right)^{\frac{D}{4}} \tag{36}$$

Thus, we may equally well consider α_1 and Δ to be the non-linear parameters of the problem.

In addition to the two non-linear parameters, we have of course the M linear parameters $c_1, c_2, \ldots, c_M$. The way to determine the parameters is to use the variational method. We have done this by pointwise solution of an $M \times M$ secular problem for the coefficients $c_1, c_2, \ldots, c_M$ and the energy, in the two-dimensional parameter space defined by α_1 and β. The (α_1, β) point leading to the lowest energy determines, then, the best approximation to $\psi_0(\mathbf{x})$ for the chosen M-value.

We have found that a linear combination of 9 gaussians gives a very good approximation to the true wavefunction. This is borne out by the numbers in Table 1, which compares the $M = 9$ energies with the exact ones from Eq. (10). In addition, the table shows the best energies that can be obtained by using only a single gaussian ($M = 1$). The orbital exponent of that gaussian is also shown. It is seen that a single gaussian leads to a reasonable estimate of the energy for large D. This is because the factor r^{D-1} in the volume element (4), for large D, puts less emphasis on the regions of space where the gaussian differs from the exact function. It does not, of course, imply that the exact function approaches a gaussian for large D.

To perform the variational calculation described above, one must evaluate the matrix elements of the Hamiltonian of the problem. The Hamiltonian is

$$\hat{H} = \hat{T} + \hat{V} \tag{37}$$

where $\hat{T}$ is the kinetic-energy operator:

$$\hat{T} = -\frac{1}{2}\left(\frac{\partial^2}{\partial x_1^2} + \frac{\partial^2}{\partial x_2^2} + \cdots + \frac{\partial^2}{\partial x_D^2}\right) \tag{38}$$

and $\hat{V}$ is the potential energy:

$$\hat{V} = -\frac{1}{r} \tag{39}$$

The evaluation of the matrix elements of $\hat{T}$ and $\hat{V}$ is quite straightforward and we get, in a notation similar to that of Eq. (20):

$$T_{ij} \equiv \langle\psi(\mathbf{x};\alpha_i)|\hat{T}|\psi(\mathbf{x};\alpha_j)\rangle = D\Delta_{ij}\frac{\alpha_i\alpha_j}{\alpha_i + \alpha_j} \tag{40}$$

and

$$V_{ij} \equiv \langle\psi(\mathbf{x};\alpha_i)|\hat{V}|\psi(\mathbf{x};\alpha_j)\rangle = -\Delta_{ij}\sqrt{\alpha_i + \alpha_j}\frac{\Gamma(\frac{D-1}{2})}{\Gamma(\frac{D}{2})} \tag{41}$$

The results of the variational calculations with $M = 9$ are presented in Table 2, except for the energies which are given in Table 1. Table 2 also gives the α-value corresponding to a single-gaussian approximation. To get this value, we put $\alpha_i = \alpha_j = \alpha$ in the expressions above and minimize the sum of (40) and (41) with respect to α. This gives the optimal values

$$\alpha = \frac{2}{D^2}\left(\frac{\Gamma(\frac{D-1}{2})}{\Gamma(\frac{D}{2})}\right)^2 \tag{42}$$

and

$$E_0(M = 1) \equiv \langle\psi(\mathbf{x};\alpha)|\hat{H}|\psi(\mathbf{x};\alpha)\rangle = -\frac{1}{D}\left(\frac{\Gamma(\frac{D-1}{2})}{\Gamma(\frac{D}{2})}\right)^2 \tag{43}$$

The $M = 1$ results in Table 1 and Table 2 were evaluated from these expressions.

Table 1. Exact and approximate ground-state energy of the D-dimensional hydrogen atom.

D	$-E_0(exact)$	$-E_0(M = 9)$	$-E_0(M = 1)$
2	2.00000000	1.99976239	1.57080
3	0.50000000	0.49998898	0.42441
10	$2.46913580 \times 10^{-2}$	$2.46913549 \times 10^{-2}$	2.3489×10^{-2}
25	$3.47222222 \times 10^{-3}$	$3.47222221 \times 10^{-3}$	3.4035×10^{-3}
100	$2.04060810 \times 10^{-4}$	$2.04060810 \times 10^{-4}$	2.0304×10^{-4}

Table 2. Gaussian expansion of the ground-state wavefunction for the D-dimensional hydrogen atom. α is the orbital exponent for $M = 1$. The remainder of the table defines the $M = 9$ expansion in the notation used in the text.

	$D = 2$	$D = 3$	$D = 10$	$D = 25$	$D = 100$
α	1.57080	0.28204	0.0046978	0.00027228	0.0000040609
α_1	0.445570	0.0801937	0.00151377	0.000103992	0.00000251296
β	3.375874	2.702590	1.728769	1.419096	1.240861
Δ	0.839766	0.836802	0.831111	0.82658	0.74788
c_1	0.3249767	0.2331289	0.0516525	0.0092818	0.0199668
c_2	0.4723280	0.4734663	0.2997391	0.1197102	0.2181261
c_3	0.2209662	0.2827043	0.4281817	0.3639203	0.5059905
c_4	0.0757559	0.1066784	0.2596936	0.3964247	0.3464624
c_5	0.0233621	0.0342184	0.0989864	0.2096995	0.0862627
c_6	0.0068552	0.0099565	0.0273806	0.0664064	0.0103179
c_7	0.0022419	0.0032368	0.0071666	0.0149571	0.0003370
c_8	0.0004442	0.0006104	0.0011818	0.0023644	0.0001305
c_9	0.0003251	0.0004477	0.0004987	0.0004905	0.0000313

Contour Maps

By means of the gaussian expansions listed in Table 2 we have prepared contour maps of the radial Wigner function $F_0(r, k)$ for $D = 1, 2, 10, 25, 100$. These maps are presented in the Figures 1 to 5. The upper map in each figure shows the 'exact' function derived from the $M = 9$ expansion, the lower map shows the single-gaussian approxi-

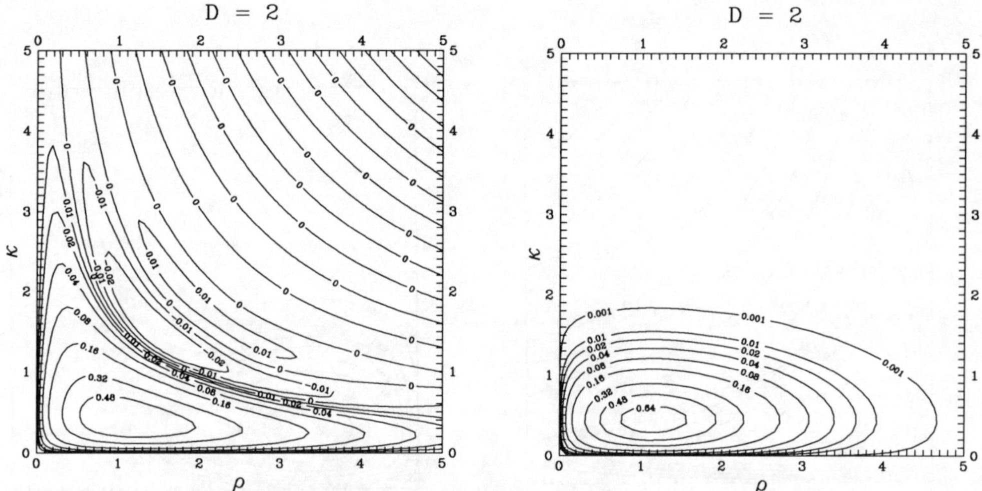

Figure 1. Contour maps of the Wigner function for $D = 2$.
Left: $M = 9$. Right: $M = 1$.

mation. The maps are referred to the scaled position coordinate

$$\rho = \frac{r}{n_0^{\frac{3}{2}}} \tag{44}$$

and the corresponding momentum coordinate

$$\kappa = n_0^{\frac{3}{2}} k \tag{45}$$

with n_0 as given by (9). The figures show that this type of scaling does equal justice to the position and momentum coordinates, in the sense that each map extends over comparable ranges of ρ and κ values. It is also seen from the figures that the region of phase space in which $F_0(r, k)$ finds its major support is of similar size in all the maps, although the region is displaced towards larger (ρ, κ) values with increasing values of D.

The single-gaussian approximation to $F_0(r, k)$ is everywhere positive for any D. The exact function is, on the other hand, characterized by a dominant region in which it is positive, and by an exterior region in which it oscillates between positive and negative values. The

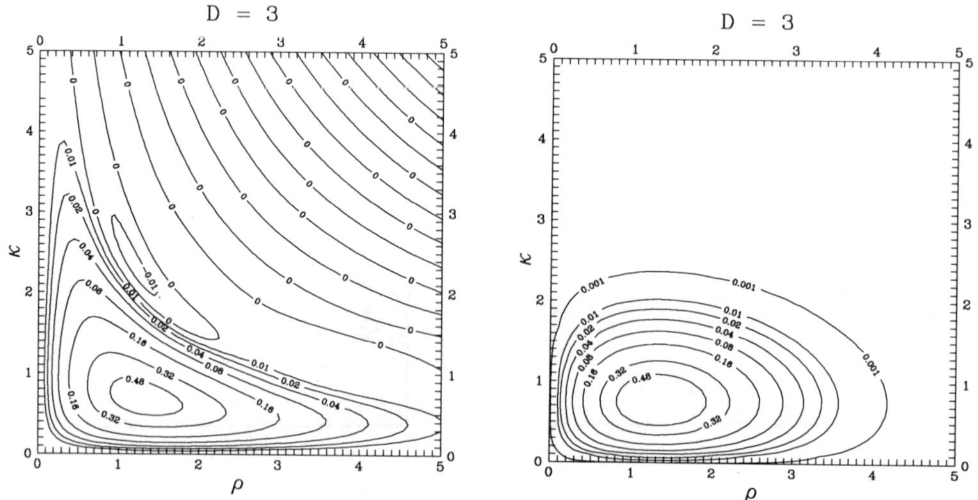

Figure 2. Contour maps of the Wigner function for $D = 3$.
Left: $M = 9$. Right: $M = 1$.

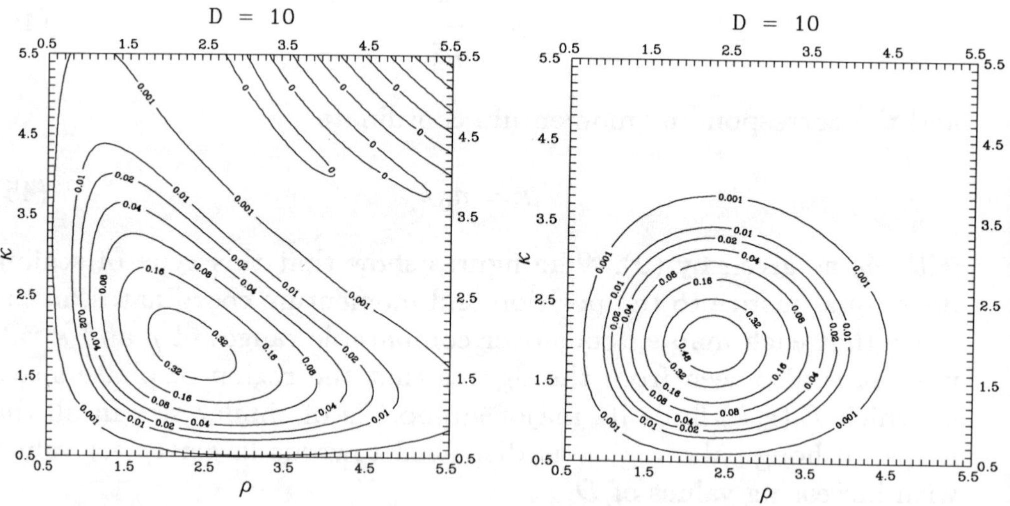

Figure 3. Contour maps of the Wigner function for $D = 10$.
Left: $M = 9$. Right: $M = 1$.

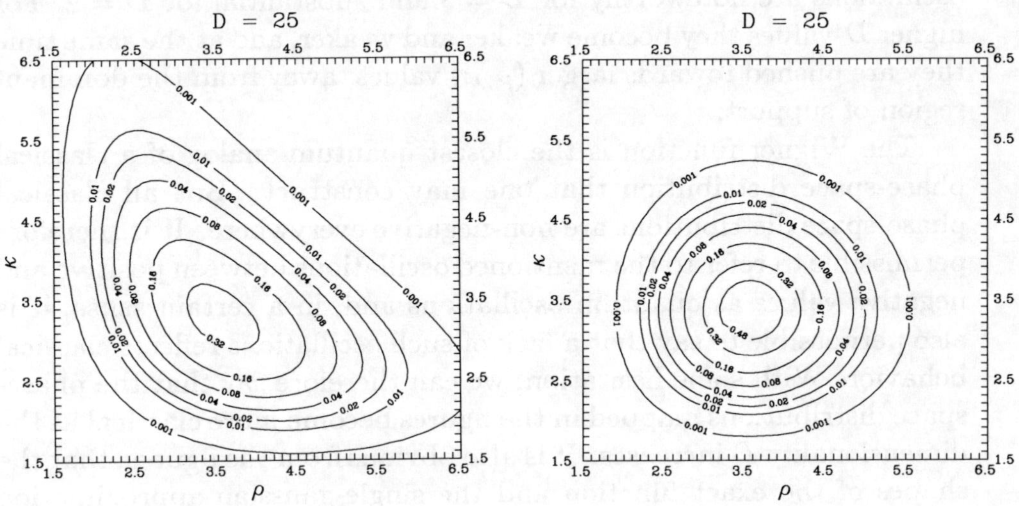

Figure 4. Contour maps of the Wigner function for $D = 25$.
Left: $M = 9$. Right: $M = 1$.

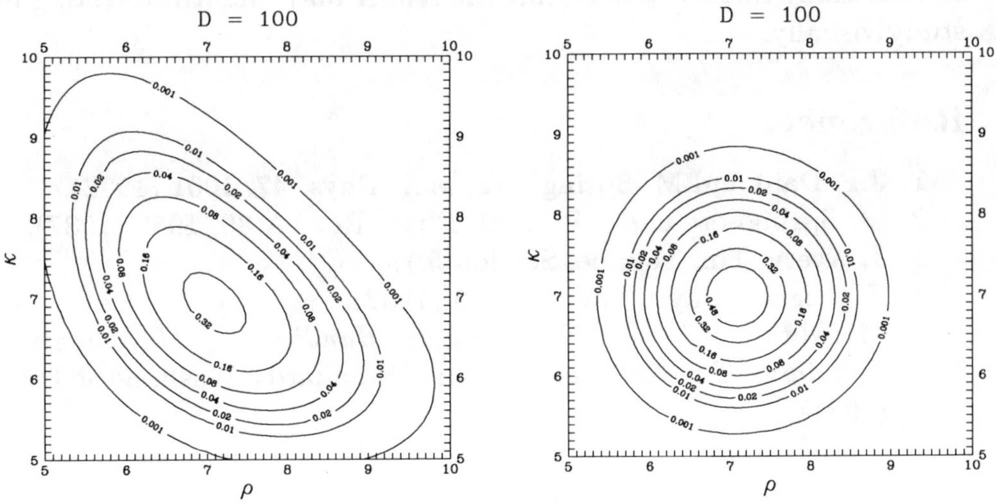

Figure 5 Contour maps of the Wigner function for $D = 100$.
Left: $M = 9$. Right: $M = 1$.

oscillations are noteworthy for $D = 3$ and substantial for $D = 2$. For higher D-values they become weaker and weaker, and at the same time they are pushed towards larger (ρ, κ) values, away from the dominant region of support.

The Wigner function is the closest quantum analog of a classical phase-space distribution that one may construct. But all classical phase-space distributions are non-negative everywhere. It is therefore permissible to refer to the mentioned oscillations between positive and negative values as quantum oscillations and, in a certain sense, it is also permissible to say that a lack of such oscillations reflects classical behavior. With some hesitation, we can therefore say that the phase-space distributions mapped in the figures become more classical as the dimensionality D increases. It is also obvious from the figures, that the shapes of the exact function and the single-gaussian approximation become more and more similar for higher D-values. This is to a large extent a consequence of the fact that the radial Wigner function includes the factor $(rk)^{D-1}$.

With these remarks, we shall close our discussion of the Wigner function for the D-dimensional hydrogen atom. The maps display, of course, many further details that the reader may find it interesting to study visually.

References

1. J.P. Dahl and M. Springborg, Mol. Phys. 47, 1001 (1982).
2. M. Springborg and J.P. Dahl, Phys. Rev. A 36, 1050 (1987).
3. J. Avery, This volume, Section 5.1.
4. E. Wigner, Phys. Rev. 40, 749 (1932).
5. M. Abramowitz and I.A. Stegun, *Handbook of Mathematical Functions*, U.S. National Bureau of Standards, Washington D.C. (1964).

5.3 1/n Expansions for Quasistationary States

V.S. Popov
Institute of Theoretical and
Experimental Physics
Moscow, 117 259, Russia

Abstract

Applications of 1/n expansions to calculations of energies and wave functions are considered, including quasistationary states. As illustrations we have examined the power-law, funnel, and Yukawa potentials. It has been shown that in many cases the method ensures high accuracy, even for small quantum numbers. The connection of these results with the properties of coherent states is briefly discussed.

Introduction

A family of methods, often referred to as the $1/N$ expansion [1-7], have proven to be effective in quantum mechanics and field theory and are widely used at present in various physical calculations (see, for example, refs. [8-10]). Versions differ in the choice of the expansion parameter; among these choices are: $N = D$; or $N = l + D/2$, where $l = 0, 1, 2, \ldots$ is the angular momentum and D is the space dimension; $N = l + D/2 - a$ (the shifted $1/N$ expansion); $N = [l(l+1)]^{1/2}$; and $N = n$, the principal quantum number. Here we consider the version of this method introduced in refs. [11-22], which uses $N = n$. An essential   aspect of this approach is the possibility of using it not only in the case of the discrete spectrum, but also in calculating energies and widths of resonances (with complex energy $E = E_r - i\Gamma/2$). This problem frequently occurs in atomic and nuclear physics, scattering theory, etc.. As examples, we consider quasistationary states in short-range potentials: the power-law, funnel, and Yukawa potentials. The structure of high orders of the $1/N$ expansion is discussed, as well as its divergence.

179

D. R. Herschbach et al. (eds.), Dimensional Scaling in Chemical Physics, 179–195.
© 1993 *Kluwer Academic Publishers. Printed in the Netherlands.*

Description of the Method

Consider a particle of mass M, subject to an attractive spherically symmetric potential characterized by a length parameter R; thus,

$$V(r) = -g\frac{\hbar^2}{MR^2}v(x) \tag{1}$$

where $x \equiv r/R$ is a dimensionless length and g a dimensionless coupling constant. We adopt units such that $\hbar = M = 1$, so energy is in units of R^{-2}. The radial wave equation is then

$$\left[-\frac{1}{2}\frac{d^2}{dx^2} + \frac{l(l+1)}{2x^2} - gv(x) \right]\Psi = ER^2\Psi \tag{2}$$

Scaling is introduced via $x = n^2\rho$, which gives

$$\left[-\frac{1}{2n^2}\frac{d^2}{d\rho^2} + \frac{l(l+1)}{2n^2\rho^2} - gn^2v(n^2\rho) \right]\psi = \epsilon\psi \tag{3}$$

where $\epsilon \equiv n^2ER^2$. As usual, $n = p + l + 1$ is the principal quantum number, $p = 0, 1, 2, ...$ is the radial quantum number (sometimes denoted by n_r), and l is the orbital angular momentum quantum number. If we fix p and let $n \to \infty$, since also $l \to \infty$ and $l(l+1)/n^2 \to 1$, the radial equation reduces to

$$\frac{1}{2\rho^2} - gn^2v(n^2\rho) = \epsilon \tag{4}$$

The condition for a minimum gives

$$-\frac{1}{\rho^3} - gn^2\frac{dv(n^2\rho)}{d\rho} = 0 \tag{5}$$

and since $dv/d\rho = (dv/dx)(dx/d\rho) = n^2v'(x)$ this yields [11,12]

$$x^3v'(x) = -\nu \tag{6}$$

where $\nu \equiv n^2/g$ represents the scaled coupling constant. Eq.(6) determines the classical equilibrium point, $x = x_0$, near which the particle is localized in the large n limit. The energy is given by

$$E_{nl} = \frac{g^2}{2n^2R^2}\epsilon_{nl} \tag{7}$$

where the scaled energy ϵ_{nl} is expanded as

$$\epsilon = \epsilon^{(0)} + \epsilon^{(1)}/n + \epsilon^{(2)}/n^2 + \ldots \tag{8}$$

with $\epsilon^{(0)}$ the limiting result obtained by evaluating Eq.(4) at the classical equilibrium point, $x_0 = n^2 \rho_0$. Although we have $D = 3$, our $1/n$-expansion is equivalent to the dimensional scaling procedure used elsewhere in this volume, because for central forces $D \to \infty$ corresponds to $l \to \infty$ and $n \to \infty$.

To develop systematically the $1/n$-expansion, we note that comparison of the Schrödinger equation at $r \approx r_0 = Rx_0$ with the equation describing the harmonic oscillator shows that n is analogous to $M/\hbar$. The amplitude of the vacuum fluctuation of this oscillator is proportional to $(\hbar/M\omega)^{1/2} \sim n^{1/2}$. Accordingly we introduce

$$x = x_0(1 + \xi n^{-1/2}) \tag{9}$$

and expand all quantities in powers of $n^{-1/2}$. (In contrast to r, the variable ξ remains finite as $n \to \infty$). In this way, coefficients of the expansion shown in equation (8), and corresponding coefficients of the wave function expansion in powers of $n^{-1/2}$ may be successively determined. For instance [11,12],

$$\begin{aligned}
\epsilon^{(0)} &= (1 + v_0)\nu^2/x_0^2 \\
\epsilon^{(1)} &= (2p + 1)(\omega - 1)\nu^2/x_0^2 \\
\epsilon^{(2)} &= \left[\frac{\nu}{\omega^2 x_0}\right]^2 \left[\frac{3}{2}(2p+1)^2\omega(\sigma - \omega^2) + \frac{1}{2}\omega^2(\tau + 1)\right. \\
&\quad \left. -\frac{11}{16}\sigma^2 + p(p+1)[\omega^4 + \omega^2(\tau - 1) - \frac{15}{8}\sigma^2]\right]
\end{aligned} \tag{10}$$

where $\omega = [3(1 + v_2)]^{1/2}$, $\sigma = 4(1 - v_3)$, $\tau = \frac{15}{2}(v_4 + \frac{3}{5})$ and

$$v_k \equiv \frac{2x^{k-1}}{(k+1)!}\left[\left(\frac{d^k v}{dx^k}\right) \Big/ \left(\frac{dv}{dx}\right)\right]_{x=x_0} \tag{11}$$

(Note that ω is the frequency of small oscillations around the equilibrium point x_0). Analytical expressions for coefficients $\epsilon^{(k)}$ with $k \geq 3$ are cumbersome, but may be obtained with ease with the help of recurrence relations convenient for a computer.

Special consideration is required for the case $E = 0$ (i.e. for the moment of bound state appearance in a short-range potential). As a rule, the "reduced" energy, $\epsilon(\nu)$, increases with increasing ν, and at some value, $\nu = \nu_{nl}^{cr}$, the level escapes to the continuous spectrum. In the lowest approximation ($n \to \infty$),

$$\nu_{cr} = 2\bar{x}_0^2 v(\bar{a}_0) \qquad \left[\frac{d\epsilon^{(0)}}{d\nu}\right]_{\nu=\nu_{cr}} = 2v(\bar{x}_0) \tag{12}$$

where $\bar{x}_0$ is the root of the equation $xv' + 2v = 0$. Corrections to order $1/n$ can be obtained from equation (4). Denoting the coupling constant corresponding to the zero energy of the nl-level by g_{nl}, we have:

$$g_{nl} = \frac{n^2}{2\bar{x}_0^2 v(\bar{x}_0)}\left[1 + \frac{\alpha_1}{n} + \frac{\alpha_2}{n^2} + O\left(\frac{1}{n^3}\right)\right] \tag{13}$$

where

$$\alpha_1 \equiv (2p+1)(\omega - 1) \tag{14}$$

$$\begin{aligned}
\alpha_2 \equiv \;& \frac{1}{2}\left[\omega^2 - \omega - 5 + \frac{1}{\omega^2}(3\sigma + \tau + 3) - \frac{11\sigma^2}{8\omega^4}\right] \\
& + p(p+1)\left[2\omega^2 - 2\omega - 9 + \frac{1}{\omega^2}(6\sigma + \tau + 3) - \frac{15\sigma^2}{8\omega^4}\right]
\end{aligned} \tag{15}$$

and where the quantities ω, σ, and τ are evaluated at $x = \bar{x}_0$.

With a further increase of ν, the frequency ω becomes zero at $\nu = \nu_*$. At this point the classical solution loses its stability and a collapse of two solutions occurs, one corresponding to stable and the other to unstable equilibrium points in the effective potential $U(r)$, which includes the centrifugal energy. At $\nu > \nu_*$, the potential $U(r)$ possesses no minimum at real r, the equilibrium point escapes to the complex plane, and the coefficients of the $1/n$ expansion become complex. Such solutions, although they have no physical meaning in terms of classical mechanics, are of special importance in quantum mechanics: They are just the solutions determining, within the $1/n$ expansion, not only the positions but also the widths of the resonance levels.

Wave functions and the $|\psi(0)|^2$ value

The evaluation of wave functions and, in particular, the determination of their asymptotic form at $r \to 0$ and $r \to \infty$ constitute a problem of considerable importance in physics. Using only two terms of the $1/n$ expansion, it is possible to obtain analytical expressions which are asymptotically exact at $n \to \infty$ for an arbitrary smooth potential.

We perform, for this purpose, the substitution of equation (8), and take into account the fact that, in the region where $|\xi| \ll n^{1/2}$, the function $\chi_n \equiv r R_{nl}(r)$ coincides with the wave function of the p-th level of an harmonic oscillator with frequency ω. Continuing this function (by means of the WKB method) into the subbarrier region, $\xi < \xi_-$ or $\xi > \xi_{\pm}$, where $\xi_{\pm} \equiv +[(2p+1)/\omega]^{1/2}$, and expanding all quantities at the point $x = x_0$ in powers of $1/n$, we come to the final result. Here we present only an expression for the coefficient giving the asymptotic behaviour at the origin:

$$\chi_{nl} = c_{nl} R^{-(l+3/2)} r^{l+1} + \ldots \qquad r \to 0 \qquad (16)$$

$$\tilde{c}_{nl} = \left[\frac{n\omega^3}{\pi(p!)^2} \right]^{1/4} (2\pi\omega)^{p/2} x_0^{p-1/2} exp\left(-[nJ_0 + (2p+1)J_1] \right) \qquad (17)$$

where the tilde indicates that this expression is approximate. Here

$$J_0 \equiv ln(x_0) + \int_0^{x_0} dx[Q_0(x) - x^{-1}] \qquad (18)$$

$$J_1 \equiv \frac{1}{2} \int_0^{x_0} dx \left[\frac{x_0}{x(x_0 - x)} - \frac{P_0(x)}{Q_0(x)} \right] \qquad (19)$$

$$Q_0(x) \equiv \left[\frac{1}{x^2} - \frac{2}{\nu} v(x) - \frac{\epsilon^{(0)}}{\nu^2} \right]^{1/2} \qquad (20)$$

and

$$P_0(x) \equiv x^{-2} + (\omega - 1)x_0^{-2} \qquad (21)$$

Analogous expressions can be obtained for $r > r_+$, including the asymptotic coefficient at $r \to \infty$ and the effective radius, r_s [18,19].

Let us consider the wave function at finite r values. For simplicity, we shall confine ourselves to the nodeless states ($p = 0$, $n = l + 1$).

Then

$$\chi_{n,n-1} = \left[\frac{n\omega}{\pi x_0^2}\right]^{1/4} exp\left(-\frac{1}{2}\omega\xi^2\right)[1+(\frac{1}{3}a\xi^3+b\xi)n^{-1/2}+O(n^{-1})] \quad (22)$$

where $a = 2(1 - v_3)\omega^{-1}$, $b = (a - 1)\omega^{-1}$ and $\int_0^\infty \chi^2 dx = 1 + O(1/n)$. This expression is valid in the vicinity of the equilibrium point x_0, including the turning points, $x_\pm$. Terms proportional to $n^{-1/2}$ and to n^{-1} take into account the anharmonicity corrections and considerably improve the agreement with numerical calculations (see ref. [19]).

Small quantum numbers

For fixed p, the above expressions are asymptotically exact at $n \to \infty$. However, in applications, the case $n \sim 1$ is the most frequent. It is not immediately obvious how useful the $1/n$ expansion is for small quantum numbers. To examine this question, we consider three examples.

(a) For *power-law* potentials,

$$V(r) = \frac{g}{s}r^s \quad (23)$$

we have

$$E_{nl}(g) \sim g^{2/(s+2)} \quad (24)$$

and

$$c_{nl} \sim g^{(l+3/2)(s+2)} \quad (25)$$

Therefore it is enough to take $g = 1$. This case corresponds to

$$\begin{aligned} v(x) &= -\frac{x^s}{s} \\ v_k &= 2\frac{(s-1)!}{(k+1)!(s-k)!} \\ x_0 &= n^{2/(s+2)} \end{aligned} \quad (26)$$

and the energy expansion is:

$$E_{nl} = \frac{1}{2}n^{2s/(s+2)}\left(\alpha_0 + \frac{\alpha_1}{n} + \frac{\alpha_2}{n^2} + ...\right) \quad (27)$$

where

$$
\begin{aligned}
\alpha_0 &\equiv 1 + 2s^{-1} \\
\alpha_1 &\equiv (2p + 1)/(\omega - 1)
\end{aligned}
\tag{28}
$$

and

$$
\alpha_2 \equiv -\frac{(s-2)(\omega-1)}{48(s+2)}[(2p+1)^2(\omega^3+\omega^2-12\omega+12)-\frac{1}{3}(\omega^3+\omega^2)] \tag{29}
$$

with

$$
\begin{aligned}
\omega &\equiv (s+2)^{1/2} \\
\sigma &\equiv 4 - \frac{1}{3}(s-1)(s-2) \\
\tau &\equiv -\frac{3}{4} + \frac{(s-1)(s-2)(s-3)}{8}
\end{aligned}
\tag{30}
$$

For $-1 \le s \le 4$, the coefficients α_k rapidly decrease with increasing k. At $s = -1$ (the Coulomb potential) and $s = 2$ (the harmonic oscillator), $\alpha_k = 0$ for all $k \ge 2$ and the series in equation (8) is then truncated and coincides with the exact solutions:

$$
\epsilon_{nl} = \begin{cases} -1 & \text{at } s = -1 \\ 2n^2(2p+1+3/2) & \text{at } s = 2 \end{cases}
\tag{31}
$$

Let us now investigate the accuracy of the asymptotic relationship shown in equation (12). For power-law potentials, a comparison of equation (12) with numerical calculations is given in ref. [18]. Even for the ground state ($l = 0$), the precision of this simple expression is surprisingly good, especially at $s = 1$ and $s = 4$ (the anharmonic oscillator). Note that in this case, the coefficients change by many orders of magnitude:

$$
c_{n,n-1} \sim const \cdot a^n \frac{n^\beta}{(n\alpha)!} \qquad n \to \infty
\tag{32}
$$

where $\alpha = 2/(s+2)$, $\beta = (3s-2)/[4(s+2)]$ and $0.368 < a < 0.737$ for $-1 < s < \infty$. The factorial decrease of $c_{n,n-1}$ is due to the centrifugal barrier.

For $s = -1$ and $s = 2$ these coefficients are known explicitly: For $s = -1$,

$$c_{n,n-1} = 2^n / n^{n+1} [2n - 1]^{1/2} \tag{33}$$

while for s=2

$$c_{n,n-1} = \left[\frac{2}{\Gamma(n + 1/2)} \right]^{1/2} \tag{34}$$

Thus, for power-law potentials, the asymptotic expression with $p = 0$ has good precision for a wide range of s values; but the precision decreases as as the number of nodes, p, increases.

(b) The *funnel* potential

$$V(r) = -\frac{\chi}{r} + \frac{r}{a^2} \tag{35}$$

is frequently used to describe the quarkonium and multiquark systems [23,24]. Here

$$\begin{aligned} v(x) &= x^{-1} - x \\ R &= (\chi a^2)^{1/2} \\ \omega &= \left[\frac{1 + 3x_0^2}{1 + x_0^2} \right]^{1/2} \end{aligned} \tag{36}$$

and x_0 is determined by the equation

$$x^3 + x = \nu \equiv n^2 \chi^{-3/2} a^{-1} = \frac{n^2}{R\chi} \tag{37}$$

The first three terms of the series in equation (3) are determined explicitly in this case (see refs. [12,20]). A comparison of the $1/n$ expansion with the results of numerical computations is presented in Table 1, where the values of $\xi_{nl}^{(k)}$ are given, where

$$\xi_{nl}^{(k)} \equiv \xi^{(0)} + \xi^{(1)} n^{-1} + \dots + \xi^{(k)} n^{-k} \tag{38}$$

The Schrödinger equation with the funnel potential was first transformed ($r = const \cdot \rho$) to the standard form:

$$\frac{d^2 u_{nl}}{d\rho^2} + \left[\xi + \frac{\lambda}{\rho} - \rho - \frac{l(l+1)}{\rho^2} \right] u_{nl} = 0 \tag{39}$$

which determines the eigenvalues, $\xi = \xi_{nl}(\lambda)$. Here

$$\int_0^\infty u_{nl}^2(\rho)d\rho = 1 \tag{40}$$

and

$$u_{nl}(\rho) = c_{nl}\rho^{l+1} + \dots \tag{41}$$

as $\rho \to 0$.

It can be seen from Table 1, that the description of the energy spectrum using partial sums $(k+1 \geq 3)$ of the $1/n$ expansion is fairly accurate for the funnel potential. The accuracy is enhanced as l increases, and it is especially high for nodeless states (1S, 2P,...). In the latter case, only three terms of the $1/n$ expansion ensure percent accuracy in energy and $\psi(0)$ computations and properly reproduce [20] the charmonium spectrum if equation with Cornell potential parameters [23] is used. It is important that the above expressions allow one to perform with ease calculations for other potentials arising from QCD.

(c) For the *Yukawa* potential,

$$V(r) = -r^{-1}exp(-\mu r) \tag{42}$$

we have

$$v(x) = \frac{e^{-x}}{x} \tag{43}$$

and

$$\nu = n^2\mu \tag{44}$$

where μ is the screening parameter. The dependence of x_0 and $\epsilon^{(0)}$ on ν is determined by

$$\nu = (x^2 + x)e^{-x} \tag{45}$$

and

$$\epsilon^{(0)} = (x^2 - 1)e^{-2x} \tag{46}$$

In this case, $\nu_{cr} = 2e^{-1} = 0.736$ and $\nu_* = 0.834$. For $\nu \to \nu_*$ the coefficients $\epsilon^{(k)}$ have singularities:

$$\begin{aligned}
\epsilon^{(0)} &= c_0 + c_1(\nu - \nu_*) + c_2(\nu_* - \nu)^{3/2} + \dots \\
\epsilon^{(1)} &= c_0' + c_1'(\nu_* - \nu)^{1/4} + \dots \\
\epsilon^{(2)} &\sim (\nu_* - \nu)^{-1}
\end{aligned} \tag{47}$$

Table 1. Accuracy of $1/n$ expansion for funnel potential.
The values of $\xi_{nl}^{(k)}$, equation (38), are given for $2M = m_c = 1.84 GeV$, $\chi = 0.52$, and $a = 2.34 GeV^{-1}$, which corresponds to the Coulomb parameter $\lambda = \chi(2Ma)^{2/3} = 1.37623$. Exact eigenvalues ξ_{nl} were obtained by numerical solution of the Schrödinger equation [25]. The coefficients c_{nl} at $r \to 0$ correspond to the normalization condition (40).

k	1S (n=1)	1P (n=2)	2S (n=2)	1D (n=3)	2P (n=3)	3S (n=3)
2	0.97932	2.61144	3.26476	3.69578	4.30275	4.89442
3	0.97985	2.61101	3.24627	3.69557	2.29871	4.87627
5	0.98029	2.61114	3.23182	3.69560	4.29692	4.85651
exact ξ_{nl}	0.980366	2.611131	3.228853	3.695599	4.296697	4.842092
c_{nl}	2.0833	0.8970	1.6634	0.3103	1.0701	1.5362
$\tilde{c}_{nl}/c_{nl}$	0.9977	0.9978	1.40	0.9983	1.22	-

with $\epsilon^{(k)} \to \infty$ for $k \geq 2$. Therefore the $1/n$ expansion is not useful in the neighborhood of $\nu = \nu_*$. However, at $\nu > \nu_*$, when the equilibrium point moves into the complex plane, the $1/n$ expansion is applicable again. As for the funnel potential, only three terms of the series of equation (8) provide acceptable accuracy for energy calculations for quasistationary states, and the greater the value of l, the higher is the accuracy. Figures 1 and 2 illustrate this performance.

Application of the $1/n$ expansion to the problem of an atom in a strong electric and/or magnetic field is discussed in section 6.2.

The $1/n$ expansion and coherent states

As can be seen from the examples considered above, the $1/n$ expansion has a gratifyingly high accuracy for $n \sim 1$ (in the case of nodeless states, $p = 0$). We shall now give a qualitative explanation of this fact. It will be shown that the above states are closest to classical mechanics. Since the first term, $\epsilon^{(0)}$ of the series (8) corresponds to a classical particle at rest at the minimum of the effective potential, the high accuracy of the $1/n$ expansion is thus explained by a lucky choice of the initial approximation.

For the states with $l =\mid m \mid= n - 1$ in a hydrogen atom, we have

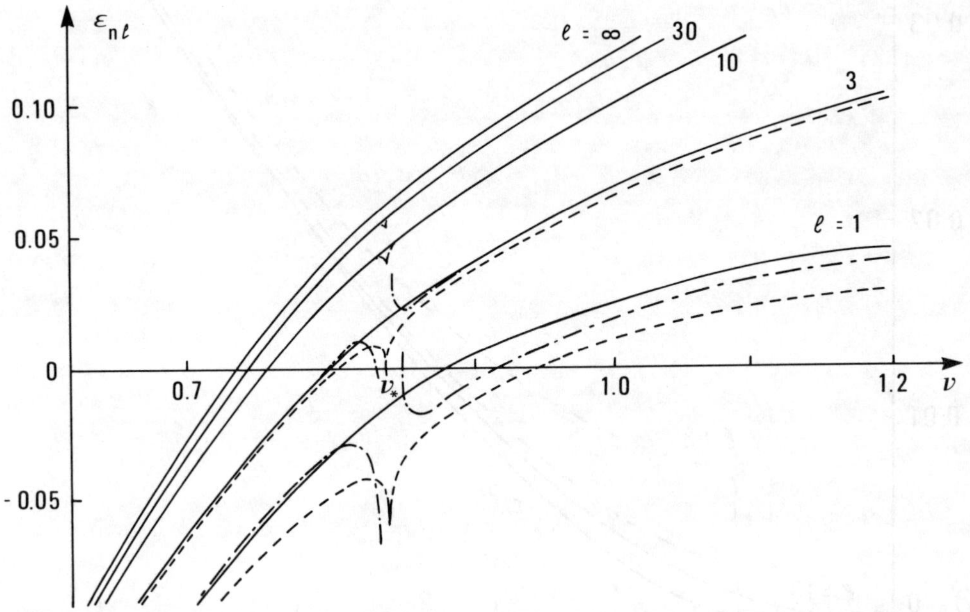

Figure 1. Energies of the nodeless states $(l = n - 1, p = 0)$ in the Yukawa potential; For $\nu > \nu_{cr}$ we show $Re(\epsilon_{nl})$. The solid, dashed and dash-dot curves correspond to one, two and three terms of the $1/n$ expansion. The curves are labelled by the l values.

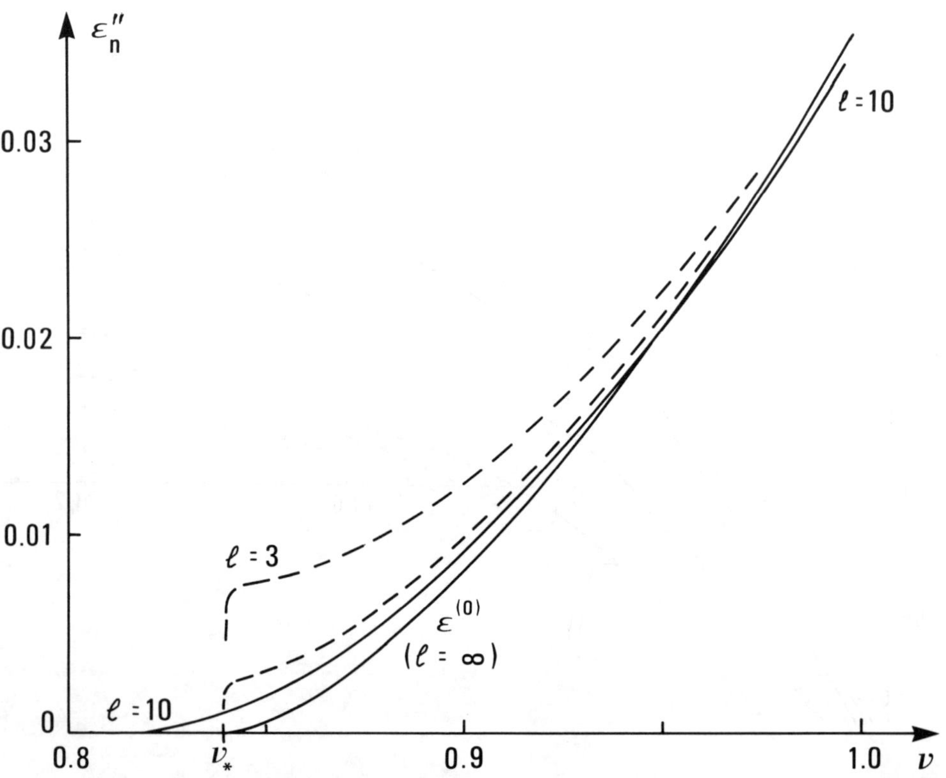

Figure 2. The Yukawa potential: $\epsilon_n = -Im(\epsilon_{nl}) = n^2\Gamma_{nl}$, $l = n - 1$. The solid curve represents the exact energy ($l = 10$), while the dashed curves correspond to two terms, $\epsilon^{(0)} + \epsilon^{(1)}n^{-1}$, of the $1/n$ expansion.

[1]:

$$\bar{r} = n(n + \frac{1}{2})$$

$$\bar{p}_r^2 = [(2n - 1)n^2]^{-2} \tag{48}$$

Hence,

$$\Delta p_r \Delta r = \frac{\hbar}{2} \left[\frac{2n + 1}{2n - 1}\right]^{1/2}$$

$$\Delta p_z \Delta z = \frac{\hbar}{2} \left[\frac{2n + 2}{2n + 1}\right]^{1/2} \tag{49}$$

At $n \gg 1$, such a state, corresponding to the circular electron orbit perpendicular to the z axis, minimizes the uncertainty relations of the radial and transverse (to the orbit plane) components of p and r. Moreover, the quantum fluctuations of the radius of the orbit and the angle of inclination of the orbit plane to the z axis decrease as $n^{-1/2}$. This makes the applicability of the semiclassical approach quite natural.

For all other states, $\Delta p_i \Delta q_i$ exceeds $\hbar/2$. Thus, if $n \to \infty$ and the quantum numbers $p \equiv n_r$ and $q = l - |m|$ are fixed, then

$$\Delta p_r \Delta r = \hbar(p + \frac{1}{2} + \frac{a}{n} + ...)$$

$$\Delta p_z \Delta_z = \hbar(q + \frac{1}{2} + \frac{b}{n} + ...) \tag{50}$$

where

$$a = \frac{1}{4}[1 - 3p(p + 1)]$$

$$b = \frac{5}{4}(2pq + p + q) - \frac{1}{8}(4q^2 - 1) \tag{51}$$

If, on the other hand, p and q are both large, then $\Delta p_i \Delta q_i \sim n$. For example, for ns-states in the power-law potential

$$V(r) = gr^s/s \tag{52}$$

we have:

$$\Delta p_r \sim n^{s/(s+2)}$$

$$\Delta r \sim n\frac{2}{s+2}$$

$$\Delta p_r \Delta r = \hbar n f(\rho) \tag{53}$$

$$\rho = [l(l+1)]^{1/2} n^{-1}$$

Here $n \gg 1$ and the function $f(\rho)$ can be calculated analytically for the Coulomb potential and harmonic oscillator [7]. The results of computations of $f(\rho)$ for the power potentials (53) are given in Figure 3, which shows that at $n \to \infty$, the product $\Delta p \Delta q = \hbar$ only at l near to $l_{max} = n-1$, i.e. at $p \sim 1$. In other cases, the quantum fluctuations of the orbit radius grow infinitely at $n \to \infty$; and thus the notion of the classical orbit becomes meaningless.

In conclusion, the following should be noted. For the harmonic oscillator, the Glauber coherent states are introduced

$$| \alpha \rangle = \sum_{n=0}^{\infty} \alpha^n (n!)^{-1/2} | n \rangle$$

$$\alpha(t) = \alpha(0)e^{-i\omega t} \tag{54}$$

for which $\Delta p \Delta x = \hbar/2$ for any α. (The average number of quanta, $n =| \alpha |^2$ may be arbitrarily large.)

The coherent states demonstrate most clearly the limiting transition from quantum to classical mechanics. However, this approach has a shortcoming: The coherent states (54) can be constructed only for very special models. On the other hand, the nodeless states (and the states with $p, q \ll n$) are easily constructed with the help of the $1/n$ expansion for an arbitrary potential, $V(r)$, as well as in a number of problems without spherical symmetry (for example, for the problem of two Coulomb centers [8]). Such states minimize the uncertainty relations at $n \to \infty$, and in that respect they are similar to (54). However, there is also a difference: the states $| \alpha(t) \rangle$ are nonstationary, while the states discussed above are stationary. This is due to the fact that we are now considering the motion of a wave

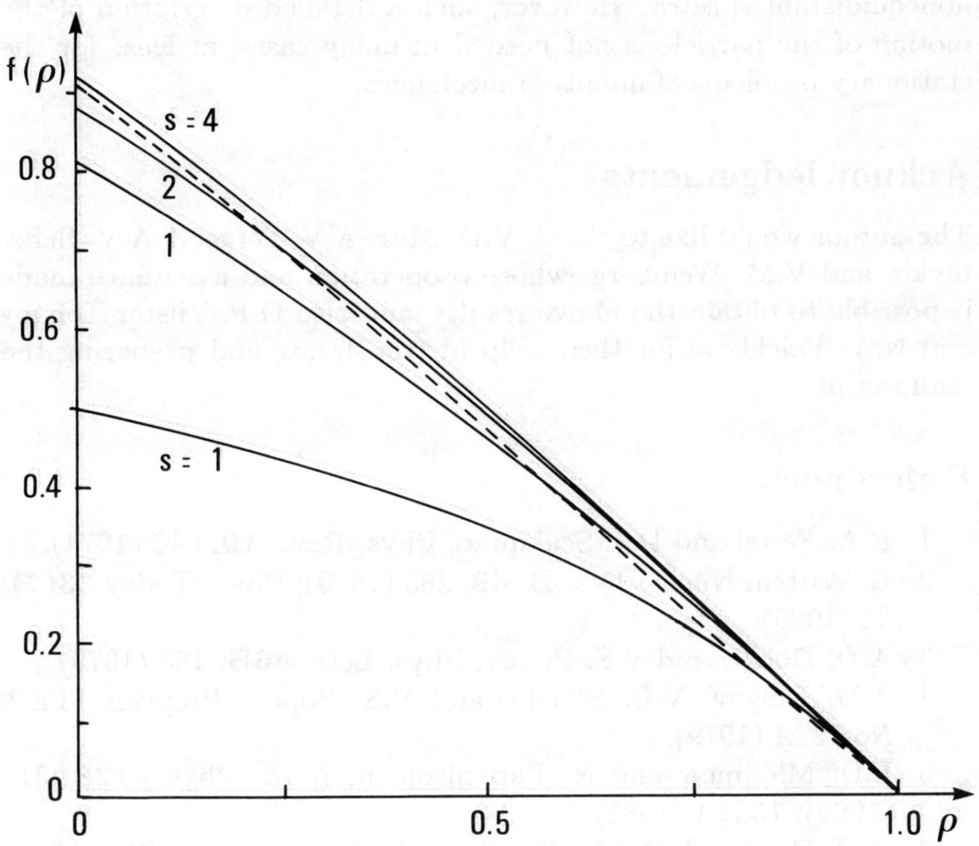

Figure 3. The function $f(\rho)$ in equation (53) for power-law potentials. The values of the index s are shown on the curves. The dashed curve corresponds to $s = \infty$, i.e. the square well potential.

packet along the orbit, averaging over many turns. In an attempt to consider in detail this type of motion, it is necessary to introduce a superposition of states with different energies, E_n, which unavoidably results in a rapid smearing out of the wave packet for systems with nonequidistant spectra. However, such a detailed description of the motion of the particle is not needed in many cases, at least for the stationary problems of quantum mechanics.

Acknowledgements

The author would like to thank V.D. Mur, A.V. Sergeev, A.V. Scheblykin and V.M. Weinberg, whose cooperation and assistance made it possible to obtain the above results, and also O.P. Vilster-Tolstoy and N.A. Volchkova for their help in translating and preparing the manuscript.

References

1. R.A. Ferrel and D.J. Scalapino, Phys. Rev. **A9**, 846 (1974).
2. E. Witten, Nucl. Phys. **B149**, 285 (1979); Phys. Today **33(7)**, 38 (1980).
3. A.D. Dolgov and V.S. Popov, Phys. Lett. **86B**, 185 (1979).
4. A.D. Dolgov, V.L. Eletsky and V.S. Popov, Preprint ITEP, No.72.M (1979).
5. L.D. Mlodinow and N. Papanicolaou, Ann. Phys. **128**, 314 (1980); **131**, 1 (1981).
6. C.M. Bender, L.D. Mlodinow and N. Papanicolaou, Phys. Rev. **A25**, 1305 (1982).
7. T. Imbo, A. Pagnamenta and U. Sukhatme, Phys. Rev. **D29**, 1669 (1984); Phys. Lett. **105A**, 183 (1984).
8. L.G. Yaffe, Rev. Mod. Phys. **54**, 407 (1982).
9. L.G. Yaffe, Phys. Today **36(8)**, 50 (1983).
10. A. Chatterjee, Phys. Reps. **186**, 249 (1990).
11. V.S. Popov, V.M. Weinberg and V.D. Mur, Pis'ma v ZhETF **41**, 4399, (1985); Preprint ITEP, No.178.M (1985).
12. V.S. Popov, V.M. Weinberg and V.D. Mur, Yad. Phys. **44**, 1103 (1986).

13. V.S. Popov et al., Doklady Akad. Nauk SSSR **289**, 1095 (1986); **293**, 851 (1988); **303**, 1102 (1988).

14. V.S. Popov, V.D. Mur, A.V. Scheblykin et al., Phys. Lett. **124A**, 77, (1987).

15. V.M. Weinberg, V.D. Mur, V.S. Popov et al., Pis'ma v ZhETF **44**, 9 (1986); ZhETF **93**, 450 (1987).

16. V.S. Popov et al., Phys. Lett. **149A**, 418, 425 (1990).

17. V.M. Weinberg et al., TMF **74**, 399 (1988).

18. V.D. Mur and V.S. Popov, Pis'ma v ZhETF **45**, 323 (1987); Yad. Fiz. **477**, 697 (1988).

19. V.D. Mur, S.G. Pozdn'akov and V.S. Popov, Yad. Fiz **51**, 390 (1990).

20. V.D. Mur, V.S. Popov and A.V. Sergeev, ZhETF **97**, 32 (1990).

21. V.D. Mur and V.S. Popov, ZhETF **97**, 1729 (1990).

22. V.M. Weinberg, V.S. Popov and A.V. Sergeev, ZhETF **98**, 847 (1990).

23. E. Eichten, K. Gottfried, T. Kinoshita et al., Phys. Rev. **D17**, 3090 (1978); **D21**, 203, 313 (1980).

24. A.M. Badalyan, B.L. Ioffe and A.V. Smilga, Nucl. Phys. **B281**, 85 (1987).

25. A.M. Badalyan, D.I. Kitoroage and D.S. Pariysky, Yad. Fiz. **46**, 226 (1987).

Chapter 6

HYPERCYLINDRICAL SYMMETRY

6.1   The Spheroidal H Atom

Stella Sung
Department of Chemistry
Harvard University
12 Oxford Street
Cambridge MA 02138, USA

Abstract

*The Schrödinger equation for a hydrogenic atom is separable in pro-
late spheroidal coordinates, as a consequence of the "hidden symmetry" stemming from the fixed spatial orientation of the classical Kepler
orbits. One focus is at the nucleus and the other a distance R away
along the major axis of the elliptic orbit. Here we evaluate the separation constant $\alpha(n, m; ZR)$ and the coefficients $g_l(\alpha)$ that specify the
spheroidal eigenfunctions as hybrids of the familiar $|nlm >$ hydrogen
atom states with fixed n and m but different l values. In the limit
$R \longrightarrow \infty$ these hybrids become the solutions in parabolic coordinates,
determined simply by geometrical Clebsch-Gordan coefficients that
account for conservation of angular momentum and the hidden symmetry. These spheroidal eigenfunctions may be used to construct exact*

D. R. Herschbach et al. (eds.), Dimensional Scaling in Chemical Physics, 197–216.
© 1993 *Kluwer Academic Publishers. Printed in the Netherlands.*

analytic solutions of two-center molecular orbitals for special values of R and the nuclear charge ratio $\frac{Z_a}{Z_b}$. Furthermore, interdimensional degeneracies may be used to relate the three-dimensional spheroidal solutions to the analogous spheroidal solutions in higher dimensions.

Introduction

As described in a previous chapter [1], the D-dimensional hydrogen atom has interdimensional degeneracies which link states having the same number of radial nodes but dimensionality differing by two units and orbital angular momentum differing by one unit. Because of the simple relationship between l and D which dictates these interdimensional degeneracies, solutions for the hydrogenic atom in any one dimension essentially encompass all alternating higher dimensions. Thus, the familiar three-dimensional hydrogen atom takes an added importance in the context of dimensional scaling.

Beyond the interdimensional degeneracies, the three-dimensional nonrelativistic hydrogen atom is characterized by the extraordinary n^2 degeneracy of its energy levels. This better-known degeneracy manifests a "hidden" dynamical symmetry. In addition to the Hamiltonian **H** and orbital angular momentum **l**, the Lenz vector **a** is a constant of the motion.[2 -5] Classically, for bound states the **a**-vector points along the major axis of the Kepler elliptic orbit and its length is proportional to the eccentricity. A consequence of this dynamical symmetry is that the bound state eigenfunctions specified by the complete set of commuting constants of the motion may be constructed simply as angular momentum eigenstates. The quantum Kepler problem is thereby soluble by purely geometrical means.

Although it is typically solved in spherical and parabolic coordinates, the hydrogenic Schrödinger equation is also separable in prolate spheroidal coordinates, with one focus at the nucleus and the other located along the Lenz vector at a distance R away. These coordinates are ordinarily used for two- center problems such as H_2^+. Previously, general features of the spheroidal hydrogen atom have been explored by Coulson and Robinson [6], who noted that the limits $R \longrightarrow 0$ and $R \longrightarrow \infty$ yield the spherical and parabolic solutions, respectively. Demkov [7] used the spheroidal eigenfunctions to construct analytic solutions (exact for the Born -Oppenheimer problem) for an electron interacting with two nuclei for certain special values of the internuclear distance R and the charge ratio $\frac{Z_a}{Z_b}$. Other aspects of the spheroidal eigenfunctions were elucidated by Judd [8], particularly the connection to the four- dimensional spherical harmonics.

Here we outline a study [9] applying and extending these results to evaluate explicitly the spectrum of the separation constant $\alpha(n, m, ZR)$ for the hydrogen atom in spheroidal coordinates and the corresponding eigenfunctions. These are obtained from a secular equation of order $n - m$ which provides coefficients $g_l(\alpha)$ for expansion of the spheroidal eigenfunctions in the usual $|nlm >$ states. In effect, the $g_l(\alpha)$ functions interpolate between the coupled and uncoupled representations and thereby play the role of generalized Clebsch-Gordan coefficients. We plot these coefficients and the probability distributions for the hybrid wavefunctions, $\sum_l g_l(\alpha)|nlm >$, for all states up through $n = 4$. We also briefly discuss applying the spheroidal eigenfunctions to further exact solutions akin to those of Demkov [7] for special diatomic molecular orbitals.

Two-Center Spheroidal Eigenstates

For an electron at distances r_a and r_b from two fixed Coulomb centers with charges Z_a and Z_b a distance R apart, the wavefunction is separable in the form $L(\lambda)M(\mu)e^{\pm im\phi}$, with $\lambda = \frac{(r_a+r_b)}{R}$ and $\mu = \frac{(r_a-r_b)}{R}$ the spheroidal coordinates and ϕ the azimuthal angle about the line between the centers. The separated equations for the $L(\lambda)$ and $M(\mu)$ factors are

$$\mathcal{L}_\lambda L(\lambda) + [A + (Z_a + Z_b)R\lambda - p^2\lambda^2]L(\lambda) = 0 \tag{1}$$

and

$$\mathcal{L}_\mu M(\mu) + [-A - (Z_a - Z_b)R\mu + p^2\mu^2]M(\mu) = 0 \tag{2}$$

where A is the separation constant and $p^2 = \frac{-1}{2ER^2}$ is an energy parameter. The operator

$$\mathcal{L}_\lambda = \frac{d}{d\lambda}(\lambda^2 - 1)\frac{d}{d\lambda} - \frac{m^2}{\lambda^2 - 1} \tag{3}$$

and $\mathcal{L}_\mu$ has the same form with the factor $(\lambda^2 - 1)$ replaced by $(1 - \mu^2)$. The pair of equations for $L(\lambda)$ and $M(\mu)$ is commonly referred to as the two-center equations.[10] For a hydrogenic atom, we take $Z_a = Z$ and $Z_b = 0$, and since $E = \frac{-Z^2}{2n^2}$ for the bound states of interest here, $p^2 = \frac{Z^2 R^2}{4n^2}$. The pair of two-center equations then become the same equation with different ranges for the variables,

$$\mathcal{L}_x F(x) + [A + 2pnx - p^2 x^2] F(x) = 0. \tag{4}$$

When $x = \lambda$ (range 1 to ∞), $F(x) = L(\lambda)$ and when $x = \mu$ (range -1 to +1), $F(x) = M(\mu)$; in either case $\mathcal{L}_x$ has the form of Equation (3). For simplicity, henceforth we write $m \equiv |m|$. When $p = 0$, the solutions are $P_l^m(x)$, the associated Legendre polynomials, with $A = -l(l+1)$. It is only necessary to extend Rodrigues' formula,

$$P_l^m(\mu) = \frac{(1 - \mu^2)^{\frac{m}{2}}}{2^l l!} \frac{d^{l+m}}{d\mu^{l+m}} (l - \mu^2)^l \, for |\mu| \le 1, \tag{5}$$

by replacing the $(1 - \mu^2)$ factors with $(\lambda^2 - 1)$ for the range $|\lambda| > 1$. For $p = 0$ a solution of either two-center equation has the form [6,10]

$$F(x) = e^{-px} \sum_{l=m}^{n-1} c_l P_l^m(x). \tag{6}$$

On substituting this into Equation (4) and simplifying with the aid of identities linking polynomials of different l, we obtain a three-term recursion relation for the coefficients:

$$[A - p^2 + l(l+1)]c_l \; + \; \frac{2p(l + n + 1)(l + m + 1)}{2l + 3} c_{l+1}$$
$$- \; 2p \frac{(l - n)(l - m)}{2l - 1} c_{l-1} = 0. \tag{7}$$

For each pair of values of n and m, the recursion relation gives a secular determinant of order $n - m$. The roots determine the eigenvalues of the separation constant A and the corresponding $\{c_l\}$ which specify the eigenfunctions.

Figure 1 shows the resulting pattern of spheroidal eigenstates up through $n = 4$. For each energy level there are n degenerate spheroidal

Table 1. Secular equations for Lenz vector eigenvalues. $\alpha_m \equiv \alpha + m(m+1); \alpha \equiv A - p^2$, where $-A$ is the eigenvalue of the Lenz vector. $\beta_l = \beta_l(n,m) \equiv \frac{(l^2-m^2)(n^2-l^2)}{(2l-1)(2l+1)}$.

$n-m$	*Secular Equation*
1	$\alpha_m = 0$
2	$\alpha_m \alpha_{m+1} = 4p^2$
3	$\alpha_m \alpha_{m+1} \alpha_{m+2} = 4p^2[t a_{m+2}\alpha_m + t a_{m+1}\alpha_{m+2}]$
4	$\alpha_m \alpha_{m+1}\alpha_{m+2}\alpha_{m+3} = 4p^2[t a_{m+3}\alpha_m \alpha_{m+1}$ $+ t a_{m+2}\alpha_m \alpha_{m+3} + t a_{m+1}\alpha_{m+2}\alpha_{m+3}] - 144p^4$

states with different m (denoted by $\sigma, \pi, \delta, ...$ for $m = 0, 1, 2, ...$), each comprised of a linear combination of the $n - m$ terms in Equation (6) for $l = m$ to $l = (n-1)$. Also listed for each of the eigenstates are the spheroidal quantum numbers n_λ and n_μ, which specify the number of nodes in the λ- and the μ-coordinate, respectively. Table 1 gives the secular equations in terms of $\alpha = A - p^2$.

Although the secular determinant associated with the $\{c_l\}$ coefficients is sufficient for calculating the separation constants, it would be even more useful to have a set of coefficients that relate the eigenstates in spheroidal coordinates, designated as $|n\alpha m >$, and the customary states in spherical coordinates, designated by $|nlm >$, in the following simple manner:

$$|n\alpha m > = \sum_l g_l |nlm >, \qquad (8)$$

and

$$|nlm > = \sum_\alpha g_l |n\alpha m >, \qquad (9)$$

where, for given n and m, the sum in Equation (8) extends from $l = m$ to $l = n-1$ and that in Equation (9) over the $n - m$ distinct α-values. The set of coefficients $\{g_l\}$ thus would define a unitary transformation between the spheroidal and spherical eigenstates, and would have the role of generalized Clebsch-Gordan coefficients,

$$g_l = < n\alpha m|nlm > = < nlm|n\alpha m >, \tag{10}$$

in analogy to the familiar angular momentum recoupling.[11]

In fact, such a set of coefficients $\{g_l\}$ may be obtained by symmetrizing the secular determinants for the spheroidal eigenstates [8] such that $c_l = Q_l g_l$, where

$$Q_l = (-1)^l \left[\frac{(2l+1)(l-m)!}{(l+m)!(l+n)!(n-l-1)!} \right]^{\frac{1}{2}} \tag{11}$$

The recursion relation for the new coefficients g_l is then

$$\frac{\alpha + l(l+1)}{2p} g_l \; + \; -\left\{ \frac{[(l+1)^2 - m^2][n^2 - (l-1)^2]}{(2l+1)(2l+3)} \right\}^{\frac{1}{2}} g_{l+1}$$

$$+ \; -\left\{ \frac{[l^2 - m^2][n^2 - l^2]}{(2l-1)(2l+1)} \right\}^{\frac{1}{2}} g_{l-1} = 0. \tag{12}$$

Figure 2 plots values of g_l^2 (normalized to sum to unity) as functions of ZR for all eigenstates up through $n = 4$.

Simple results are obtained for both the small R and large R limits. For $R \longrightarrow 0$, where $\alpha \longrightarrow -l(l+1)$, only one g_l coefficient is nonzero for each eigenstate (so the normalized $g_l^2 \longrightarrow 1$), and the corresponding values of n, l, and m are used to label that state. For $R \longrightarrow \infty$, where $\alpha \longrightarrow -2p(m_+ - m_-)$, the eigenvalue becomes proportional to that for the z-component of the Lenz vector [9] and the g_l become Clebsch-Gordan coefficients.[9] In Fig. 3, the limiting values are marked on the ordinate axis.

Explicit Eigenfunctions and Probability Distributions

The eigenstates in spheroidal coordinates (λ, μ, ϕ) are given explicitly by

$$|n\alpha m> = e^{-p(\lambda+\mu)}[(\lambda^2 - 1)(1 - \mu^2)]^{\frac{m}{2}} f_{n\alpha m}(\lambda) f_{n\alpha m}(\mu) e^{\pm im\phi}, \tag{13}$$

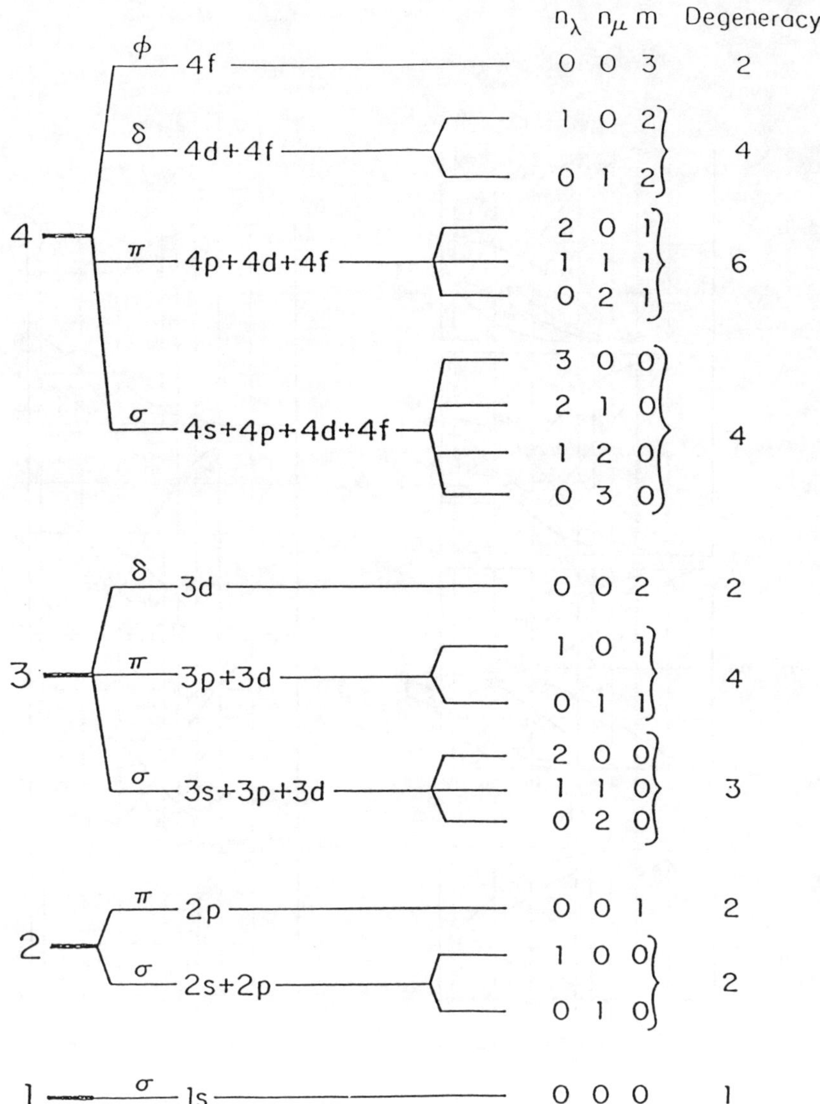

Figure 1. Degenerate energy eigenstates for each n sorted according to $m = 0, 1, 2, ... (\sigma, \pi, \delta, \phi)$ and according to eigenvalues of the separation constant A (designated by the spheroidal quantum numbers n_λ and n_μ), thereby specifying spheroidal eigenstates as hybrids of the usual spherical eigenstates.

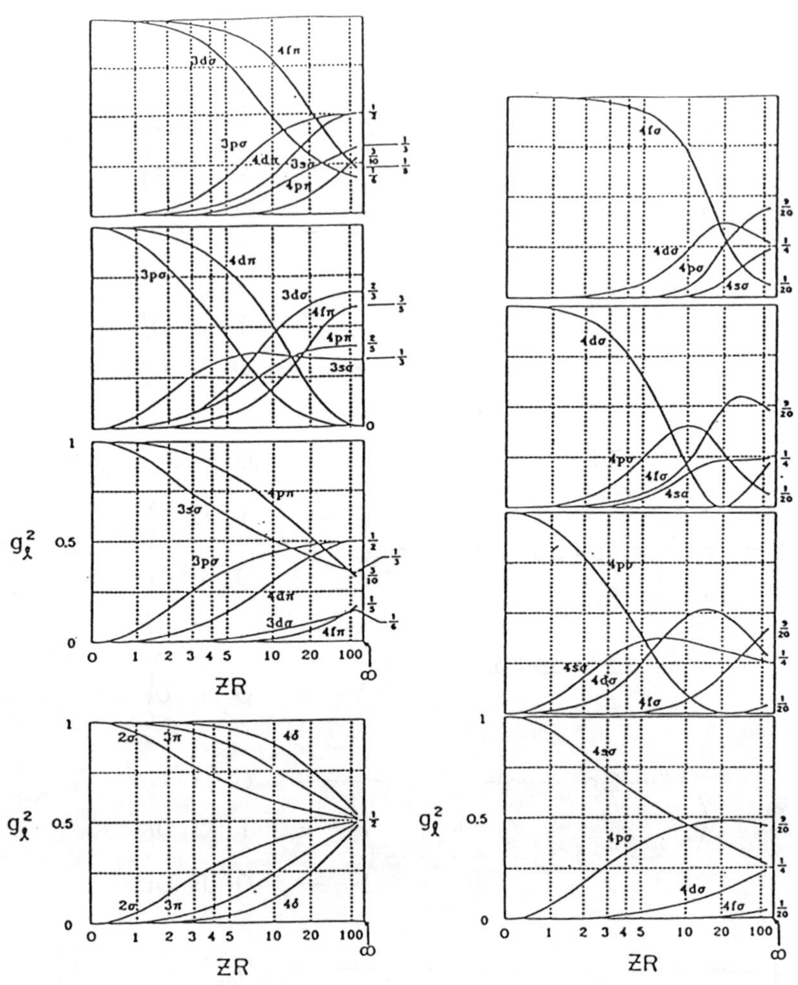

Figure 2. Squares of transformation coefficients g_l as functions of ZR. States are labeled as the $|nlm>$ states to which the spheroidal $|n\alpha m>$ reduce for $R \longrightarrow 0$. The limiting values of g_l^2 for $R \longrightarrow \infty$ are indicated along the ordinate at the right.

aside from normalization.[6] The polynomial factors in $|n\alpha m>$ are given by

$$f_{n\alpha m}(x) = \sum_l g_l Q_l C_{l-m}^{m+\frac{1}{2}}(x), \tag{14}$$

aside from normalization. $C_{l-m}^{m+\frac{1}{2}}(x)$ is a Gegenbauer polynomial. Again, for given n and m, there are $n-m$ distinct values of α, and the sum extends from $l = m$ to $l = n-1$. Table 2 lists the roots of the polynomials for these eigenstates for several values of ZR.

From the relationships between the various coordinates, one can find that the analogous eigenstates in spherical coordinates $|nlm>$ and in parabolic coordinates $|n\tau m>$ have the same form as Equation (13). The only difference is that the $f_{n\alpha m}(\lambda)f_{n\alpha m}(\mu)$ polynomials are replaced by

$$\rho^{l-m} L_{n-l-1}^{2l+1}(\rho) C_{l-m}^{m+\frac{1}{2}}(cos\theta) \tag{15}$$

or by

$$L_s^m(\xi)L_t^m(\eta) \tag{16}$$

where $L_\nu^q(x)$ is an associated Laguerre polynomial.

In the limit $R \longrightarrow 0$, each spheroidal eigenfunction $|n\alpha m>$ reduces to a particular spherical function $|nlm>$, specified by $\alpha \longrightarrow -l(l+1)$. Likewise, in the limit $R \longrightarrow \infty$, each $|n\alpha m>$ becomes a particular parabolic function $|n\tau m>$, which in turn can be obtained from Equation (8) as a linear combination of spherical functions with the g_l given by the Clebsch-Gordan coefficients.[9] Regardless of the internuclear separation, and, consequently, of the coordinate system chosen, the total number of nodes of the eigenfunctions is conserved at $n-m-1$. Other aspects of the transition to these limits have been examined and illustrated by Coulson and Robinson.[6]

For the probability distributions, a format analogous to that customary for spherical functions can be obtained from Equation (8). In spherical coordinates the joint distribution obtained from the squared modulus $<n\alpha m|n\alpha m>$ is not separable,

$$\mathcal{P}_{n\alpha m}(\rho,\theta,\phi) = \sum_l \sum_{l'} g_l g_{l'} \rho R_{nl}(\rho)\rho R_{nl'}(\rho) Y_{lm}(\theta,\phi) Y_{l'm}(\theta,\phi), \tag{17}$$

but radial and angular distributions can be derived by integrating over either the spherical angles or radius, yielding

$$\mathcal{P}_{n\alpha m}(\rho) = \sum_l g_l^2 |\rho R_{nl}(\rho)|^2 \tag{18}$$

and

$$\mathcal{P}_{n\alpha m}(\theta) = \sum_l g_l^2 |Y_{lm}(\theta, \phi)|^2 \tag{19}$$

where $R_{nl}(\rho)$ is the usual normalized radial wavefunction and $Y_{lm}(\theta, \phi)$ is a spherical harmonic.[11] Figure 3 shows the radial probability distributions for all states up through $n = 4$ for $ZR = 0, 5, 15, 50,$ and ∞. Of course the small R and large R limits represent the familiar spherical and parabolic results, respectively. From this figure, one can see that for a series with the same values of n and m, the probability distributions change shape with varying ZR more noticeably for states associated with higher values of the united atom quantum number l. This is because increasing l for fixed m corresponds to increasing the number of angular nodes, given by $l - m = n_\mu$, and to decreasing the number of radial nodes, given by $n - l - 1 = n_\lambda$. Radial nodes are less sensitive to change in internuclear distance than angular nodes. Also, since R scales in the same manner as the principal quantum number n for the Coulomb problem, the radial probability distributions begin to change shape at larger values of ZR for states with higher principal quantum number n. Figure 4 shows corresponding angular probability distributions for $ZR = 0, 15,$ and ∞. The various knobs and lobes that emerge or retreat as ZR is varied reflect the shifting pattern of nodes as the weighting factors g_l^2 change. However, the probability distributions do not take their simplest or most revealing form in these spherical coordinates, since ρ and θ involve sums and products of the spheroidal coordinates. In particular, in $\mathcal{P}_{n\alpha m}(\theta)$ the symmetry about the plane $\theta = 90°$ conceals the asymmetry about the $\mu = 0$ plane which exists in the actual spheroidal eigenstates.

In terms of spheroidal coordinates, the probability distribution as obtained directly from Equation (13) is separable,

$$\mathcal{P}_{n\alpha m}(\lambda, \mu) = \mathcal{P}_{n\alpha m}(\lambda)\mathcal{P}_{n\alpha m}(\mu) \tag{20}$$

with

$$\mathcal{P}_{n\alpha m}(x) = e^{-2px}|x^2 - 1|m[f_{n\alpha m}(x)]^2 \tag{21}$$

where again x represents μ in the interval (-1, 1) and represents λ in the interval $(1, \infty)$. Figure 5 shows some examples of these probability distributions for both $x = \mu$ and $x = \lambda$. In these coordinates, the nodal structure is rather simple and stable. The number of zeros in λ or μ is given by the spheroidal quantum numbers, n_λ or n_μ, respectively, and remains the same as ZR is varied. Although these distributions may be constructed from the explicit expressions for $f_{n\alpha m}(x)$, one may also obtain quick estimates of the spheroidal probability distribution from tables [9] of the zeros of the $f_{n\alpha m}(x)$ polynomials for a few values of ZR. From the zeros $x_1, x_2, \ldots$ the polynomial can of course be obtained as $f(x) = (x - x_1)(x - x_2)\ldots$ The zeros located at values of $x \geq 1$ pertain to the λ- coordinate; those at $-1 \leq x \leq 1$ pertain to the μ-coordinate.

Figure 6 plots the zeros of the $f_{n\alpha m}$ polynomials as a function of the scaled internuclear distance, $\frac{10R}{5+R}$, for $n = 4$ a nd $m = 0$. The state labeled $4s\sigma$ and associated with the united atom quantum numbers $n = 4, l = 0, m = 0$ must have three λ nodes and no μ nodes for the entire range of internuclear separations, since $n_\lambda = n - l - 1 = 3$, and $n_\mu = l - m = 0$. Likewise, the s tate labeled $4p\sigma$ has two λ nodes and one μ node, independent of R. $4d\sigma$ has a single λ node and two μ nodes; $4f\sigma$ has all three nodes in the μ region. As the internuclear distance and, hence, the scaled distance increase, the λ nodes go to the limiting value of 1, and the μ nodes approach the value of -1.

Application to Simple Analytic Solutions of the Generalized Hydrogen Molecule Ion

The hydrogenic spheroidal eigenstates can be used to construct exact analytic solutions for special configurations of the general two-center Coulombic system comprised of a charged particle interacting with a pair of fixed charges Z_a and Z_b a distance R apart.[10] As seen in Equations (1) and (2), for this general problem the $L(\lambda)$ and $M(\mu)$ factors of the eigenfunctions each satisfy similar equations, with the same values for p^2 and A but different parameters, $(Z_a + Z_b)$ and $(Z_a - Z_b)$, respectively. Thus, as noted by Coulson and Robinson [6], both factors may have hydrogenic forms if for some pair of principal quantum numbers n_λ and n_μ the energy eigenvalues are equal and

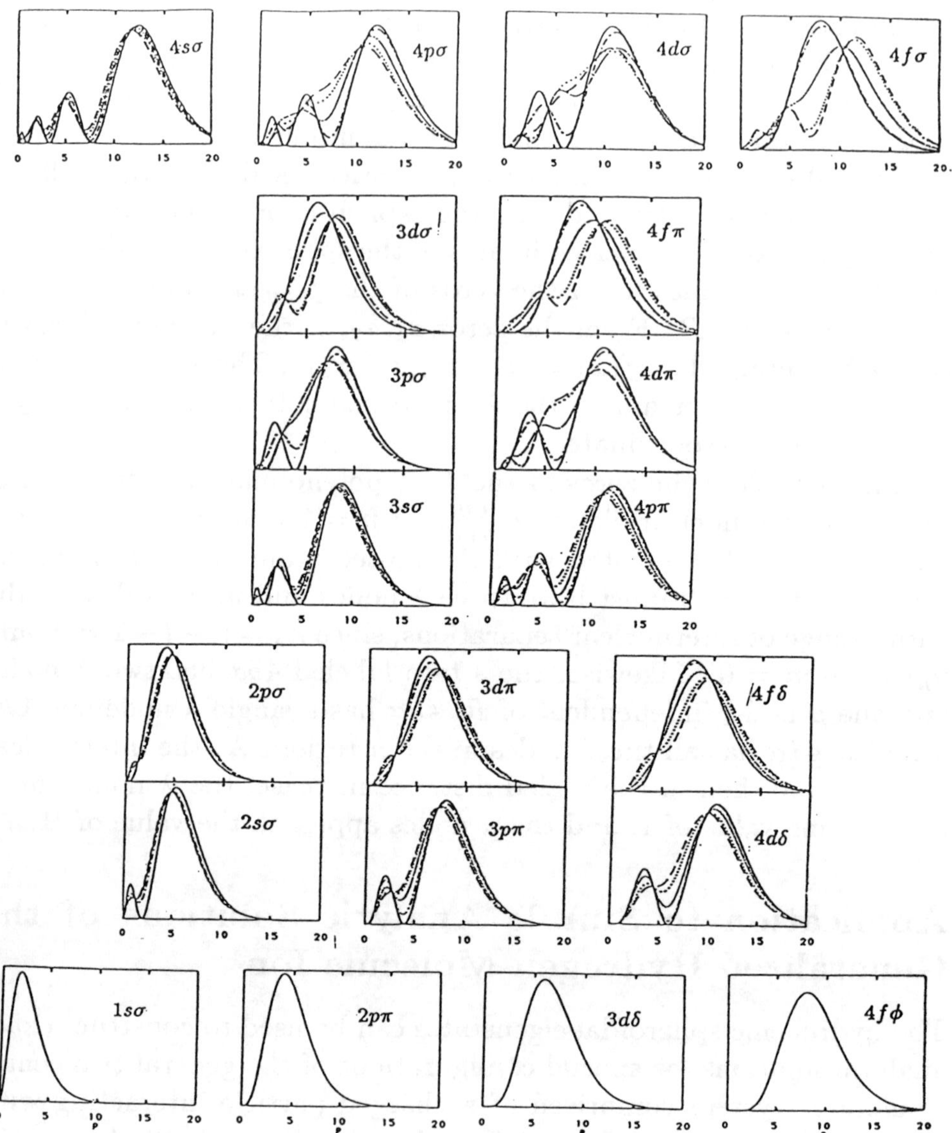

Figure 3. Radial probability distributions for the spheroidal hybrids up through $n = 4$, for $ZR = 0$ (solid line), 5 (_ . _), 15 (_ . . . _), 50 (. . . .), and ∞(dashed line). $\rho = \frac{2Zr}{n}$.

Figure 4. Polar plots of angular probability distributions for the spheroidal hybrids up through $n = 4$, for $ZR = 0$ (solid line), 15 (. . . .), and ∞ (dashed line). The z-axis is horizontal.

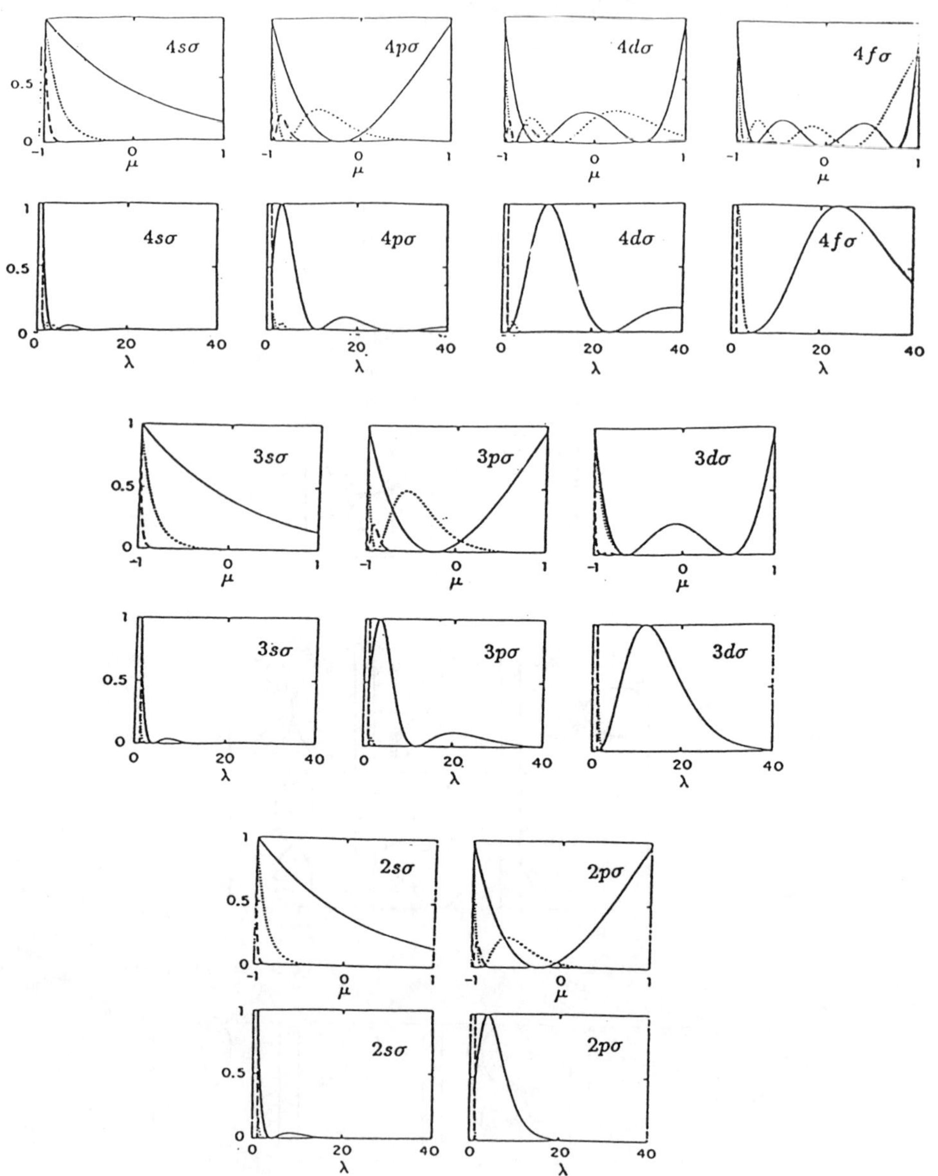

Figure 5. Probability distributions for the $m = 0$ spheroidal hybrids. In each case, the λ- and μ-factors are plotted separately, for $ZR = 1$ (solid line), 15 (. . . .), and 100 (dashed line).

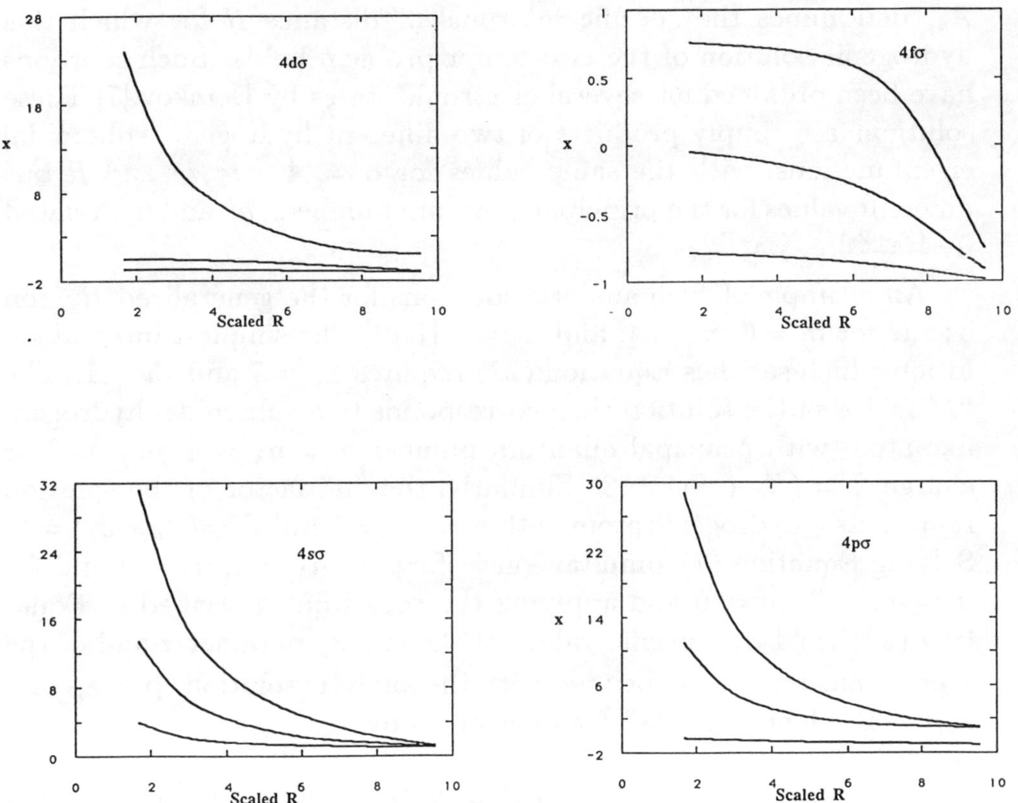

Figure 6. Zeros of the $f_{n\alpha m}$ polynomials for $n = 4$, $m = 0$ as a function of scaled internuclear distance $= \frac{10R}{5+R}$. States are labelled by the corresponding united atom quantum numbers.

the nodal structures correspond to hydrogenic states. The energy condition requires

$$E = -\frac{1}{2}\frac{(Z_a + Z_b)^2}{n_\lambda^2} = -\frac{1}{2}\frac{(Z_a - Z_b)^2}{n_\mu^2} \qquad (22)$$

The nodal condition for $L(\lambda)$ requires that the number of nodes does not exceed $n_\lambda - m - 1$ in the interval $(1, \infty)$ and for $M(\mu)$ that it not exceed $n_\mu - m - 1$ in the interval $(-1, +1)$. The requirement that the separation constant must be the same for both equations, $A_\lambda = A_\mu$, determines the specific internuclear distance R for which this hydrogenic solution of the two-center problem holds. Such solutions have been obtained for several electronic states by Demkov.[7] These solutions are simply products of two different hydrogenic spheroidal eigenfunctions, with the same values for $\alpha = A - p^2, m$ and R but different values for the principal quantum numbers, n_λ and n_μ, related by $\frac{(Z_a+Z_b)}{n_\lambda} = \frac{(Z_a-Z_b)}{n_\mu}$.

An example of such analytic solutions for the generalized H_2^+ ion occurs for $m = 0$, $n_\lambda = 4$, and $n_\mu = 3$.[7,12] The simplest integral solution which satisfies Equation (22) requires $Z_a = 7$ and $Z_b = 1$. The "λ" factor of the solution then corresponds to a spheroidal hydrogen-like atom with principal quantum number $n = n_\lambda = 4$ and nuclear charge $Z = (Z_a + Z_b) = 8$. Similarly, the "μ" factor of the solution represents a hydrogenic atom with $n = n_\mu = 3$ and $Z = (Z_a - Z_b) = 6$. Solving Equation (7) simultaneously for $n = n_\lambda = 4$, $m = 0$ and for $n = n_\mu = 3$, $m = 0$ and applying the constraint described in Equation (22) yield the specific values of the energy parameter and of the separation constant associated with the analytic solution: $p = \frac{\sqrt{7}}{3}$ and $A = -\frac{53}{9}$. Hence, the "λ" factor is given by

$$L_{n_\lambda \alpha m}(\lambda) = e^{-\frac{\sqrt{7}}{3}\lambda}[-7\lambda^3 + 12\sqrt{7}\lambda^2 + 33\lambda - 4\sqrt{7}], \qquad (23)$$

and the "μ" factor is

$$M_{n_\mu \alpha m}(\mu) = e^{-\frac{\sqrt{7}}{3}\mu}[3\sqrt{7}\mu^2 + 6\mu - \sqrt{7}]. \qquad (24)$$

The exact (unnormalized) eigenfunction for an electron between two charges $Z_a = 1$ and $Z_b = 1$, which are separated by a distance of

$R = \frac{\sqrt{7}}{3}$ a.u., is then

$$e^{-\frac{\sqrt{7}}{3}(\lambda+\mu)}[-7\lambda^3 + 12\sqrt{7}\lambda^2 + 33\lambda - 4\sqrt{7}][3\sqrt{7}\mu^2 + 6\mu - \sqrt{7}], \quad (25)$$

with corresponding energy eigenvalue $E = -2$ a.u.

Interdimensional Degeneracies

The hydrogen atom in spheroidal coordinates may be viewed in terms of two different sets of interdimensional degeneracies. In the united atom limit, the spheroidal solutions reduce to the familiar spherical states $|nlm\rangle$. When these states are generalized to arbitrary dimension, interdimensional degeneracies arise between states having the same number of radial nodes but dimensionality differing by two units and orbital angular momentum differing by one unit [13]. Away from the united atom limit, the spheroidal solutions have non-zero internuclear separation, and the system becomes a two-center problem rather than a one-center one. Hence, the cylindrical symmetry remains, but the spherical symmetry is destroyed. This implies that m, the orbital angular momentum projection on the internuclear axis, is a good quantum number for the whole range of R, while l is only valid for the united atom limit. Because of this cylindrical symmetry, the D-dimensional spheroidal solutions of the hydrogen atom have the same kind of interdimensional degeneracies found in the D-dimensional hydrogen molecule ion [14]. These interdimensional degeneracies link states which have the same number of radial nodes but differ in dimension by two units and in m by one unit.

Discussion

Beyond providing an unconventional molecular perspective for the hydrogen atom, the spheroidal eigenfunctions have practical utility. As is well known, the parabolic eigenstates provide the correct zeroth order linear combinations of the degenerate spherical eigenstates for treating the Stark effect of the hydrogen atom in a uniform electric field [15]. Although little known, the spheroidal eigenstates likewise provide the correct zeroth order hybrids, specified by the g_l coefficients, for treating the Stark effect induced by a point charge or a point dipole [16]. In the $R \longrightarrow \infty$ limit, this perturbation becomes equivalent to a uniform field and the spheroidal functions indeed reduce to the parabolic eigenstates.

Other applications arise when an hydrogenic atom is subject to a two- center perturbation. In these instances, the spheroidal eigenfunctions provide an appropriate basis for treating the system. This is exemplified by the recent work of Pont and Gavrila on hydrogen in a circularly polarized, high intensity and high frequency laser field [17].

Similarly, spheroidal eigenstates are appropriate for the exchange or tunneling of an electron between a pair of protons. As shown in a later chapter, [18] this process has been treated extensively for the g, u pairs of electronic states that stem from separated atoms with $n = 1$ and for all excited states with maximal values of $m = l = n - 1$ as well. These correspond to the $ls\sigma, 2p\pi, 3d\delta, 4f\phi, ...$ states, as depicted in Figs. 1, 3, and 4, that do not hybridize with others of the same n when subject to a two-center perturbation; in effect, such states remain spherical. The spheroidal eigenstates offer a natural basis for treating the many other excited states that have less than maximal m-values and strong hybridization. The spheroidal basis is implicit in asymptotic expansions in $\frac{1}{R}$ developed by Damburg and Propin [19] and by Cizek, *et al* [20] for the eigenvalues of H_2^+ states.

Furthermore, the two-center interdimensional degeneracies and the special solutions of the generalized hydrogen molecule ion may be useful in conjunction with electronic structure calculations based on perturbation expansions in $\frac{1}{D}$ [21]. Summation procedures for such expansions [22] can be substantially improved by exploiting information about singularity structure at large-D and low-D. For the hydrogen molecule ion there appears to be an essential singularity at $\delta = 0$, the point about which the expansion is performed, and this induces divergence of the perturbation expansion. Due to interdimensional degeneracies, there will be ground states of the generalized hydrogen molecule ion in certain higher dimensions which correspond to the excited states in three dimensions that have hydrogenic solutions. Hence, these simple analytic solutions of the generalized H_2^+ ion may be used to identify specific internuclear distances and charge ratios at which singularities disappear and the $\frac{1}{D}$ series converges. By studying the behavior of the $\frac{1}{D}$-series as R varies from the calculated hydrogenic distances, these simple hydrogenic solutions could then elucidate understanding of the singularity structure inherent in $\frac{1}{D}$ perturbation expansion calculations.

Overall, the hydrogen atom in spheroidal coordinates may be regarded as "an atom in a molecule's clothing." By allowing us to view the hydrogen atom as a diatomic molecule in the limiting case when one nucleus has zero charge, the spheroidal solutions serve as a connection between the three-body and the two-body problem. Such unconventional perspectives are often useful, whether manifested as a bridge between the simplest molecular system and its atomic archetype or as a link connecting hypothetical dimensions to the comfortable familiarity of our three-dimensional world.

Acknowledgements

I am grateful to Dudley Herschbach for his collaboration and for personifying a generalized Clebsch-Gordan coefficient by helping me interpolate from a mildly confused, "uncoupled" state to one of "coupled" understanding and enjoyment. I am indebted to Jan-Michael Rost for stimulating discussions of the three-body problem and for his help in preparing Figure 4. John Briggs was also very helpful, with or without the benefit of "øl." A graduate fellowship from the National Science Foundation and a conference grant from Radcliffe College are gratefully acknowledged.

References

1. D.R. Herschbach, this volume, Chapter 2.
2. W. Pauli, Z. Phys. **36** , 336 (1926).
3. L.I. Schiff, *Quantum Mechanics* , 3rd Ed. (McGraw-Hill, New York, 1968) , pp. 236-239.
4. G. Baym, *Lectures on Quantum Theory* (Benjamin/Cummings Publishing Co., Inc., Reading, MA, 1969), pp. 175-179.
5. M.J. Englefield, *Group Theory and the Coulomb Problem* (Wiley-Interscience, New York, 1972).

6. C.A. Coulson and P.D. Robinson, Proc. Phys. Soc. London **71** , 815 (1958).

7. Yu. N. Demkov, Zh. Eksp. Teor. Fiz. Pis'ma Red. **7** , No. 3, 101 (1968); JETP Lett. **7** , 76 (1968).

8. B.R. Judd, *Angular Momentum Theory for Diatomic Molecules* (Academic Press, New York, 1975), pp. 56-58, 67-73, 81-84.

9. S.M. Sung and D.R. Herschbach, J. Chem. Phys., **95** , in press (1991).

10. E. Teller and H.L. Sahlin, in *Physical Chemistry–An Advanced Treatise*, H. Eyring, D. Henderson, and W. Jost, Eds. (Academic Press, New York, 1970), Vol. 5, pp. 35-124.

11. R.N. Zare, *Angular Momentum* (Wiley, New York, 1988).

12. S.M. Sung, D.D. Frantz, and D.R. Herschbach, to be submitted.

13. D.R. Herschbach, J. Chem. Phys. **84** 838 (1986).

14. D.D. Frantz and D.R. Herschbach, J. Chem. Phys. **92** , 6669 (1990).

15. L.D. Landau and E.M. Lifshitz, *Quantum Mechanics* (Pergamon Press, London, 1958), pp. 121-125, 130-132, 251-256.

16. P.D. Robinson, Proc. Phys. Soc. London **71** , 815 (1958).

17. M. Pont, Phys. Rev. A **40**, 5659 (1989); M. Pont and M. Gavrila, Phys. Rev. Letts. **65** , 2362 (1990).

18. S. Kais and D.R. Herschbach, this volume, Chapter 6.

19. R.J. Damburg and R.K. Propin, J. Phys. B **1**, 681 (1968).

20. J. Cizek, R.J. Damburg, S. Graffi, V. Grecchi, E.M. Harrell, J.G. Harris, S. Nakai, J. Paldus, R.K. Propin, and H.J. Silverstone, Phys. Rev. A**33**, 12 (1986).

21. M. López-Cabrera, D.Z. Goodson, D.R. Herschbach and J.D. Morgan III, to be submitted.

22. D.Z. Goodson and M. Lópex-Cabrera, this volume, Chapter 7.

6.2   1/n Expansion for a Hydrogen Atom in Strong External Fields

V.S. Popov
Institute of Theoretical and
Experimental Physics
Moscow, Russia

Abstract

The Stark shifts and widths of atomic states in a strong electric field and the energy of a hydrogen atom in electric and magnetic fields are calculated using the 1/n expansion. The asymptotic behavior of large orders of the expansion is briefly discussed.

Introduction

In a previous section [1] we considered a version of the $1/n$-expansion which is applicable not only in the case of the discrete spectrum, but also for quasistationary states. Here we apply the same method to the energy spectrum of a hydrogen atom in electric and/or magnetic fields that are strong relative to the atomic field at the position of the electron orbit. Note that in the presence of an electric field, all atomic states are not stable, but quasistationary (with complex energy $E = E_r - i\Gamma/2$); so the appearance of complex equilibrium points is quite natural.

Throughout this paper we shall use atomic units. $n = n_1 + n_2 + |m| + 1$ is the principal quantum number of the $|n_1, n_2, m\rangle$ state, and n_1, n_2, and m are parabolic quantum numbers [2].

The Stark Effect in a Strong Field

We will briefly discuss here the application of the $1/n$-expansion to this problem, restricting ourselves to $|\,0, 0, n-1\rangle$ states of a hydrogen atom, and using the approach and notation of reference [3]. These states have no "radial" excitations (in parabolic coordinates ξ and η).

217

D. R. Herschbach et al. (eds.), Dimensional Scaling in Chemical Physics, 217–229.
© 1993 *Kluwer Academic Publishers. Printed in the Netherlands.*

In the limit $n \to \infty$, the classical equilibrium point (ξ_0, η_0) and the energy E are determined from the system of equations:

$$\frac{dU_1}{d\xi} = \frac{dU_2}{d\eta} = 0$$
$$U_1(\xi_0) = U_2(\eta_0) = \frac{1}{4}E \tag{1}$$
$$\beta_1 + \beta_2 = 1$$

where $\xi = r + z$ and $\eta = r - z$ are parabolic coordinates, β_i are separation constants and U_i are effective potentials [1]:

$$U_1(\xi) = -\frac{\beta_1}{2\xi} + \frac{m^2 - 1}{8\xi^2} + \frac{1}{8}\mathcal{E}\xi$$
$$U_2(\eta) = \frac{\beta_2}{2\eta} + \frac{m^2 - 1}{8\eta^2} - \frac{1}{8}\mathcal{E}\eta \tag{2}$$

(where atomic units are used). Substituting equations (2) into (1) we arrive at a system of five equations. Their solution may be considerably simplified if we take into account that, at $n \gg 1$, the Bohr atom model is applicable. The state $|\, 0, 0, m \rangle$ corresponds to a circular electron orbit, perpendicular to the z axis. When an electric field $\mathcal{E}$ is turned on, this orbit shifts along the z axis and changes its radius, remaining circular. From the condition of equilibrium of forces applied to the electron in its rest frame, we get:

$$\mathcal{E} = -zr^{-3}$$
$$\rho r^{-3} = v^2 \rho^{-1}$$
$$\rho v = m \approx n \tag{3}$$
$$E = \tfrac{1}{2}v^2 - r^{-1} + \mathcal{E}z$$

where $\rho = (r^2 - z^2)^{1/2}$. Performing the scaling transformation:

$$r \to n^2 r$$
$$v \to v/n$$
$$\epsilon = \epsilon' - i\epsilon'' = 2n^2 E \tag{4}$$
$$F = n^4 \mathcal{E}$$

and substituting $\rho = (r^2 - z^2)^{1/2}$, we finally arrive at

$$\epsilon^{(0)} = -(1 - \tau^2)^2(1 + 3\tau^2)$$
$$\epsilon^{(1)} = (1 - \tau^2)^3[(1 + 3\tau)^{1/2} + (1 - 3\tau)^{1/2} - 2] \tag{5}$$

where τ is defined implicitly by $\tau(1 - \tau^2)^4 = F$. The next coefficients $\epsilon^{(k)}$ can be, in principle, found analytically, but the expression for $\epsilon^{(2)}$ is already rather cumbersome [4]. We computed $\epsilon^{(k)}$ numerically by recurrence relations, described in reference [4]. As long as $0 < \tau < 1/3$ (or $F < F_* = 2^{12} \times 3^{-9} = 0.2081$), all coefficients $\epsilon^{(k)}$ are real. In this region, the $1/n$-expansion determines only Stark shifts of the levels. At $F < F_*$, the root of equation (5) escapes to the complex plane:

$$\tau = \frac{1}{3} \pm i(F - F_*)^{1/2} + \dots \tag{6}$$

This corresponds to barrier disappearance in the potential $U_2(\eta)$. Since $\epsilon^{(k)}$ is complex, it opens the possibility of computation within the $1/n$-expansion not only the shifts, but also the widths of atomic levels.

The computed complex energy eigenvalues $E = E_r - i\Gamma/2$ for different states $| \, n_1, n_2, m \rangle$ of a hydrogen atom are presented in Table 2, where $(1/n)$ denotes the results of the $1/n$-expansion and HPA marks the values obtained by perturbation series summation with the help of Hermite-Padé approximations [6]. The agreement between different computational methods is quite good.

The results for $| \, 0, 0, n - 1 \rangle$ Stark resonances are presented in Figures 1 and 2, where the curve $n = \infty$ corresponds to the initial term $\epsilon^{(0)}$ of the $1/n$-expansion. Note that in very strong fields, Stark shifts change their sign. For further details, we refer the reader to [4-6].

Table 1. Energies and widths of hydrogenic states $| n_1, n_2, m \rangle$ in an electric field. Here $\varepsilon_n = 2n^2(E_r - i\Gamma/2)$, $F = n^4\mathcal{E}$, $\hbar = e = m_e = 1$, and $\nu = (-2E_r)^{-1/2}$, $\nu = n$ for a free atom ($\mathcal{E} = 0$).

a) $| 0, 0, n-1 \rangle$ states:

n	F	$-\varepsilon_n$	method
1	0.1	1.05484+i0.01454	HPA
		1.054836+i0.014538	ref.[7]
1	0.5	1.25+i0.55	HPA
		1.246+i0.559	$1/n$
3	0.5	1.224+i0.317	HPA
		1.22393+i0.31685	$1/n$
1	1.0	1.248+i1.294	$1/n$
		1.2487+i1.2936	ref.[8]
3	1.0	1.27891+i0.83875	$1/n$
10	1.0	1.2851+i0.6739	HPA
		1.28518+i0.67388	$1/n$

b) $| \frac{1}{2}(n-1), \frac{1}{2}(n-1), 0 \rangle$ states:

n	$\mathcal{E}$(a.u.)	F	ν	$10^6\Gamma$	method
5	1.8×10^{-4}	0.1125	4.92402	2.283	HPA
			4.9239	2.22	$1/n$
			4.9240	2.282	ref.[3]
11	1.0×10^{-5}	0.1464	10.713	2.83	HPA
			10.7128	2.82	$1/n$
			10.688	2.815	ref.[3]
15	3.0×10^{-6}	0.1519	14.577	1.35	HPA
			14.5766	1.338	$1/n$
			14.5771	1.338	ref.[3]

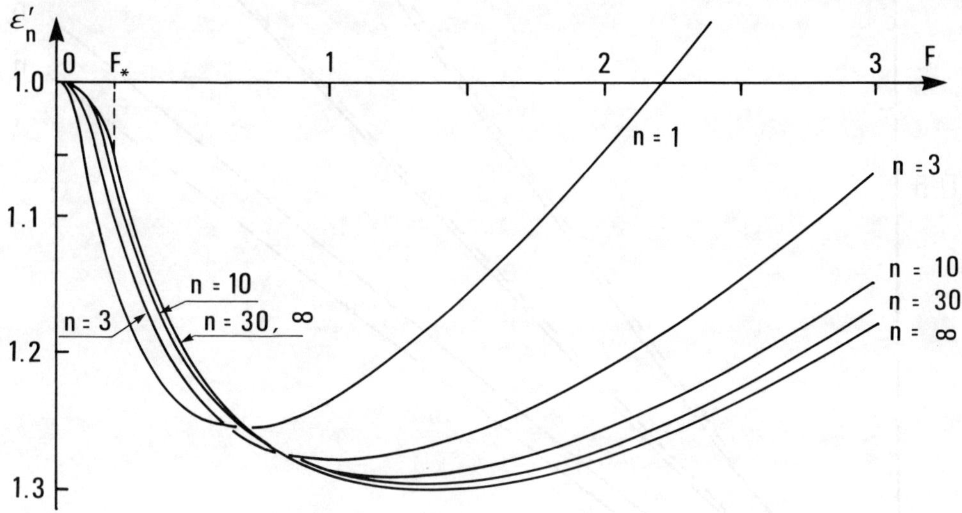

Figure 1. The Stark shifts for the $|0, 0, n-1\rangle$ states of a hydrogen atom. We use the reduced variables (3).

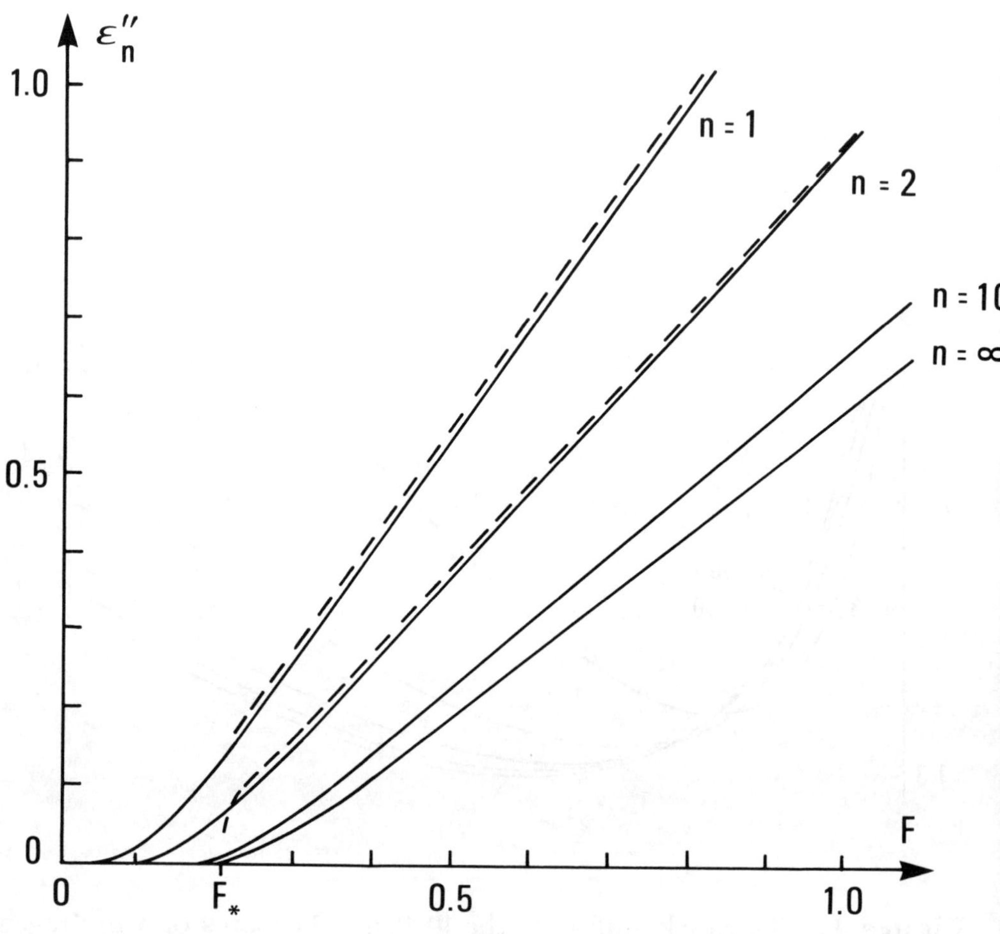

Figure 2. The electric field dependence of ε''_n for the $|\,0,0,n-1\rangle$ states. The dashed curves correspond to two terms, $\varepsilon^{(0)} + \varepsilon^{(1)}n^{-1}$, of the $1/n$-expansion.

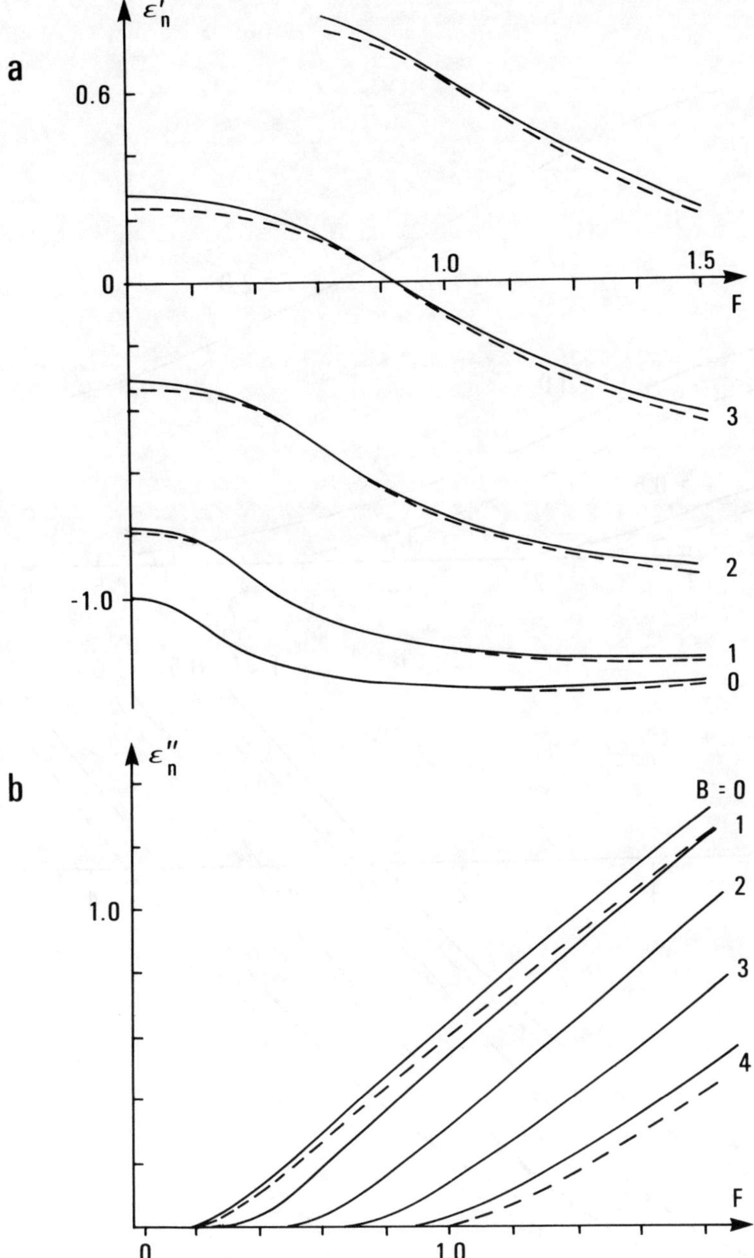

Figure 3. A hydrogen atom in parallel electric and magnetic fields: (a) $\epsilon'_n = 2n^2 Re(E_n)$, (b) $\epsilon''_n = n^2 \Gamma_n$, $(n_1 = n_2 = 0, n = m + 1 = 10)$. The values of $B = n^3 \mathcal{H}$ are shown on the curves.

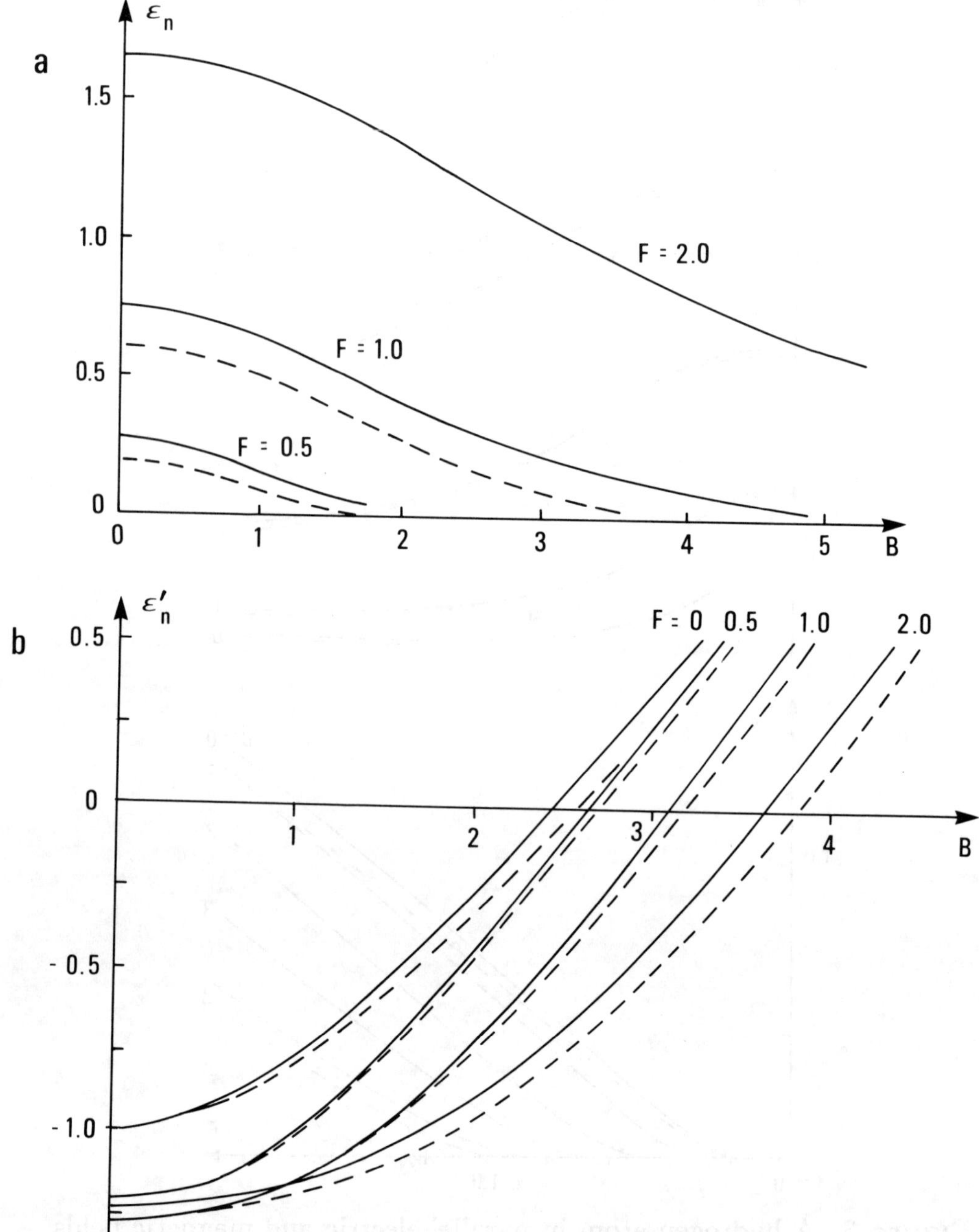

Figure 4. The same as in Figure 3 for the state with $n = m + 1 = 5$.

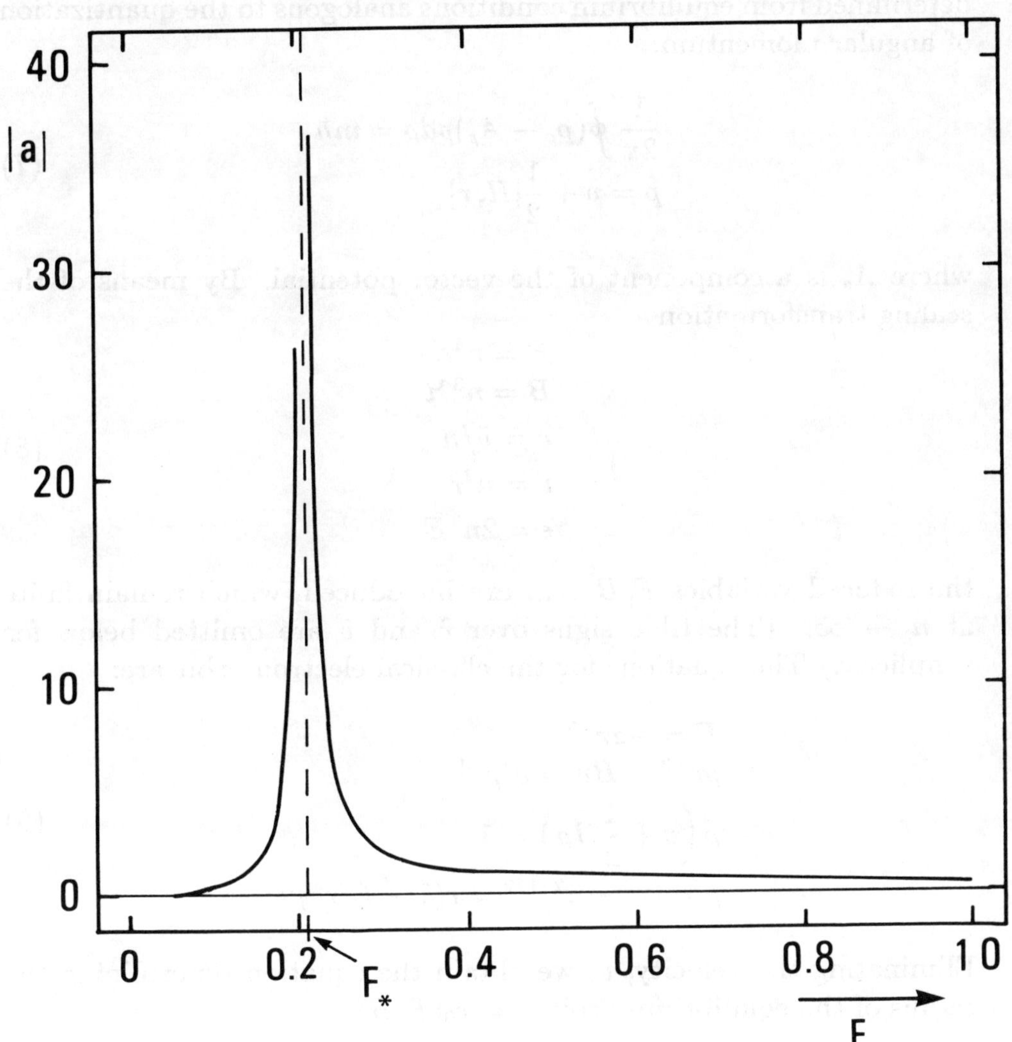

Figure 5. The dependence of $|a(F)|$ on F for the $|0,0,n-1\rangle$ states.

A hydrogen atom in external fields [6]

Here we restrict ourselves to the case $\mathcal{E} \parallel \mathcal{H}$ and to the states with $m = n-1$, i.e. the states with maximal possible values of the magnetic quantum number m. Parameters of the classical electron orbit are determined from equilibrium conditions analogous to the quantization of angular momentum:

$$\frac{1}{2\pi} \oint (p_\rho - A_\rho)\rho d\rho = m\hbar$$
$$p = v + \frac{1}{2}[H, r] \tag{7}$$

where A_ρ is a component of the vector potential. By means of the scaling transformation

$$F = n^4\mathcal{E}$$
$$B = n^3\mathcal{H}$$
$$v = \tilde{v}/n \tag{8}$$
$$r = n^2\tilde{r}$$
$$\epsilon = 2n^2 E$$

the reduced variables $F, B, \epsilon, \ldots$ are introduced, which remain finite at $n \to \infty$. (The tilde signs over $\tilde{r}$ and $\tilde{v}$ are omitted below for simplicity.) The equations for the classical electron orbit are:

$$F = -zr^{-3}$$
$$\rho r^{-3} - Bv = v^2\rho^{-1}$$
$$\rho\left(v + \frac{1}{2}B\rho\right) = 1 \tag{9}$$
$$\rho = (r^2 - z^2)^{1/2} = r(1 - F^2r^4)^{1/2}$$

Eliminating the velocity, v, we obtain the equation determining the radius of the equilibrium orbit, $r = r_0(F, B)$:

$$r(1 - F^2r^4)^2\left(1 + \frac{1}{4}B^2r^3\right) = 1 \tag{10}$$

The first term of the $1/n$-expansion, $\epsilon^{(0)}$, is equal to the energy of the electron in this orbit, and the second is determined by small oscillations around it:

$$\epsilon^{(0)} = 2U(r_0) = 2\rho_0^{-2} + \frac{\rho_0^2}{r_0^3} - \frac{4}{r_0}$$

$$\epsilon^{(1)} = 2(n_1 + 1)\omega_1 + (2n_2 + 1)\omega 2 - 2(n_1 + n_2 + 1)\rho_0^{-2} \tag{11}$$

where $n_1 + n_2 + 1 = n - |m|$, $n_j = 0, 1, ...$ are oscillator quantum numbers, and where the frequencies are given by:

$$\omega_{1,2} = \left(\frac{1}{r_0^3} + \frac{1}{2}B^2 \pm \left[9\frac{F^2}{r_0^2} + 3F^2 B^2 r_0 + \frac{1}{4}B^4\right]^{1/2}\right)^{1/2} \tag{12}$$

Higher orders of the $1/n$-expansion ($k \geq 2$) take into account anharmonicity of the effective potential $U(r)$, and are computed by means of recurrence relations. The effective potential, $U(r)$, has a minimum only for small enough electric fields, $F < F_*(B)$. At $F = F_*$, $\omega_2 \to 0$, which corresponds to the collapse of the classical solutions. At $F > F_*(B)$, the radius of the classical orbit and the coefficients $\epsilon^{(k)}$ become complex. Such a solution is meaningless in classical mechanics, but in quantum mechanics it makes it possible to describe within the $1/n$-expansion not only the shifts, but also the widths of the levels in strong fields.

Part of the results obtained are presented in Figures 3 and 4 (for the states with $n_1 = n_2 = 0$, $m = n - 1$). The solid lines represent the real and imaginary parts of the reduced energy; while dashed lines pertain to the classical energy, $\epsilon^{(0)}$. It can be seen that $Re[\epsilon^{(0)}]$ qualitatively represents the dependence of the energy on the field. As to $Im[\epsilon^{(0)}]$, it is a rather rough approximation; i.e. summation of the series

$$\epsilon = \epsilon^{(0)} + \epsilon^{(1)}/n + \epsilon^{(2)}/n^2 + ... \tag{13}$$

is necessary in this case.

The above examples clearly show that the $1/n$-expansion is quite an effective calculational method, especially for Rydberg ($n \gg 1$) states. The investigation of such states, including those in external fields, is a problem of much interest in atomic physics at present. Due to the development of laser techniques, the region of strong fields has become accessible, including fields which are comparable with the atomic field in an electron orbit. In these case, the $1/n$-expansion is the most adequate calculational method, since it does not assume that the perturbation applied to the free atomic hamiltonian is small.

Asymptotic behaviour of the $1/n$-expansion

It is known that certain perturbation series in quantum mechanics and field theory are divergent, due to the instability of the vacuum state when the coupling constant changes its sign (Dyson's phenomenon [10]). In the case of the $1/n$-expansion, such an argument is unknown to us, and therefore a numerical investigation of the asymptotic behaviour of $\epsilon^{(k)}$ at $k \to \infty$ was performed.

It turned out that as $k \to \infty$

$$\epsilon^{(k)} \sim k!! a^k k^\beta \tag{21}$$

where a and β are computable constants (for their computation, 40-50 initial coefficients appeared to be significant). As an example, in Figure 5 we give $\mid a(F) \mid$ for the $\mid 0,0,n-1\rangle$ state in the Stark effect problem. Until $F < F_* = 0.2081$, the parameter $a > 0$. When $F \to F_*$, $a(F)$ has a pole. This accounts for the above-mentioned sharp decrease in the accuracy of the energy, determined by means of the $1/n$-expansion, at $F \approx F_*$. When $F > F_*$, the parameter $a(F)$ is complex, but $arg[a(F)] \to 0$ as $F \to F_*$.

From equation (21), it follows that the radius of convergence of the $1/n$-expansion (13) is zero. However, the series can be successfully summed with the help of Padé approximants, and this procedure determines both the Stark shift and the level width. (In the region $0 < F < F_*$, where $\epsilon^{(k)}(F) < 0$, the Padé-Hermite approximants [7] can be used).

Acknowledgements

The author would like to thank V.D. Mur, A.V. Sergeev, A.V. Scheblykin and V.M. Weinberg, for their cooperation and assistance. Thanks are also due to O.P. Vilster-Tolstoy and N.A. Volchkova for their help in translating and preparing the manuscript.

References

1. V.S. Popov, this volume, Section 5.3.
2. H.A. Bethe and E.E. Salpeter, *Quantum Mechanics of One- and Two-Electron Systems*, Springer Verlag, Berlin (1957), pages 27-32.
3. R.J. Damburg and V.V. Kolosov, J. Phys. **B9**, 3149 (1976).
4. V.S. Popov, V.D. Mur, A.V. Scheblykin et al., Phys. Lett. **124A**, 77, (1987).
5. V.M. Weinberg et al., TMF **74**, 399, (1988).
6. V.S. Popov et al., Phys. Lett. **149A**, 418, 425, (1990).
7. V. Franceschini, V. Grecchi and H.J. Silverstone, Phys. Rev. **A32**, 1338, (1985).
8. L. Benassi and V. Grecchi, J. Phys. **B13**, 911 (1980).
9. L.D. Landau and E.M. Lifshitz, *Quantum Mechanics*, Addison-Wesley, Reading, (1985).
10. F.J. Dyson, Phys. Rev. **85**, 631, (1952).

6.3 Simple Molecules and Variant Scalings

A. L. Tan[1] and J. G. Loeser
Department of Chemistry
Oregon State University
Corvallis, OR 97331-4003.

Abstract

The D-dimensional N-body hamiltonian is written in terms of the interparticle distances r_{ij}, and recast to show the natural D-dependent partitioning of the kinetic energy into centrifugal and differential components; modifications required for some alternative coordinate systems are also described. As a simple application, the $D \to \infty$ solution is worked out for a collection of N equivalent gravitating bosons, and combined with the previously obtained $D \to 1$ solution to obtain quantitative predictions at $D = 3$. We also present a simple procedure for scaling dimensionally generalized Coulomb problems which is both finitizing and uniform for $1 \leq D \leq \infty$. This scaling allows one to use dimensional limit results for any Coulomb problem to construct approximate solutions at $D = 3$. We use the uniform scaling in conjunction with dimensional interpolation and with the $1/D$ expansion to treat H_2^+ and H_2; these problems also suggests a possible connection between $D = 3$ equilibrium geometries and $D \to \infty$ symmetry breaking points. Finally, we describe a procedure for modifying the classical $D \to \infty$ hamiltonian using $D \to 1$ solutions without destroying the classical character of the former, thereby allowing it to provide quantitative classical models for $D = 3$ electronic structures.

---

[1]Graduate student, Department of Chemistry, Harvard University, Cambridge, MA 02138. Currently based at Oregon State University.

D. R. Herschbach et al. (eds.), Dimensional Scaling in Chemical Physics, 230–255.

Many-body Hamiltonian in r_{ij} form

Derivation

In atomic units, the non-relativistic Schrödinger equation of a system of N particles in D-dimensional Euclidean space is

$$\left(-\frac{1}{2}\sum_{i=1}^{N}\frac{1}{m_i}\nabla_i{}^2 + V\right)\Psi = E\Psi, \tag{1}$$

where

$$\nabla_i{}^2 = \sum_{\nu=1}^{D}\frac{\partial^2}{\partial x_{i\nu}{}^2}.$$

Our aim is to recast the equation in a form appropriate for analysis in the $D\to\infty$ limit. Since this limit is a "classical" one, characterized by interparticle distances which are fixed relative to one another, we will use the interparticle distances,

$$r_{ij} = \left[\sum_{\nu=1}^{D}(x_{i\nu} - x_{j\nu})^2\right]^{1/2}, \tag{2}$$

as our internal coordinate system. This choice keeps the analysis as general as possible. We note that for $D \geq N-1$ the $\frac{1}{2}N(N-1)$ coordinates thus generated are independent (though still restricted by triangle inequalities).

If we transform to the center of mass and restrict attention to spherically symmetric states, then the Laplacians can be rewritten in terms of the internal coordinates r_{ij}:

$$\nabla_i{}^2 = \sum_{j\neq i}\left(\frac{\partial^2}{\partial r_{ij}{}^2} + \frac{D-1}{r_{ij}}\frac{\partial}{\partial r_{ij}} + \sum_{k\neq i,j}\frac{r_{ij}{}^2 + r_{ik}{}^2 - r_{jk}{}^2}{2r_{ij}r_{ik}}\frac{\partial^2}{\partial r_{ij}\partial r_{ik}}\right). \tag{3}$$

As noted elsewhere [1,2], we can derive and interpret solutions more easily if we make a transformation which removes first derivative terms. This may be done by incorporating a factor related to the volume element, which we shall denote by J, into the wave function:

$$\Psi = J^{-1/2}\Phi. \tag{4}$$

Expanding $\nabla_i^2 \Psi$ and demanding that first derivatives of Φ vanish gives the condition on J that

$$\frac{D-1}{r_{ij}} - \frac{1}{J}\left(\frac{\partial J}{\partial r_{ij}} + \sum_{k \neq i,j} \frac{r_{ij}^2 + r_{ik}^2 - r_{jk}^2}{2 r_{ij} r_{ik}} \frac{\partial J}{\partial r_{ik}}\right) = 0 \qquad (5)$$

for all r_{ij}.

In analogy to earlier results [1,2], we predict that the factor J will be of the form

$$J = v^{D-1}, \qquad (6)$$

where v is the internal volume of the parallelotope defined by the N particles in D-dimensional space. The square of v can be expressed in terms of the determinant W [3]:

$$v^2 = -(-\tfrac{1}{2})^{N-1} W,$$

$$W = \begin{vmatrix} 0 & 1 & 1 & \cdots & 1 \\ 1 & 0 & r_{12}^2 & \cdots & r_{1N}^2 \\ 1 & r_{12}^2 & 0 & \cdots & r_{2N}^2 \\ \vdots & \vdots & \vdots & \ddots & \vdots \\ 1 & r_{1N}^2 & r_{2N}^2 & \cdots & 0 \end{vmatrix}.$$

Now $J = v^{D-1}$ satisfies the condition given by Eq. (5). Hence we have obtained an appropriate expression for J.

We now write the Schrödinger equation in terms of the probability amplitude Φ:

$$\left\{-\frac{1}{2}\sum_{i=1}^{N} \frac{1}{m_i}\left[\sum_{j \neq i}\left(\frac{\partial^2}{\partial r_{ij}^2} + \sum_{k \neq i,j} \frac{r_{ij}^2 + r_{ik}^2 - r_{jk}^2}{2 r_{ij} r_{ik}} \frac{\partial^2}{\partial r_{ij} \partial r_{ik}}\right)\right.\right.$$
$$\left.\left. - \left(\frac{D-1}{2}\right)\left(\frac{D-2N+1}{2}\right)\frac{v_i^2}{v^2}\right] + V\right\}\Phi = E\Phi.$$
$$(7)$$

Here v_i is the volume of the parallelotope defined by all but the ith particle, and is defined exactly like v, except with row and column i in W deleted (and with the exponent of the prefactor decremented by 1).

We note that $v/v_i = h_i$, where h_i is the perpendicular distance of the ith particle from the flat ("base") specified by the other $N-1$ particles. Therefore the nondifferential component of the kinetic energy is a generalized centrifugal potential T_{cent} which has the form

$$T_{\text{cent}} \propto \sum_i \frac{v_i{}^2}{2v^2} = \sum_i \frac{1}{2h_i{}^2}. \tag{8}$$

The casting of the Laplacian as outlined here is general to all many-body problems, including non-Coulombic systems, and is particularly useful for systems with a high degree of symmetry.

The large-D limit

In absolute units, the solutions to Eq. (7) will become unbound in the $D \to \infty$ limit, because of the quadratically divergent centrifugal potential. However, we can obtain a finite result as a useful perturbation limit by means of a suitable scaling. For problems with an inverse distance potential such a scaling is given by

$$\begin{aligned} \hat{\rho}_{ij} &= r_{ij}/\beta^2, \\ \mathcal{E} &= \beta^2 E, \\ \beta &= \tfrac{1}{2}(D-1). \end{aligned} \tag{9}$$

From this, one obtains immediately

$$(T_{\text{deriv}}^{(D)} + T_{\text{cent}}^{(D)} + V_{\text{sys}})\Phi = \mathcal{E}\Phi, \tag{10}$$

where

$$\begin{aligned} T_{\text{deriv}}^{(D)} &= -\frac{1}{2\beta^2} \sum_{i=1}^{N} \frac{1}{m_i} \sum_{j \neq i} \left(\frac{\partial^2}{\partial \hat{\rho}_{ij}^2} + \sum_{k \neq i,j} \frac{\hat{\rho}_{ij}^2 + \hat{\rho}_{ik}^2 - \hat{\rho}_{jk}^2}{2\hat{\rho}_{ij}\hat{\rho}_{ik}} \frac{\partial^2}{\partial \hat{\rho}_{ij} \partial \hat{\rho}_{ik}} \right), \\ T_{\text{cent}}^{(D)} &= \frac{1}{2} \sum_{i=1}^{N} \frac{1}{m_i} \frac{\hat{v}_i^2}{\hat{v}^2} \left(1 - \frac{N-1}{\beta} \right), \\ V_{\text{sys}} &= \sum_{i \neq j} \frac{a_{ij}}{\hat{\rho}_{ij}}, \end{aligned}$$

where for a system of charges, $a_{ij} = Z_i Z_j$. For the $D \to \infty$ limit, the β-dependent portion of $T_{\text{cent}}^{(D)}$ and the entirety of $T_{\text{deriv}}^{(D)}$ drop out of the problem, which thereby reduces to one of finding the minimum of the effective potential $T_{\text{cent}} + V_{\text{sys}}$. Here T_{cent} denotes the large-D limit for the centrifugal potential, namely

$$T_{\text{cent}} = \frac{1}{2} \sum_{i=1}^{N} \frac{1}{m_i} \frac{\hat{v}_i^2}{\hat{v}^2}.$$

So in principle at least, we can find the $D \to \infty$ solution for any Coulombic system. However, for molecules it is usually desirable to utilize a different coordinate system (Section 1.4) and a different scaling (Section 2.1) from the one described above.

A simple application

As a simple application of the many-body hamiltonian in r_{ij} form, we consider the problem of N equivalent mutually attracting (gravitating) bosons. That is, we consider the above with all $m_i = m$, and all $a_{ij} = -g$; for convenience we set $m = 1$, but retain g through the calculations. This problem is interesting since it is a true many-body problem which can be solved exactly (and easily) for $D = 1$ and for $D \to \infty$, though apparently not for $D = 3$. Here we will solve the problem in the large-D limit both exactly and in the Hartree approximation.

We assume that the minimum of the effective potential $T_{\text{cent}} + V_{\text{sys}}$ will in this case have complete symmetry (that of the symmetric group, S_N). Thus, we set all $\hat{\rho}_{ij}$ to the common value of $\hat{\rho}$. With this simplification, $W = (-1)^N N \hat{\rho}^{2N-2}$, from which $h_i^2 = N\hat{\rho}^2/(2N-2)$ and

$$\mathcal{H}_\infty = N\left(\frac{N-1}{N\hat{\rho}^2}\right) - \frac{N(N-1)}{2}\left(\frac{g}{\hat{\rho}}\right). \tag{11}$$

Taking the minimum with respect to $\hat{\rho}$ gives $\mathcal{E}_\infty = -\frac{1}{16}N^2(N-1)g^2$.

In the Hartree approximation, the ground state wavefunction takes the form $\varphi(\hat{\varrho}_1)\cdots\varphi(\hat{\varrho}_N)$, where the $\hat{\varrho}_i$ are distances with respect to a common center. Again we symmetrize, assuming that minimization of the Hartree effective potential will give all $\hat{\varrho}_i$ the same value $\hat{\varrho}$. The interparticle distances will then all be $\sqrt{2}\hat{\varrho}$, since in the $D \to \infty$ limit the Hartree approximation orthogonalizes the particle vectors. Thus, the symmetrized Hartree effective potential is

$$\mathcal{H}_\infty^{\mathrm{H}} = N\left(\frac{1}{2\hat{\varrho}^2}\right) - \frac{N(N-1)}{2}\left(\frac{g}{\sqrt{2}\hat{\varrho}}\right). \tag{12}$$

Taking the minimum gives $\mathcal{E}_\infty^{\mathrm{H}} = -\frac{1}{16}N(N-1)^2 g^2$. Combining this with the exact result, the correlation energy is given by $\mathcal{E}_\infty^{\mathrm{corr}} = -\frac{1}{16}N(N-1)g^2$.

What makes this an especially interesting problem to study is the fact that the $D=1$ version can also be solved both exactly and in the Hartree approximation [4]. Using the same energy scaling as in Eq. (9), but a distance scaling $\check{\rho}_i = r_i/\beta$, the $D=1$ problem becomes

$$\mathcal{H}_1 = -\frac{1}{2}\sum_{i=1}^{N}\frac{\partial^2}{\partial\check{\rho}_i^2} - g\sum_{i\neq j}\delta(\check{\rho}_i - \check{\rho}_j), \tag{13}$$

for which the exact and Hartree eigenvalues are $\mathcal{E}_1 = -\frac{1}{24}N(N^2-1)g^2$ and $\mathcal{E}_1^{\mathrm{H}} = -\frac{1}{24}N(N-1)^2 g^2$. The eigenfunctions can also be written down explicitly. (We note that the use of different distance scalings in the low-D and high-D limits can be avoided by means of the procedure described in Sec. 2, but is not necessary for this problem.)

Finally, we can apply these results to the real (though perhaps not too real) $D=3$ problem. The problem is mentioned elsewhere [5], but the authors are not aware of any quantitative treatment. However, the above solutions yield predictions via simple linear interpolation in $1/D$,

$$\mathcal{E}_{3\,(\mathrm{DI})} = \tfrac{1}{3}\mathcal{E}_1 + \tfrac{2}{3}\mathcal{E}_\infty. \tag{14}$$

The results for the exact, Hartree, and correlation energies are summarized in Table 1.

One sees that the N-dependences of the Hartree energies are the same in the dimensional limits (and therefore in the interpolated result as well). In fact, it isn't hard to show that for this problem the N-body Hartree energy is just $\frac{1}{2}N(N-1)^2$ times the corresponding two-body energy for any D.

Table 1. Exact and Hartree energies for the system of N equivalent gravitating particles for $D = 1$ and $D \to \infty$, and the dimensionally interpolated estimates at $D = 3$.

D	$\mathcal{E}_D$	$\mathcal{E}_D^{\mathrm{H}}$	$\mathcal{E}_D^{\mathrm{corr}}$
1	$-\frac{1}{24}(N+1)N(N-1)g^2$	$-\frac{1}{24}N(N-1)^2g^2$	$-\frac{1}{12}N(N-1)g^2$
3(DI)	$-\frac{1}{18}(N+\frac{1}{4})N(N-1)g^2$	$-\frac{1}{18}N(N-1)^2g^2$	$-\frac{5}{72}N(N-1)g^2$
∞	$-\frac{1}{16}N^2(N-1)g^2$	$-\frac{1}{16}N(N-1)^2g^2$	$-\frac{1}{16}N(N-1)g^2$

Coordinate systems with distinguished axes

Results like those obtained in Section 1.1, particularly equations (6) and (8), pertain to coordinate systems in which all the original Cartesian coordinates are involved in the transformation to internal coordinates, such as the set of interparticle coordinates employed above, or the set of radial coordinates and cosines of angles [2]. It is worth noting that it is possible to utilize these or any other suitable sets of coordinates, together with the associated results, in a context where it is desirable to have one or more of the original Cartesian axes separated out. Regarding the latter, in the case of axially symmetric systems, we leave the z-coordinate alone. That is, the dimensional generalization is performed only with respect to variables orthogonal to the z-axis. Similarly, we can construct coordinate systems in which 2, 3 or more coordinates are not involved in the coordinate transformation.

As an example of the above generalization, we consider the simplest case of analogues of the radial coordinate (with origin defined) of a one-particle system. We can envisage a coordinate perpendicular to an axis, to a plane, to a 3-dimensional flat, and so on.

For these coordinate systems, derivations like those in Section 1.1 can be carried out without change in the space "left over" after the distinguished axes have been separated out. Thus the derivative-removing factor is now

$$J = v^{D-\eta-1},$$

where η is the number of coordinates separated out. Similarly, in the D-dependent factor of the centrifugal operator in Eq. (7), we replace

D with $D-\eta$; the proportionality expressed in Eq. (8) is not affected except that v, v_i and h_i refer to the "compressed" analogues of volume and height in $D-\eta$ coordinate space.

It should be noted that for the set of interparticle coordinates, if, in the course of distinguishing axes, we fix the positions of two or more particles (*e.g.* nuclei) in the distinguished subspace, the number of particles N in Eq. (7) needs to be reinterpreted. To avoid confusion, in such cases we shall henceforth exclude the particles fixed in the subspace in our definition of N. The centrifugal factor is then $\frac{1}{4}(D-\eta-1)(D-\eta-2N-1)$, which renders this formulation equivalent to treatment using the generalization of the set of radial coordinates and cosines of angles.

The virial theorem and the large-D limit

The virial theorem [8] applies to a system of particles whose positions and momenta are both bounded. For Coulombic systems in equilibrium, in the absence of external forces, the virial theorem states that

$$2\langle T \rangle = -\langle V \rangle. \tag{15}$$

The dimensional continuation treatment suggests a further breakdown of contributions to the energy, since the transformations described above lead to a well-defined separation of the kinetic operator into differential and centrifugal components. The relative contributions of these pieces depend on the dimensionality. For example, for the hydrogen atom, we can express the kinetic operator as follows:

$$T = -\frac{1}{2}\frac{d^2}{dr^2} + \frac{(D-1)(D-3)}{8r^2}. \tag{16}$$

In order to finitize the large-D limit, we scale according to Eq. (9). The virial theorem then reads

$$\left\langle -\frac{1}{\beta^2}\frac{d^2}{d\hat{\rho}^2} + \frac{1}{\hat{\rho}^2}(1 - 1/\beta) \right\rangle = \left\langle \frac{1}{\hat{\rho}} \right\rangle \tag{17}$$

As $D \to \infty$, the entire derivative term and the entry in $1/\beta$ in the centrifugal term vanish, and for the kinetic portion we are left with twice the centrifugal energy. Since the system is localized at the minimum of the effective potential in this limit, we can discard the angled brackets and write simply

$$\frac{1}{\hat{\rho}_0{}^2} = \frac{1}{\hat{\rho}_0},$$

It isn't hard to see that for any Coulombic system whose effective potential possesses a global minimum, the relationship

$$2\bar{T}_{\text{cent}} = -\bar{V}_{\text{coul}}. \tag{18}$$

will hold for $D \to \infty$. This is always the case for atoms.

For molecules handled by the Born-Oppenheimer approximation, we have to take into account some external forces which hold the nuclei at rest [9]. In the case of diatomic molecules, where R denotes the internuclear distance, the general form of the virial theorem is

$$2\langle T \rangle + \langle V \rangle + R\frac{\partial E}{\partial R} = 0, \tag{19}$$

or, in the $D \to \infty$ limit, after scaling,

$$2\bar{T}_{\text{cent}} + \bar{V}_{\text{coul}} + \hat{R}\frac{\partial \mathcal{E}}{\partial \hat{R}} = 0. \tag{20}$$

As a result, Eq. (15) (or (18), in the large-D limit) does not hold in general, except when no external forces are needed to hold the nuclei at rest. The latter case arises in two important circumstances: at infinite nuclear separation, and at the minimum of the potential energy curve, corresponding to an equilibrium configuration.

It is also worth noting that since the nuclear repulsion term is included in both V and E, its contribution to V is exactly cancelled out in the term $R(\partial E/\partial R)$. Hence we have for $D \to \infty$

$$2\bar{T}_{\text{cent}} + \bar{V}_{\text{e}} + \hat{R}\frac{\partial \mathcal{E}_{\text{e}}}{\partial \hat{R}} = 0, \tag{21}$$

where V_{e} is the contribution to the Coulomb energy from the electrons alone, and $\mathcal{E}_{\text{e}}$ the total electronic energy.

An alternative scaling scheme

Uniform scaling procedure

The straightforward dimensional generalization of an electronic structure hamiltonian is accomplished by reinterpreting Laplacians and Coulomb terms as being D-dimensional quantities:

$$\nabla_i^2 = \sum_{\nu=1}^{D} \frac{\partial^2}{\partial x_{i\nu}^2},$$

$$\frac{1}{r_i} = 1 \Big/ \left(\sum_{\nu=1}^{D} x_{i\nu}^2 \right)^{1/2}.$$

$$(22)$$

Since the dimensional generalization introduces D "upstairs" in the Laplacians, and "downstairs" in the Coulomb terms, one might expect the two sets of terms to get out of balance as D is varied, with the Laplacians dominating at high D, and the Coulomb terms at low D. This is of course a rather simplistic perspective, but it does describe qualitatively what happens to the solutions as D is varied. A ground state hydrogen atom ($E = -\frac{1}{2}[\frac{2}{D-1}]^2$), for example, falls apart as $D \to \infty$, and falls into the nucleus as $D \to 1$.

It is of course desirable to work with solutions that remain finite. Usually in the dimensional scaling approach this is done by means of scalings applied to the distance and energy variables. The aim of these scalings is essentially to finitize the effects of the various terms in the hamiltonian. In particular, the aim is to cancel the singular behavior associated with the Laplacians in the $D \to \infty$ limit, and with the Coulomb terms in the $D \to 1$ limit. The problem with the Laplacians at large D is the fact that the centrifugal terms grow quadratically in D, so that they tend to push the system apart. The one-particle central-field Laplacian in radial form, for example, is

$$\nabla^2 = \frac{\partial^2}{\partial r^2} - \frac{\left(\ell + \frac{D-3}{2}\right)\left(\ell + 1 + \frac{D-3}{2}\right)}{r^2}.$$

$$(23)$$

In general, Laplacians need to divided by factors quadratic in D in order to finitize their effects as $D \to \infty$. Similarly, the problem with the Coulomb terms at $D = 1$ is that they are too strong; without centrifugal barriers, systems bound by Coulomb potentials collapse. However, Coulomb singularities can be tamed in the $D \to 1$ limit through multiplication by factors of $D-1$, since

$$\lim_{D \to 1} \frac{D-1}{2r} = \delta(r), \tag{24}$$

and $\delta(r)$ yields a finite ground state energy.

Often, the scalings needed to finitize the $D \to \infty$ and $D \to 1$ limits have been considered independently of one another. For example, for simple electronic structure problems, the above aims can be (and usually have been) met by means of the scalings

$$E = \left(\frac{2}{D-1}\right)^2 \mathcal{E} \qquad r_i = \left(\frac{D-1}{2}\right) \check{\rho}_i \qquad \text{for } D \to 1,$$

$$E = \left(\frac{2}{D-1}\right)^2 \mathcal{E} \qquad r_i = \left(\frac{D-1}{2}\right)^2 \hat{\rho}_i \qquad \text{for } D \to \infty. \tag{25}$$

These scalings do finitize the dimensional limits. For example, for the hydrogen atom they give the limiting forms

$$\mathcal{H}_1 = -\frac{1}{2}\frac{d^2}{d\check{\rho}^2} - \delta(\check{\rho}),$$

$$\mathcal{H}_\infty = \frac{1}{2\hat{\rho}^2} - \frac{1}{\hat{\rho}}, \tag{26}$$

both of which yield finite expectation values for distances and energies. (In fact, $\mathcal{E} = -\frac{1}{2}$ and $\langle \check{\rho} \rangle = \frac{1}{2}$ for $D \to 1$, and $\mathcal{E} = -\frac{1}{2}$ and $\langle \hat{\rho} \rangle = 1$ for $D \to \infty$.) However, the inconsistency of the two distance scalings in Eq. (26) renders the relationship of the $D=3$ solutions to the dimensional limit solutions unclear. The aim of this section is to show how one can construct a single scaling which simultaneously finitizes both dimensional limits, thereby allowing one to use the limiting solutions for purposes of quantitative prediction at $D=3$.

Because scalings can be confusing (especially when there are several different varieties being used by the practitioners of dimensional scaling!), we choose to describe the uniform scaling in a different, but ultimately equivalent, way. Specifically, instead of viewing the scaling as one performed on the distances and energies, we will view it as being performed on the terms of the hamiltonian itself. That is, we will write for a typical electronic structure problem

$$\mathcal{H} = c_1 \begin{pmatrix} \text{Lapla-} \\ \text{cians} \end{pmatrix} + c_2 \begin{pmatrix} \text{Coulomb} \\ \text{terms} \end{pmatrix}, \tag{27}$$

and choose the scale factors c_1 and c_2 in such a way as to finitize the effects of the terms which they multiply for $1 \leq D \leq \infty$. Of course, with the hamiltonian written in this form, it is clear that we must also have $c_1 = c_2 = 1$ for $D = 3$.

As discussed above, the general behavior required of c_1 and c_2 in order for them to finitize the dimensional limits while leaving the $D = 3$ version of the hamiltonian unaltered, are

$$c_1 \begin{cases} \text{is finite} & \text{for } D \to 1 \\ = 1 & \text{for } D = 3 \\ \sim 1/D^2 & \text{for } D \to \infty, \end{cases} \qquad c_2 \begin{cases} \sim D-1 & \text{for } D \to 1 \\ = 1 & \text{for } D = 3 \\ \text{is finite} & \text{for } D \to \infty. \end{cases} \tag{28}$$

Although these conditions could be met in a variety of ways, probably the simplest is

$$c_1 = \left(\frac{3}{D}\right)^2, \qquad c_2 = \left(\frac{3}{D}\right)\left(\frac{D-1}{2}\right). \tag{29}$$

This defines the scaling that we will be using.

We now consider two examples of this scaling procedure. First consider a two-electron atom of nuclear charge Z. The dimensionally generalized hamiltonian, with Laplacians and Coulomb terms scaled by the procedure just described, is

$$\mathcal{H}_D^u(\vec{\rho}; Z) = \left(\frac{3}{D}\right)^2 \left(-\tfrac{1}{2}\nabla_1^2 - \tfrac{1}{2}\nabla_2^2\right)$$
$$+ \left(\frac{3}{D}\right)\left(\frac{D-1}{2}\right)\left(-\frac{Z}{\rho_1} - \frac{Z}{\rho_2} + \frac{1}{\rho_{12}}\right). \tag{30}$$

The limiting forms of this hamiltonian at low D and high D are readily evaluated:

$$\mathcal{H}_1^u(\vec{\rho}; Z) = -\frac{9}{2}\frac{d^2}{d\rho_1^2} - \frac{9}{2}\frac{d^2}{d\rho_2^2} - 3Z\delta(\rho_1) - 3Z\delta(\rho_2) + 3\delta(\rho_1 - \rho_2),$$

$$\mathcal{H}_\infty^u(\vec{\rho}; Z) \;=\; \frac{9}{4}\,\frac{\rho_1{}^2 + \rho_2{}^2}{\rho_1{}^2\rho_2{}^2 + \rho_2{}^2\rho_{12}{}^2 + \rho_{12}{}^2\rho_1{}^2 - \tfrac{1}{2}\left(\rho_1{}^4 + \rho_2{}^4 + \rho_{12}{}^4\right)} \\ -\,\frac{3Z}{2\rho_1} - \frac{3Z}{2\rho_2} + \frac{3}{2\rho_{12}}.$$

$$(31)$$

The superscript ^{u} is used to indicate that these limiting forms were obtained by means of a uniform scaling. The large-D hamiltonian has been written, as usual, as one for the probability amplitude.

If instead one had dimensionally generalized the helium hamiltonian by means of Eq. (22), and then treated the dimensional limits independently by means of Eq. (25), then one would have obtained the more familiar-looking dimensional limits

$$\mathcal{H}_1^n(\vec{\rho}; Z) \;=\; -\frac{1}{2}\frac{d^2}{d\check{\rho}_1{}^2} - \frac{1}{2}\frac{d^2}{d\check{\rho}_2{}^2} - Z\delta(\check{\rho}_1) - Z\delta(\check{\rho}_2) + \delta(\check{\rho}_1 - \check{\rho}_2),$$

$$\mathcal{H}_\infty^n(\vec{\rho}; Z) \;=\; \frac{\hat{\rho}_1{}^2 + \hat{\rho}_2{}^2}{\hat{\rho}_1{}^2\hat{\rho}_2{}^2 + \hat{\rho}_2{}^2\hat{\rho}_{12}{}^2 + \hat{\rho}_{12}{}^2\hat{\rho}_1{}^2 - \tfrac{1}{2}\left(\hat{\rho}_1{}^4 + \hat{\rho}_2{}^4 + \hat{\rho}_{12}{}^4\right)} \\ -\,\frac{Z}{\hat{\rho}_1} - \frac{Z}{\hat{\rho}_2} + \frac{1}{\hat{\rho}_{12}}.$$

$$(32)$$

The superscript ^{n} denotes that the dimensional limits have been treated by means of a nonuniform scaling. If one is only interested in computing eigenvalues, it doesn't matter for *this* problem whether one uses the uniform or nonuniform scaling, since they are rendered equivalent by simple changes of units for the dynamical variables:

$$\mathcal{H}_1^u(\vec{\rho}; Z) \;=\; \mathcal{H}_1^n(\tfrac{1}{3}\vec{\rho}; Z),$$

$$\mathcal{H}_\infty^u(\vec{\rho}; Z) \;=\; \mathcal{H}_\infty^n(\tfrac{2}{3}\vec{\rho}; Z).$$

$$(33)$$

Consider now a hydrogen molecular ion with bond length R. The dimensionally generalized hamiltonian, with Laplacians and Coulomb terms scaled by the procedure described above, is

$$\mathcal{H}_D^u(\vec{\rho}; R) = \left(\frac{3}{D}\right)^2\left(-\tfrac{1}{2}\nabla^2\right) - \left(\frac{3}{D}\right)\left(\frac{D-1}{2}\right) \times$$

$$\left(\frac{1}{\sqrt{\rho_\perp{}^2 + (\rho_\parallel + \frac{1}{2}R)^2}} + \frac{1}{\sqrt{\rho_\perp{}^2 + (\rho_\parallel - \frac{1}{2}R)^2}} \right). \tag{34}$$

The dimensional limits may be written

$$\mathcal{H}_1^u(\vec{\rho}; R) = -\frac{9}{2}\frac{d^2}{d\rho_\parallel{}^2} - 3\delta(\rho_\parallel + \tfrac{1}{2}R) - 3\delta(\rho_\parallel - \tfrac{1}{2}R),$$

$$\mathcal{H}_\infty^u(\vec{\rho}; R) = \frac{9}{8\rho_\perp{}^2} - \frac{3}{2\sqrt{\rho_\perp{}^2 + (\rho_\parallel + \frac{1}{2}R)^2}} - \frac{3}{2\sqrt{\rho_\perp{}^2 + (\rho_\parallel - \frac{1}{2}R)^2}}.$$

$$\tag{35}$$

In this case, the use of the independent dimensional limit scalings, Eq. (25), leads to

$$\mathcal{H}_1^n(\vec{\rho}; R) = -\frac{1}{2}\frac{d^2}{d\check{\rho}_\parallel{}^2} - \delta(\check{\rho}_\parallel + \tfrac{1}{2}R) - \delta(\check{\rho}_\parallel - \tfrac{1}{2}R),$$

$$\mathcal{H}_\infty^n(\vec{\rho}; R) = \frac{1}{2\hat{\rho}_\perp{}^2} - \frac{1}{\sqrt{\hat{\rho}_\perp{}^2 + (\hat{\rho}_\parallel + \frac{1}{2}R)^2}} - \frac{1}{\sqrt{\hat{\rho}_\perp{}^2 + (\hat{\rho}_\parallel - \frac{1}{2}R)^2}}.$$

$$\tag{36}$$

For this problem, the two sets of limits are *not* related by changing the units of the dynamical variable alone. Rather, the parameter of the problem (the bond length) must be brought into play:

$$\mathcal{H}_1^u(\vec{\rho}; R) = \mathcal{H}_1^n(\tfrac{1}{3}\vec{\rho}; \tfrac{1}{3}R),$$

$$\mathcal{H}_\infty^u(\vec{\rho}; R) = \mathcal{H}_\infty^n(\tfrac{2}{3}\vec{\rho}; \tfrac{2}{3}R). \tag{37}$$

Dimensional interpolation

The import of the above observation is that for molecules (and more generally for any problem which possesses a length scale independent of that defined by the electronic motions themselves), it is only from the dimensional limits obtained by the uniform scaling that one can hope to obtain quantitative results by means of simple procedures such as dimensional expansions and interpolation.

With regard to the latter, we choose $1/D$ as the interpolation parameter, since it is the natural perturbation parameter from our choice of scaling coefficients. By simple proportion, we construct a linear interpolating scheme as follows:

$$E_3(R) \approx \tfrac{1}{3}E_1^u(R) + \tfrac{2}{3}E_\infty^u(R) \qquad (38)$$

It should be noted that one *can* use the more familiar independent scalings, as long as one evaluates the energy at suitably modified values for parameters such as the bond length. For example, from Eq. (37), one sees that if one uses the non-uniform scalings, then linear interpolation requires energy eigenvalues obtained at $\tfrac{1}{3}$ and $\tfrac{2}{3}$ of the actual bond length:

$$E_3(R) \approx \tfrac{1}{3}E_1^n(\tfrac{1}{3}R) + \tfrac{2}{3}E_\infty^n(\tfrac{2}{3}R) \qquad (39)$$

Note that the scalings of the bond length are merely undoing the inconsistency introduced by the separate $D=1$ and $D \to \infty$ scalings of Eq. (25). Although one can use either Eq. (38) or (39) for purposes of interpolation, the former is recommended as being more direct and intuitive. However, an equation like (39) may be useful for purposes of salvaging quantitative results from dimensional limits which happen to have been evaluated using the conventional scalings.

Table 2 lists some results obtained for H_2^+ and H_2 using linear interpolation between the dimensional limits. We observe that the electronic energies are accurate to within $\tfrac{1}{2}\%$ (bond energies to within $2\tfrac{1}{2}\%$), while the bond lengths are accurate to within 6%. In Figure 1, the total energy of H_2^+ as calculated from interpolation is plotted as a function of the internuclear distance. The agreement with the exact energy is quite good in the region of the equilibrium bond length and for smaller R.

We note that the uniform scaling procedure has been applied to a number of other model problems as well [7]. These include the hydrogen atom in a spherical cavity, HF two-electron atoms in electric and magnetic fields, and perturbed one-body Coulomb problems (*e.g.* the Yukawa potential).

Symmetry breaking

Quite generally, the molecular electronic structure obtained in the large-D limit (by minimizing effective potentials) will exhibit abrupt

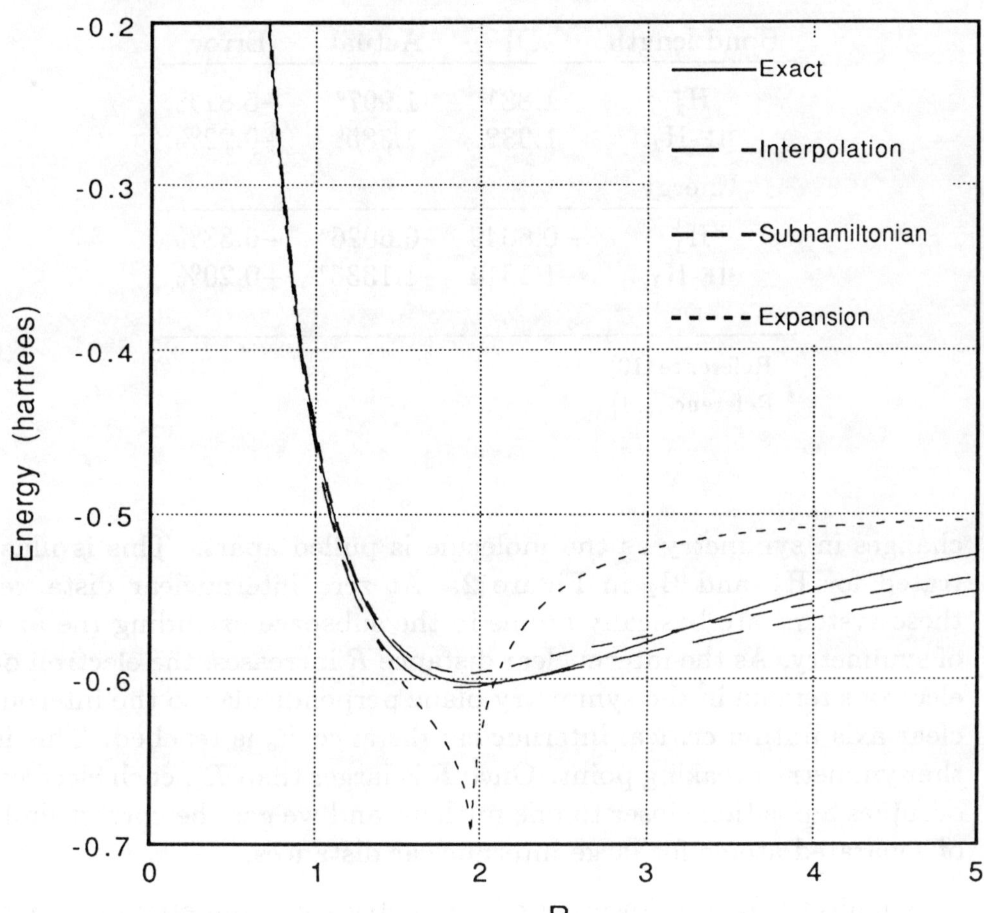

Figure 1. H_2^+ energies calculated using various dimensional continuation methods: linear interpolation between uniformly scaled $D \rightarrow 1$ and $D \rightarrow \infty$ limits; expansion to first order in $1/D$ about the uniformly scaled $D \rightarrow \infty$ limit; and minimization of subhamiltonians constructed using uniformly scaled $D \rightarrow 1$ and $D \rightarrow \infty$ energies.

Table 2. Estimates of bond length and ground state energy for H_2^+ and Hartree-Fock H_2 as calculated by dimensional interpolation; actual values are listed for comparison.

Bond length	DI	Actual	Error
H_2^+	1.881	1.997[a]	−5.81%
HF-H_2	1.388	1.385[b]	+0.22%
Energy			
H_2^+	−0.6049	−0.6026[a]	−0.38%
HF-H_2	−1.1314	−1.1336[b]	+0.20%

[a] Reference [10]

[b] Reference [11]

changes in symmetry as the molecule is pulled apart. This is illustrated for H_2^+ and H_2 in Figure 2. At zero internuclear distance, these systems are basically atoms in the subspace excluding the axis of symmetry. As the internuclear distance R increases, the electron or electrons remain in the symmetry plane perpendicular to the internuclear axis until a critical internuclear distance R_c is reached. This is the symmetry breaking point. Once R is larger than R_c, each electron occupies a position closer to one nucleus, and we get the correct limit of separated atoms for large internuclear distances.

Qualitatively, the lowering of symmetry (or symmetry breaking) which occurs as the bond is lengthened are the large-D equivalents of the change from molecular orbital to valence bond description [6]. If one works with a uniform scaling, however, the symmetry breakings may have quantitative significance as well. This is illustrated in Table 3, where $D = 3$ equilibrium bond lengths R_e are compared with the $D \to \infty$ critical internuclear distances R_c (those at which symmetry breaking occurs). One sees that for unapproximated problems there is very good agreement between the two values. (On the other hand, imposition of the Hartree-Fock approximation in H_2 en-

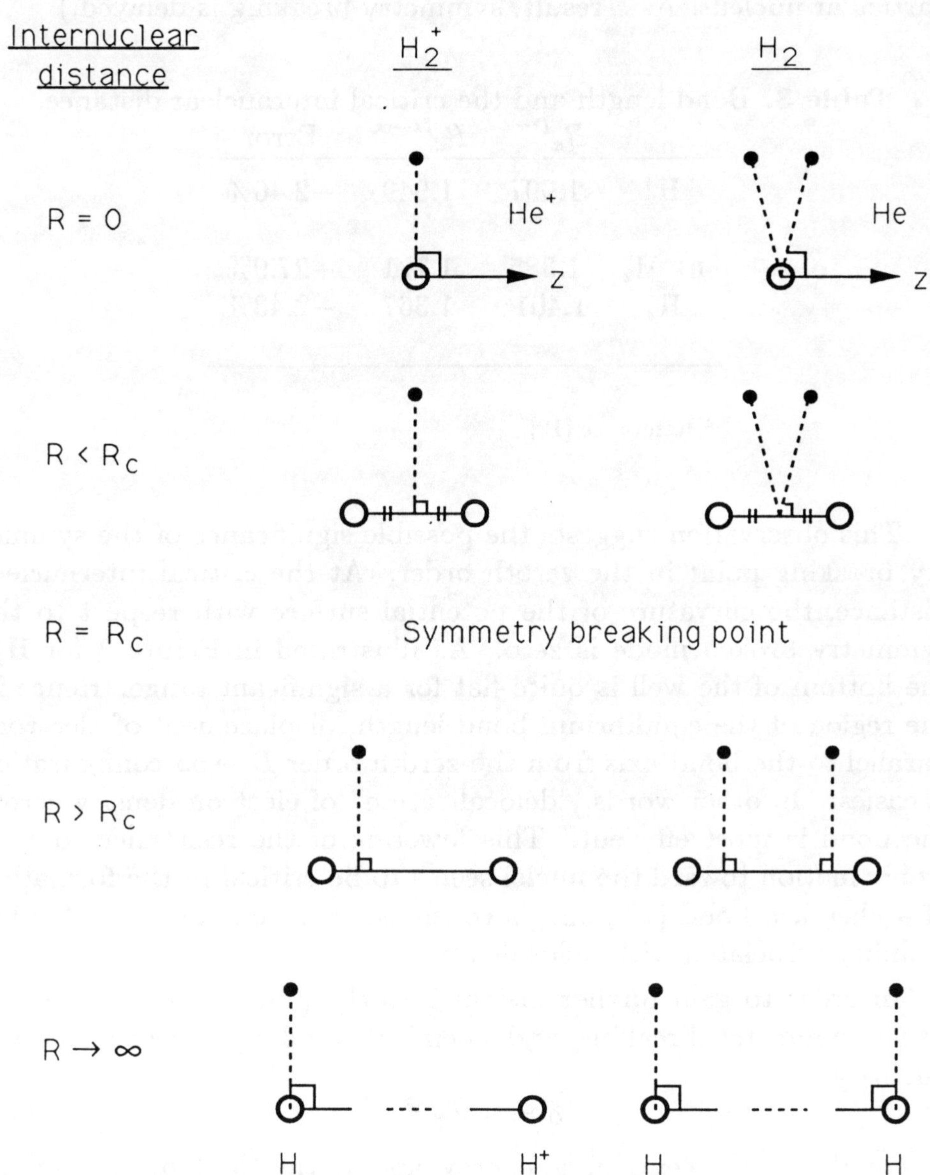

Figure 2. $D \to \infty$ electronic structures of H_2^+ and H_2.

tails a molecular orbital description of the electronic structure for all R. This involves contributions from ionic structures, which are represented in the large-D limit by both electrons being closer to one particular nucleus. As a result, symmetry breaking is delayed.)

Table 3. Bond length and the critical internuclear distance.

	$R_e^{D=3}$	$R_c^{D\to\infty}$	Error
H_2^+	1.997	1.949	-2.40%
HF-H_2	1.385	1.771	$+27.9\%$
H_2	1.401^c	1.367	-2.43%

^{c} Reference [12]

This observation suggests the possible significance of the symmetry breaking point in the zeroth order. At the critical internuclear distance, the curvature of the potential surface with respect to the symmetry broken mode is zero. As illustrated in Figure 3 for H_2^+, the bottom of the well is quite flat for a significant range. Hence in the region of the equilibrium bond length, displacement of electrons parallel to the bond axis from the zeroth order $D \to \infty$ configuration is easiest. In other words, "delocalization" of electron density across the bond is most efficient. This lowering of the resistance to electronic motion toward the nuclei seems to be critical in the formation of a chemical bond [13], and is consistent with our notions of extra stability associated with delocalization.

In order to gain further insight into the possible connection between symmetry breaking and chemical bonding, we consider the quantity

$$\alpha = -\bar{V}_e/\bar{T}_{\text{cent}} \tag{40}$$

where $\bar{T}_{\text{cent}}$ is the centrifugal energy, and $\bar{V}_e$ the Coulomb energy attributed to the electrons alone, *i.e.* without nuclear repulsion.

At zero internuclear distance we find that $\alpha = 2$, which we recognize as the virial theorem for atoms, Eq. (18). Empirically, we find that

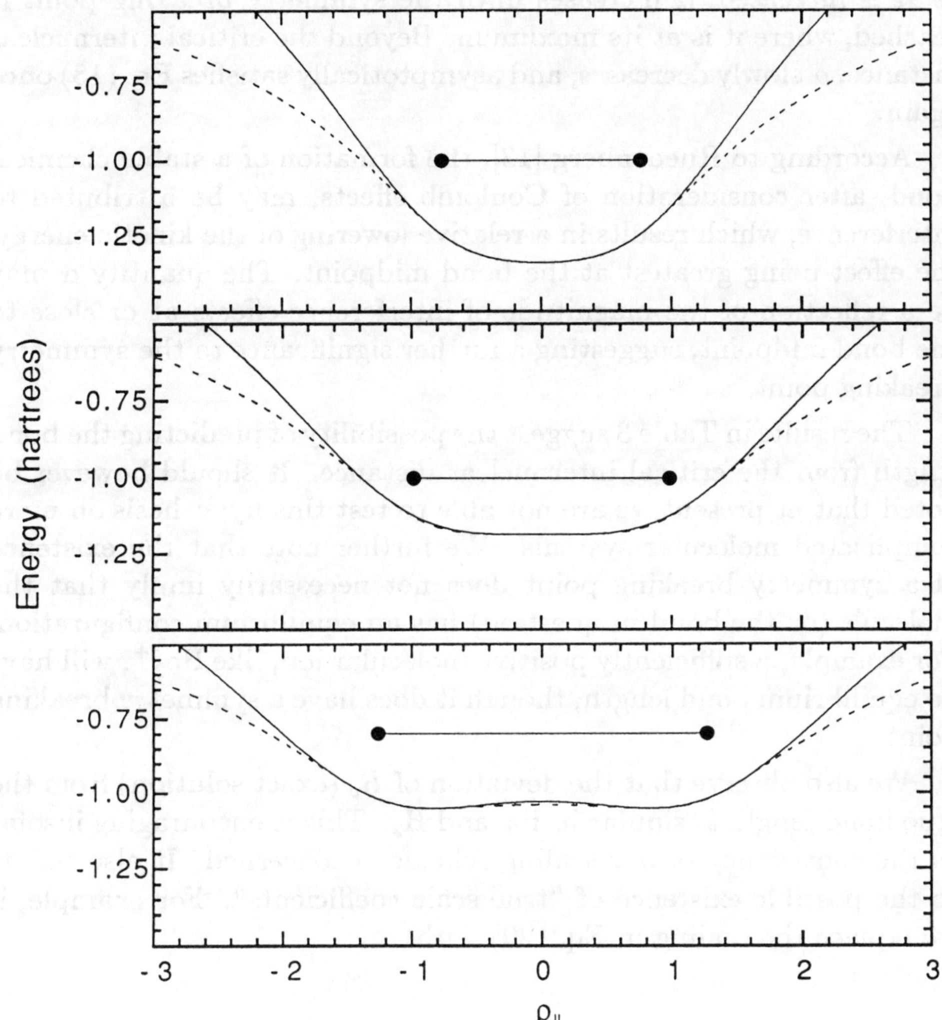

Figure 3. Effective potential for the electron in H_2^+ with respect to axial motion $\rho_\parallel$ before, at, and after symmetry breaking ($R = 1.5$, 1.949, and 2.5, respectively). Heavy figures indicate nuclear positions. Curves at both constant $\rho_\perp$ (that assumed at the global minimum, solid) and minimizing $\rho_\perp$ (that which minimizes the potential for each $\rho_\parallel$, dashed), are shown.

as R is increased, α increases until the symmetry breaking point is reached, where it is at its maximum. Beyond the critical internuclear distance, α slowly decreases, and asymptotically satisfies Eq. (18) once again.

According to Ruedenberg [13], the formation of a stable chemical bond, after consideration of Coulomb effects, may be attributed to interference, which results in a relative lowering of the kinetic energy, the effect being greatest at the bond midpoint. The quantity α may be a reflection of the magnitude of interference effects at or close to the bond midpoint, suggesting a further significance to the symmetry breaking point.

The results in Table 3 suggest the possibility of predicting the bond length from the critical internuclear distance. It should however be noted that at present we are not able to test this hypothesis on more complicated molecular systems. We further note that the existence of a symmetry breaking point does not necessarily imply that the molecule (or the bond in question) has an equilibrium configuration. For example, a sufficiently positive molecular ion, like He_2^{3+}, will have no equilibrium bond length, though it does have a symmetry breaking point.

We also observe that the deviation of R_c (exact solution) from the true bond length is similar in H_2^+ and H_2. This is encouraging insofar as the consistency of our scaling scheme is concerned. It also points to the possible existence of "true scale coefficients". For example, if we replace the scaling in Eq. (29) with

$$c_1 = \left(\frac{3+\xi}{D+\xi}\right)^2; \qquad c_2 = \left(\frac{3+\xi}{D+\xi}\right)\left(\frac{D-1}{2}\right) \qquad (41)$$

(which is still completely consistent with Eq. (28)), then the assignment

$$\xi = 0.075 \qquad (42)$$

gives symmetry breaking points at $1.997a_0$ for H_2^+ and $1.401a_0$ for H_2. These values reproduce the known bond lengths for both species.

Expansion from the large-D limit

In this and the following section we consider two alternatives to dimensional interpolation for obtaining the $D = 3$ results from the dimensional limit calculations. First is the $1/D$ expansion. The zeroth-order term in this expansion is just the $D \to \infty$ limit solution, as given by the minimum of an effective potential, such as Eq. (35). The first-order or $1/D$ term arises from harmonic vibrations about the minimum of the effective potential. Together, the first two terms of the $1/D$ expansion for axially symmetric systems take the form

$$E_D = E_\infty + \frac{1}{D}\left[3\sum_\mu (n_\mu + \tfrac{1}{2})\lambda_\mu - (2N+4)\bar{T}_{\text{cent}} - \bar{V}_e\right] + \dots, \quad (43)$$

Here the index μ runs over the normal modes, while $3\lambda_\mu/D$ is the frequency and n_μ is the number of quanta associated with mode μ. The remaining entries in the $1/D$ term come from the expansions of the centrifugal and potential terms, namely

$$\left(\frac{3}{D}\right)^2 \left(\frac{D-2}{2}\right)\left(\frac{D-2N-2}{2}\right) = \frac{9}{4}\left(1 - \frac{2N+4}{D} + \dots\right),$$

$$\left(\frac{3}{D}\right)\left(\frac{D-1}{2}\right) = \frac{3}{2}\left(1 - \frac{1}{D} + \dots\right).$$

If, prior to vibrational analysis, the dynamical variables are rescaled according to Eq. (37), the only modification to Eq. (43) would be replacing the factor of 3 associated with the vibrational frequencies by 2. We note that for atoms, this reformulation is equivalent to Loeser's [2] 2-term approximation, since for such cases $\alpha = -\bar{V}_e/\bar{T}_{\text{cent}} = 2$ as a consequence of the virial theorem.

In Figure 1, the total energies obtained from the first-order $1/D$ expansion for H_2^+ are plotted as a function of the internuclear distance. We observe that the eigenvalues are underestimated for $R < R_c$, and overestimated for $R > R_c$. The former can be explained by the fact that the curvature at the bottom of the well with respect to axial motion, from which the associated frequency is obtained, does not reflect the overall shape of the well, and in particular ignores the steep potential rise at the sides of the well (see, for example, Figure 3). On the other hand, beyond symmetry breaking, there are two equivalent minima; vibrational analysis does not take into account the existence of the second minimum, and hence does not address overlap and/or tunneling effects which result in a lowering of the energy. This is in contrast to linear interpolation of energies which actually underestimates the energy for larger internuclear distances.

Interpolated subhamiltonians

The procedure of using subhamiltonians is based on the premise that the effects of the kinetic energy operator for finite dimensionalities may be modeled by a modified centrifugal potential in a zeroth-order-like Hamiltonian [14]. For example, for H_2^+ we write

$$\mathcal{H}_D^{\text{sub}}(R) = \frac{9a_D}{8\rho_\perp{}^2} - \frac{3}{2} \left[\frac{1}{\sqrt{\rho_\perp{}^2 + (\rho_\parallel + \frac{1}{2}R)^2}} + \frac{1}{\sqrt{\rho_\perp{}^2 + (\rho_\parallel - \frac{1}{2}R)^2}} \right].$$

$$(44)$$

where a_D is a parameter. Given an explicit expression for a_D, one can minimize this subhamiltonian for any D (with respect to $\rho_\perp$ and $\rho_\parallel$), thereby obtaining an energy and a classical model for the electronic structure at that D. Here we will model a_D in a very simple manner. Specifically, we will approximate it as a linear function of $1/D$. Since from Eq. (35) we must clearly have $a_D \to 1$ as $D \to \infty$, we can write

$$a_D = 1 + b/D. \tag{45}$$

Then, since the $D=1$ solution to H_2^+ is available [15], we can solve for b by matching the subhamiltonian minimum with the $D=1$ energies.

The $D=3$ energies for H_2^+ obtained by minimizing $\mathcal{H}_D^{\text{sub}}$ with this prescription for a_D are plotted in Figure 1. The potential energy curve is comparable in accuracy to that obtained by dimensional interpolation, though it is somewhat better at large R and in the prediction of the equilibrium bond length ($1.92a_0$ $vs.$ $1.88a_0$, as compared to the actual bond length of $2.00a_0$).

The subhamiltonian approach is especially appealing because it allows one to use simple classical models based on those of the large-D limit, but corrected quantitatively for the effects of finite D. The structures obtained by minimizing the subhamiltonians might be considered as optimal localized representations of real electronic structures.

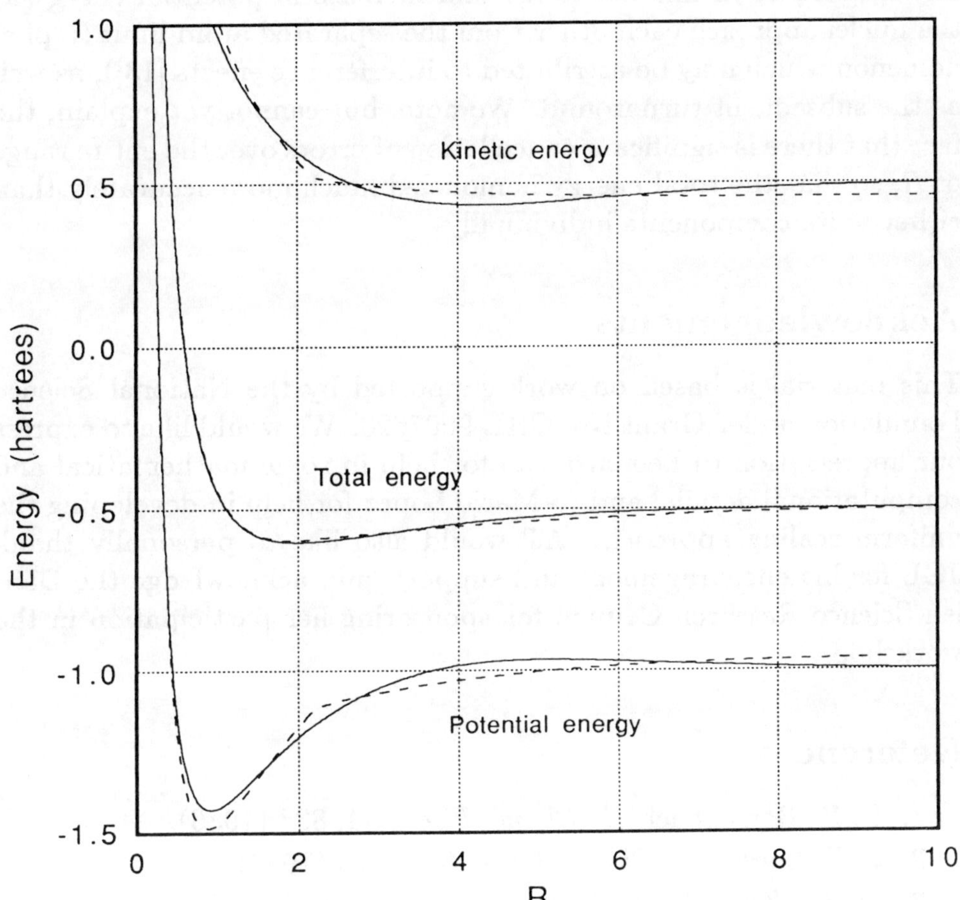

Figure 4. Variation of H_2^+ total, kinetic, and potential energies with internuclear distance R, as calculated from subhamiltonians (dashed curves) and exact eigenvalues (solid curves).

In addition, use of the subhamiltonians affords a convenient breakdown of energy contributions. As shown in Figure 4, the centrifugal energies and total Coulomb potentials thus obtained do model in a rough sense the variation of the actual $D = 3$ kinetic and potential energies with the internuclear distance. In particular, they model the initial decrease in kinetic energy and increase in potential energy as the nuclei approach each other from the separated atom limit (a phenomenon which may be attributed to interference effects [13]), as well as the subsequent turnaround. We note, but cannot yet explain, the fact that there is significant cancellation of errors over the entire range of R, so that the total energy is modeled much more accurately than either of its components individually.

Acknowledgements

This material is based on work supported by the National Science Foundation under Grant No. CHE-9007620. We would like to express our appreciation to Leonard Tan for help in some mathematical and computational details, and to Mario Lopez for help in developing the uniform scaling approach. AT would also like to personally thank JGL for his encouragement and support, and acknowledge the Danish Science Research Council for sponsoring her participation in the workshop.

References

1. D. R. Herschbach, *J. Chem. Phys.* **84**, 838 (1986).

2. J. G. Loeser, *J. Chem. Phys.* **86**, 5635 (1987).

3. M. G. Kendall, *A Course in the Geometry of n Dimensions* (Charles Griffin, London, 1962), pp. 38-40.

4. F. Calogero and A. Degasperis, *Phys. Rev.* A **11**, 265 (1975).

5. W. Thirring, *A Course in Mathematical Physics* (Springer-Verlag, Vienna, 1980), Vol. 4, p. 18.

6. D. D. Frantz and D. R. Herschbach, *Chem. Phys.* **126**, 59 (1988).

7. M. M. Lopez, A. L. Tan and J. G. Loeser, in preparation.

8. L. I. Schiff, *Quantum Mechanics*, second edition (McGraw-Hill, New York, 1955), p. 140.

9. J. C. Slater, *Quantum Theory of Molecules and Solids*, Vol. 1 (McGraw-Hill, 1963), pp. 32-33.

10. H. Wind, *J. Chem. Phys.* **42**, 2371 (1965).

11. A. Szabo and N. S. Ostlund, *Modern Quantum Chemistry*, revised first edition (McGraw-Hill, 1989), pp. 190-205.

12. W. Kolos and L. Wolniewicz, *J. Chem. Phys.* **49**, 404 (1968).

13. K. Ruedenberg, *Revs. Mod. Phys.* **34**, 326 (1962); M. J. Feinberg and K. Ruedenberg, *J. Chem. Phys.* **54**, 1495 (1971).

14. Z. Zhen and J. G. Loeser, *Large-D limit for N-electron atoms*, this book.

15. F. C. Goodrich, *A Primer of Quantum Chemistry* (Wiley, 1972), pp. 161-163.

6.4 Symmetry-breaking and Tunneling in H_2^+

Sabre Kais
Department of Chemistry
Harvard University
12 Oxford Street
Cambridge MA 02138, USA

Abstract

In the large-D regime the effective potential for electronic structure often undergoes symmetry breaking for certain ranges of nuclear charges or geometries and thus acquires multiple minima. Tunneling among such minima is akin to resonance among valence bond structures. Here we treat the D-dimensional H_2^+ molecule-ion as a prototype case. We evaluate the energy splitting ΔE between the lowest pair of states $(1s\sigma_g$ and $2p\sigma_u)$ by two semiclassical techniques, the asymptotic and instanton methods. This treatment is markedly simplified by use of the large-D limit.

Introduction

In the application of dimensional scaling to electronic structure, symmetry breaking and tunneling are major features [1]. Typically, the effective potential W_∞ for the $D \to \infty$ limit acquires multiple minima when the nuclear charges or geometrical parameters are varied. For instance, W_∞ for the He atom ($Z = 2$) has a single minimum, with the electrons equidistant from the nucleus. If the nuclear charge drops below a critical value ($Z_o = 1.2279...$), the symmetric configuration becomes a saddle point and W_∞ acquires two equivalent unsymmetrical minima; these have one electron much closer to the nucleus than the other, as in the H^- ion [2]. An analogous symmetry breaking transition occurs in W_∞ for H_2^+ when the internuclear distance R is varied [3]. When R is small, W_∞ has a single minimum, with the electron midway between the nuclei. When R becomes large enough($R_c >1.2990...$), however, W_∞ has a pair of equivalent minima,

256

D. R. Herschbach et al. (eds.), Dimensional Scaling in Chemical Physics, 256–273.
© 1993 *Kluwer Academic Publishers. Printed in the Netherlands.*

with the electron localized on one or the other nucleus. For finite D, tunneling

between these double minima becomes prominent; it is tantamount to resonance among valence bond structures. Since this produces energy splitting that depend exponentially on D and so vanish more rapidly than any power of $1/D$, a nonperturbative technique such as the instanton method is required to evaluate the tunneling contributions.

Our aim is to develop and test practical, numerically stable means of treating nonseparable potentials in which tunneling occurs in two or more degrees of freedom. The H_2^+ system is particularly suitable for this purpose. Exact numerical calculations [4] are available for comparison over a wide range of R. Also, in spheroidal coordinates the double-minimum potential is separable and tunneling occurs in only one coordinate [5] whereas in cylindrical coordinates [6] the potential is nonseparable and tunneling occurs in two coordinates. This offers an opportunity to compare approximation methods for separable and nonseparable versions of the same system.

Here we will briefly recapitulate pertinent features of H_2^+ in D-dimensions, including the W_∞ potential and interdimensional degeneracies, and show that the major D-dependence of the energy splitting ΔE_D produced by tunneling has a simple, generic form. We employ two semiclassical techniques, the asymptotic and instanton methods. The asymptotic method blithely uses an asymptotic approximation for the wavefunction to evaluate flux through the median plane midway between the nuclei [7]. The instanton method examines the evolution of the system in imaginary time, equivalent to motion in real time in an inverted potential $(V \to -V)$ [8].

Our results agree well with numerical calculation. It is remarkable that use of the effective potential for large-dimension, which is exactly calculable from classical electrostatics, yields quantitative results for electronic tunneling, an intrinsically quantal phenomenon.

The H_2^+ Molecule-Ion in D-dimensions

We consider H_2^+ with "clamped nuclei," corresponding to the Born-Oppenheimer approximation, for which very accurate numerical solutions at $D = 3$ are available [4]. As shown in chapter 2, in D-dimension the Hamiltonian can be cast into the same form as $D = 3$, with the addition of a centrifugal potential that depends quadratically on D as a parameter. For general D, solutions can be obtained by exploiting an exact interdimensional degeneracy: $D \to D + 2$ is equivalent to $m \to m + 1$, increasing by one unit the projection m of the electronic angular momentum on the internuclear axis [4]. Our results pertain to the tunnel effect splitting ΔE between the lowest g, u states of the D-dimensional system. However, by virtue of the exact interdimensional degeneracies, this yields ΔE for all pairs of g, u states of the $D = 3$ molecule that stem from separated atom states with $m = l = n - 1$, for $n = 1 \to \infty$.

Although the H_2^+ problem for D-dimensions is separable in spheroidal coordinates, just as for $D = 3$, since we want to examine the nonseparable situation, we employ cylindrical coordinates. In these coordinates the nuclei are located on the z-axis at $-R/2$ and $+R/2$, respectively, and the electron is at (ρ, z). Dimensional scaling is introduced by using units of κ^2/Z bohr radii for distance and Z^2/κ^2 hartrees for energy, with Z the nuclear charge and $\kappa = (D - 1)/2$. The scaled Schrödinger equation for H_2^+ then takes a simple form,

$$[-\frac{1}{2\kappa^2}(\frac{\partial^2}{\partial\rho^2} + \frac{\partial^2}{\partial z^2}) + W_D]\Phi = E_D(R)\Phi, \tag{1}$$

where κ^2 has the role of an effective mass, $\Phi(\rho, z; R)$ is the Jacobian-weighted wavefunction, and $E_D(R)$ the electronic energy . The effective potential energy is given by

$$W_D(\rho, z; R) = \frac{f_D}{2\rho^2} - \frac{1}{r_a} - \frac{1}{r_b} \tag{2}$$

The first term is the scaled centrifugal potential, for $m = 0$; this contains a D-dependent coefficient,

$$f_D = \frac{(D - 2)(D - 4)}{(D - 1)^2} \tag{3}$$

Note $f_D \to 1$ in the $D \to \infty$ limit; we shall chiefly use the effective potential W_∞ for this limit. The Coulombic terms are specified by the electron-nucleus distances, given by

$$r = [\rho^2 + (z \pm R/2)^2]^{1/2},$$ (4)

where the $+$ sign pertains for r_a and the $-$ sign for r_b.

At the united atom limit ($R = 0$), the potential surface W_∞ has a single well, but at distances near the equilibrium bond length at ($R \sim 2$) double minima become prominent [3]. At large R these evolve into a pair of isolated wells in the separated atom limit. The critical point at which the symmetry breaking transition from single to double wells occurs is determined from the conditions $\frac{\partial W}{\partial \rho} = 0$ and $\frac{\partial^2 W}{\partial^2 z} = 0$, both evaluated at $z = 0$. At that point: $R_c = (27/16)^{1/2} = 1.299038$; $\rho_c = (27/32)^{1/2} = 0.918559$; $W_c = -32/27 = -1.185185$. Figure 1 shows the variation with R of the coordinates r_{am}, and r_{bm} that correspond to the minima of W_∞.

Throughout the domain $R_c < R < \infty$, tunneling through the barrier between the two minima occurs. However, as D and hence the effective mass κ^2 increases, tunneling diminishes markedly. Our chief aim is to evaluate the splitting $\Delta E_D(R)$ between the lowest two eigenvalues of Eq.(1), produced by tunneling in the double well domain.

Spheroidal coordinates $\lambda = (r_a + r_b)/R$ and $\mu = (r_a - r_b)/R$ are related to the cylindrical coordinates by $z = R\lambda\mu/2$ and $\rho^2 = R^2(\lambda^2 - 1)(1 - \mu^2)/4$. In these coordinates, Eq.(1) separates into a pair of equations [5] with effective potentials $W_D(\lambda; R)$ and $W_D(\mu; R)$ given by

$$W_D(\lambda) = \frac{1}{2(\lambda^2 - 1)}\Big(\frac{f_D}{\lambda^2 - 1} - 2R\lambda - A'\Big)$$ (5)

and

$$W_D(\mu) = \frac{1}{2(1 - \mu^2)}\Big(\frac{f_D}{1 - \mu^2} + A'\Big)$$ (6)

where $f_D = \frac{D-5}{D-1}$ and $A' = A - p^2$, with A the usual separation constant [4], and $p^2 = -\frac{1}{2}R^2 E$. The corresponding energy is $\epsilon = -\frac{1}{2}p^2 = \frac{1}{4}R^2 E$. The double well structure appears only in the μ potential [5].

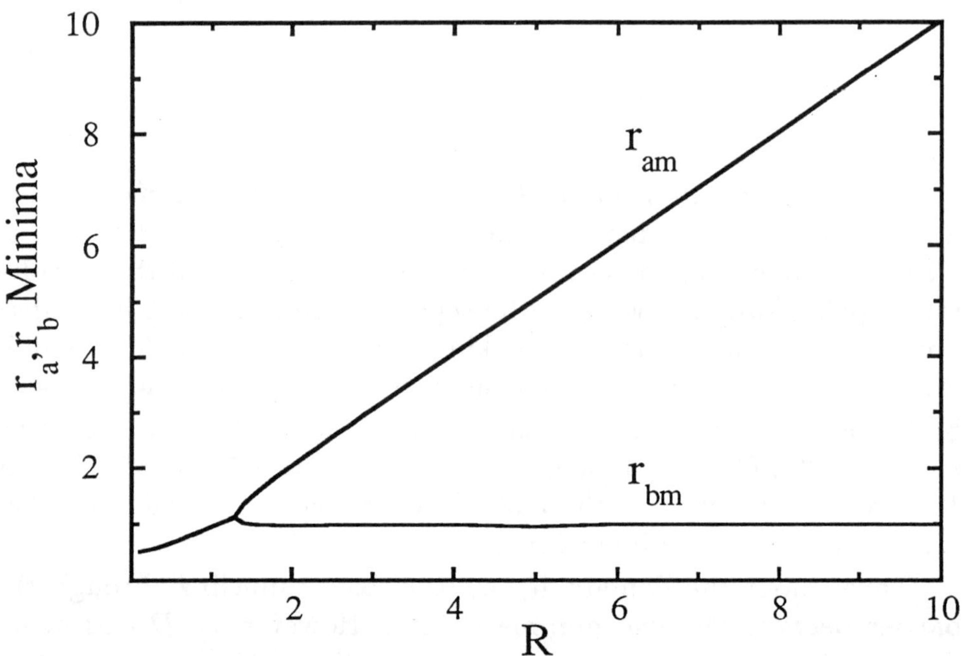

Figure 1. Variation with R of coordinates r_{am} and r_{bm} corresponding to minima of effective potential curves for H_2^+.

Asymptotic Method

For any two eigenvalues $E_\pm$ for which the eigenfunctions $\Psi_\pm$ are respectively even and odd under reflection in some coordinate, $x_1 \rightarrow -x_1$, the splitting $\Delta E = E_- - E_+$ is rigorously [9] proportional to the probability current flowing across the $x_1 = 0$ plane:

$$\Delta E = \frac{C}{2}\frac{I_S}{I_V} \tag{7}$$

in terms of surface and volume integrals given by

$$I_S = \int_{x_1=0} \Psi_+ \frac{\partial \Psi_-}{\partial x_1}\, dx_2 dx_3 \cdots dx_D \tag{8}$$

and

$$I_V = \int_{x_1 \geq 0} \Psi_+ \Psi_-\, dx_1 dx_2 \cdots dx_D \tag{9}$$

The constant C depends on the choice of units; $C = \kappa^{-2}$ in our dimension-scaled units. For one-degree of freedom, the splitting formula reduces to

$$\Delta E = \frac{C\Psi_+ \frac{d\Psi_-}{dx}\big|_{x=0}}{2\int_{x\geq 0}\Psi_+\Psi_-\,dx} \tag{10}$$

In our application to H_2^+, we have $x_1 = \mu$. The wavefunctions factor in the form $\Psi_\pm = M_\pm(\mu)L_\pm(\lambda)\Phi(\phi)$. In setting up the integrals I_S and I_V we used the gradient operator and the surface and volume elements in spheroidal coordinates to obtain [5]

$$\Delta E = \Delta E_\mu/F, \tag{11}$$

where the factor F involves the expectation value of a Jacobian. Aside from this factor, the result corresponds exactly to tunneling in one degree of freedom, the separable $\mu-$coordinate. The factor F arises because the Jacobian in the normalization integral is not separable. For our purposes, we need to evaluate F only in the large-D limit, Where the coordinates take the fixed values λ_m and μ_m; thus we obtain simply

$$F \to \frac{R^2}{4}(\lambda_m^2 - \mu_m^2) \simeq R \tag{12}$$

At the large D-limit the wavefunction $M(\mu)$ has the form

$$M(\mu) \simeq \frac{1}{(1-\mu^2)^{1/2}}\left(\frac{1-\mu}{1+\mu}\right)^{\kappa/2}\exp\left[-\frac{\kappa A'_\infty}{2}\mu\right] \tag{13}$$

and

$$M_\pm = \frac{1}{\sqrt{2}}[M(\mu) \pm M(-\mu)] \tag{14}$$

Substituting $M_\pm$ into Eq.(10) leads to the standard formula

$$\Delta E_\mu = \frac{2}{\kappa^2}\frac{M(0)M'(0)}{\int M^2\,d\mu} \tag{15}$$

Using Eqs.(12) and (13) we can show that [5]

$$\Delta E_D(R) \simeq 2\frac{\kappa^{\kappa-1}}{\Gamma(\kappa)}\exp[-\kappa(R+1-\ln 2R)] \tag{16}$$

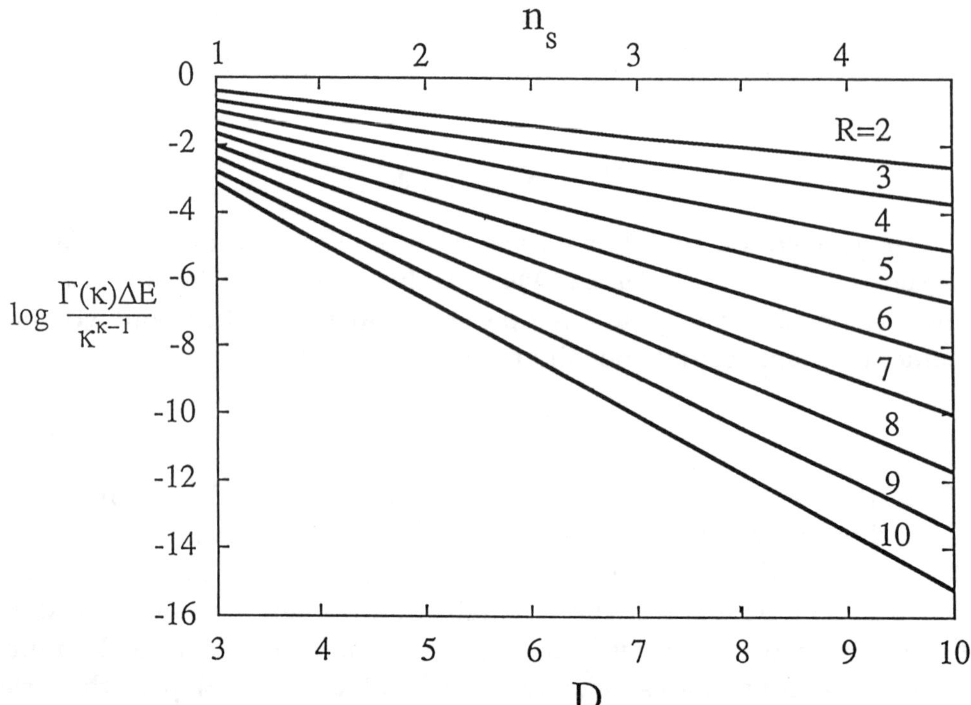

Figure 2. Semilog plot of scaled tunneling splitting $\Delta E_D(R)$ versus D for $R = 2$ to 10.

This simple result agrees exactly with the simple transcription [4] derived from interdimensional degeneracies,

$$\Delta E_D(R) = \kappa^2 \Delta E_3(R/\kappa^2) \tag{17}$$

For large D, and hence large κ, the bracketed quantity in the preexponential factor becomes simply $(2\pi\kappa)^{-1/2}\exp(\kappa)$, and $\Delta E_D(R)$ becomes

$$\Delta E_D(R) \simeq (2/\pi\kappa)^{1/2}\exp[-\kappa(R - \ln 2R)] \tag{18}$$

Figure 2 shows that plots of $\ln\frac{\Gamma(\kappa)\Delta E_D}{\kappa^{\kappa-1}}$ versus κ at constant R are quite linear even at small D and R.

Instanton Method

We consider motion along a Cartesian coordinate x, governed by a double-well potential $V(x)$ with minima at $x = \pm a$. The energy zero is chosen so that $V(\pm a) = 0$, and units adopted such that the kinetic energy operator is just $-\frac{1}{2}\frac{d^2}{dx^2}$. In its canonical version [10], the instanton method pertains to potentials adequately approximated as parabolic near the minima, with $\omega = [V''(\pm a)]^{1/2}$ the frequency for harmonic oscillation about either minimum. To leading order in $\hbar$, the splitting ΔE of the lowest pair of energy levels has the form

$$\Delta E = \hbar^{1/2} A \exp(-S_o/\hbar) \tag{19}$$

where $\hbar$ is Planck's constant and S_o is the classical action for motion along the instanton path, the classical trajectory of zero-energy between the two maxima in the inverted potential. The pre-exponential factor A involves contributions from fluctuations about the instanton path.

The instanton action is given by

$$S_o = \int_{-a}^{a} [2V(x)]^{1/2}\, dx \tag{20}$$

The great advantage of the instanton method is that S_o can be immediately generalized to more degrees of freedom [11]. We need only rep lace the scalar position variable x with a vector, and the scalar displacements $x \pm a$ with magnitudes of the vector displacements. These simple substitutions suffice to determine the exponential factor $\exp(\frac{-S_o}{\hbar})$, the leading contribution to energy level splitting produced by tunneling for any number of degrees of freedom.

For the double-well problem, the calculation [10] of the instanton fluctuation factor A can reduced [6] to evaluating a limit involving a singular quadrature,

$$A = 2\omega(\hbar\omega/\pi)^{1/2}(S_o)^{1/2}K \tag{21}$$

where

$$K = \lim_{\bar{x}\to a}(a - \bar{x})\exp(\omega \int_{o}^{\bar{x}} (2V(x))^{-1/2}\, dx) \tag{22}$$

This provides, to leading order in $\hbar$, an explicit prescription for computing ΔE from the potential function. Note that, the transcription $\hbar \leftrightarrow 1/\kappa$ holds for D-scaling to this order.

Application to H_2^+ in Spheroidal Coordinates

The application to H_2^+ in spheroidal coordinates [5] involves special features; here we only outline the chief results. Already in the $D \to \infty$ limit, the instanton potential is given by

$$V(x) = V_o \frac{[1 - (x/a)^2]^2}{(1 - x^2)^2} \tag{23}$$

with $x = \mu$ and $V_o = \frac{1}{2}\frac{a^4}{(1-a^2)^2}$. Except for the denominator, $(1-x^2)^{-1}$, this function resembles the double-quartic potential . However, the shape is strongly affected by the constraint $\mid x \mid \le 1$. This imposes sharp cusps at the minima, especially when $a \to 1$, as occurs for large R. Both the barrier height, $V_o = \frac{1}{4}\omega a^3$, and the harmonic frequency, $\omega = \frac{2a}{(1-a^2)^2}$, then grow very large.

From Eq.(20) the instanton action is given by

$$S_o = \frac{2a}{(1 - a^2)} - \ln \frac{(1 + a)}{(1 - a)} \tag{24}$$

This simple result is in full agreement with the action evaluated from the exact numerical results [12]. However, in evaluating the prefactor A from Eq.(21) we encounter an instinctive difficulty. The requisite quadrature is readily carried out and gives

$$K = 2a \exp[\omega a(1 - a^2)] \tag{25}$$

This K factor grows very large when $a \to \pm 1$, as occurs for large R. The unphysical divergence occurs because the usual instanton technique comes to grief for a potential with sharp, nonquadratic cusps [13]. The instanton path then does not linger long enough near the classical minima and the fluctuations diverge while approaching these minima. We find that this malady is easily cured by rescaling the time variable. As suggested by a procedure applied to H_2^+ classical trajectories [14], we define a new time variable τ by $dt = \frac{R^2}{4}(\lambda^2 - \mu^2)d\tau$. This rescaling is required to conform to the separation of variables in the Hamiltonian, which involves multiplying the original Schrödinger equation by the factor $\frac{R^2}{4}(\lambda^2 - \mu^2)$. The net effect is that in the quadrature of Eq.(22) can use a weighted potential,

$$V(x) = V_o[1 - (x/a)^2]^2 \tag{26}$$

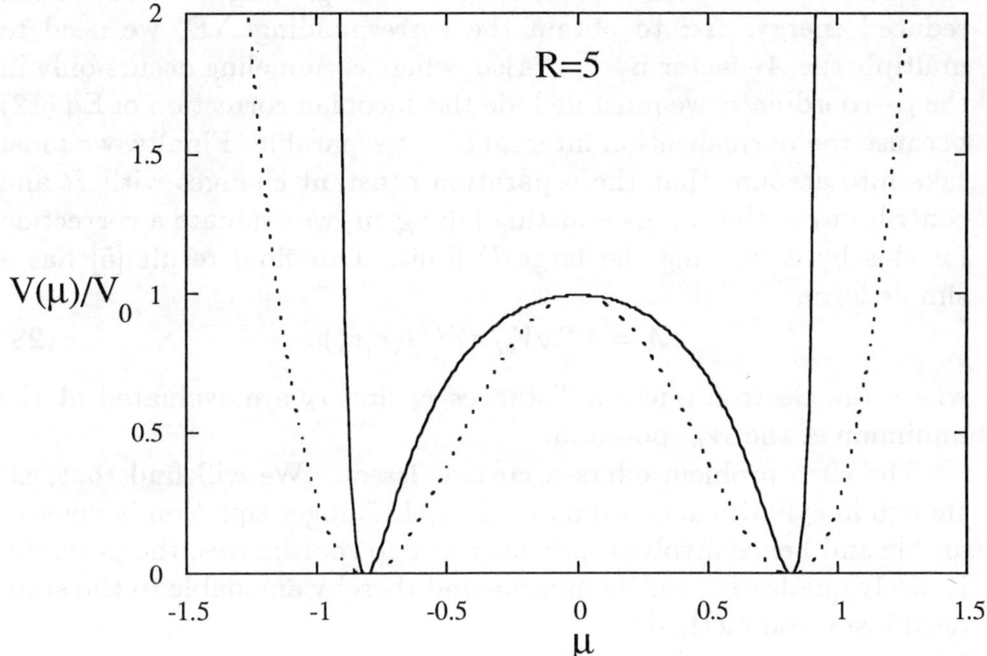

Figure 3. Comparison of double-minimum potentials $V(\mu)$ for the $D \to \infty$ limit at $R = 5$ evaluated without (full curve) and with (dotted curve) rescaled time transformation.

where V_o is the same as in Eq.(23). This is a simple double-quartic potential, for which ΔE can be evaluated exactly [5]. Rescaling the time thus removes the factor $(1 - x^2)^{-2}$ that appeared in Eq.(23) and produced the sharp cusps which induced divergence of the fluctuations about the instanton path. Figure 3 compares the double-minimum potential $V(x)$ at $R = 5$ evaluated without and with the rescaled time transformation. Rescaling the time does not affect the calculation of the action S_o, whereas we now obtain simply $K = 2a$ from the quadrature. Accordingly, we find from Eq.(21) that the fluctuation factor is given by

$$A = 4a\omega(\omega/\pi)^{1/2} \tag{27}$$

Some adjustments are required to obtain our final result. Since our calculation began with Eq.(6), it gives the splitting in terms of the reduced energy, $\Delta\epsilon$; to obtain the corresponding ΔE, we need to multiply the $A-$factor by $\frac{4}{R^2}$. Also, whereas tunneling occurs only in the $\mu-$coordinate, we must include the Jacobian correction of Eq.(12) because the normalization integral is not separable. Finally, we must take into account that the separation constant changes with R and contributes to the energy splitting [5]; again, we evaluate a correction for this by employing the large-D limit. Our final result [5] has a simple form,

$$A = 4(2\omega V_o/\pi)^{1/2}/(r_a^2 r_b^2)_m \tag{28}$$

where the electron-nucleus distances r_a and r_b are evaluated at the minimum of the W_∞ potential.

The cusp problem offers a curious lesson. We will find that, although in cylindrical coordinates the Schrödinger equation is nonseparable and hence involves tunneling in two coordinates, the potential is nicely quadratic near its minima and thereby amenable to the standard instanton methods.

Instanton Method for Two Degrees of Freedom

The ground state energy splitting given by the instanton method has the general form of Eq.(19) regardless of the number of dimensions or degrees of freedom. However, to evaluate the instanton action for a multivariable potential, we must find a particular zero energy path between the maxima of the inverted potential. For H_2^+ in cylindrical coordinates we deal with two degrees of freedom. At a fixed internuclear distance R beyond the critical point for symmetry breaking, the instanton path can be determined by finding $\rho(z)$ such that it minimizes the integral

$$S_o = \int_{-z_m}^{z_m} F(\rho, z)[1 + (\partial\rho/\partial z)^2]^{1/2}\, dz \tag{29}$$

where $F(\rho, z) = [2V(\rho, z)]^{1/2}$ and $V(\rho, z) = W(\rho, z) - W(\rho_m, z_m)$, with W the effective potential of Eq.(2) for the $D \to \infty$ limit. The subscript m refers to the potential minima, located at $(\rho_m, \pm z_m)$. From the calculus of variations [15], we have derived an efficient al-

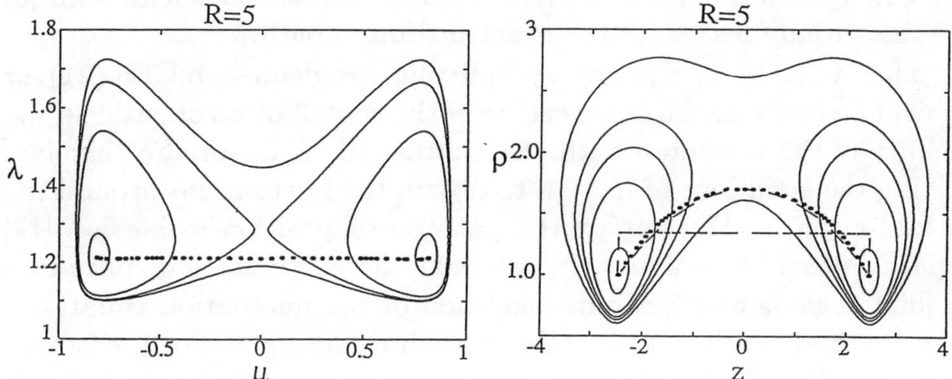

Figure 4. Instanton path for tunneling between pairs of minima in effective potential at $D \to \infty$ limit for H_2^+ with $R = 5$. Panel at left show contour maps of effective potential in spheroidal coordinates, at right maps in cylindrical coordinates. Heavy dots show exact instanton paths; long-dashes indicate parabolic approximation; short-dashes straight-line approximation.

gorithm for numerical evaluation of the instanton path and action [6].

Figure 4 shows contour plots of the effective potential and compares the exact instanton path obtained from our algorithm with the straight-line approximation [16] and with a parabolic approximation.

We see that the exact path deviates markedly from the straight-line but does closely resemble the parabolic approximation. This result is readily understood

as a consequence of the separability in spheroidal coordinates. The action computed for tunneling in two degrees of freedom (cylindrical coordinates) proves to agree closely with the analytical result of Eq.(24) obtained for one degree of freedom

(spheroidal coordinates). In the range examined, the action is a nearly linear function of R; relative to the result for the exact instanton path, as R increases the action for the straight-line approximation path becomes too large and that for the parabolic path too small. The agreement with the action derived from the numerical calculations [6] is substantially better for the exact instanton path.

The A-factor in the energy splitting, as defined by Eq.(21) or equivalent expressions, is a measure of the contribution of paths in the vicinity of the instanton path. Evaluating the A-factor thus involves solving the equations of motion to obtain the fluctuations around the instanton path. We employ the path decomposition expansion [17] a path integral technique which breaks the configuration space into disjoint regions and facilitates solution of the fluctuation equations. This method specifies two surfaces, each enclosing one of the potential wells. For H_2^+ in cylindrical coordinates the wells have quadratic minima and we can use the harmonic oscillator wavefunctions within the enclosed regions. With this approximation, the net result for the pre-exponential factor takes the form

$$A = 2\Lambda(\omega_a\omega_b/\pi\kappa)^{1/2} \tag{30}$$

where ω_a and ω_b are the harmonic frequencies for vibrational normal modes at the minima of the effective potential and $\kappa = (D-1)/2$ is again the dimensional scaling factor. The factor Λ comes from numerical integration of the fluctuation equations [6]. Agreement with the accurate numerical results is less good than for the action integral. However, Eq.(30) from the instanton treatment in two degrees of freedom (cylindrical coordinates) proves to be appreciably better than the analytical formula of Eq.(28) obtained for one degree of freedom (spheroidal coordinates). Both instanton versions are much better than the straight-line approximation. Figure 5 compare the tunneling splitting obtained from the asymptotic, instanton, and uniform semiclassical approximations [14] with practically exact numerical results [4,12].

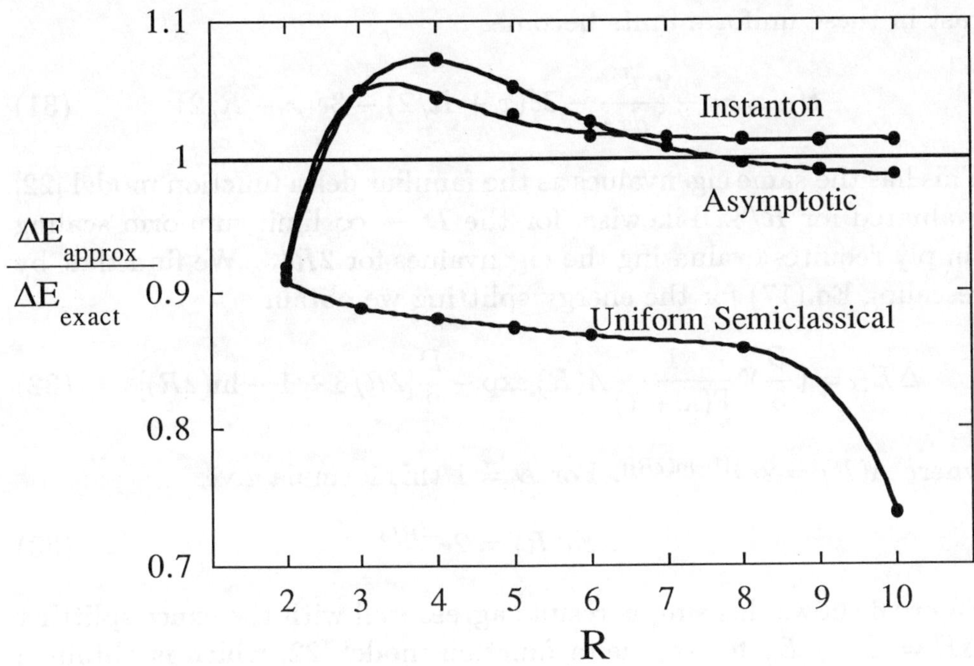

Figure 5. Comparison of tunneling splitting from the asymptotic, instanton, and uniform semiclassical approximations with practically exact numerical result.

Tunneling and Uniform Scaling

In Eq.(1) and subsequently, we used units of κ^2/Z bohr radii for distance and Z^2/κ^2 hartrees for energy. This scaling ensures that the energy remains finite for $D \to \infty$, but not for the $D \to 1$ limit. In the treatment of atoms, the $D \to 1$ limit has prove extremely useful [1,2]; combined with the $D \to \infty$ limit, this allows quite accurate results for $D = 3$ to be obtained by resummation or interpolation techniques [18]. Finite solutions can be obtained by use of uniform scaling [19]. This again employs Z^2/κ^2 hartrees for energy, but uses $(D/3)\kappa$ bohr radii for distances. With this adjustment, it is found that interpolation between the $D \to \infty$ and $D \to 1$ limits yields a good first approximation for the ground state H_2^+ potential energy curve [3,19,20]. The electronic Hamiltonian for the $D \to 1$ limit when cast in these uniform units becomes

$$H_{D\to1} = -\frac{9}{2}\frac{d^2}{dx^2} - 3\delta(x + R/2) - 3\delta(x - R/2) \tag{31}$$

This has the same eigenvalues as the familiar delta function model [22] evaluated for $R/3$. Likewise, for the $D \to \infty$ limit, uniform scaling simply requires evaluating the eigenvalues for $2R/3$. We find that by rescaling Eq.(17) for the energy splitting we obtain

$$\Delta E_D = (\frac{D}{3})^\kappa \frac{1}{\Gamma(\kappa + 1)} A(R) \exp -\frac{D}{2}[2R/3 + 1 - \ln(2R)] \tag{32}$$

where $A(R) = 2e^{\frac{1}{2}[1-\ln(2R)]}$. For $D = 1$ this formula gives

$$\Delta E_1(R) = 2e^{-R/3} \tag{33}$$

Figure 6 shows this simple results agrees well with the exact splitting $\Delta E = E_- - E_+$ for the delta function model [22] which is obtained by solving numerically a pair of transcendental equations,

$$E_\mp = -\frac{1}{2}(1 \pm e^{-\sqrt{(-2E_\mp)}R/3})^2. \tag{34}$$

Discussion

For a wide range of internuclear distance, our results for ΔE, which pertain to the leading order in $1/D$, gives good agreement with comparable semiclassical methods as well as with exact numerical calculations.

The instanton method is particularly congenial for the large-D limit, in that the zero-energy trajectory in imaginary time corresponds precisely to tunneling between minima in the effective potential with $D \to \infty$. Despite the infinite effective mass, tunneling still occurs because quantum fluctuations persist in this limit. As usual with semiclassical methods, the dynamical aspects are evaluated by classical mechanics. However, the unusual and striking aspect of our

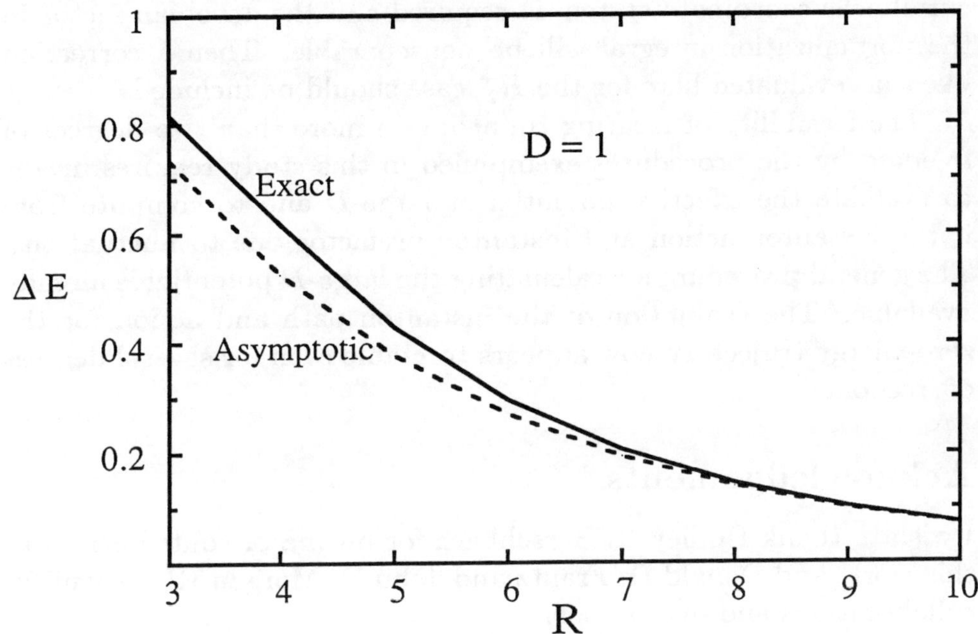

Figure 6. Comparison of tunneling splitting from the asymptotic approximation at $D = 1$ with the exact numerical result from the delta function model.

treatment of tunneling in the large-D limit is that the effective potential for the electronic motions is also calculated classically. The form of the effective potential changes markedly with D, and tunneling is very sensitive to the potential. Despite this, we find that the effective mass factors involving $\kappa = (D - 1)/2$ adequately describe the major D-dependence of the tunneling. Accordingly, we can exploit the simplifications available in the pseudoclassical large-D limit, yet obtain quantitative results in good agreement with conventional treatments for $D = 3$.

Another feature of general interest is the Jacobian correction to ΔE, which proves to be inversely proportional to R and thus is a significant factor. It is extremely easy to evaluate this factor in the large-D limit. Many treatments of tunneling assume or approximate the tunneling path to be a separable coordinate. However, usually a curvilinear coordinate system is required and the Jacobian factor in the normalization integral will be nonseparable. Then a correction like that evaluated here for the H_2^+ case should be included.

The feasibility of treating tunneling in more than two degrees of freedom by the procedures exemplified in this study requires means to evaluate the effective potential at large-D and to compute from it the instanton action and instanton prefactor due to fluctuations. The general procedure for calculating the large-D potential is already available. The evaluation of the instanton path and action for the zero-energy trajectory now appears tractable even for several degrees of freedom.

Acknowledgements

I wish to thank Dudley R. Herschbach for his major contributions to this work, and Donald D. Frantz and John D. Morgan III for fruitful collaborations and discussions.

References

1. D.R. Herschbach, Faraday Discuss. Chem. Soc. **84**, 465 (1988).
2. D.J. Doren and D.R. Herschbach, J. Phys.Chem. **92** , 1816 (1988).
3. D.D. Frantz and D.R. Herschbach, Chem.Phys. **126**, 59 (1988).
4. D.D. Frantz and D.R. Herschbach, J. Chem. Phys. **92**, 6668 (1990).
5. S. Kais, D.D. Frantz, and D.R. Herschbach, Chem. Phys. (in press, 1992).
6. S. Kais, J.D. Morgan III and D.R. Herschbach, J. Chem. Phys. (in press, 1991).
7. C. Herring, Rev. Mod. Phys. **34**, 631 (1962); L.D. Landau and E.M. Lifshitz, *Quantum Mechanics* (Academic Press, New York, 1965); R.J. Damburg and R.Kh. Propin, J. Phys. **B1**, 681 (1968).
8. J.S. Langer, Ann. Phys. **41**, 108 (1967); W.H. Miller, J. Chem. Phys. **55**, 3146 (1971); G. t'Hooft, Phys. Rev. Lett. **37**, 8 (1976); A.M. Polyakov, Nucl. Phys. **B 121**, 429 (1977).

9. E.M. Harrele, Commun. Math. Phys. **75**, 239 (1980).

10. S. Coleman, Phys. Rev. **D15**, 2929 (1977); C.G. Gallen and S. Coleman, Phys. Rev. **D16**, 1762 (1977); *The Uses of Instantons, in The Whys of Subnuclear Physics*, A. Zichichi Ed. (Plenum, New York, 1979); S. Coleman, *Aspects of Symmetry* (Cambridge Univ. Press, 1985), pp. 265-350.

11. H.J. DeVega, J.L. Gervais, and S.Sakita, Nucl. Phys. **B 139**, 20 (1978).

12. D.D. Frantz, D.R. Herschbach, and J.D. Morgan III, Phys. Rev. **A40**, 1175 (1989) and work cited therein.

13. P. Van Baal and A. Auerbach, Nucl. Phys. **B275**, 93 (1986).

14. M.P. Strand and W.P. Reinhardt, J. Chem. Phys. **70**, 3812 (1979).

15. R. Weinstock, *Calculus of Variations* (McGraw-Hill, New York 1963)

Chapter 7

GENERAL COMPUTATIONAL STRATEGIES

7.1 Singularity Analysis and Summation of 1/D Expansions

David Z. Goodson and Mario López-Cabrera
Department of Chemistry
Harvard University
12 Oxford Street
Cambridge, MA 02138, USA

Abstract

The poor convergence of the partial sums of the $1/D$ expansion for atomic and molecular energies is due to the fact that the energy function $E(\delta)$, $\delta \equiv 1/D$, is not a polynomial. The large-D limit appears to be an excellent qualitative model, but in order to accurately continue the solution from $\delta = 0$ to $\delta = 1/3$ it is necessary to take into account the functional form of $E(\delta)$ in that general region of the complex plane. The most important feature in this functional form is a second-order

D. R. Herschbach et al. (eds.), Dimensional Scaling in Chemical Physics, 275–314.

pole at $\delta = 1$. Incorporating this pole into the form of the summation approximants, using either of three methods that we describe, accounts for over 99% of the solution at $\delta = 1/3$ for ground-state energies of H_2^+ and helium. Almost all the remaining error can be accounted for by including a rather complicated singularity at $\delta = 0$. Through an analysis of the large-order behavior of the δ expansion, we construct a functional form for this singularity. The functional form suggests that the energy expansion can be summed most effectively using Borel summation, with the integrand of the Borel sum approximated by Padé approximants or, better yet, by approximants in the form of Darboux functions. We present numerical results that support this conclusion.

Introduction

The theoretical description of atoms and molecules has to rely on approximate solutions to Schrödinger's equation. For the standard methods in current use, the starting approximation treats the electrons as if they were independent particles. The advantage of this approach is the ease with which it can be formulated even for very large systems [1]. However, the correlation of electronic motion often has a major role, particularly in chemical bonding and reactivity. The independent-electron approximation does not provide a qualitative model for correlation effects, nor an efficient basis for evaluating numerical contributions from correlation.

The large-dimension limit offers an approach that explicitly includes electron correlation [2,3]. It seems to give an excellent qualitative description of correlation effects that are difficult to understand in terms of independent-electron models [3,4]. This is because all the coulombic forces are retained in their exact form. The dimensionality of space is treated as an arbitrary and continuous parameter, D. For any atom or molecule, an exact solution can be obtained in the limit $D \to \infty$. The problem then becomes how to exploit this limiting result in constructing an approximate solution for the actual physical system with $D = 3$. Our aim is twofold: to find an efficient way to calculate accurate expectation values, and to develop reliable qualitative interpretations of the $D \to \infty$ results that elucidate the correlation effects at $D = 3$.

The most straightforward way to evaluate corrections to the large-dimension limit is to compute a $1/D$ expansion [5]. The energy eigenvalues each have a unique asymptotic expansion in the form

$$E = D^{-2} \sum_{k=0}^{\infty} E_k D^{-k}. \tag{1}$$

The exact solutions for the expansion coefficients, E_k, are given by a straightforward recursion relation [6-8]. Unfortunately, the convergence properties of this expansion are very poor. Eq. (1) is not useful in practice for accurate numerical calculations.

Let us consider the energy to be a function $E(\delta)$ of $\delta \equiv 1/D$. The prefactor D^{-2} in Eq. (1) is an overall scaling, which can be absorbed into the energy units. Then the leading term, $E_0 \equiv E(0)$, is given exactly by the large-dimension limit. We wish to study the behavior of $E(\delta)$ as δ goes from 0 to 1/3. Eq. (1), when truncated at any finite order, expresses E as a polynomial in δ. Its failure indicates that the true energy function is not a polynomial. Of course, any *analytic* function can be approximated to arbitrary precision through a Taylor series. The failure of Eq. (1) can therefore be more precisely attributed to the presence of singularities in E in the complex δ plane.

Our strategy is to locate and characterize these *dimensional singularities*, and then infer from them an approximate functional form for $E(\delta)$. We then design summation methods for the energy expansion that incorporate the singularities. In this paper we report on our recent attempts to implement this strategy [7,9-12]. We consider the energy expansions of two prototypical systems: the ground state of the hydrogen molecule cation and the ground state of the two-electron atom. Our general conclusions are as follows: 1) $E(\delta)$ has poles at $\delta = 1$ and a rather complicated singularity at $\delta = 0$. 2) A functional form for the singularity at the origin that is consistent with the large-order behavior of the δ expansion can be constructed from the expansion coefficients using methods based on the Borel sum. 3) At very low order in the expansion the convergence behavior is dominated by the poles at $\delta = 1$; the most effective summation methods at low order appear to be methods that concentrate on fitting these poles, without attempting to fit the singularities at the origin. 4) At medium to large order it is necessary to include the singularity at $\delta = 0$ as well as the poles at $\delta = 1$; we recommend in particular the use of summation methods based on the Borel sum, which can yield an accuracy of greater than eight significant figures.

Dimensional Scalings

As a preliminary, we here briefly discuss the question of dimensional scalings. The Schrödinger equation can be solved exactly at $\delta = 0$ and at $\delta = 1$; however, these solutions will be useful models for the physical solution at $1/3$ only if appropriate dimension dependent scale factors are introduced into the wavefunction and into the units of energy and distance [9,13-16]. For eigenstates with zero total angular momentum, the scaling of the wavefunction allows us to write the Schrödinger equation in the form

$$(T + D^2 U + V - E)\Psi = 0, \tag{2}$$

where V is the usual potential energy and $T + D^2 U$ is the kinetic energy. (For states with nonzero angular momentum the equation is more complicated to write down, but the analysis that follows remains essentially the same [17].) T is an effective kinetic energy operator that consists of terms proportional to second derivatives, while $D^2 U$ is a centrifugal potential. T and V are independent of D and U has only a weak dimension dependence that goes to zero as D^{-1} at large D. Therefore,
Eq. (2) does not provide a useful large dimension limit, since the centrifugal term becomes infinite.

T and U depend on distance to the power -2, while V depends on distance to the power -1. Consider the two-electron atom. Our coordinates are r_1, r_2 and θ and the potential energy is

$$V = Z\left[r_1^{-1} + r_2^{-1} + \lambda(r_1^2 + r_2^2 - 2r_1 r_2 \cos\theta)^{-1/2}\right], \tag{3}$$

where Z is the nuclear charge and $\lambda = 1/Z$. Suppose we change the distance units through the substitution $r_i \rightarrow (D^2/Z)r_i$ and then multiply Eq. (2) by D^2/Z^2. Then the Schrödinger equation becomes

$$(\delta^2 T + U + \tilde{V} - \tilde{E})\Psi = 0, \tag{4}$$

where $\tilde{V}$ is identical to V of Eq. (3) except with the factor of Z omitted and $\tilde{E} = (D^2/Z^2)E$. The scaling by Z is simply a matter of convenience. The D^2 scaling, however, is significant; it causes Eq. (4) to have the desired large-dimension limit of the vibrating Lewis structure. Any scale factor that is proportional to D^2 in the limit of large D will have this effect. In practice, it is usually more convenient to replace D in the scale factors with $(D/3)$, which goes to unity at $D = 3$.

The one-dimension limit of Eq. (2) as written is also not well-defined. The problem now is that the expectation value $\langle V \rangle$, and hence the energy, becomes infinite at particle coalescences [18], as we discussed in Chapter 4.1. $\langle V \rangle$ can be rendered finite through a distance scaling $r_i \to [(D-1)/Z]r_i$. The Schrödinger equation becomes

$$[T + D^2 U + (D-1)\tilde{V} - \bar{E}]\Psi = 0 \tag{5}$$

where $\bar{E} = [(D-1)^2/Z^2]E$. The factor $(D-1)$ exactly cancels the divergence of $\langle \tilde{V} \rangle$ so that the scaled eigenvalue $\bar{E}$ remains finite in the limit $D \to 1$. Any other scale factor proportional to $(D-1)$ in the one-dimension limit would work as well.

For atoms, these dimensional scalings amount to nothing more than a simple change of variables. We are free to use one scaling at $\delta = 0$ and another at $\delta = 1$; no inconsistency results from this. In the case of molecules, however, there is an additional subtlety, due to the Born-Oppenheimer approximation. For example, the potential energy for H_2^+, in cylindrical coordinates, is

$$V = -[\rho^2 + (z - \tfrac{1}{2}R)^2]^{-1/2} - [\rho^2 + (z + \tfrac{1}{2}R)^2]^{-1/2}, \tag{6}$$

where z is the distance along the internuclear axis, measured from the midpoint, ρ is the perpendicular distance from the axis, and R is the internuclear distance. Within the Born-Oppenheimer approximation R is treated as a parameter, constant with respect to the electronic coordinates. Now suppose we introduce a dimensional scaling $z \to fz$ and $\rho \to f\rho$, where f is some function of D. Then

$$V = f^{-1}\left\{-[\rho^2 + (z - \tfrac{1}{2}R/f)^2]^{-1/2} - [\rho^2 + (z + \tfrac{1}{2}R/f)^2]^{-1/2}\right\}. \tag{7}$$

In order to obtain well defined limits we choose $f \propto D^2$ as $\delta \to 0$ and $f \propto D-1$ as $\delta \to 1$, as with the atom. Note, however, that in the former limit the terms R/f in Eq. (7) go to zero and in the latter limit they become infinite. Thus, the large-dimension limit we obtain in this way is equivalent to what we would have found if we had set $R = 0$ at the outset, and the one-dimension limit is equivalent to what we would have found had we set R infinite. $R = 0$ corresponds to the atomic ion He^+ and infinite R corresponds to a single hydrogen atom, neither of which is a particularly good model for H_2^+.

More useful dimensional limits can be obtained if one treats R/f as a parameter that is independent of dimension [19]. Thus, we replace R/f in Eq. (7) with a parameter $\tilde{R}$, which we assume to be dimension-independent, whose numerical value we choose to equal R/f using the value for f that corresponds to $D = 3$. Complications of this type can be expected in general when dimensional continuation is applied to systems that are not purely coulombic. Previous studies of model central potentials [20,21] and of atoms in an external field [9,22,23] have resorted to this technique of parameter scaling, with good results. Of course $\tilde{R}$ is truly dimension-dependent, whether or not we admit it, and this fact can reassert itself by affecting the convergence of the $1/D$ expansion. (We will briefly address this point in the next Section.) Nevertheless, the dimensional limits for H_2^+ obtained in this way prove to be an excellent qualitative model for the three-dimensional solution [19], so we will use parameter scalings here.

However, this approach introduces an important additional consideration that is not present in the treatment of atoms [9,24]. With the molecule we are no longer free to choose different scalings at the two limits. The scaling of the distance units is no longer just a change of variables, but also implies a particular dimension dependence in the unscaled R. Each choice of f implies a different function $E(\delta)$. For $\delta = 1/3$ the two functions will have the same numerical value, but elsewhere they will differ. Therefore, at both limits we must choose the scaling

$$f \propto D(D - 1), \tag{8}$$

which goes as D^2 when $\delta \to 0$ and as $D - 1$ when $\delta \to 1$. This restriction on the scale factor is due to the Born-Oppenheimer approximation. (Dimensional perturbation theory is not as dependent on the Born-Oppenheimer approximation as are independent-electron approaches [16], and if R is treated as a dynamical variable, then it is not necessary to use the same scale factor at the two dimensional limits.)

In what follows, we will express the energy of the two-electron atom in units of hartree atomic units multiplied by Z^2/D^2. Instead of Eq. (1), which gives the form of the energy expansion for the unscaled energy, we now have

$$E = \sum_{i=0}^{\infty} E_i \delta^i. \tag{9}$$

For H_2^+ we will use the energy expansion calculated [9,10] using the scale factor $D(D-1)/6$. (The factor $1/6$ is included for the sake of convenience, so that $\tilde{R} = R$ when $D = 3$.) However, we have expanded the factor of $D(D-1)$ in the energy units in powers of δ so that this expansion will also have the form of Eq. (9) with energy units of D^{-2}.

Dimensional Singularities

A characteristic feature of $E(\delta)$ for systems with coulomb potentials is the presence of a second-order pole at $\delta = 1$. Indeed, the exact solution for the ground state of the one-electron atom is simply [25]

$$E(\delta) = -2(1 - \delta)^{-2}. \tag{10}$$

These poles can have a strong effect on the convergence of the δ expansion.

Consider the sequence of approximants S_n,

$$S_n(\delta) = \sum_{i=0}^{n} E_i \delta^i, \tag{11}$$

which are called the *partial sums* of the δ expansion. We would like for the S_n to steadily converge to E as n increases. Unfortunately, one finds in practice that there are values of δ for which the partial sums converge very slowly, or not at all, due to the presence of singularities in the function $E(\delta)$ in the complex δ plane. Suppose that $E(\delta)$ is analytic within some closed region of the complex plane that includes the origin. Then the values of δ for which the partial sums ultimately converge comprise the region of the complex plane bounded by a circle centered at the origin. The radius of this circle is called the *radius of convergence.* Taylor's theorem states that the radius of convergence is equal to the distance in the complex plane between the origin and the nearest singularity of $E(\delta)$ [26]. If there is no nonzero δ at which the partial sums converge, then the radius of convergence is said to be zero; this would imply the presence of a singularity at $\delta = 0$. For the energy of the one-electron atom, Eq. (10), the nearest singularity to the origin is a second-order pole at $\delta = 1$, so the radius of convergence is 1 and the physical solution $\delta = 1/3$ is within the circle of convergence. However, $1/3$ is sufficiently close to the circle of convergence that the convergence is quite slow.

The origin of the pole in Eq. (10) has been explained by Doren and Herschbach [18] in terms of an analysis of the dimension dependence of the Schrödinger equation at particle coalescences. They developed a systematic technique that allows one to predict the locations and types of a certain class of dimensional singularities without having to actually solve for $E(\delta)$ [18,27]. (We have described this technique in some detail in Chapter 4.1.) For problems with coulomb potentials, the analysis predicts singular behavior at $\delta = 1$ of the form

$$E(\delta) \sim \frac{a_{-2}}{(1 - \delta)^2} + \frac{a_{-1}}{1 - \delta}. \tag{12}$$

We call these singularities *coulombic poles*, since they result from the divergence of expectation values of the coulomb potentials [18]. In the case of the one-electron atom the residue a_{-1} is zero, but for the ground states of helium and H_2^+ it is nonzero.

If the Doren-Herschbach singularity analysis for helium and for H_2^+ is carried out with the wavefunction expressed as a linear combination of hydrogenic basis functions, then other singularities are also predicted [18]. In addition to the poles in Eq. (12), one finds as well poles at $\delta = -(2i - 1)^{-1}$, $i = 1, 2, 3, \ldots$. This is an infinite sequence of poles between -1 and 0 converging at 0, which would seem to imply the existence of an essential singularity at $\delta = 0$. If so, then the radius of convergence of the δ expansion for these systems is zero and the partial sums diverge for any nonzero value of δ.

This analysis, however, is not mathematically rigorous. Even if a truncated basis expansion for the wavefunction leads to an accurate numerical value for the energy, it will not necessarily give the correct functional dependence of the energy on other parameters in the problem. It is easy to construct counterexamples in which this approach yields incorrect results [27,28].

There is in fact a singularity in $E(\delta)$ at $\delta = 0$ for the ground states of H_2^+ and helium, but the characterization of the singularity and the determination of its cause is less straightforward than was the case for the coulombic poles. In what follows we will introduce several techniques for analyzing this singularity. Our immediate purpose is to better understand the nature of the singularity and its effect on the convergence properties of the δ expansion; in addition, however, each of these techniques will prove to be useful in the development of summation procedures in Section IV.

A. Large-Order Behavior of the δ Expansion

An advantage of the δ expansion over other types of perturbation theory used in electronic structure calculations is that it is relatively easy to calculate the expansion coefficients, the E_i of Eq. (9). They can be calculated exactly and to high order using a recursive algorithm in terms of moments of the coordinates [6-10]. E_i are now available for the following systems: the ground state of H_2^+ (within the Born-Oppenheimer approximation) for various values of the internuclear distance [9,10], several eigenstates of the two-electron atom [7,8,12], and the hydrogen atom in an external magnetic or electric field [22,23]. They have also been calculated for a wide variety of model central potentials [6,20,21,29-32]. In this paper we will consider the ground-state of H_2^+, a prototype for molecules, and the ground-state of the helium atom, a prototype for many-electron systems. In the case of H_2^+, we will use the δ expansion calculated with the unscaled internuclear distance $R = 1$, which implies the scaled internuclear distance $\tilde{R} = D(D-1)/6 = 1$, as discussed in Section II. The application of dimensional perturbation theory to excited states of helium is described in Chapter 8.1.

The simplest method for determining the radius of convergence, r_c, of an asymptotic expansion is the ratio test: If the ratios $R_i \equiv$

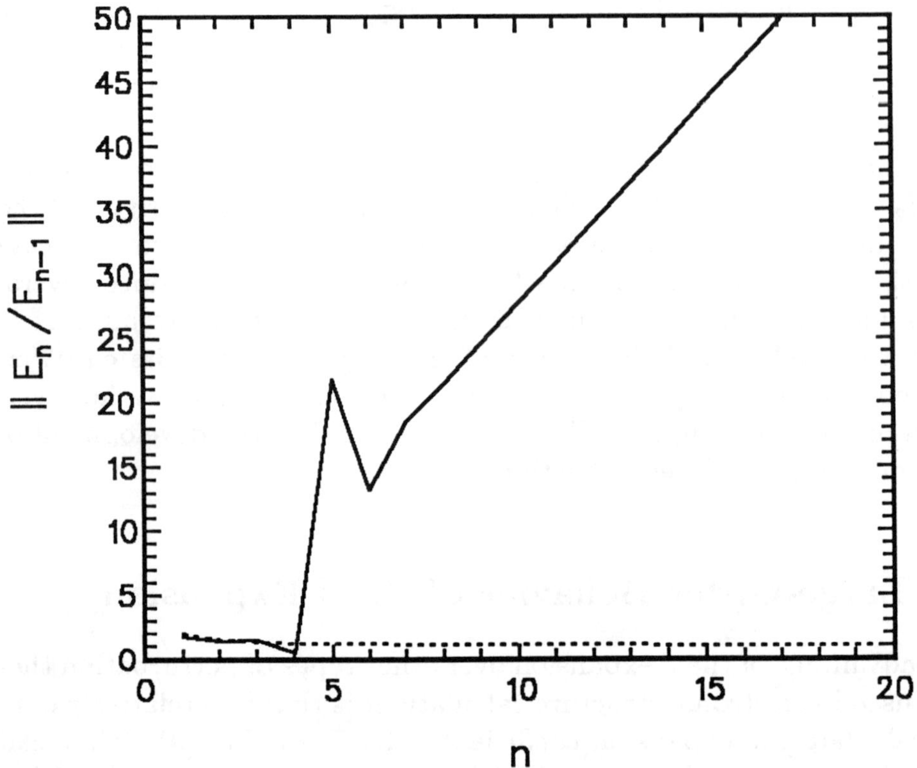

Figure 1. Ratio test for expansion coefficients of the ground state energy of H_2^+ with $R = 1$ (solid curve) and the ground state energy of the hydrogen atom (dotted curve).

$|E_i/E_{i-1}|$ converge in the limit of large i, then their limit is $1/r_c$,

$$\lim_{i \to \infty} |E_i/E_{i-1}| = r_c^{-1}. \tag{13}$$

Fig. 1 shows the ratios, R_i, plotted as a function of i for H_2^+. For comparison, the ratios for the ground state of the H atom are also shown, as a dotted curve. We know from Eq. (10) that for the latter case we have the finite nonzero radius of convergence $r_c = 1$, and we see from the Figure that the R_i rapidly converge to $1/r_c$. At low order the ratios for the molecule also seem to converge, and in fact are almost identical to those for the hydrogen atom, as if r_c were determined by the coulombic poles at $\delta = 1$ according to Eq. (12). However, the apparent convergence is only asymptotic; at high orders the R_i diverge to infinity, implying $r_c = 0$.

A zero radius of convergence from the ratio test implies the *existence* of a singularity at $\delta = 0$, but it does not indicate the nature of the singularity. In what follows, our basic aim will be to determine a functional form that describes $E(\delta)$ in the neighborhood of the singularity. This would allow us, in effect, to sum the highly divergent large-order behavior, leaving for the remainder an asymptotic series that has better convergence properties.

It is interesting to note that the divergence of the R_i for H_2^+ seems to be linear in i. A linear rate of increase in the R_i implies that $E_i \sim i!$. This factorial divergence appears to be a characteristic feature of δ expansions for ground states; it is the large-order behavior that has been found for most systems whose expansion coefficients have been calculated to high enough order [20,32]. Factorial divergence is not at all uncommon for perturbation theories of practical interest.

We consider next the large-order behavior for helium. Fig. 2 shows the R_i for this case, with the results for hydrogen given by the dotted curve. At low order, once again, the ratios are quite similar to those for hydrogen, converging toward $r_c^{-1} = 1$, but at larger orders the helium ratios are much more difficult to characterize than the H_2^+ ratios. Fig. 3 shows the results of the nth root test, which gives the radius of convergence in terms of the limit of the nth roots of the expansion coefficients,

$$\lim_{n \to \infty} |E_n|^{1/n} = r_c^{-1}. \tag{14}$$

This test tends to be more robust than the ratio test. The solid curve in Fig. 3 corresponds to helium, the dashed curve to H_2^+, and the dotted curve to hydrogen. It would seem from this Figure that the large-order behavior of helium is qualitatively similar to that of H_2^+ except that an oscillation with period of approximately 10 is superimposed on the helium curve. For other values of Z or R the results are similar except that the slopes of the linear increase shift somewhat [9,10,12]. Note that the large-order behavior n^n implied by a linear result from the nth root test is equivalent, according to Stirling's formula, to the $n!$ behavior from the ratio test as long as n is sufficiently

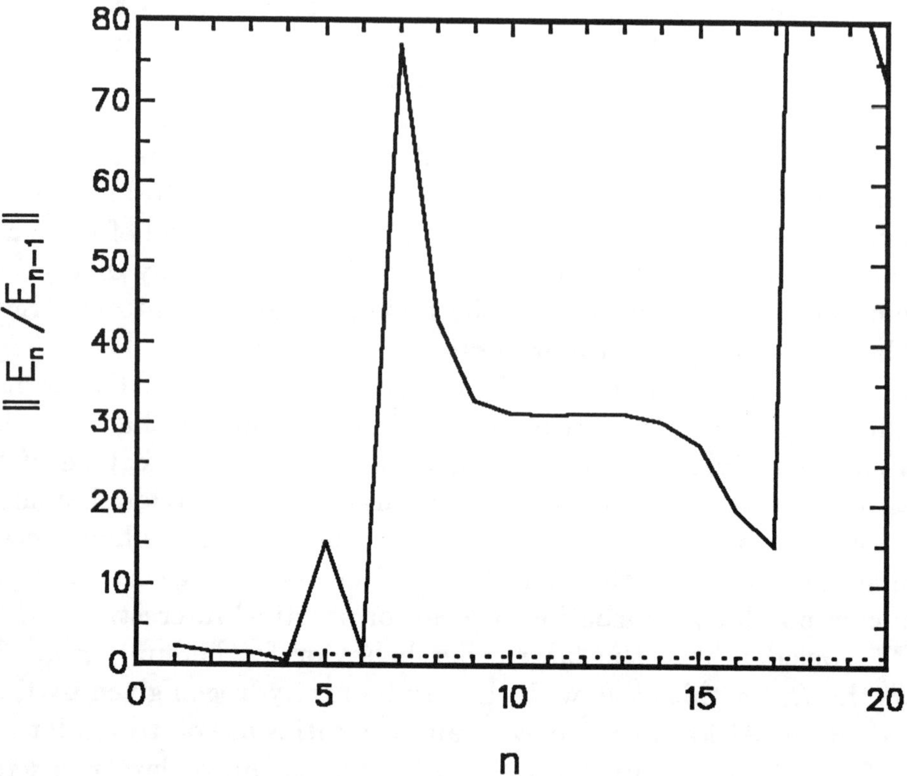

Figure 2. Ratio test for expansion coefficients of the ground state energy of helium (solid curve) and the ground state energy of the hydrogen atom (dotted curve).

large.

B. Padé Approximants

A *Padé approximant* is an approximant to a power series in the form of a ratio of polynomials,

$$S_{[L/M]}(\delta) = \frac{p_0 + p_1\delta + p_2\delta^2 + \cdots + p_L\delta^L}{1 + q_1\delta + q_2\delta^2 + \cdots + q_M\delta^M}. \tag{15}$$

The p_i and q_i are assigned by expanding Eq. (15) in powers of δ and then equating it order by order with the energy expansion, Eq. (9). L

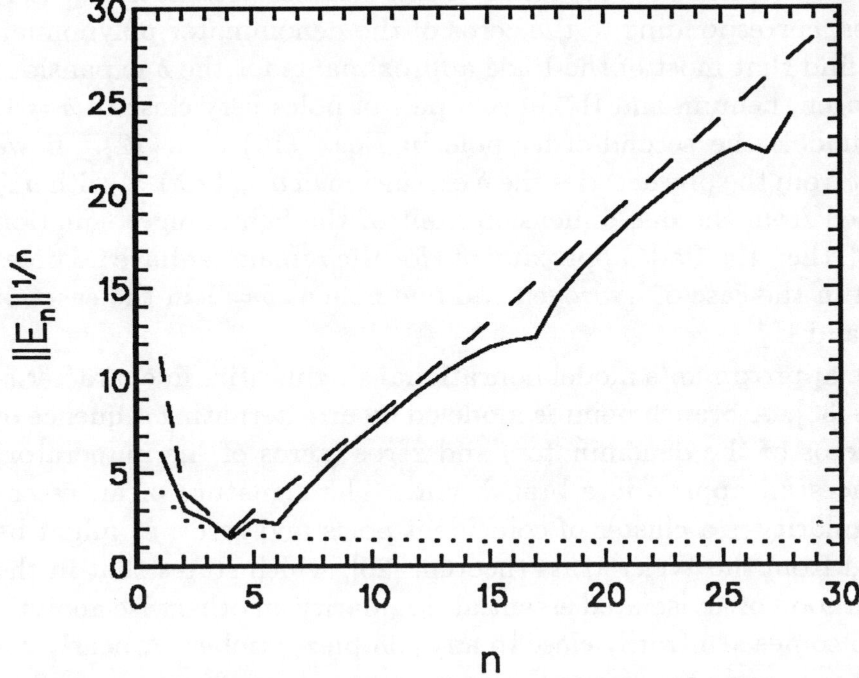

Figure 3. nth-root test for expansion coefficients of the ground state energy of helium (solid curve), the ground state energy of H_2^+ with $R = 1$ (dashed curve), and the ground state energy of the hydrogen atom (dotted curve).

and M are nonnegative integers chosen such that $L + M = n$, where n is the order of the power series. The function $S_{[L/M]}(\delta)$ has poles at the zeros of the denominator polynomial. Thus, Padé approximants provide a simple method for constructing approximants that have some flexibility for describing functions with singularities.

The usefulness of Padé approximants for summing δ expansions will be discussed in Section IV. For present purposes, we will use them simply to gain information about the singularity structure of $E(\delta)$. The mathematical theory of the accuracy with which the singularity structure of a function is described by its Padé approximants is not well developed, although practical experience provides some general guidelines, as described, for example, in Baker's books on the subject

[33]. It is clear that the only kinds of singularities present in Eq. (15) are poles, corresponding to the zeros of the denominator polynomial, and we find that most of the Padé approximants for the δ expansions of hydrogen, helium, and H_2^+ have a pair of poles very close to $\delta = 1$, which models the second-order pole in Eqs. (10) and (12). If we subtract from the power series the δ expansion of $a_{-2}(1-\delta)^{-2}$, with a_{-2} calculated from the one-dimension limit of the Schrödinger equation [9,10,18], then the Padé approximants for the remainder have no poles at $\delta = 1$ in the case of hydrogen and one pole at $\delta = 1$ in the cases of helium and H_2^+.

Padé approximants model nonrational singularities in characteristic ways [33]. A branch point is modeled by an alternating sequence of poles (zeros of the denominator) and zeros (zeros of the numerator) that traces an appropriate branch cut. The signature of an essential singularity is a cluster of coincident poles and zeros, as might be expected from the Weierstrass theorem [26], which states that in the neighborhood of an isolated essential singularity an otherwise analytic function comes arbitrarily close to any complex number. A nearly coincident pole and zero causes the approximate to go from $-\infty$ to 0 to $+\infty$ over a very small range of δ.

Figs. 4 and 5 show the singularity structure of the [12/13] Padé approximants for H_2^+ and helium, respectively. Poles of the approximants are designated by × and zeros are designated by o. The second-order pole at $\delta = 1$ is clearly discernible. There is also a structure on the negative real axis that seems to combine the features of a branch point and of an essential singularity: there is a cluster of coincident poles and zeros near the origin, which could suggest an essential singularity at $\delta = 0$, consistent with the prediction of Doren and Herschbach, but there is also a sequence of coincident or nearly coincident pole-zero pairs along the negative real axis, suggestive of a branch point. Finally, H_2^+ seems to have additional singularities in the negative half-plane, at approximately $-0.3 \pm 0.3\,i$.

We can rule out a simple square-root branch point for the singularity at the origin since that would lead to half-integer powers of δ in the asymptotic expansion, Eq. (9); the δ expansions of the wavefunction and of the Hamiltonian do have half-integer powers, but they cancel exactly leaving Eq. (9) as the correct form for the energy ex-

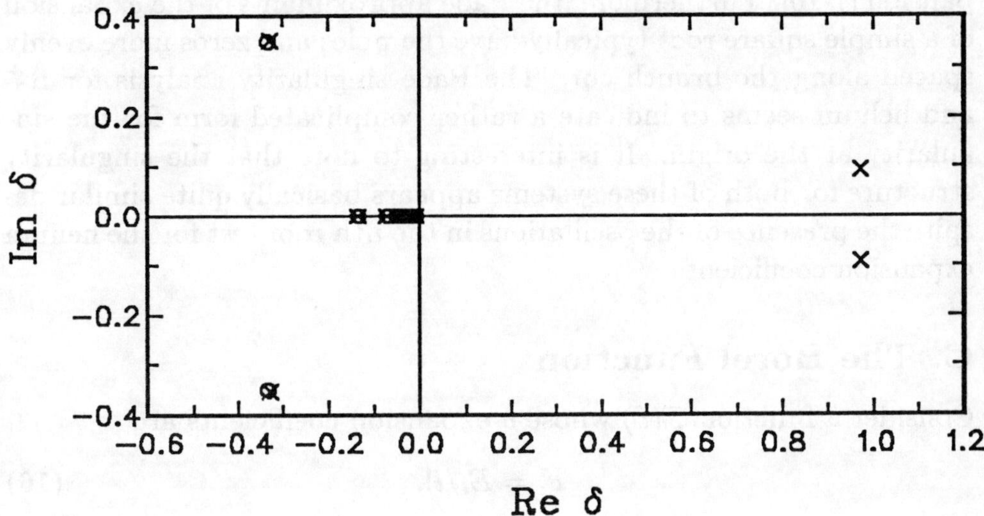

Figure 4. Poles ($\times$) and zeros (o) of the [12/13] Padé approximant for the ground state energy of H_2^+ with $R = 1$.

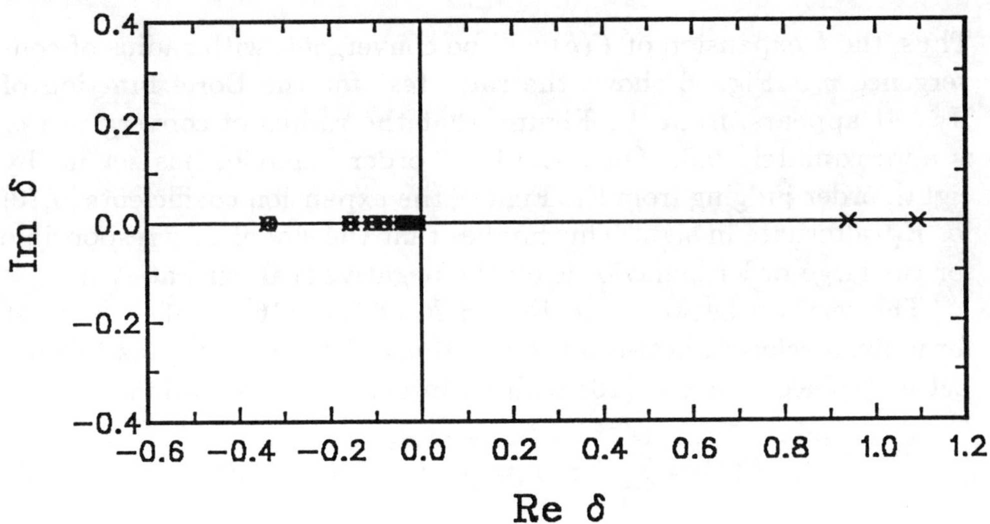

Figure 5. Poles ($\times$) and zeros (o) of the [12/13] Padé approximant for the ground state energy of helium.

pansion [7-10]. Furthermore, the Padé approximants of the expansion of a simple square root typically have the poles and zeros more evenly spaced along the branch cut. The Padé singularity analysis for H_2^+ and helium seems to indicate a rather complicated form for the singularity at the origin. It is interesting to note that the singularity structure for both of these systems appears basically quite similar despite the presence of the oscillations in the nth root test for the helium expansion coefficients.

C. The Borel Function

Consider a function $F(\delta)$ whose δ-expansion coefficients are

$$F_i = E_i/i!. \tag{16}$$

We will refer to F as a *Borel function* corresponding to E [34]. If the E_i increase as $i!$, as suggested by the ratio test for H_2^+, then the ratios $|F_i/F_{i-1}|$ will converge to a nonzero constant,

$$|F_i/F_{i-1}| = |E_i/(iE_{i-1})| \sim 1/p_c. \tag{17}$$

Thus, the δ expansion of $F(\delta)$ will be convergent, with radius of convergence p_c. Fig. 6 shows the ratio test for the Borel function of H_2^+. It appears, from the Figure, that the radius of convergence p_c is approximately 0.3. Once the large-order behavior has set in, by eighth order judging from the Figure, the expansion coefficients [9,10] for H_2^+ alternate in sign. This implies that the singularity responsible for the large-order behavior is on the negative real axis, at −0.3.

The relation between the E_i and F_i of Eq. (16) implies, at least formally, a relation between the functions $E(\delta)$ and $F(\delta)$, as follows. Let us replace $i!$ in Eq. (16) with its integral representation. Then

$$E(\delta) = \sum_{i=0}^{\infty} i!\, F_i\, \delta^i = \sum_{i=0}^{\infty} \int_0^{\infty} e^{-t}(\delta t)^i F_i dt. \tag{18}$$

Now suppose that we can interchange the order of summation and integration. Then

$$E(\delta) = \int_0^{\infty} e^{-t} \sum_{i=0}^{\infty} (\delta t)^i F_i dt = \int_0^{\infty} e^{-t} F(\delta t) dt. \tag{19}$$

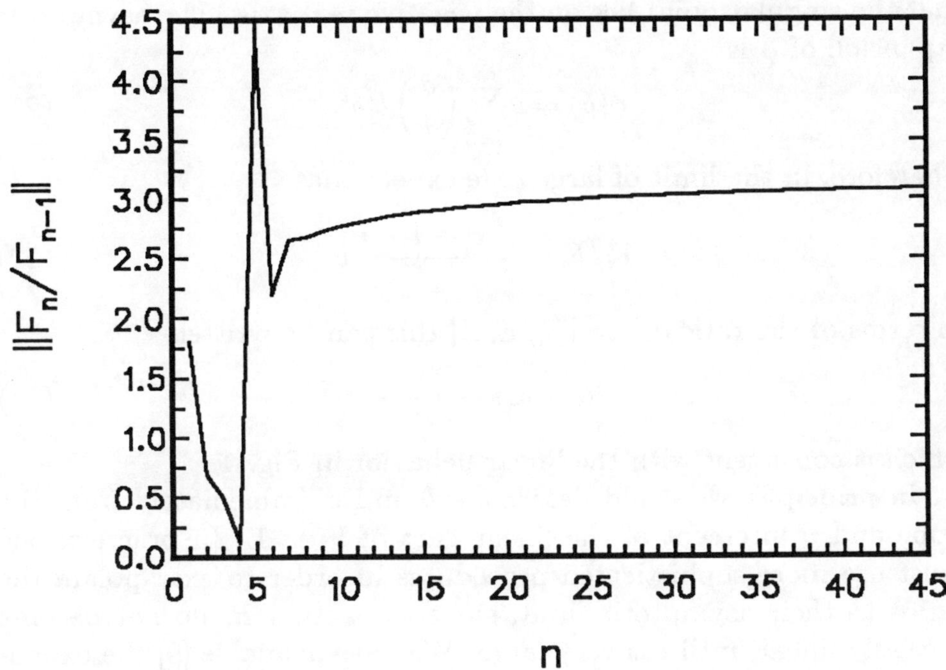

Figure 6. Ratio test for the expansion coefficients of the Borel function corresponding to the ground state energy of H_2^+ with $R = 1$.

Eq. (19) is called the *Borel sum* of the expansion of E.

The fact that the expansion of F is convergent makes it easier to construct an explicit functional form for the singularity responsible for the large-order behavior. We will first treat H_2^+ since it lacks the complication of the superimposed oscillation. Consider the following generic form for a singularity:

$$\phi(\delta) = a\,(1 + b\delta)^\sigma. \tag{20}$$

We would like to fit the parameters b and σ so that the δ expansion of ϕ is consistent with the large-order behavior of the ratios F_i/F_{i-1}. If σ is not a positive integer, then $\phi(\delta)$ represents a singularity at $\delta = -b^{-1}$. Let us assume that this is the form of the singularity that is nearest to the origin in the complex δ plane, so that the radius of convergence is $p_c = |b^{-1}|$. The fact that the F_i beyond $i = 3$ strictly alternate in sign [9,10] implies that b is a positive real number, so that the singular point lies on the negative real axis. The asymptotic expansion of ϕ is

$$\phi(\delta) = a \sum_{i=0}^{\infty} \binom{\sigma}{i} b^i \delta^i. \tag{21}$$

Therefore, in the limit of large i we expect that

$$F_i/F_{i-1} \sim \frac{\sigma + 1 - i}{i} b. \tag{22}$$

In terms of the ratios $R_i = |E_i/E_{i-1}|$ this can be written

$$R_i \sim b(i - \sigma - 1), \tag{23}$$

which is consistent with the linear behavior in Fig. 1.

In principle, we could determine b and σ immediately from the slope and y intercept of the linear part of Fig. 1. In practice, one must use more sophisticated procedures in order to extrapolate the ratios to their asymptotic limit, since the actual R_i do not become precisely linear until i is very large. We have available [9] the expansion coefficients for H_2^+ through order 49, although the higher-order values suffer some from roundoff error. Using Neville-Richardson extrapolation, we have been able to establish [9,10] that $\sigma = 1/2$, so that ϕ represents a square-root branch point,

$$\phi(\delta) = a(1 + b\delta)^{1/2}, \tag{24}$$

with $b^{-1} \approx 0.3138411$.

We have been able to fit the parameter b even more precisely using a modified version of the Padé singularity analysis [10]. Padé, in his original treatise [35], suggested a generalization of his approximants that can explicitly model branch-point singularities. Let $P(\delta)$ and $Q(\delta)$ be the numerator and denominator polynomials, respectively. Then Eq. (15) can be written in the form of a linear equation in E,

$$QE - P \sim 0. \tag{25}$$

Suppose we introduce a third polynomial, $R(\delta)$, and replace Eq. (25) with a quadratic equation,

$$QE^2 - PE + R \sim 0. \tag{26}$$

Then $E \sim S_{[L/M/N]}$, with the *quadratic Padé approximants* given by

$$S_{[L/M/N]} = \frac{1}{2}\frac{P}{Q} + \frac{1}{2}\frac{(P^2 - 4QR)^{1/2}}{Q}, \qquad (27)$$

where the index N is the order of the polynomial R. The advantage of Eq. (27) is that it contains square-root branch points at zeros of the the discriminant $P^2 - 4QR$. For the Borel function of H_2^+ we find [9,10] that the quadratic approximants consistently place a branch point on the negative real axis in the neighborhood of -0.3. This is the nearest singularity to the origin, and as the order of the approximant increases, the location of the branch point converges quite well giving the result $b^{-1} = 0.31384121$.

We can use the same basic approach to determine ϕ for the helium, although a minor modification is required. The general rate of increase from the nth root test is linear but there is a superimposed oscillation with a period of approximately 10, which cannot be accounted for by Eq. (20). This behavior suggests [36] a pair of complex-conjugate branch points slightly displaced from the negative real axis,

$$\phi(\delta) = a\left[(1 + be^{i\zeta}\delta)^{1/2} + (1 + be^{-i\zeta}\delta)^{1/2}\right] \qquad (28),$$

with the displacement angle ζ defined as in Fig. 7. The period of oscillation in the expansion coefficients of Eq. (28) is $2\pi/\zeta$. The quadratic Padé approximants of the Borel function for helium do indeed give as the singularities nearest the origin a complex-conjugate pair of branch points corresponding to $\zeta = 34.5°$ and $b^{-1} = 0.3388$.

The Borel sum for H_2^+ is

$$\eta(\delta) = a\int_0^\infty e^{-t}(1 + b\delta t)^{1/2}dt \qquad (29a)$$

and for helium is

$$\eta(\delta) = a\int_0^\infty e^{-t}\left[(1 + be^{i\zeta}\delta t)^{1/2} + (1 + be^{-i\zeta}\delta t)^{1/2}\right] dt \qquad (29b)$$

Presumably, these functions represent the singularity in the energy function, $E(\delta)$, at $\delta = 0$. It can be proved [36], both for Eq. (29a) and for Eq. (29b), that η has the form

$$\eta(\delta) = \eta_1(\delta) + \delta^{1/2}\eta_2(\delta), \qquad (30)$$

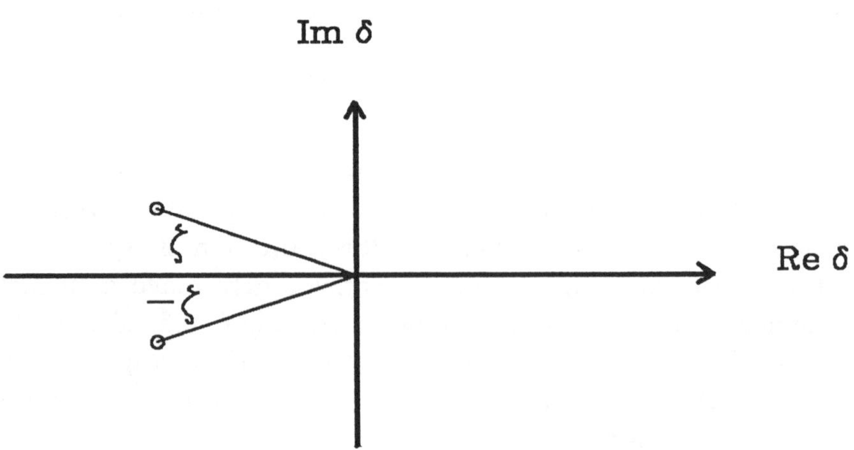

Figure 7. Locations of the square-root branch points of the Borel function corresponding to the ground state energy of helium.

where η_1 is analytic at $\delta = 0$ while η_2 has an essential singularity there. Thus, the singularity in $E(\delta)$ at the origin appears to have the form of a square-root branch point multiplied by an essential singularity. This is consistent with the behavior of the poles and zeros of the Padé approximants, shown in Figs. 4 and 5, which seemed to combine the usual features branch points and removable singularities.

We conclude from this analysis that in the neighborhood of the origin of the complex δ plane $E(\delta)$ has the form

$$E(\delta) = \eta(\delta) + \bar{E}(\delta), \tag{31}$$

with the large-order behavior of the δ expansion of E determined by η, which is given by Eqs. (29). This does not necessarily imply that the function $\bar{E}(\delta)$ is regular at the origin. It could be the case that the Borel function corresponding to $\bar{E}$ has singularities that are farther from the origin than the branch points we found for the ϕ, and it is quite possible that these additional singularities would also be mapped to the origin by the Borel integral. However, the expansion coefficients of η will converge to the E_i for i sufficiently large.

This singularity structure from the Borel sum differs somewhat from the singularity structure predicted from the coulombic poles of the hydrogenic basis expansion of the wave function, but it is consistent with the results from the Padé analysis. We noted above that the hydrogenic expansion is not a mathematically rigorous procedure for determining δ dependence of E. Unfortunately, the Borel analysis is not rigorous either. It is easy to construct a counterexample. Consider the function

$$\mathcal{E}(\delta) = E(\delta) + e^{-1/\delta}. \tag{32}$$

The exponential term goes to zero for small (but positive) δ faster than any power. Therefore, the expansion coefficients of $\mathcal{E}$ are exactly equal to the expansion coefficients of E. Since the Borel analysis is based solely on the E_i, it gives the same result for $\mathcal{E}$ as for E.

A function that can be added to E without affecting the δ expansion is said to be *subdominant* [37]. It is obvious that one needs additional information beyond E_i to characterize subdominant contributions. For some other types of perturbation theory it has proved possible to derive subdominant terms from a direct, but rather complicated, analysis of the Schrödinger equation [38,39]. Such an analysis is probably possible in our case as well, but has not yet been carried out. Fortunately, summation methods based on the Borel sum, which will be described in Section IV, yield very accurate values for the energy eigenvalues in the cases we have studied. This would seem to indicate that subdominant corrections to the Borel sum, if they exist at all, are numerically insignificant at $\delta = 1/3$.

D. Divergence Due to Dimension-Dependent Parameters

The large-dimension limit is formally equivalent to a system of classical particles each confined by a potential well [13]. The effective masses of the particles are proportional to D^2. For δ small but nonzero the particles oscillate about the minima of their wells, and in the limit $\delta \to 0$ they all come to rest. The scaled energy eigenvalue at $\delta = 0$ is equal to the potential energy of the particles at their rest configuration.

A possible explanation for the existence of a singularity in $E(\delta)$ at the origin is that the extremum of the effective potential about which we are expanding does not accurately reflect the physics of the problem at nonzero δ. The formal expressions for the effective potentials for helium and for H_2^+ have stable wells at the extremum, and the solutions that correspond to small oscillations about the bottom of the well seem to be excellent qualitative models for the three-dimensional solutions. However, the dimensional scaling of the internuclear distance in H_2^+, described in Section II, introduces an implicit dimension dependence into the effective potential, which must be included in our analysis.

Consider first a simpler system: the Schrödinger equation for an isotropic harmonic oscillator with a quartic perturbation, which can be written in the (unscaled) form [20]

$$\left[-\frac{1}{2}\frac{d^2}{dr^2} + \frac{(D-1)(D-3)}{r^2} + \frac{1}{2}r^2 + gr^4 - E \right] \Psi(r) = 0, \qquad (33)$$

where g is a positive constant. In the limit $D \to \infty$ the centrifugal term, which is proportional to D^2, becomes infinite. However, if we introduce dimension-scaled distance units with the substitution $r \to D^{1/2}r$, then

$$\left[-\frac{\delta^2}{2}\frac{d^2}{dr^2} + V_{eff}(r) - \tilde{E} \right] \Psi(r) = 0, \qquad (34)$$

with the effective potential

$$V_{eff}(r) = (1 - 4\delta + 3\delta^2)r^{-2} + \tfrac{1}{2}r^2 + \tilde{g}r^4 \qquad (35)$$

and the dimension-scaled coupling constant and energy

$$\tilde{g} = \delta^{-1}g, \qquad \tilde{E} = \delta E. \qquad (36)$$

As long as δ is positive, the scaled coupling constant $\tilde{g}$ is also positive and V_{eff} is a stable well, in the shape of the solid curve in Fig. 8. Since the behavior of the asymptotic expansion depends on the singularity structure within an infinitely small circle in the δ plane centered at the origin, we must also consider what happens as δ approaches zero from below. If δ is negative, then $\tilde{g}$ is also negative and V_{eff} looks like the dashed curve in Fig. 8. The dashed curve is not stable at the bottom of the well since the particle can tunnel out to ∞. Hikami and Brézin

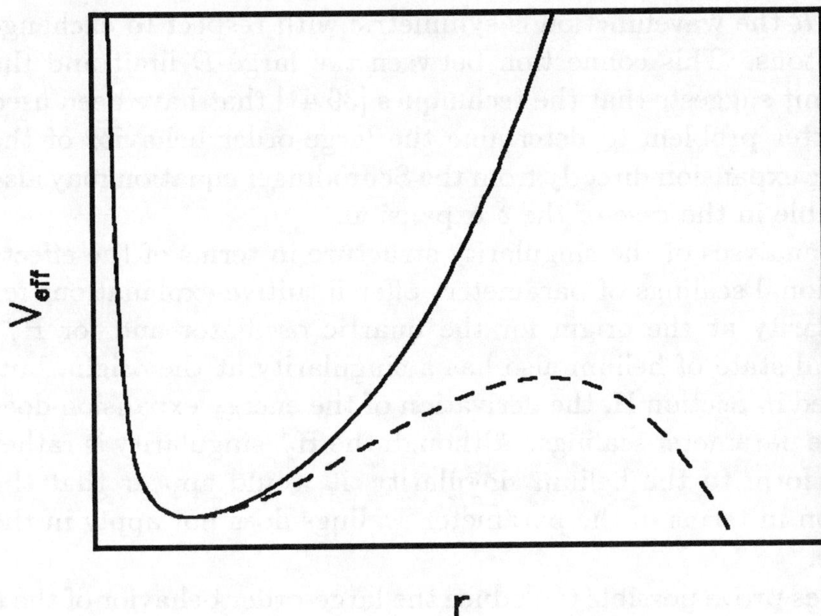

Figure 8. Shape of the large-dimension effective potential for the harmonic oscillator with quartic perturbation, with positive coupling constant (solid curve) and negative coupling constant (dashed curve).

[20] demonstrated that the rate of divergence of the δ expansion for the energy can be calculated from the tunneling rate at negative δ.

We can apply a somewhat analogous argument to H_2^+, in which case we use the distance scaling $r \to fr$ with $f \propto D(D-1)$, which leads to the scaled internuclear distance $\tilde{R} = R/f$. $\tilde{R}$ is by assumption independent of dimension. Therefore, our dimensional continuation of the Schrödinger equation corresponds to a dimension dependent internuclear distance

$$R(\delta) \propto \delta^{-2}(1 - \delta). \tag{37}$$

Thus, our solution at $\delta = 0$ corresponds to the solution of the unscaled Schrödinger equation in the limit $R \to \infty$. The energy as a function of $1/R$ is known to have a complicated branch point singularity in this limit [39], which has been attributed [40] to the fact that at infinite R the electron is localized at only one of the protons, while at any finite R the wavefunction is symmetric with respect to exchange of the protons. This connection between the large-D limit and the large-R limit suggests that the techniques [39,41] that have been used for the latter problem to determine the large-order behavior of the asymptotic expansion directly from the Schrödinger equation may also be applicable in the case of the δ expansion.

These analyses of the singularity structure in terms of the effects of dimensional scalings of parameters offer intuitive explanations for the singularity at the origin for the quartic oscillator and for H_2^+. The ground state of helium also has a singularity at the origin, but, as discussed in Section II, the derivation of the energy expansion does not involve parameter scalings. Although the H_2^+ singularity is rather similar in form to the helium singularity, it would appear that the explanation in terms of the parameter scalings does not apply in the latter case.

If it does prove possible to deduce the large-order behavior of the δ expansion directly from an analysis of the Schrödinger equation, then it might be feasible to calculate exactly the parameters a, b and ζ of Eqs. (29). The expansions of the functions $\eta(\delta)$ could be subtracted from the energy expansion, and the resulting expansion would have much better convergence properties [42]. Very accurate values of the parameters would be needed on account of the factorial divergence. The direct analysis of the differential equation could also yield a more rigorous derivation of the singularity at the origin; it would not depend on the validity of the Borel sum and it could identify subdominant terms missed by an analysis based only on the expansion coefficients [38].

Summation Methods

The partial sums S_n of Eq. (11) are not suitable approximants for summing dimensional expansions for atoms and molecules since they cannot model the dimensional singularities. More effective approximants can be developed by incorporating what is known in advance about the singularity structure of the energy function $E(\delta)$ into the form of the approximants.

Another important consideration is the fact that different singularities may manifest themselves at different orders in the expansion. In the last Section, we showed that the ground state energies of H_2^+ and helium have poles at $\delta = 1$ and a rather complicated singularity at $\delta = 0$. Furthermore, we suggested that the large-order behavior of the δ expansion could be accounted for if the singularity at $\delta = 0$ had the form of $\eta(\delta)$ of Eqs. (29). Fig. 9 shows, for H_2^+, the nth roots of the expansion coefficients of the poles, with residues calculated from the one-dimensional limit [9,10], and the ratios of the expansion coefficients of η, with its parameters fit to the large-order behavior of the E_n [9,10]. The poles correspond to the dashed curve and η corresponds to the dash-dot curve. The dotted curve corresponds to the sum of these two expansions. Finally, the ratios of the exact E_n are shown, for comparison, as the solid curve. It is clear that for $0 \leq n \leq 3$ the E_n are modeled quite well by the poles. The expansion coefficients of η in that range of n are relatively insignificant. This suggests that if one only had available the first few terms in the expansion then it would not be helpful to try to include the singularity at $\delta = 0$ in the approximants; its contribution would not yet be manifest, so the parameters needed to specify η could not be determined from the E_n. Hence, we will need different summation methods, depending on how many of the expansion coefficients we have calculated. For $n \geq 5$ the exact result is modeled very well by η while the poles are insignificant. The sum of the poles and η are in excellent agreement with the exact result everywhere except $n = 4$, where the transition occurs.

In Section A, we will describe several methods that only take into account poles and are therefore only appropriate at low orders. The low-order terms of dimensional expansions are relatively easy to derive, so these methods are useful for a quick qualitative analysis. This level of approximation corresponds to a simple physical model: E_0 is the energy of the infinite-D Lewis configuration, E_1 corresponds to the harmonic Langmuir oscillations, and E_2 represents cubic and quartic anharmonicities [13].

It is straightforward to calculate the E_n to reasonably large order [7-10], but some effort is required, since the computer codes that are needed are rather complicated. These higher expansion coefficients are useless unless the summation method can model the singularity at $\delta = 0$. With appropriate methods highly accurate results can be

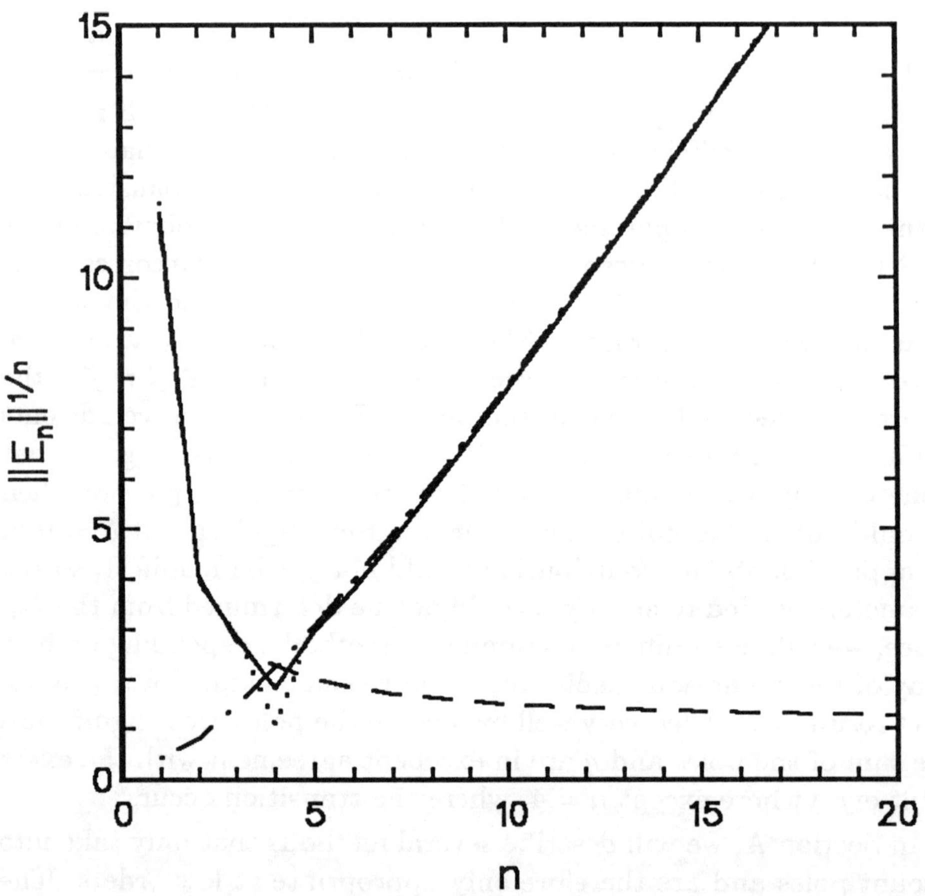

Figure 9. nth-root test for the expansion coefficients of the ground state energy of H_2^+ with $R = 1$ and for the expansion coefficients corresponding to the following models for the dimensional singularity structure: the coulombic poles, Eq. (12) (dashed curve); the singularity at $\delta = 0$, Eq. (29a) (dash-dot curve); and the sum of both of these singularities (dotted curve).

obtained. This is the subject of Section B.

A. Summation of Low-Order Terms

According to Eq. (12), the coulombic singularities have the form of a second-order pole and a confluent first-order pole. The residues of the poles can in principle be calculated exactly, from the $D \to 1$ limit of the Schrödinger equation, as we discussed in Chapter 4.1. This suggests that we use an approximant of the form

$$S_n = \frac{a_{-2}}{(1-\delta)^2} + \frac{a_{-1}}{1-\delta} + \sum_{i=0}^{n} E_i' \delta^i, \tag{38}$$

where

$$E_i' = E_i - (i+1)a_{-2} - a_{-1}. \tag{39}$$

The summation in Eq. (38) is equivalent to the nth partial sum of the function

$$E'(\delta) \equiv E(\delta) - a_{-2}(1-\delta)^{-2} - a_{-1}(1-\delta)^{-1}. \tag{40}$$

This function is regular at $\delta = 1$, so we can expect that it will be more accurately modeled by partial sums than will the original function E. This summation technique was suggested by Doren and Herschbach [43], who called it the *hybrid expansion*, since it can be thought of as the sum of a truncated Laurent expansion about $\delta = 1$ and a truncated Taylor expansion about $\delta = 0$.

Suppose, however, that the residues a_{-2} and a_{-1} are not known. It may be the case for more complicated systems than those studied thus far that the $D \to 1$ limit of the Schrödinger equation is very difficult to solve. Even if the solution is possible it would be useful to have a quick procedure that could yield a reasonably accurate result from the expansion coefficients alone. A particularly simple method for doing this is the *rescaled expansion*, which was suggested several years ago by Herschbach [13]. Suppose that we multiply and divide the partial sums by $(1-\delta)^{-2}$ as follows:

$$S_n = (1-\delta)^{-2}(1-\delta)^2(E_0 + E_1\delta + E_2\delta^2 + \cdots + E_n\delta^n)$$

$$\sim (1-\delta)^{-2}[E_0 + (E_1 - 2E_0)\delta + (E_2 - 2E_1 + E_0)\delta^2$$

$$+ \cdots + (E_n - 2E_{n-1} + E_{n-2})\delta^n]. \qquad (41)$$

The series within the brackets represents a function that is regular at $\delta = 1$, and which therefore can be more accurately modeled by a polynomial in δ than can the original function. Indeed, this procedure has been found to greatly accelerate the low-order convergence of δ expansions for coulombic systems [2,11,13,15].

Another technique, that is more flexible than the rescaled expansion and appears to converge slightly better, is *weighted truncation* [11]. This method is based on the idea of an optimal asymptotic approximation [36]. A typical characteristic of divergent asymptotic expansions is that their partial sums at first steadily approach the correct value but then, after a certain point, become steadily worse. An estimate of the error in the nth partial sum is given by the term in the expansion of order $n+1$, that is $E_{n+1}\delta^{n+1}$. The partial sum for which this error estimate is smallest is called the *optimal asymptotic approximation*. If δ is sufficiently small, then the optimal asymptotic approximation will be the point of closest approach to the correct value [37]. We can use the asymptotic error estimate as a criterion for fitting the residue a_{-2} in Eq. (38). Let S_n be an approximant of the form

$$S_n = \frac{\mu}{(1 - \delta)^2} + \sum_{i=0}^{n}[E_i - (i + 1)\mu]\delta^i, \qquad (42)$$

with μ treated as a variable parameter. The error estimate for S_{n-1} is

$$\Delta S_{n-1} = [E_n - (n + 1)\mu]\delta^n, \qquad (43)$$

which suggests that we choose $\mu = E_n/(n+1)$ so as to set $\Delta S_{n-1} = 0$. In this way we can construct a new sequence of approximants,

$$S_n = \frac{E_n/(n + 1)}{(1 - \delta)^2} + \sum_{i=0}^{n-1}\left(E_i - \frac{i + 1}{n + 1}E_n\right)\delta^i. \qquad (44)$$

We call this technique *weighted truncation*, since each of these approximants is the optimal asymptotic approximation of a series that has been weighted by subtracting out the expansion of a pole with the residue chosen so as to minimize the asymptotic error of the truncation.

Padé approximants were used in Section 3.B. as a tool for singularity analysis, but their primary function is as a summation method. Unfortunately, their convergence with dimensional expansions tends to be rather slow and uneven. They can be useful if the expansion is known to high order, but they are not appropriate as a low-order summation technique. However, their rate of convergence can be improved considerably if they are used in conjunction with the hybrid expansion, of Eq.(38). The Padé approximants of the E' converge quite well even at low order. We will refer to this technique as *hybrid Padé summation.*

If the residues are not known, then they too can be calculated from Padé approximants [7]. The residues can be expressed in terms of the δ expansion according to

$$a_{-2} = \lim_{\delta \to 1}(1 - \delta)^2 E(\delta), \tag{45a}$$

$$a_{-1} = \lim_{\delta \to 1}(1 - \delta)\left[E(\delta) - a_{-2}(1 - \delta)^{-2}\right]. \tag{45b}$$

We can expand Eqs. (45) in powers of δ and then calculate the residues from Padé approximants evaluated at $\delta = 1$. The values for the residues obtained in this way are not extremely accurate, but they are usually accurate enough to remove the poles at $\delta = 1$ from the Padé approximants for the E'. In cases where the exact value for a_{-2} is available, so that only a_1 needs to be fit, then this procedure appears to be the method of choice at low order [11].

Finally, we mention one other method, the *shifted expansion*, which was discovered empirically [44] before the dimensional singularity structure was understood. This method consists of reexpanding the expansion parameter $\delta \equiv 1/D$ in terms of $(D-\sigma)^{-1}$, where σ is an arbitrary shift parameter. For the ground state of He the optimal shift parameter was found to be $\sigma = 1$, which leads to approximants of the form

$$S_n = (1 - \delta)^{-2} \sum_{i=0}^{n} E_i^{(Sh)} (D - 1)^{-i}, \tag{46}$$

where

$$E_i^{(Sh)} = \sum_{i=0}^{n} (-1)^i \binom{n+1}{i} E_{n-i}. \tag{47}$$

For $n = 0$ the shifted expansion is identical to the rescaled expansion, Eq. (41), with a second-order pole at $\delta = 1$, but they differ at higher order since the shifted expansion adds additional higher-order poles at $\delta = 1$. We know from dimensional singularity analysis [18] that there are no poles at $\delta = 1$ of order higher than two, so we can expect that the shifted expansion will be less and less accurate than the rescaled expansion as n increases. This is indeed what happens in practice [11]. The shifted expansion procedure has been rather popular [31], but now that we have information about the dimensional singularity structure of the problem this method should probably be considered obsolete.

Figs. 10 and 11 compare the convergence of the summation methods just discussed, using expansion coefficients through fifth order. Fig. 10 shows results for the ground state of H_2^+ and Fig. 11 shows results for the ground state of helium. Points beyond the optimal asymptotic approximation are omitted. (There is no optimal asymptotic approximation for the methods that use Padé summation. In principle, these should continue indefinitely to improve with increasing order.) The dotted curves correspond to hybrid Padé summation with the exact value for a_{-2} [9,10,18]. The ordinate in these Figures is the number of accurate digits, which we define as $-\log_{10} |(E - S_n)/E|$, where E is the exact value for the energy. It is clear that the methods that explicitly incorporate a second-order pole at $\delta = 1$—the rescaled expansion, weighted truncation, and hybrid Padé summation—are to be preferred over those that do not. In between is the shifted expansion, which does place a singularity at $\delta = 1$ but it is a singularity of the wrong type.

The relative merits of these low-order summation methods have been compared using several other δ expansions in addition to the two considered here and the conclusions are as follows [11]: If the residue a_{-2} is known exactly then hybrid Padé summation is recommended; if a_{-2} is not known then at very low order weighted truncation generally appears to be the method of choice, but beyond about third order the hybrid Padé summation again seems to be best. An additional advantage of weighted truncation and hybrid Padé summation is the fact that both of these methods can be applied to singularities that are more complicated than poles. For example, relativistic problems

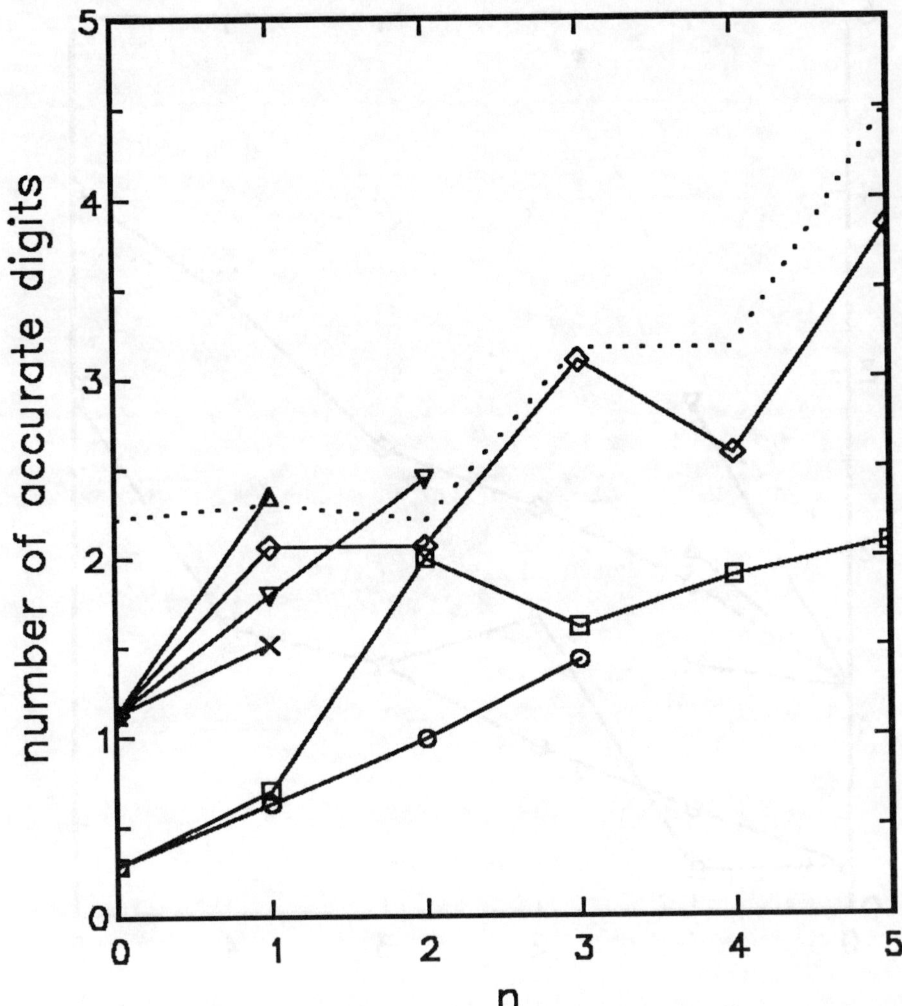

Figure 10. Accuracy of various summation methods for the expansion of the ground state energy of H_2^+ with $R = 1$. The curves are labeled as follows: o, partial sums, Eq. (11); ×, shifted expansion, Eq. (46); □, conventional Padé approximants; △, rescaled approximants, Eq. (41); ▽, weighted truncation, Eq. (44); ◇, hybrid Padé approximants with residues from Eqs. (45). The dotted curve corresponds to biased hybrid Padé summation with the exact value for the residue a_{-2}. The number of accurate digits is defined as $-\log_{10}|(S_n - E)/E|$, where E is the exact energy.

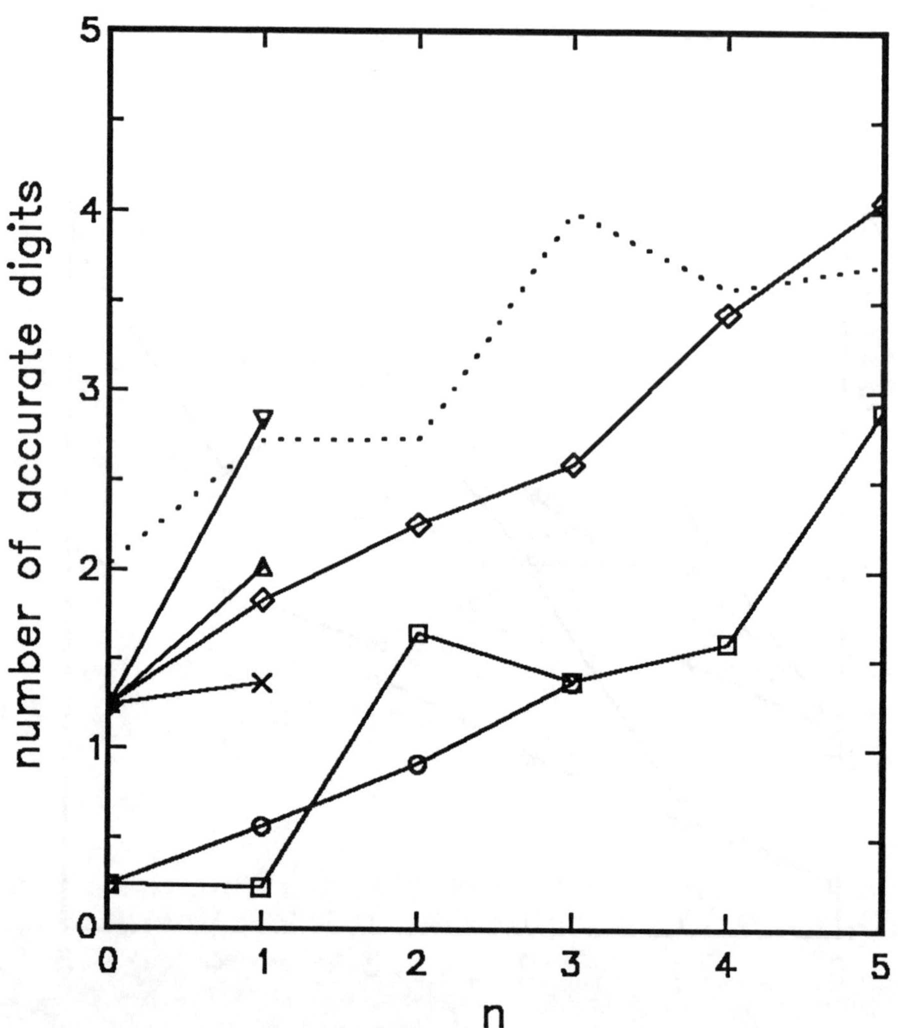

Figure 11. Accuracy of various summation methods for the expansion of the ground state energy of helium. The curves are labeled as follows: $\circ$, partial sums, Eq. (11); $\times$, shifted expansion, Eq. (46); $\square$, conventional Padé approximants; $\triangle$, rescaled approximants, Eq. (41); $\triangledown$, weighted truncation, Eq. (44); $\diamond$, hybrid Padé approximants with residues from Eqs. (45). The dotted curve corresponds to biased hybrid Padé summation with the exact value for the residue a_{-2}. The number of accurate digits is defined as $-\log_{10}|(S_n - E)/E|$, where E is the exact energy.

have branch points on the positive real axis instead of poles, and it is straightforward to modify Eq. (38) to incorporate these singularities [27].

B. Large-Order Summation

The presence of a singularity in $E(\delta)$ at $\delta = 0$ implies that any of the summation methods described in Section A that employ partial sums will eventually diverge. The optimal asymptotic approximations are reached quite early, and these methods cannot utilize the expansion coefficients beyond that point. A major advantage of dimensional perturbation theory over other types of perturbation theory is the relative ease with which the expansion coefficients can be calculated [6-10]. For this reason, the development of summation methods that continue to work at large order is a central goal.

Padé summation is an obvious first choice on account of its computational simplicity and its ability, in principle, to model functions with singularities. Although the only singularities in these approximants are poles, they can model more complicated singularities with groupings of poles and zeros, as we discussed in Section 3. The Padé approximants do indeed continue to improve as the order increases. Fig. 12 shows the convergence of Padé summation for H_2^+ through 35th order. The dash-dot curve shows results from the conventional kind of Padé approximants while the other curves correspond to various forms of hybrid Padé summation. This Figure demonstrates that the hybrid methods add from one to four significant figures in accuracy. The only difference between the hybrid summation, on the one hand, and conventional Padé summation, on the other, is that the former explicitly include the coulombic poles. It is interesting that the improvement from the hybrid methods continues at large order, even though the numerical contribution of the poles to the expansion coefficients becomes insignificant compared to the contribution from the branch point. The solid curve corresponds to hybrid summation with approximate values for both of the residues of the poles at $\delta = 1$ as determined from Padé summation of Eqs. (45). The dashed curve was calculated with the exact value for the residue a_{-2} and the approximate value for a_{-1}. The dotted curve was calculated with exact

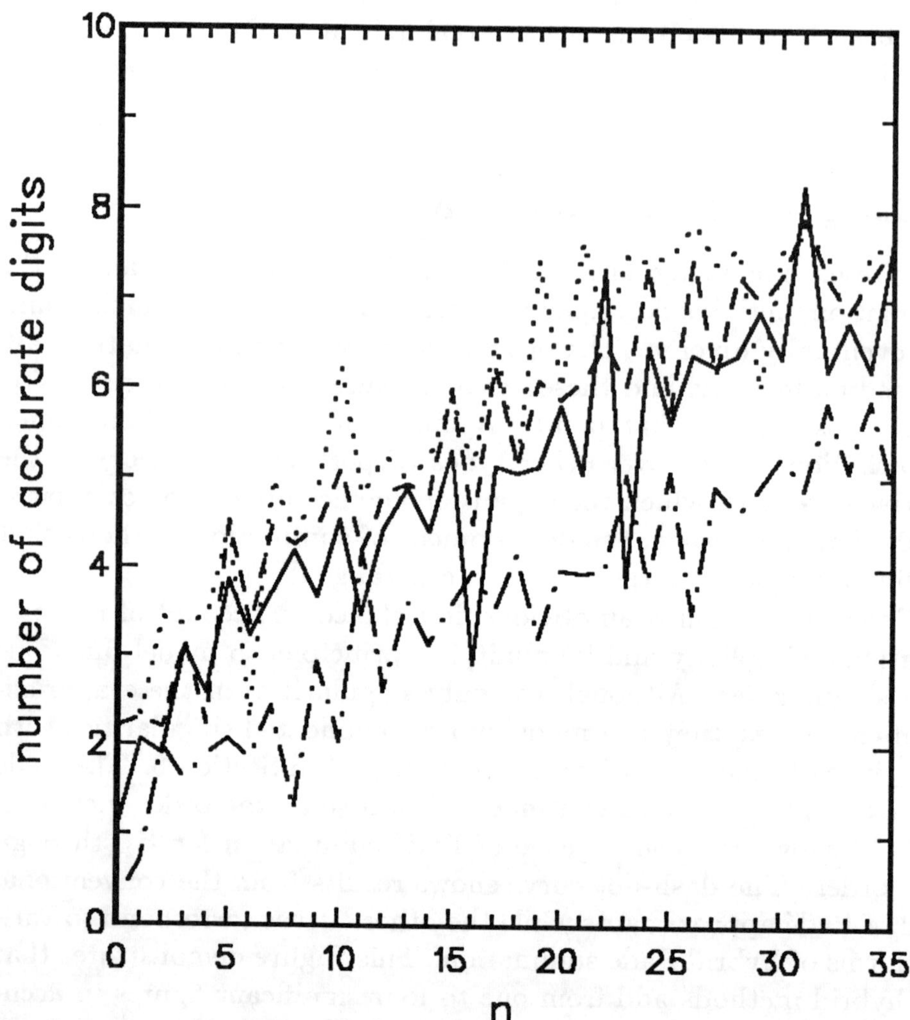

Figure 12. Accuracy of various types of Padé summation for the ground state energy of H_2^+ with $R = 1$. The curves are labeled as follows: dash-dot curve, conventional Padé approximants; solid curve, hybrid Padé approximants with both residues estimated by Padé summation; dashed curve, biased hybrid Padé approximants with exact value for a_{-2}; dotted curve, biased hybrid Padé approximants with exact values for both residues. The number of accurate digits is defined as $-\log_{10}|(S_n - E)/E|$, where E is the exact energy.

values for both residues. The exact values for a_{-2} are the ground-state energy eigenvalues of dimension-scaled Schrödinger equations in the limit $D \rightarrow 1$. The coulomb potentials in this limit become delta functions [25], and the ground-state energy can be computed to arbitrary numerical accuracy both for H_2^+ [9,10,45] and for helium [18,46]. a_{-1} can, in principle, also be determined exactly for these two systems using perturbation theory about the $D = 1$ solutions [9,18] but these calculations have not yet been carried out. Here we use $a_{-1} = -0.9830$, which is the value to which Padé summation of the 35th-order expansion of Eq. (45b) appears to converge. The convergence of Padé summation of the helium expansion [12] is very similar to that of the H_2^+ expansion.

Our analysis in Section 3 of the large-order behavior of the expansion coefficients suggested that the singularity structure of the Borel function $F(\delta)$ is more manageable than that of the energy function. We found that $F(\delta)$ has a finite radius of convergence while $E(\delta)$ has a radius of convergence of zero. Furthermore, we found that the singularity responsible for determining the radius of convergence of $F(\delta)$ is simply a square-root branch point (or a pair of square-root branch points in the case of helium), as opposed to the rather more complicated singular point in $E(\delta)$. Therefore, we can expect that Padé approximants will be able to sum the F_i, of Eq. (16), more effectively than the E_i. Let $S_{[L/M]}[F(\delta)]$ represent a Padé approximant of F. Substituting these into the Borel sum, we get a sequence of *Padé-Borel approximants* [47] for the energy,

$$S_{[L/M]}^{(PB)} = \int_0^\infty e^{-t} S_{[L/M]}[F(\delta t)]dt. \tag{48}$$

The integral is evaluated by quadrature.

A disadvantage of these approximants is the fact that the Padé approximant in the integrand needs to be evaluated along the entire real axis, not just at $\delta = 1/3$, and as the distance from the origin increases it will become less accurate. Furthermore, spurious poles on the real axis can wreak havoc with the quadrature. These problems are balanced by the facts that the integrand approximants converge quite rapidly and the exponential factor weights small values of t much more heavily than large values. The net result is an improvement over the hybrid Padé approximants, at large orders, of about one or two significant figures. Padé-Borel summation of the helium expansion gives a similar improvement [12]. The solid curve in Fig. 13 shows the accuracy of Padé-Borel approximants for H_2^+. Actually, these are hybrid Padé-Borel approximants, with Eq. (48) applied to the expansion coefficients $F_i' = E_i'/i!$, where the E_i' are given by Eq. (39). Failure to subtract out the poles before applying the Padé-Borel summation results in poles in the Borel function on the positive real axis. The curves in Fig. 13 were calculated using the exact values for both residues. For a few of the approximants the quadrature did not converge, due to poles on the integration path. We have omitted these points. (In principle, however, these poles are integrable through a partial-fraction decomposition of the Padé approximant [9].) The dotted curve, which is repeated from Fig. 12 for the sake of comparison, corresponds to hybrid Padé summation using exact values for both residues.

We can improve the Padé-Borel summation by explicitly incorporating a square root in the approximant for the integrand [9,10]. We determined in the last Section that the large-order behavior of $F(\delta)$ for H_2^+ was determined by a singularity of the form $a(1 + b\delta)^{1/2}$. This implies that F has the functional form [48]

$$F(\delta) = g(\delta)(1 + b\delta)^{1/2} + h(\delta), \qquad (49)$$

where g and h have no singularities in the region $|\delta| \leq b^{-1}$. An approximant consistent with Eq. (49) is

$$S_{[L,M/N]} = \frac{P(\delta)}{R(\delta)} + \frac{Q(\delta)}{R(\delta)}(1 + b\delta)^{1/2}, \qquad (50)$$

where P, Q and R are polynomials of orders L, M and N, respectively, and b has the value we determined from the singularity analysis of the quadratic Padé approximants. Setting Eq. (50) equal to the expansion of F and then multiplying both sides of the equation by R, we obtain a set of linear equations that determine the three polynomials. The functional form of Eq. (49) is one of a class known as Darboux functions [48], so we refer to the Borel sums of the approximants defined by Eq. (50) as *Darboux-Borel approximants*. The results for

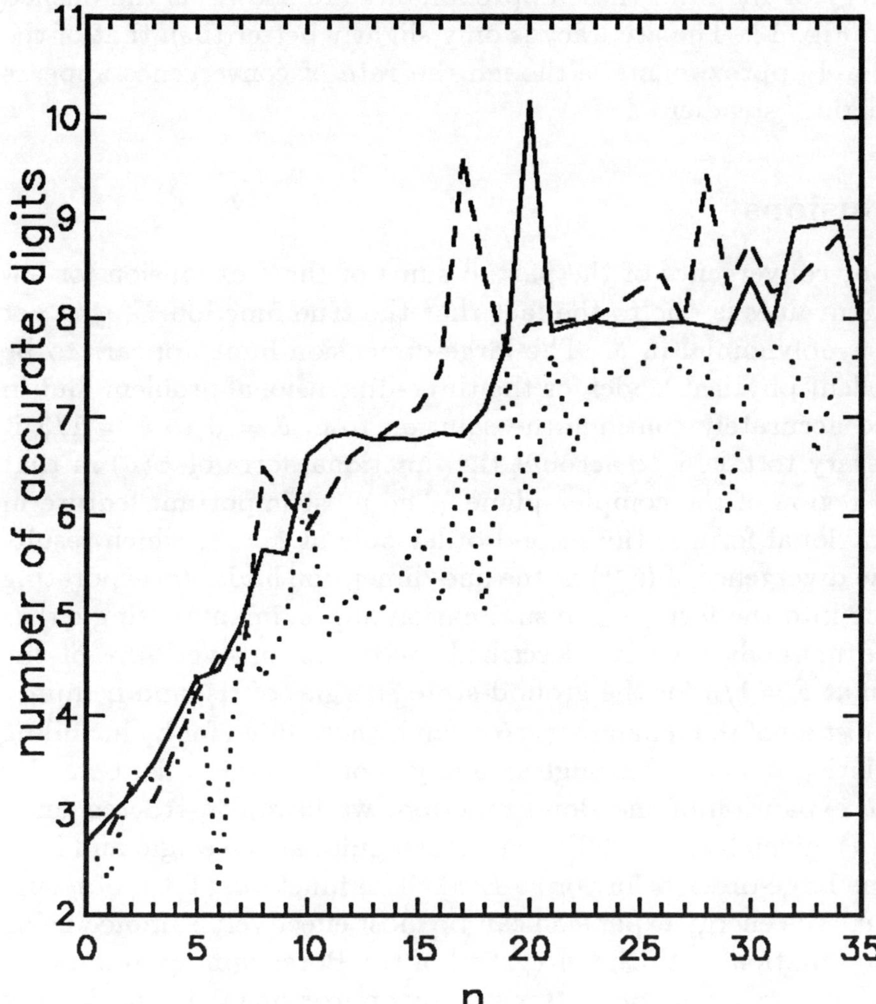

Figure 13. Accuracy of Borel summation for the ground state energy of H_2^+ with $R = 1$. The curves are labeled as follows: solid curve, Padé-Borel approximants; dashed curve, Darboux-Borel approximants. Hybrid Padé summation is shown, for comparison, as the dotted curve. All three curves were calculated using exact values for both residues. The number of accurate digits is defined as $-\log_{10} |(S_n - E)/E|$, where E is the exact energy.

the energy of H_2^+ from these approximants are shown as the dashed curve in Fig. 13. The accuracy is only slightly better than that of the Padé-Borel approximants, although the rate of convergence appears to be slightly steadier.

Conclusions

The poor convergence of the partial sums of the δ expansion for energy eigenvalues is due to the fact that the true function $E(\delta)$ is not simply a polynomial in δ. The large-dimension limit appears to be an excellent physical model for the three-dimensional problem, but in order to accurately continue the solution from $\delta = 0$ to $\delta = 1/3$ it is necessary to take into account the functional form of $E(\delta)$ in that general region of the complex plane. The most important feature in this functional form is the second-order pole at $\delta = 1$, which results from the divergence of $\langle r^{-1} \rangle$ in the one-dimension limit. Incorporating this pole into the form of the summation approximants, using either of three methods we have described, accounts for over 99% of the solution at $\delta = 1/3$ for the ground-state energies of H_2^+ and helium.

Almost all of the remaining error can be accounted for by including a singularity at $\delta = 0$. Through an analysis of the large-order behavior of the δ expansion of the Borel function, we have constructed functions $\eta(\delta)$, given by Eqs. (29), that are singular at the origin and have the same large-order behavior as $E(\delta)$. The functional form of η suggests that the energy expansion can be most effectively summed using Borel summation, with the integrand of the Borel sum approximated by Padé approximants or, better yet, by approximants in the form of Darboux functions. Our numerical results indicate that summation methods of these types can yield an accuracy of at least eight or nine significant figures. By taking advantage of interdimensional degeneracies that relate the energies of ground states at higher values of D to excited states at $D = 3$ [25], it is possible to obtain even higher accuracy for certain excited states.

It should be straightforward to use this same approach—summing the δ expansion with a summation method designed to fit the appropriate singularity structure—to calculate expectation values other than the energy. It is possible to compute a δ expansion for the wave-

function [6], although because the expansion coefficients would each
be functions of the coordinates such an expansion would be rather
awkward to use directly if it proved to be divergent. A better approach would be to calculate a δ expansion for the desired expectation value. The singularity structure will probably vary depending on
the operator in question. Indeed, a study of the type of dimensional
singularities that result from a given type of operator might shed light
on the cause of the branch point in the energy function.

We expect that dimensional perturbation theory will continue to
be applicable with many-electron systems. A D-dimensional generalization of the Schrödinger equation is now available for any arbitrary
N-body problem [15,16], although coefficients of the δ expansion have
not yet been calculated beyond first order [15]. A potential difficulty
in the treatment of larger systems is the fact that the moment method
[6,7], which we have used here to compute the expansion coefficients,
is somewhat unstable to roundoff error. This results in a decrease
in the number of terms that can be computed as the number of degrees of freedom increases. To treat large systems, other methods for
calculating the expansion coefficients will have to be developed.

It will be interesting to see what type of dimensional singularity
structure results when there are more than two electrons. The poles
at $\delta = 1$ still ought to be present. In Chapter 4.1 we pointed out
that atoms in the $D \to 1$ limit only bind $1s$ electrons, which suggests
that the presence of electrons with higher principal quantum numbers will not affect the singularity structure at $\delta = 1$. The coulombic
poles result from the cusp behavior of the wavefunction, which is essentially a two-body phenomenon. In contrast, the $D \to \infty$ limit
emphasizes electron correlation, which is sensitive to the number of
electrons in the system. More complicated correlations could lead to a
more complicated singularity structure at $\delta = 0$, which would increase
the importance of carrying out a dimensional singularity analysis before trying to sum the δ expansion. If it proves possible to discern
a generic singularity structure at $\delta = 0$ for coulombic systems, then
this structure could perhaps be used as a probe of the accuracy with
which an approximate wavefunction models electron correlation. Conventional approaches to electronic structure begin with a separable
approximation for the wavefunction. Rather complicated techniques

are then used to improve it so that it can describe correlation effects. One could calculate the δ expansion within a given approximation to the wavefunction and then deduce from it a singularity structure. The extent to which the resulting singularities are consistent with a known generic behavior could serve as a criterion for the quality of the wavefunction.

Finally, we note that excited states can have a dimensional singularity structure that appears to be qualitatively different from that found for the ground state. In Chapter 8.1 the δ expansions of three excited states of helium are analyzed. The $1s2s\ ^3S$ state appears to have the same structure as the ground state, while the $1s2s\ ^1S$ and $2p^2\ ^1S$ states seem to have a rather different structure.

Acknowledgements

We thank Dudley Herschbach, John D. Morgan III, Don Frantz and John Loeser for many helpful discussions. This work was supported by grants from the U. S. National Science Foundation and the Office of Naval Research, and by a Cray Research and Development Grant.

References

1. C. C. J. Roothaan, Rev. Mod. Phys. **23**, 69 (1951).
2. J. G. Loeser and D. R. Herschbach, J. Phys. Chem. **89**, 3444 (1985).
3. D. Z. Goodson and D. R. Herschbach, J. Chem. Phys. **86**, 4997 (1987).
4. D. R. Herschbach, J. G. Loeser and D. K. Watson, Z. Phys. D **10**, 195 (1988).
5. L. D. Mlodinow and N. Papanicolaou, Ann. Phys. (N.Y.) **128**, 314 (1980); **131**, 1 (1981).
6. J. Ader, Phys. Lett. **97A**, 178 (1983).
7. D. Z. Goodson and D. R. Herschbach, Phys. Rev. Lett. **58**, 1631 (1987)

7.2 Dimensional Scaling and Spectral Properties

Lotten Hägg and Osvaldo Goscinski
Department of Quantum Chemistry
Uppsala University
Box 518, S-751 20 Uppsala, Sweden

Abstract

A review is made of the calculation of multipole polarizabilities in three dimensions, for hydrogen-like ions with nuclear charge Z, in which the solutions of the inhomogeneous equations in perturbation theory are used. The procedure is extended to arbitrary dimension D, and an explicit expression for exact multipole polarizabilities is obtained. A calculation of an approximate ground state energy using $1/D$ expansions is made, leading to a classic harmonic oscillator problem. The possibility to estimate the dipole polarizability in the large D limit from the obtained expressions is used. A comparison with the exact result yields quantitative agreement in the coefficient as well as in the D dependence, when using a physically motivated form of the perturbation.

Inequalities for oscillator strengths, previously used for estimating dipole polarizabilities in three dimensions, are generalized to D dimensions, and expressions for the dipole polarizability in the large D limit are obtained. The exact results, the dimensional scaling calculations, and the expressions obtained from inequalities are compared and evaluated. It is shown that the exact first order correction to the unperturbed wave function reduces to one term in the sum over states expression. The asymptotic result for the dipole polarizabilities is, in atomic units, $\alpha_2 = (64Z^4)^{-1}D^6$.

Introduction

In the past few years the method of dimensional scaling [12,22,23] has become increasingly more important in quantum theory. Using this technique one can solve the many-particle Schrödinger equation in a space of arbitrary dimension D [17]. By taking the limit of infinite

315

D. R. Herschbach et al. (eds.), Dimensional Scaling in Chemical Physics, 315–335.
© 1993 *Kluwer Academic Publishers. Printed in the Netherlands.*

D, which in some sense is a classical limit, one can find exact solutions of the equation. In reality the procedure of dimensional scaling is reduced to finding a local minimum for a harmonic potential. Using perturbation series based on the parameter $1/D$, and sometimes interpolation procedures, it is possible to obtain wave functions and energies in the physical three-dimensional space [15,17,21]. As shown elsewhere in this volume [13], one can calculate reasonable approximations to *low-lying* eigenvalues and eigenfunctions with dimensional scaling. In particular this is done for hydrogen-like atoms since the exact solutions for such atoms are known, so an evaluation of the method is facilitated. So far most applications of the method have concerned static properties like energies and wave functions for individual levels [12,15,17], but not properties like polarizabilities, characterizing the *response* of a system to a perturbation. In terms of conventional perturbation theory polarizabilities involve sums over all levels.

A basic aim here is to compare exact calculations for arbitrary D with approximations usually made in dimensional scaling where one uses the large D limit. We start by presenting *exact* calculations of the atomic polarizability in three dimensions as done earlier by Dalgarno [7]. Thereafter these exact calculations are generalized to D dimensions, using hyperspherical coordinates and a generalized multipole operator. The choice of the perturbation is discussed. How to calculate approximate energies with the method of dimensional scaling is described, and from the previously mentioned harmonic potential we can estimate a polarizability in the large D limit to be compared with the exact analytic expression. The large D limit for the dipole polarizability is also studied carefully using sum over states expressions as well as sum rules and known bounds for dipole polarizabilities.

Atomic Polarizabilities in Three Dimensions

Perturbation Theory

To calculate polarizabilities, which describe how a system responds, one can use perturbation theory [7,20]. The system one wishes to study is assumed to be characterized by a Hamiltonian $\hat{H}^{(0)}$ with eigenfunction $\Psi^{(0)}$ and energy $E^{(0)}$ before the perturbation is applied.

We shall restrict our attention to a *bound* ground state. If the perturbation can be described by the Hamiltonian $\hat{H}^{(1)}$ the new, perturbed system has the Hamiltonian:

$$\hat{H} = \hat{H}^{(0)} + \lambda \hat{H}^{(1)} \tag{1}$$

where λ is a small parameter, signifying the order of magnitude of the perturbation. The new wave function and the new energy of the system can be formally expanded in powers of λ.

$$\Psi = \Psi^{(0)} + \lambda \Psi^{(1)} + \lambda^2 \Psi^{(2)} + \dots \tag{2}$$

$$E = E^{(0)} + \lambda E^{(1)} + \lambda^2 E^{(2)} + \dots \tag{3}$$

The time-independent Schrödinger equation describing any system is:

$$\hat{H}\Psi = E\Psi \tag{4}$$

If the expressions for $\hat{H}$, Ψ and E from equations (1) - (3) are used in equation (4), and terms with the same powers of λ are collected, one obtains a set of equations, one for each power of λ. The zeroth order equation is

$$(\hat{H}^{(0)} - E^{(0)})\Psi^{(0)} = 0 \tag{5}$$

and the inhomogeneous equation for the first order wave function is

$$(\hat{H}^{(0)} - E^{(0)})\Psi^{(1)} + (\hat{H}^{(1)} - E^{(1)})\Psi^{(0)} = 0 \tag{6}$$

The first order correction to the energy is:

$$E^{(1)} = <\Psi^{(0)}|\hat{H}^{(1)}|\Psi^{(0)}> \tag{7}$$

and the second order correction is

$$E^{(2)} = <\Psi^{(1)}|(\hat{H}^{(1)} - E^{(1)})|\Psi^{(0)}> \tag{8}$$

The quantities appearing in equations (1)-(8) can be used to study the response of a system to perturbations when $\hat{H}^{(1)}$ is specified.

Multipole Perturbations

The conventional solutions of the inhomogeneous equations in perturbation theory are based on expanding the wave functions in terms of eigenfunctions of the unperturbed Hamiltonian $\hat{H}^{(0)}$. In such an approach one must include both the wave functions for the discrete levels and for the continuum. Because of the difficulties in finding a correct description of the continuum one seeks alternatives to this procedure, two of which are widely used. One procedure, which goes back to Schrödinger [19], is to use what are known as Sturmian functions [14] and this has for instance been done by Avery and Herschbach [3]. The solutions of the inhomogeneous equation (6) can be expressed in terms of propagators and the Coulomb propagator plays a crucial role [5]. It is possible to express the Coulomb propagator in terms of a complete *and* discrete set of basis functions (Sturmian functions). The other procedure introduced by Dalgarno and Lewis [8], is based on a factorization of the first order wave function:

$$\Psi^{(1)} = F\Psi^{(0)} \tag{9}$$

One substitutes this expression into the inhomogeneous first order equation (6) and obtains a differential equation for F. It turns out that for a number of perturbations, F assumes a simple form.

In a number of cases we can write the specific perturbation as [7]:

$$\hat{H}^{(1)} = \hat{V}_k(\mathbf{r}) = r^k P_k(\cos\theta) \tag{10}$$

where P_k is a Legendre polynomial and where $z = r\cos\theta$ is the axis of quantization. The form of perturbation in (10) yields special forms of equations (5)-(9) in which the wave functions are denoted $\Psi_k(\mathbf{r})$, $\Psi^{(0)}(\mathbf{r})$, $\Psi_k^{(1)}(\mathbf{r})$, $\Psi_k^{(2)}(\mathbf{r})$ etc. and the energies E_k, $E^{(0)}$, $E_k^{(1)}$, $E_k^{(2)}$ etc. The first order correction to the energy (from equation (7)) vanishes in the case of multipole perturbations ($k \geq 1$):

$$E_k^{(1)} = <\Psi^{(0)}|\hat{V}_k|\Psi^{(0)}> = 0 \tag{11}$$

We also have the second order correction to the energy from equation (8),(10) and (11) as

$$E_k^{(2)} = <\Psi_k^{(1)}|\hat{V}_k|\Psi^{(0)}> \tag{12}$$

It is important to notice once again that the wave functions $\Psi^{(0)}$ and $\Psi_k^{(1)}$ describe *bound* states.

From Dalgarno [7] we have that the 2^k-polarizability is, in atomic units,

$$\alpha_{2^k} = -2E_k^{(2)} = -2 < \Psi_k^{(1)}|\hat{V}_k|\Psi^{(0)} > \qquad (13)$$

This is the starting point for a calculation of the multipole polarizability. One uses equation (13) with the perturbation $\hat{V}_k$ from equation (10). The factorization of $\Psi_k^{(1)}$ (equation (9)) by Dalgarno and Lewis [8] is used, R becomes a two term polynomial and explicit expressions for the polarizabilities are found.

Atomic Polarizabilities in Three Dimensions

We now present the calculations for three dimensions as done by Dalgarno [7] using the scheme described above. The normalized ground state wave function for a hydrogen-like atom in three dimensions is

$$\Psi^{(0)}(\mathbf{r}) = \frac{\Psi^{(0)}(r)}{\sqrt{4\pi}} = \sqrt{\frac{Z^3}{\pi}}\, e^{-Zr} \qquad (14)$$

If we now use that $E_k^{(1)} = 0$ because of symmetry (equation (11)) and the explicit perturbation from equation (10) then equation (6) i reduced to

$$(\hat{H}^{(0)} - E^{(0)})\Psi_k^{(1)}(\mathbf{r}) + r^k P_k(\cos\theta)\Psi^{(0)}(r) = 0 \qquad (15)$$

Instead of the conventional sum over states expressions, obtained by expanding the wave function in terms of eigenfunctions of $\hat{H}^{(0)}$, we factorize the solution of the inhomogeneous equation. Following Dalgarno [7] as well as Dalgarno and Lewis [8] we write the solution in the form

$$\Psi_k^{(1)}(\mathbf{r}) = F_k(\mathbf{r})\Psi^{(0)}(\mathbf{r}) \qquad (16)$$

where

$$F_k(\mathbf{r}) = f_k(r)P_k(\cos\theta) \qquad (17)$$

we get another form of $\Psi_k^{(1)}(\mathbf{r})$ which is inserted into equation (15) giving

$$(\hat{H}_k^{(0)} - E^{(0)})f_k(r)\psi^{(0)}(r) + r^k\psi^{(0)}(r) = 0 \qquad (18)$$

where

$$\hat{H}_k^{(0)} = -\frac{1}{2}\frac{\partial^2}{\partial r^2} - \frac{1}{r}\frac{\partial}{\partial r} + \frac{k(k+1)}{2r^2} - \frac{Z}{r} \qquad (19)$$

The radial eigenfunction $\psi^{(0)}(r)$ for the unperturbed system obeys equation (5), that is

$$\left\{ -\frac{1}{2}\frac{\partial^2}{\partial r^2} - \frac{1}{r}\frac{\partial}{\partial r} - \frac{Z}{r} - E^{(0)} \right\} \psi^{(0)}(r) = 0 \qquad (20)$$

If the Hamiltonian $\hat{H}_k^{(0)}$ is applied to the product $f_k(r)\psi^{(0)}(r)$ in equation (18) and then $\psi^{(0)}(r)$ is factored out, a differential equation for $f_k(r)$ is obtained:

$$-\frac{1}{2}\frac{\partial^2 f_k}{\partial r^2} + \left(Z - \frac{1}{r} \right)\frac{\partial f_k}{\partial r} + \left(\frac{k(k+1)}{2r^2} + r^k \right) f_k = 0 \qquad (21)$$

One can try a polynomial approximation for $f_k(r)$, substituting the following form into the differential equation (21):

$$f_k(r) = a_k r^{k+1} + b_k r^k \qquad (22)$$

This substitution yields a sum of the form $c_{-2}r^{k-2} + c_{-1}r^{k-1} + c_0 r^k = 0$ and the fact that each one of the coefficients in front of the terms has to be zero gives a_k and b_k:

$$a_k = -\frac{1}{Z(k+1)} \qquad (23)$$

$$b_k = -\frac{1}{kZ^2} \qquad (24)$$

Using these values in equation (22) for $f_k(r)$ we have from equations (16) and (17) the wave function

$$\Psi_k^{(1)}(\mathbf{r}) = P_k(\cos\theta)f_k(r)\psi^{(0)}(\mathbf{r}) \qquad (25)$$

The results above with a two term polynomial imply that only *two* Sturmian functions are necessary when expressing the Coulomb propagator in the first procedure described above on how to solve the inhomogeneous equation.

After having carried out these calculations, which give us an explicit expression for the wave function $\Psi_k^{(1)}(\mathbf{r})$, we are ready to calculate the matrix element $< \Psi_k^{(1)}|\hat{V}_k|\Psi^{(0)} >$ needed to get the polarizability (equation (13)). Since we have $\Psi^{(0)}(\mathbf{r})$ from equation (14), $\Psi_k^{(1)}(\mathbf{r})$ from equation (25) and $\hat{V}_k(\mathbf{r})$ from equation (10), all separated into radial and angular parts, we can factorize integral into a product of an angular integral I_a and a radial integral I_r.

$$< \Psi_k^{(1)}|\hat{V}_k|\Psi^{(0)} >= I_a \times I_r \tag{26}$$

The angular integral is:

$$I_a = \int | P_k(\cos\theta) |^2 \, d\Omega = \frac{4\pi}{2k+1} \tag{27}$$

The radial integral is:

$$I_r = \frac{1}{4\pi} \int_0^\infty r^{k+2} f_k(r) \, | \psi^{(0)}(r) |^2 \, dr = -\frac{(2k+2)!}{4\pi(2Z)^{2k+2}} \left(\frac{2k+3}{k+1} + \frac{2}{k} \right) \tag{28}$$

With the radial part in equation (28) and the angular part in equation (27) the matrix element in equation (26) is known. The multipole polarizability α_{2k} is given by equation (13) and we obtain in three dimensions

$$\alpha_{2k} = \frac{(2k+2)!(k+2)}{2^{2k+1}Z^{2k+2}k(k+1)} \tag{29}$$

This expression yields the known result $\alpha_2 = 4.5$ for the ground state hydrogen atom ($Z = 1$).

Atomic Polarizabilities in D Dimensions

The scheme for calculating polarizabilities in D dimensions follows the calculations for three dimensions presented above. The perturbation expressions are applicable independent of the dimension of the system and we can therefore use equation (13) for the polarizability. In analogy with three dimensions the unperturbed wave function (equation (14), [4]) is (for $D > 2$)

$$\Psi^{(0)}(\mathbf{r}) = \frac{\Psi^{(0)}(r)}{\sqrt{S_D}} = \mathcal{N}e^{-K_0 r} \tag{30}$$

where ([4], pages 15,79,80)

$$K_0 \equiv \frac{2Z}{(D-1)}$$

$$\mathcal{N} \equiv \left(\frac{(2K_0)^D}{(D-1)!S_D}\right)^{1/2} \tag{31}$$

$$S_D \equiv \int d\Omega = \frac{2\pi^{D/2}}{\Gamma(D/2)}$$

In the theory of hyperspherical harmonics in a D-dimensional space, the Gegenbauer polynomials, C_λ^α, with $\alpha = D/2-1$, play a role analogous to that of the Legendre polynomials in the theory of three-dimensional spherical harmonics [4]. Since the Gegenbauer polynomials are the D-dimensional generalizations of Legendre polynomials, a possible choice of a D-dimensional multipole perturbation analogous to $\hat{H}^{(1)}$ of equation (10), is

$$\hat{H}^{(1)} = \hat{V}_k = r^k C_k^\alpha(\cos\theta) \tag{32}$$

The equations corresponding to (16) and (17) are

$$\Psi_k^{(1)}(\mathbf{r}) = G_k(\mathbf{r})\Psi^{(0)}(\mathbf{r}) \tag{33}$$

and

$$G_k(\mathbf{r}) = g_k(r)C_k^\alpha(\cos\theta) \tag{34}$$

while the D-dimensional generalization of (19) is

$$\hat{H}_k^{(0)} = -\frac{1}{2}\frac{\partial^2}{\partial r^2} - \frac{(D-1)}{2r}\frac{\partial}{\partial r} + \frac{k(k+D-2)}{2r^2} - \frac{Z}{r} \tag{35}$$

Thus we obtain the following D-dimensional counterpart of equation (21):

$$-\frac{1}{2}\frac{\partial^2 g_k}{\partial r^2} + \left(K_0 - \frac{(D-1)}{2r}\right)\frac{\partial g_k}{\partial r} + \left(\frac{k(k+D-2)}{2r^2}\right)g_k + r^k = 0 \tag{36}$$

With the form of the polynomial $g_k(r)$ similar to the form of $f_k(r)$ in equation (22):

$$g_k(r) = a_k r^{k+1} + b_k r^k \tag{37}$$

we now get the coefficients

$$a_k = -\frac{1}{K_0(k+1)}$$

$$b_k = -\frac{(2k+D-1)}{2k(k+1)K_0^2}$$

(38)

In accordance with the calculations for three dimensions of the matrix element (equations (26)-(28)) we now wish to calculate $E_k^{(2)}$ by calculating one radial integral and one angular integral. The angular integral (analogous to (27)) is equal to ([4], page 43):

$$I_a = \int |\, C_k^\alpha(\cos\theta)\,|^2\, d\Omega = \frac{(D-2)(k+D-3)!S_D}{k!(D+2k-3)(D-3)!} \tag{39}$$

The radial integral in D dimensions (corresponding to equation (28)) is:

$$I_r = \frac{1}{S_D}\int_0^\infty |\psi^{(0)}(r)|^2 g_k(r) r^{k+D-1}\, dr \tag{40}$$

With $\psi^{(0)}(r)$ from equation (30) and $g_k(r)$ in equations (37)-(38), and using K_0 from equation (31), the integral I_r becomes:

$$I_r = -\frac{(D-1)^{2k+1}[k(2k+D)! + (2k+D-1)(2k+D-1)!]}{2^{4k+3}Z^{2k+2}k(k+1)(D-2)!S_D} \tag{41}$$

The matrix element in equation (13) is the angular integral (39) times the radial integral (41) and hence we have the polarizability in D dimensions , if $D \geq 3$, as

$$\alpha_{2k} =$$

$$\frac{(D-1)^{2k+1}(k+D-3)!(D-2)(2k+D-1)!(2k^2+kD+2k+D-1)}{2^{4k+2}Z^{2k+2}k(k+1)(D-3)!(D+2k-2)k!(D-2)!}$$

(42)

If the expression in equation (42) is used, the 2^k-polarizability for $D = 3$ is

$$\alpha_{2k} = \frac{(2k+2)!(k+2)}{2^{2k+1}Z^{2k+2}k(k+1)} \tag{43}$$

which agrees with equation (29).

One can study the asymptotic expansions for large D which follow from equation (42) giving a dipole polarizability ($k = 1$)

$$\alpha_2 = \frac{D^8}{64Z^4} \tag{44}$$

The coefficient $1/64Z^4$ is of major interest in this work as well as the D-dependence to which we will return later in the text.

One should observe though, that the dipole ($k = 1$) perturbation using equation (32) is of the form

$$\hat{V}_1(\mathbf{r}) = r\mathcal{C}_1^\alpha(\cos\theta) \tag{45}$$

which, using standard forms of Gegenbauer poynomials (Avery [4] page 27) with $\alpha = D/2 - 1$, yields

$$\hat{V}_1(\mathbf{r}) = r(D - 2)(\cos\theta) = r(D - 2)P_1(\cos\theta) \tag{46}$$

A *physically* motivated multipole expansion of a coulombic test charge leads to a perturbing dipole potential of *Legendre form*, even for D dimensions. The appropriate perturbing potential has to be corrected by a $(D - 2)$ factor and this in turn modifies equation (42) yielding for large D:

$$\alpha_2 = \frac{D^6}{64Z^4} \tag{47}$$

This simple proportionality for dipole perturbations does not carry over to other symmetries, i.e. the quadrupole term is a linear combination of two Gegenbauer polynomials.

Dimensional Scaling and Polarizabilities of Hydrogen-like Atoms

Large D Limit as a Harmonic Oscillator

The approximations used in dimensional scaling convert, for large D, the original potential into a harmonic oscillator problem. It is of interest to calculate approximate polarizabilities using large D expansions. The dimensional scaling procedure developed by Witten [22] and Herschbach [12], for S-state atoms, can be considered to be based on the procedure described below.

Given the generalized equation for a hydrogen-like atom in D dimensions

$$\left\{ -\frac{1}{2}\frac{\partial^2}{\partial r'^2} + \frac{(D-1)(D-3)}{8r'^2} - \frac{Z}{r'} - E_0 \right\} \psi_0(r') = 0 \qquad (48)$$

a scaling of the coordinates

$$r' = rM^2 \qquad (49)$$

and thereafter a multiplication by M^2 yields

$$\left\{ -\frac{1}{2M^2}\frac{\partial^2}{\partial r^2} + \frac{(D-1)(D-3)}{8M^2r^2} - \frac{Z}{r} - E_0 M^2 \right\} \psi_0(rM^2) = 0 \qquad (50)$$

We put $M = (D-a)$, where a is a constant to be chosen appropriately in a computational procedure in the limit of large D [15,21]. This choice can also be regarded as a choice of an effective mass $m = (D-a)^2$. The equation for a hydrogen-like atom now is

$$\left\{ -\frac{1}{2(D-a)^2}\frac{\partial^2}{\partial r^2} + \frac{(D-1)(D-3)}{8(D-a)^2r^2} - \frac{Z}{r} - E_0(D-a)^2 \right\} \psi_0(r(D-a)^2) = 0 \qquad (51)$$

with an effective potential

$$V^{eff}(r) = \frac{(D-1)(D-3)}{8(D-a)^2r^2} - \frac{Z}{r} \qquad (52)$$

which in the limit $D \to \infty$ is both a and D independent:

$$V^{eff}(r) = \frac{1}{8r^2} - \frac{Z}{r} \qquad (53)$$

The new energy is

$$E = E_0(D-a)^2 \qquad (54)$$

For infinite D the ground state energy $\overline{E}$ is given as a classical limit (no quantum fluctuations) and it is identical to the minimum of the potential $V^{eff}(r)$. Taking the derivative of $V^{eff}(r)$ we get

$$(V^{eff}(r))' = -\frac{1}{4r^3} + \frac{Z}{r^2} \qquad (55)$$

and the point of the minimum of $V^{eff}(r)$ is $\bar{r} = 1/4Z$. The system, in the infinite mass limit, is now classical and has a ground state energy:

$$\overline{E} = V^{eff}(\bar{r}) = \frac{1}{8(1/4Z)^2} - \frac{Z}{(1/4Z)} = -2Z^2 \qquad (56)$$

Converting to the original physical units, using equation (54), the energy is:

$$E_0 = \frac{E}{(D-a)^2} = -\frac{2Z^2}{(D-a)^2} \qquad (57)$$

The second derivative of $V^{eff}(r)$, which will be of interest later in the calculations, is

$$(V^{eff}(r))'' = \frac{3}{4r^4} - \frac{2Z}{r^3} \qquad (58)$$

and in the point $\bar{r} = 1/4Z$ the second derivative is

$$(V^{eff}(\bar{r}))'' = \frac{3}{4(1/4Z)^4} - \frac{2Z}{(1/4Z)^3} = 64Z^4 \qquad (59)$$

In conventional dimensional scaling the constant a is taken to be zero [22]. Various resuming considerations (work by Imbo $et.\,al.$ on shifted $1/D$ expansions [15]) lead to $a = 1$ which in the above expression (57) would yield the exact result for the ground state energy for all $D \neq 1$.

The significance of the underlying coherent state approach as described by Witten [22] motivates a harmonic form of $V^{eff}(r)$ for any central potential provided one scales appropriately [23,24]. One can expand the potential about the local minimum in order to introduce fluctuations, obtaining a quantum harmonic oscillator from the "classical" harmonic oscillator obtained for large D:

$$
\begin{aligned}
V(r) &= V^{eff}(\bar{r}) + (r - \bar{r})\{V^{eff}(r)\}'_{r=\bar{r}} + \frac{(r - \bar{r})^2}{2!}\{V^{eff}(r)\}''_{r=\bar{r}} \\[2mm]
&= -2Z^2 + \frac{1}{2}\left(r - \frac{1}{4Z}\right)^2 64Z^4
\end{aligned}
$$

$$\qquad (60)$$

Scaling back to the original coordinate r' we get

$$V(r') = -2Z^2 + \frac{1}{2}\left(\frac{r'}{(D-a)^2} - \frac{1}{4Z}\right)^2 64Z^4$$

$$= -2Z^2 + \frac{1}{2}\left(r' - \frac{(D-a)^2}{4Z}\right)^2 \frac{64Z^4}{(D-a)^4} \tag{61}$$

Comparing with the usual equation for a harmonic oscillator where we have the term $\frac{1}{2}m\omega^2$ we see that

$$\frac{1}{2}m\omega^2 = \frac{1}{2}\frac{64Z^4}{(D-a)^4} \tag{62}$$

Using $m = (D-a)^2$ we have that for a finite mass the system is described by a quantum oscillator with frequency

$$\omega = \frac{8Z^2}{(D-a)^3} \tag{63}$$

The fact that this term does not equal zero for finite D gives rise to what is called quantum fluctuations. Notice that from the minimum of the potential in equation (53) and from equation (49) one can make the estimate that

$$<r'> \simeq \frac{(D-a)^2}{4Z} \tag{64}$$

For $D = 3$ and $a = 1$ this yields a first approximation $1/Z$ to the mean value of the radial coordinate to be compared with $3/2Z$ which is the exact result (in atomic units).

Polarizabilities in D Dimensions

We will first derive rough approximations to the polarizabilities using conventional dimensional scaling as described above and then compare them with independent estimates that one can make using sum rules and operator inequalities [11]. One should study the response of the oscillator given by equation (61) to an arbitrary linear perturbation in order to get the polarizability. The range of the previously considered oscillator is actually confined to $r \geq 0$ but treating it as a usual harmonic oscillator in one dimension the ground state energy becomes

$$E = \frac{-2Z^2}{(D-a)^2} + \frac{\omega}{2(D-a)^2} \tag{65}$$

Furthermore one can also consider the *polarizability* of a one-dimensional harmonic oscillator in the variable x, as done by Fermi [9]. This yields a correction to the energy when a field $-\mathcal{G}x$ is applied (where $\mathcal{G} = (D-a)^2\mathcal{F}$ relates to the physical field strength $\mathcal{F}$ and the overall scaling of the energy previously presented), as

$$E(\mathcal{G}) = \frac{m\omega^2}{2}\left(x - \frac{\mathcal{G}}{m\omega^2}\right) - \frac{\mathcal{G}^2}{2m\omega^2} \tag{66}$$

in terms of mass m and frequency ω. Using our mass $m = (D-a)^2$ and frequency (63) it follows that

$$\frac{1}{2m\omega^2} = \frac{(D-a)^4}{64Z^4} \tag{67}$$

In terms of the properly scaled energy units (dividing $E(\mathcal{G})$ by $(D-a)^2$) the second order correction to the energy becomes

$$E(\mathcal{F}) = \frac{E(\mathcal{G})}{(D-a)^2}$$

$$= -\frac{1}{2} \cdot \frac{(D-a)^4(D-a)^4}{64Z^4(D-a)^2}\mathcal{F}^2 \tag{68}$$

$$\simeq -\frac{1}{2} \cdot \frac{D^6}{64Z^4}\mathcal{F}^2$$

which yields the standard dipole polarizability for large D:

$$\alpha_2 = \frac{D^6}{64Z^4} \tag{69}$$

Comparing equation (69) with equation (47) we see that the correct coefficient, including the Z^{-4} dependence for dipole polarizabilities, is obtained. For $D = 3$, $a = 1$ and $Z = 1$ we get a dipole polarizability $\alpha_2 = 1$, which is a rough *first* approximation to the correct value 4.5. This accuracy should be compared with the corresponding approximation to the ground state energy emerging from equation (57), i.e. $-2Z^2/(D-a)^2$ which yields the exact result $-1/2$.

Polarizabilities and Oscillator Strengths for a Hydrogen Atom in the Limit of Large D

The question which now arises is to compare equations (47), formally exact for a given multipole perturbation, and (69), obtained through a gross approximation for dipole polarizabilities at large D only. It is remarkable that the results obtained by two rather different procedures are the same.

One can use other general considerations to study the problem of polarizabilities and there are two common methods for doing so. One is based on the sum over eigenstates to the unperturbed Hamiltonian and is usually slowly converging because of the contributions from the continuum [6,14]. The other one is based on operator inequalities and can yield upper and lower bounds to the polarizabilities [11]. From the theory of operator inequalities [11,16] for the dipole polarizability in three dimensions we can write the $(\alpha_2)_{zz}$ component of the polarizability tensor in the form

$$(\alpha_2)_{zz} = 2 < 0|\hat{z}\hat{P}(\hat{H}^{(0)} - E^{(0)})^{-1}\hat{P}\hat{z}|0 > \tag{70}$$

In this expression the projection operator

$$\hat{P} = \hat{I} - |0 > < 0| \tag{71}$$

enters, where $\hat{I}$ is the identity operator.

We also have

$$(\hat{H}^{(0)} - E^{(0)})|0 > = 0 \tag{72}$$

indicating the unperturbed ground state and $\hat{z}$ which is one component of the dipole operator. One of many inequalities that follow from equation (70) is [11] :

$$(\alpha_2)_{zz} \geq \frac{2| < 0|\hat{z}P\hat{z}|0 > |^2}{< 0|\hat{z}(\hat{H}^{(0)} - E^{(0)})\hat{z}|0 >} \tag{73}$$

Using the Thomas-Kuhn-Reiche sum rule for the denominator [11] one has,

$$< 0|\hat{z}(\hat{H}^{(0)} - E^{(0)})\hat{z}|0 > = 1/2 \tag{74}$$

Furthermore, for a spherically symmetric ground state one has

$$< 0|\hat{z}|0 > = < 0|\hat{y}|0 > = < 0|\hat{x}|0 > = 0 \tag{75}$$

In addition to the above, these relations holds:

$$< 0|\hat{x}^2|0 > \; = \; < 0|\hat{y}^2|0 > \; = \; < 0|\hat{z}^2|0 > \; = \; \frac{1}{3} < 0|r^2|0 > \tag{76}$$

From equations (73) - (76) the $(\alpha_2)_{zz}$ component of the dipole polarizability is

$$(\alpha_2)_{zz} \;\geq\; \frac{2| < 0|\hat{z}^2|0 > |^2}{< 0|\hat{z}(\hat{H}^{(0)} - E^{(0)})\hat{z}|0 >}$$

$$= \; \frac{2(\frac{1}{3} < 0|r^2|0 >)^2}{< 0|\hat{z}(\hat{H}^{(0)} - E^{(0)})\hat{z}|0 >} \tag{77}$$

$$= \; \frac{4}{9}(< 0|r^2|0 >)^2$$

The relations above are valid for three dimensions, but have counterparts in D dimensions. If the coordinates are denoted x_1, x_2, ... , x_D we see that equations (74) - (76) easily can be generalized. Equation (74) is valid in all dimensions and so is equation (75) while the factor 3 in (76) is replaced by D. This gives us a relation analogous to equation (77) as:

$$(\alpha_2)_{x_1 x_1} \;\geq\; \frac{2| < 0|\hat{x}_1^2|0 > |^2}{< 0|\hat{x}_1(\hat{H}^{(0)} - E^{(0)})\hat{x}_1|0 >}$$

$$= \; \frac{2(\frac{1}{D}(< 0|r^2|0 >)^2)}{1/2} \tag{78}$$

$$= \; \frac{4}{D^2}(< 0|r^2|0 >)^2$$

The radial expectation value appearing in (78) can be estimated in two ways. The first way is by calculating it exactly using the D-dimensional ground state wave function (30) thereby obtaining

$$< r^2 > = \frac{(D-1)(D+1)D(D-1)}{16Z^2} \simeq \frac{D^4}{16Z^2} \tag{79}$$

The second way is by using the large D harmonic approximation to the equilibrium "distance" $\bar{r}$ as done earlier in equation (64)

$$< r'^2 > = < r' >^2 \simeq \frac{(D-a)^4}{16Z^2} \tag{80}$$

We see that at large D the rather gross harmonic approximation in equation (80) actually becomes *exact* as compared to equation (79).

Substituting (79) into (78) we obtain

$$\alpha_2 \geq \frac{D^6}{64Z^4} \tag{81}$$

which again agrees with the results in (47) and (69). It should be noted that Au *et.al.* have studied sum rules [2] for hydrogenic systems using logarithmic perturbation theory [1] where factorizations of Dalgarno-type appear in a natural way. They have given closed form results and even sum rules in D dimensions [1,2]. However it is not straightforward to use their results directly. In the hydrogenic problem they impose $Z = (D-1)/2$ which fixes the ground state eigenvalue $-(1/2)$ for all D which is not quite the problem addressed here.

Relationships Between Exact Results and Sum Over States

It is very useful to examine the asymptotic behaviour of the solutions of the inhomogeneous equations (33) - (36). From (37) - (38) it follows that, for large D, g_k in equation (37) reduces to one term. As an example, for $k = 1$,

$$g_1(r) = -\left\{\frac{(D-1)r^2}{4Z} + \frac{(D^2-1)r}{16Z^2}\right\} \tag{82}$$

reduces to the second term only when D is large. This means in turn that the first order wave function (33) reduces for large D to one term (asymptotic behaviour of the radial part):

$$\psi_1^{(1)}(r) \simeq r \exp\left(\frac{-2Zr}{D}\right) \tag{83}$$

which coincides with the asymptotic behaviour of the first order unperturbed excited state (c.f. Avery [4] page 80)

$$R_{21}(Zr) = Zr \exp\left(\frac{-2Zr}{D+1}\right) \simeq Zr \exp\left(\frac{-2Zr}{D}\right) \tag{84}$$

One could express this as

$$\psi_1^{(1)}(r) \sim R_{21}^{(0)}(r) \tag{85}$$

namely the first order wave function becomes, for large D, identical to an unperturbed radial wave function belonging to the first excited state. The asymptotic behaviour leading to (85) is implicitly related to the asymptotics of (42)-(44). It is important though to verify the above results by explicitly evaluating the sum over states expressions from perturbation theory. One obtains for the first term, when D is large:

$$\Delta_{2p} = \frac{2| < \psi_{1s}|\hat{V}_1|\psi_{2p} > |^2}{E_{2p} - E_{1s}} = \frac{D^6}{64Z^4} \tag{86}$$

The wave functions are given by

$$\begin{aligned}
\psi_{1s} &= [S_D]^{-1/2} R_{10} \\[2ex]
\psi_{2p} &= [D/S_D]^{1/2} \cos\theta R_{21} \\[2ex]
S_D &= \int d\Omega
\end{aligned} \tag{87}$$

The radial parts are given by Avery [4] (page 80) and so are the energies (page 79). The perturbation used in (86) is the first order Legendre polynomial, i.e.

$$\hat{V}_1 = r \cos\theta \tag{88}$$

Similarly, for the next term in the sum over states ψ_{3p} with the radial part R_{31} and the same angular part as ψ_{2p}, one obtains a contribution to the polarizability

$$\Delta_{3p} = \frac{1}{2D} \cdot \frac{D^6}{64Z^4} \tag{89}$$

which does *not* contribute asymptotically and similar results are obtained for the higher terms. The significance of these results is that one can use the *first* term in the sum over states and that the operator approximations and inequalities previously used become exact for large D.

Discussion

The possibility of using Sturmian functions for expanding arbitrary radial functions has been known for some time [18]. It was used by Avery and collaborators [3,4] in order to express exact solutions of general electronic problems in terms of hyperspherical coordinates. It is recognized that dipole polarizabilities might also be calculated using Sturmian functions. The two-term-expansions of equations (22) and (37) imply that the solutions of the inhomogeneous equations in perturbation theory can be expressed in terms of finite expansions of Sturmian functions [18]. It is of significance that finite expansions remain valid for arbitrary extended dimensions and multipole perturbations. It is observed that the two terms reduce asymptotically to only one and this has the significance that the sum over states reduces to one term only. This is in counterdistinction to the calculations for $D = 3$ where the first term plays a significant but not dominating role, requiring contributions from the excited states as well as the continuum.

In this work we have calculated the polarizabilities for the hydrogen atom. The question of calculating polarizabilities of arbitrary electronic systems (many-electron atoms, molecules etc.) using $1/D$-expansions remains open. The results presented here are evidence that when using harmonic approximations and large D results one will in general require anharmonicities, that is higher terms in the expansion, for accurate evaluation of the polarizabilities. Singularity analysis and summations of $1/D$ expansions as described by Goodson and Cabrera in this volume [10] should also be useful for improved calculations of polarizabilities.

The main results of this work are the exact hyperpolarizabilities in D dimensions for a class of multipole perturbations to start with. The possibility of computing accurate approximations to them for high D using conventional dimensional scaling is then established. Finally, using operator approximations and sum rules in D dimensions, comparisons of various approaches, such as the sum over states, and scaling transformations are possible. The approximations become exact for large D. The asymptotic value for the dipole polarizability for hydrogen-like ions, for a Legendre perturbation, $\alpha_2 = (64Z^4)^{-1}D^6$ is obtained.

Acknowledgements

We wish to thank Dr. John Avery for keen interest and most valuable help in hyperspace.

References

1. Y. Aharonov and C.K. Au, Phys. Rev. Lett. **42**, 1582 (1979). C.K. Au and Y. Aharonov, Phys. Rev. A **20**, 2245 (1979).
2. C.K. Au, J. Phys. B **17**, L553 (1984). C.K. Au, Phys. Rev. A **33**, 717 (1986). C.K. Au, J. Phys. B **20**, L115 (1987). C.K. Au, K.L. Poon and K. Young, J. Phys. B **24**, 4671 (1991).
3. J. Avery and D.R. Herschbach, IJQC **41**, 673 (1992).
4. J. Avery, *Hyperspherical Harmonics; Applications in Quantum Theory*, Kluwer Academic Publishers, Dordrecht, Netherlands (1989).
5. S.M. Blinder, J. Math. Phys. **25**, 905 (1984).
6. S. Canuto and O. Goscinski, IJQC **16**, 985 (1979).
7. A. Dalgarno, Adv. Phys. **11**, 281 (1962).
8. A. Dalgarno and J.T. Lewis, Proc. Roy. Soc. A **233**, 70 (1955).
9. E. Fermi *Notes on Quantum Mechanics*, University of Chicago Press, Chicago, (1961), pp 89-91.
10. D.Z. Goodson and M. López-Cabrera, chapter 7.1 in this volume
11. O. Goscinski, IJQC **2**, 761 (1968)
12. D.R. Herschbach, J. Chem. Phys. **84**, 838 (1986).
13. D.R. Herschbach, chapter 2 in this volume.
14. E.A. Hylleraas, Z. Physik **48**, 469 (1928). See also discussion by P.-O. Löwdin, Advances in Chemical Physics **2**, 207 (1959).
15. T. Imbo, A. Pagnamenta and U. Sukhatme, Phys. Rev. D **29**, 1669 (1984).
16. P.O. Löwdin, Phys. Rev. **139 A**, 357 (1965). P.O. Löwdin, J. Chem. Phys. **43**, S175 (1965).
17. L.D. Mlodinow and M.P. Shatz, J. Math. Phys. **25**, 943 (1984).
18. M. Rotenberg, Adv. At. Mol. Phys. **6**, 233 (1970).
19. E. Schrödinger, Ann d. Physik **79**, 361 (1926).

20. T.C. Scott, R.A. Moore, G.J. Fee, M.B. Monagan, G. Labahn
    and K.O. Geddes, Int. J. Mod. Phys. C **1**, 53 (1990).
21. U. Sukhatme and T. Imbo, Phys. Rev. D **28**, 418 (1983).
22. E. Witten, Physics Today **33, no.2**, 38 (1980).
23. L.G. Yaffe, Rev. Mod. Phys. **54**, 407 (1982).
24. L.G. Yaffe, Physics Today **36, no.8**, 50 (1983).

7.3 The Dimensional Dependence of Rates of Convergence of Rayleigh-Ritz Variational Calculations on Atoms and Molecules

John D. Morgan, III
Department of Physics and Astronomy
University of Delaware
Newark, DE 19716, USA

Abstract

I shall review the types of singularities possessed by the wavefunctions of atoms and molecules and how they determine the rate of convergence of a Rayleigh-Ritz variational calculation with a given basis. I then show how these results generalize to an arbetrary number of dimensions D. There is shown to be a delicate balance between singularity effects for small D and localization effects for large D. A byproduct of this analysis is the suggestion of an alternative to the moment method for computing the coefficients of dimensional expansions, which can be expected to be significantly more efficient and to allow accurate calculations to be performed on larger atoms and molecules.

Introduction

The attainment of high accuracy in Rayleigh-Ritz variational calculations on atoms and molecules is crucially dependent on how rapidly such calculations converge to the exact limit as the number of basis functions increases. In other words, one wants to know how rapidly finite linear combinations of basis functions converge to the exact solution of the Schrödinger equation in the infinite-dimensional Hilbert space. It is thus intuitively obvious that a key issue is the magnitude of that part of the exact wavefunction which is orthogonal to the fixed finite basis which one is using in the Rayleigh-Ritz variational calculation.

D. R. Herschbach et al. (eds.), Dimensional Scaling in Chemical Physics, 336–358.

This idea made its appearance in a relatively neglected article by Charles Schwartz [1] on the rates of convergence of basis set expansions of simple functions of a single variable. A crucial idea of quite general importance which emerged from Schwartz's study is that the rate of convergence of a basis set expansion is determined by how well finite linear combinations of basis functions duplicate the singularity structure of the exact function which one is trying to approximate. The analytical study of rates of convergence of basis set expansions then received little attention for almost two decades, until it was taken up again by Bruno Klahn and me [2]. Klahn and I found a more general operator-based formalism for predicting the rate of convergence of a basis set expansion, and we showed how in the case of slowly convergent expansions the expectation values of operators containing large powers of position variables or the momenta could converge very slowly, or even diverge or converge to incorrect values as the basis approached completeness. Soon thereafter my colleague at the University of Delaware, Robert N. Hill, found a powerful method for extending the analysis of Klahn and me, and used it to obtain analytically the first two coefficients of the partial wave expansion for the helium ground state energy in terms of integrals over the exact eigenfunction [3]. (One does not, of course, know the exact eigenfunction for helium, but the use of highly accurate trial functions in Hill's formulas yielded numerical results in excellent agreement with the empirical coefficients obtained earlier by Carroll, Silverstone, and Metzger [4].) For a readable overview of the analytical theory (as of 1988) of rates of convergence of Rayleigh-Ritz variational calculations on atoms and molecules, see my contribution to the proceedings of the Versailles NATO Advanced Study Workshop on numerical approaches to the electronic structure of atoms and molecules [5].

In this article I shall discuss how spatial dimension D impacts the rates of convergence of Rayleigh-Ritz variational calculations on atoms and molecules, in which the nuclei and electrons are assumed to interact by $1/r$ potentials (rather than Coulombic potentials, which are proportional to $1/r^{D-1}$ and hence overwhelm the kinetic energy operator, which scales as $(\text{length})^{-2}$, for $D > 4$). As we shall see, regarding the spatial dimension D as a flexible parameter provides valuable insight into the theory in the practical case of $D = 3$. Ubiquitous powers of 3, whose origin appears mysterious if our thinking is confined to a three-dimensional world, turn out to arise precisely from the number of spatial dimension of the world in which we live. Furthermore, we shall see that doing an accurate correlated electronic structure calculation with the ubiquitous partial wave expansion involves avoiding on the one hand the Scylla of tight electron localisation at large D and on the other hand the Charybdis of extremely slow duplication of electron-electron cusp structure at small D. Thus it is indeed fortunate for practicing quantum chemists that they have been floating all along in the relative calm of $D = 3$ and hence $\delta = 1/3$, roughly midway between the two hitherto unperceived perils at $\delta = 0$ and $\delta = 1$.

Analytic Structure of Atomic and Molecular Wavefunctions

Let us consider the Hamiltonian for a system of N point particles, each of mass m_i and charge q_i, interacting by Coulomb potentials. In units where $\hbar = 1$, the non-relativistic N-body Hamiltonian is

$$\sum_{i=1}^{N} -\frac{1}{2m_i}\nabla_i^2 + \sum_{j>i=1}^{N} \frac{q_i q_j}{r_{ij}}, \tag{1}$$

where r_{ij} is the distance between particles i and j. Since this differential operator is elliptic, we know from a theorem of Fritz John [6] that the solutions of the Schrödinger Equation $(H - E)\psi = 0$ are analytic in Cartesian coordinates wherever the potential energy is analytic in these coordinates. However, the Coulomb potential is not analytic where two or more particles coalesce, so there is no reason why the eigenfunctions should be analytic at such coalescences. It was proved by T. Kato [7] that at two particle coalescences wavefunctions have cusps (discontinuities of derivatives with respect to Cartesion coordinates) which are described by what have come to be known as the *Kato cusp conditions*:

$$\left(\frac{\partial \hat{\psi}}{\partial r_{ij}}\right)_{r_{ij}=0} = \mu_{ij}\, q_i\, q_j\, \psi(r_{ij} = 0), \tag{2}$$

where $\mu_{ij} = m_i m_j/(m_i + m_j)$ is the reduced mass of the two-particle subsystem and $\hat{\psi}$ denotes the average of ψ over an infinitesimal sphere centered at $r_{ij} = 0$. Although Kato's fully rigorous derivation of the cusp conditions is rather long and complex, one can easily understand whence they arrive in an intuitive way by examining the Schrödinger Equation $(H - E)\psi = 0$ for the multiparticle system and transforming the variables $\mathbf{r}_i$ and $\mathbf{r}_j$ into relative and center-of-mass coordinates

$$\mathbf{R}_{ij} = \frac{m_i \mathbf{r}_i + m_j \mathbf{r}_j}{m_i + m_j}$$

$$\mathbf{r}_{ij} = \mathbf{r}_i - \mathbf{r}_j \tag{3}$$

Then as $r_{ij} \to 0$ and all other interparticle distances remain bounded away from 0, the only singularity in the potential energy term in the Schrödinger Equation is

$$\frac{q_i\, q_j}{r_{ij}}\, \psi \,, \tag{4}$$

and the kinetic energy operator will contain a term

$$-\frac{1}{2\mu_{ij}}\left(\frac{\partial^2}{\partial r_{ij}^2} + \frac{2}{r_{ij}}\frac{\partial}{\partial r_{ij}} + \frac{L_{ij}^2}{r_{ij}^2} \right)\psi \,. \tag{5}$$

If we now perform an average about an infinitesimally small sphere centered at $r_{ij} = 0$, this will eliminate from ψ all except the s-wave component, which is itself annihilated by L_{ij}^2. Assuming that $\partial^2 \hat{\psi}/\partial r_{ij}^2$ is bounded, as it is for eigenfunctions of the hydrogen atom, then the coefficients of the $1/r_{ij}$ singularities in eqq. (4) and (5) should cancel, which leads to the Kato cusp condition given by eq. (2).

This simple rationalisation of the Kato cusp conditions in 3 spatial dimensions makes it obvious how to extend them to higher dimensions, for a Hamiltonian in which the interparticle potentials retain their $1/r$ character. All the above analysis goes through unchanged, save that in eq. (5) we make the replacement

$$\frac{2}{r_{ij}}\frac{\partial}{\partial r_{ij}} \to \frac{D-1}{r_{ij}}\frac{\partial}{\partial r_{ij}} \,, \tag{6}$$

which leads to the D-dimensional Kato cusp condition

$$\left(\frac{\partial \hat{\psi}}{\partial r_{ij}} \right)_{r_{ij}=0} = \frac{2}{D-1}\mu_{ij}\, q_i\, q_j\, \psi(r_{ij} = 0) \,, \tag{7}$$

From this expression one can see that as D increases, the amount of the discontinuity in the first derivatives of ψ with respect to Cartesian coordinates decreases, which suggests that as $D \to \infty$ cusps in wavefunctions will be of diminished importance.

Let us examine two representative illustrations of cusps. At an electron-nucleus cusp, μ_{ij} is the electron-nucleus reduced mass μ and $q_i q_j = -Z$, so that $\mu_{ij} q_i q_j = -\mu Z$ is negative, and the wavefunction decreases as r_{ij} increases away from 0, as shown in the figure below:

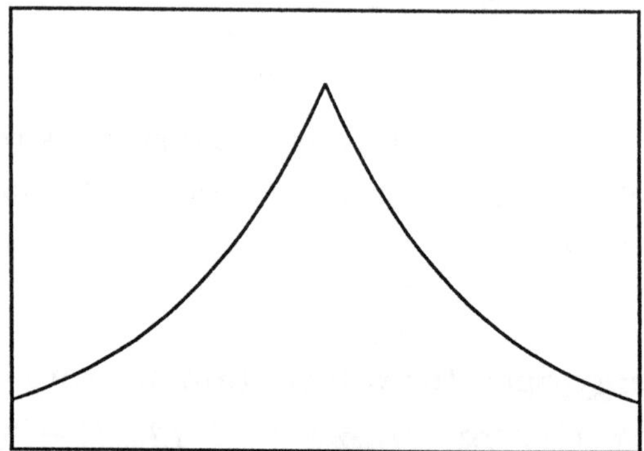

The behavior of a hydrogenic eigenfunction provides a typical example of such cusp behavior. At electron-electron coalescences of electrons with antiparallel spins (so that the wavefunction does not vanish because of the Pauli exclusion principle), $\mu_{ij} = 1/2$ and $q_i q_j = 1$ is positive, so that the wavefunction displays an inverted cusp behavior, as illustrated in the graph below:

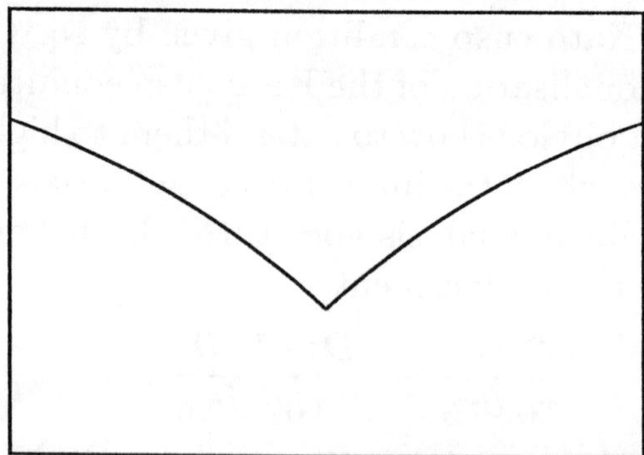

Ternary and higher coalescences of three or more particles are described by an expansion developed by V. A. Fock [8] in 1954 and subsequently studied by many atomic physicists in the former Soviet

Union [9] and the West [10-13]. The Fock expansion features an expansion

$$\sum_{j,k=0}^{\infty} R^{j/2}(\ln R)^k \psi_{j,k}(\Omega) \tag{8}$$

in half-integral powers of the hyperradius R and integral powers of $\ln R$, where

$$R = \sum_i r_i^2 \tag{9}$$

is the sum of the squares of the electron-nucleus distances, and the coefficients $\psi_{j,k}(\Omega)$ of the powers of R and $\ln R$ are functions of the hyperspherical angles. The first several $\psi_{j,k}$ are known in closed form. For singlet states (in $D = 3$ spatial dimensions), it is straightforward to derive the results

$$\psi_{0,0} = 1$$
$$\psi_{1,0} = -Z\frac{r_1 + r_2}{R^{1/2}} + \frac{1}{2}\frac{r_{12}}{R^{1/2}} \tag{10}$$
$$\psi_{2,1} = -Z\frac{\pi - 2}{3\pi}\frac{\mathbf{r}_1 \cdot \mathbf{r}_2}{R}$$

One can readily see that

$$\psi_{0,0} + R^{1/2}\psi_{1,0} = 1 - Z(r_1 + r_2) + \frac{1}{2}r_{12} , \tag{11}$$

as would be expected from the Kato cusp condition. Efforts to solve for $\psi_{2,0}$ were unsuccessful for many years, until it was obtained in closed form by Gottschalk, Abbott, and Maslen [13] in terms of Lobachevsky's function and more elementary functions.

In dimensions D different from 3, the above results are only slightly changed: the functional forms remain the same, while the coefficients are multiplied by factors of $(D - 1)/2$ and the like.

The convergence properties of Fock's expansion have been studied mathematically by Macek [10], Leray [11], and me [12], and it now is essentially certain that Fock's expansion does provide a convergent representation of many-electron wavefunctions, although a few mathematical details remain to be wrapped up.

Rates of Convergence of Rayleigh-Ritz Variational Calculations (for $D = 3$)

One of the most commonly used expansions in the theory of the electronic structure of atoms and molecules is the partial wave expansion, in which individual atomic orbitals are expressed as products of radial functions and spherical harmonics. Appropriately symmetrized sums of products of the spherical harmonics for the coordinates of each particle can be formed to yield eigenfunctions of total L^2, L_z, S^2, and S_z. To prevent lengthy expressions involving 3-j and 6-j symbols from obscuring the essential physics, I shall focus on the partial wave expansion for an S-state of the helium atom:

$$\Psi(\mathbf{r}_1, \mathbf{r}_2) \simeq \Psi_L(\mathbf{r}_1, \mathbf{r}_2) = \sum_{l=0}^{L} f_l(r_1, r_2) \, P_l(\cos\theta) \tag{12}$$

where the coefficient functions $f_l(r_1, r_2)$ are optimised to minimise the variational energy. It was first found empirically by Carroll, Silverstone, and Metzger [4] and then was proven theoretically by my Delaware colleague Robert N. Hill [3] that the variational energies obtained from the partial wave expansion converge to the exact result with errors ΔE_L which obey the asymptotic relation

$$\Delta E_L = C_3(L+1)^{-3} + C_4(L+1)^{-4} + \dots , \tag{13}$$

where

$$C_3 = 2\pi^2 \int_0^\infty dr \, r^5 |\Psi(r, r, 0)|^2$$

$$C_4 = \frac{12\pi}{5} \int_0^\infty dr \, r^6 |\Psi(r, r, 0)|^2 \tag{14}$$

Numerical evaluation of these integrals using the highly accurate 230-term function of Freund, Huxtable, and Morgan [14] to represent the exact ground state $\Psi(r_1, r_2, r_{12})$ yields

$$C_3 = 0.0247419 \dots$$

$$C_4 = 0.00774727 \dots , \tag{15}$$

in excellent agreement with the numerical estimates which Carroll, Silverstone, and Metzger [4] obtained by numerical fitting. This rate of convergence is rather slow, as can be seen from the table below, which lists the desired accuracy in the energy, the number $(L+1)$ of partial waves required to obtain it, and the total number of basis functions, which grows like $(L+1)^2$:

ΔE_L (a.u.)	$L+1$	$(L+1)^2$
10^{-2}	2	4
10^{-3}	3	9
10^{-4}	6	36
10^{-5}	13	169
10^{-6}	28	784
10^{-7}	60	3,600
10^{-8}	130	16,900

Clearly it requires an enormous number of basis functions to achieve sub-microhartree accuracy if one is using a partial wave expansion. The origin of the slow convergence is the slow representation of the electron-electron cusp at $r_{12} = 0$ by the partial wave expansion, which for any finite L yields an approximate wavefunction analytic in Cartesian coordinates at $r_{12} = 0$. Much faster convergence can be obtained by using a basis which explicitly incorporates linear terms (and higher odd powers) in r_{12}. Such a basis, for example, is provided by the Pekeris functions

$$L_p(\xi u)\, L_q(\xi v)\, L_r(\xi w)\, e^{-\xi(u+v+w)/2} \tag{16}$$

of perimetric coordinates

$$\begin{aligned}
u &= (-r_1 + r_2 + r_{12}) \\
v &= (\ r_1 - r_2 + r_{12}) \\
w &= 2(r_1 + r_2 - r_{12})\,,
\end{aligned} \tag{17}$$

where ξ is a flexible scaling parameter which can be varied to optimise the variational energy, although for convenience in setting up his equations Pekeris set it equal to its non-optimal value $\sqrt{-E}$ to eliminate the energy term from Schrödinger Equation. In his famous calculation with 1078 such basis functions [15], Pekeris obtained a variational upper bound of

$$-2.903724375 \text{ a.u.}$$

which is accurate to 2×10^{-9} a.u. The improvement in accuracy achieved with many fewer basis functions is attributable to the inclusion in the basis of terms linear in $r_{12} = (u + v)/(2\xi)$.

It helps even more to build into the basis terms which describe the singularity in the wavefunction at the 3-particle coalescence, as was first demonstrated by Frankowski and Pekeris [16], who with only 246 functions, including some suggested by the Fock expansion, obtained a variational upper bound of

$$-2.9037243770326 \text{ a.u.}$$

By including some higher-order terms from the Fock expansion which Frankowski and Pekeris should not have omitted, and including many more powers of $(r_2 - r_1)/(r_1 + r_2)$ and $r_{12}/(r_1 + r_2)$ to reflect electron-electron correlation on short to moderate length scales, Freund, Huxtable, and Morgan [14] obtained with 230 functions an upper bound of

$$-2.90372437703407 \text{ a.u.,}$$

and with 476 such functions, including some with an 'open-shell' exponential factor to account for long-range 'in-out' electron correlation, Baker, Freund, Hill, and Morgan [17] obtained an upper bound of

$$-2.9037243770341184 \text{ a.u.,}$$

in close agreement with Kleindienst and Emrich's value [18]

$$-2.903724377034119 \text{ a.u.}$$

obtained with 448 Frankowski-Pekeris functions [16] with a single length scale. It would be practically impossible to achieve such high accuracy using a partial wave expansion to represent the wave function.

Powers of 3 appear in several contexts in the analysis of rates of convergence of Rayleigh-Ritz variational calculations:

(i) As we have just seen, in the partial wave expansion

$$\Delta E_L \sim (L + 1)^{-3} \tag{18}$$

(ii) If one uses a basis of the first N spherically symmetry Hermite orthonormal functions, with a fixed length scale, in a Rayleigh-Ritz calculation of the energy levels of s-states of the hydrogen atom, it was found by Klahn and Morgan [2] that the errors ΔE_N fall off as

$$\Delta E_N \sim N^{-3/2} = (N^{1/2})^{-3}, \tag{19}$$

a result which was then proven rigourously by Hill [3].

(iii) If one uses a basis of N Gaussian basis functions with 'floating' (i.e., variable) exponents

$$\{\exp(-\zeta_j^{(N)} r^2)\}_{j=1}^N \tag{20}$$

in a Rayleigh-Ritz variational calculation of energies of s-states of the hydrogen atom, it was found numerically by Haywood, Nicholson, and Morgan [19] with further theoretical analysis by Kenneth Wilson [20], that the errors ΔE_N obey

$$\Delta E_N \sim (\zeta_{\max}^{(N)})^{-3/2} , \tag{21}$$

where $\zeta_{\max}^{(N)}$ is the largest of the $\zeta_j^{(N)}$.

The origin of these ubiquitous powers of 3 can be easily understood in terms of a simple heuristic argument. We use the Fourier transform to get from configuration space to momentum space, under which a function with a hydrogenic cusp is mapped to a function with p^{-4} fall-off at large momenta p:

$$e^{-Zr} \;\rightarrow\; \frac{1}{(Z^2 + p^2)^2} . \tag{22}$$

Suppose one's finite N-dimensional basis has little weight on a momentum scale beyond $\tilde{p}_N$. The error in the variational energy, ΔE_N, is roughly proportional to (and if the virial theorem holds is exactly equal to) the error in the kinetic energy ΔT_N:

$$\Delta E_N \sim \Delta T_N \sim \int_{\tilde{p}_N}^{\infty} dp\, p^2 \frac{p^2}{2} \left(\frac{1}{(Z^2 + p^2)^2} \right)^2 , \tag{23}$$

where the first factor of p^2 comes from the 3-dimensional Jacobian, the second factor of $p^2/2$ comes from the kinetic energy, and the third factor, which behaves as p^{-8} for large p, comes from the density in momentum space. Proceeding to the large $\tilde{p}_N$ limit and ignoring multiplicative constants, we have

$$\Delta E_N \sim \Delta T_N \sim \int_{\tilde{p}_N}^{\infty} dp\, p^2\, p^2\, p^{-8} = \int_{\tilde{p}_N}^{\infty} dp\, p^{-4} \sim (\tilde{p}_N)^{-3} . \tag{24}$$

All the previously mentioned rates of convergence can be seen to be consequences of this result:

(i) In the case of the partial wave expansion, the angular contribution to the kinetic energy of a Legendre polynomial $P_l(\cos\theta)$ can be calculated by re-expanding it in terms of 1-particle spherical harmonics:

$$\frac{1}{2}\left(\frac{l_1^2}{r_1^2} + \frac{l_2^2}{r_2^2}\right) P_l(\cos\theta) = \frac{1}{2} l(l+1)\left(\frac{1}{r_1^2} + \frac{1}{r_2^2}\right) P_l(\cos\theta) , \qquad (25)$$

so for typical electron-nucleus distances r_1 and r_2 of the order of an a.u. we see that

$$\tilde{p}_L \sim \sqrt{L(L+1)} \sim L+1 \qquad (26)$$

and thus

$$\Delta E_L \sim (L+1)^{-3} . \qquad (27)$$

(ii) In the case of the approximation of a hydrogenic s-function by a basis of the first N spherically symmetric Hermite orthonormal functions, which are eigenfunctions of the spherical harmonic oscillator Hamiltonian, the nth spherically symmetric Hermite orthonormal function ϕ_n obeys the Schrödinger Equation

$$\left(-\frac{1}{2}\nabla^2 + \frac{1}{2}r^2\right)\phi_n = (2n + 3/2)\phi_n , \qquad (28)$$

and by the virial theorem

$$\langle p^2 \rangle = E_n = (2n + 3/2), \qquad (29)$$

so

$$\tilde{p}_N \sim \sqrt{E_N} \sim N^{1/2} \qquad (30)$$

and thus

$$\Delta E_N \sim (N^{1/2})^{-3} \qquad (31)$$

(iii) In the case of the approximation of a hydrogenic s-function by a basis of N Gaussians with floating exponents, it is obvious that under Fourier transformation

$$e^{-\zeta r^2} \rightarrow e^{-\frac{1}{4\zeta}p^2} , \qquad (32)$$

and thus it is evident that the biggest kinetic energy associated with the Gaussian basis is set by $2\zeta_{\max}^{(N)}$ and that the biggest momentum scale is set by

$$\left(4\zeta_{\max}^{(N)}\right)^{1/2} = 2\left(\zeta_{\max}^{(N)}\right)^{1/2} \tag{33}$$

and hence that

$$\Delta E_N \sim \left(\left(\zeta_{\max}^{(N)}\right)^{1/2}\right)^{-3} \tag{34}$$

Rates of Convergence of Rayleigh-Ritz Variational Calculations (for arbitrary D)

We have seen that the ubiquitous powers of 3 can be explained in terms of the power of 3 in the relationship

$$\Delta E_N \sim (\tilde{p}_N)^{-3} . \tag{35}$$

To understand the origin of this power of 3 in a more fundamental way, let us consider the generalisation of our analysis to a hydrogenic wavefunction of angular momentum l in D spatial dimensions. For simplicity, we take the principal quantum number $n = l + 1$. Then such a hydrogenic eigenfunction has in configuration space the form

$$r^l \, Y_{l,\dots}(\Omega) \, e^{-\frac{2Z}{D-1}r} \tag{36}$$

which under Fourier transformation is mapped to

$$p^l \, Y_{l,\dots}(\Omega) \, \frac{1}{\left(\left(\frac{2Z}{D-1}\right)^2 + p^2\right)^{\frac{D+1}{2}+l}} . \tag{37}$$

Thus if a finite N-dimensional basis has insufficient variational flexibility on a momentum scale beyond $\tilde{p}_N$, the error ΔE_N goes as

$$\Delta E_N \sim \Delta T_N \sim \int_{\tilde{p}_N}^{\infty} dp \, p^{D-1} \frac{p^2}{2} \frac{p^{2l}}{\left(\left(\frac{2Z}{D-1}\right)^2 + p^2\right)^{D+1+2l}} . \tag{38}$$

Thus for large values of $\tilde{p}_N$, and as always ignoring inessential multiplicative constants, we see that

$$\Delta E_N \sim \int_{\tilde{p}_N}^{\infty} dp \, p^{D+1} \frac{p^{2l}}{p^{2(D+1+2l)}} \sim \int_{\tilde{p}_N}^{\infty} dp \, p^{-(D+2l+1)} \sim (\tilde{p}_N)^{-(D+2l)} , \tag{39}$$

which depends, as it should, on D and l only through the combination $(D + 2l)$. We therefore see that the power of 3 to which $(L + 1)^{-1}$ is raised in the rate of convergence of variational energies obtained using the partial wave expansion is precisely *the number of spatial dimensions of the universe in which we live.*

Let us now ponder some consequences of this fact. If our universe had 2 spatial dimensions (and charged particles still interacted by potentials of the form $1/r$), then the partial wave expansions for correlation energies would converge even more slowly, like $(L + 1)^{-2}$, than they do in our 3-dimensional universe. Elsewhere in these proceedings Pekka Pyykkö and Yongfang Zhao have provided us with a sketch, as determined by Hartree-Fock calculations, of chemistry in "Flatland", where $D = 2$. The even slower convergence of partial wave expansions for correlation energies in "Flatland", because of the increased importance of the correlation cusps, suggests that it would be extremely laborious to compute accurate correlation energies of these 2-dimensional atoms and molecules. Thus this is one more reason (besides the old familiar reasons such as inability to tie knots and thus one's own shoelaces in 2 dimensions) why people who earn their livings doing quantum chemical calculations employing truncated partial wave expansions should be glad they live in a 3-dimensional rather than a 2-dimensional universe.

One might think that correlation energies would be even easier to calculate in a universe with $D \geq 4$, and indeed the error in a truncated partial wave expansion due to the inexact representation of the electron-electron cusps becomes much less important, falling off as $(L + 1)^{-D}$. However, as D increases another problem becomes of increasing importance: the phenomenon of localisation about the minimum, or minima, of the effective potential.

The phenomenon of localisation is readily illustrated by examining the the figure below, which is taken from Dudley Herschbach's introductory article [21] on the theory of dimensional scaling of atomic properties.

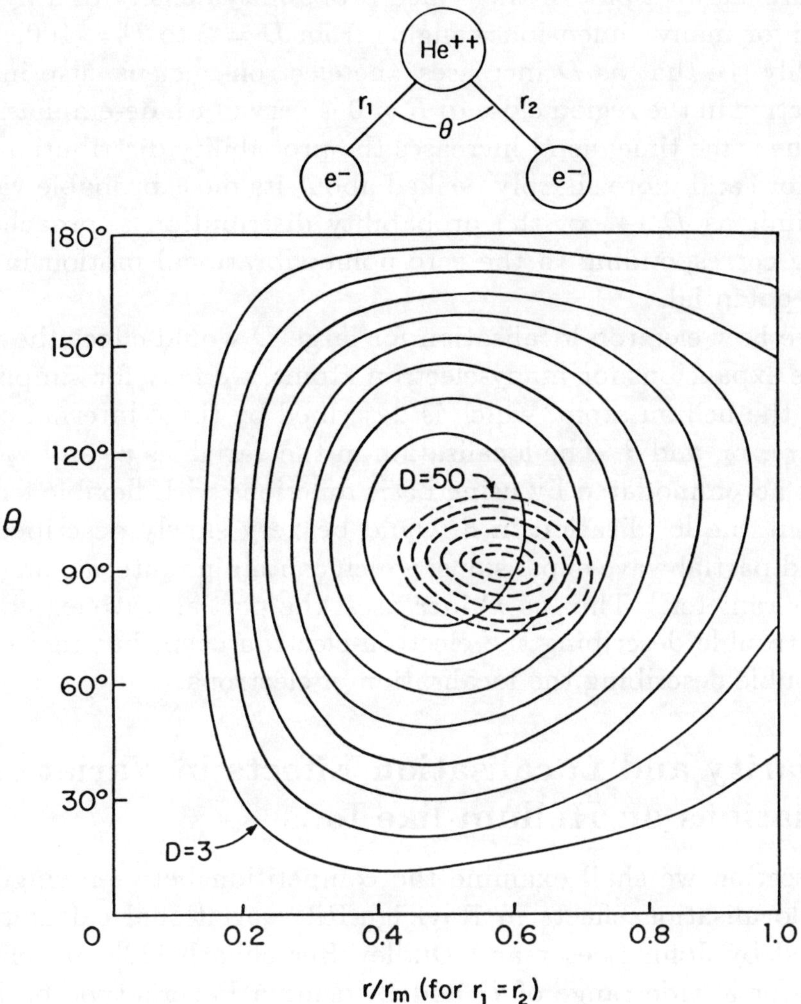

Figure 3. Contour map of Jacobian-weighted probability density , $|\Phi_D|^2 = J_D|\Psi|^2$ for ground-state helium. Solid contour pertains to 90% of maximum; others at 10% intervals. I thank Dudley Herschbach for supplying this figure.

This figure shows a plot of the scaled probability density of a hydrogenic ion for many dimensions ranging from $D = 2$ to $D = 100$. One can readily see that as D increases, the electron-nucleus cusp in the wavefunction in the region close to $R = 0$ is very much de-emphasised. But at the same time, as D increases the probability distribution becomes more and more sharply peaked about its most probable value. In the limit as $D \to \infty$, the probability distribution approaches a Gaussian corresponding to the zero-point vibrational motion in the effective potential.

To see how electron localisation for large D would effect the partial wave expansions for many-electron atoms, we may for simplicity consider the helium atom, which is described by three internal coordinates, r_1, r_2, and θ. The localisation in the variables r_1 and r_2 can easily be accommodated by using basis functions with flexible length scales, but the localisation in θ would be very slowly described by truncated partial wave expansions, which contain no internal angular scaling parameter.[1] Thus as D increases, the partial wave expansion has less trouble describing the electron-electron cusp, but more and more trouble describing the localisation of electrons.

Singularity and Localisation Effects in Variational Calculations on Helium-like Ions

In this section we shall examine the competition between singularity and localisation effects in Rayleigh-Ritz variational calculations performed by John Loeser and Dudley Herschbach [22] on helium-like ions for a wide range of D and Z, using a Pekeris-type basis of products of generalised Laguerre functions

$$L_p^\alpha(\xi u)\, L_q^\alpha(\xi v)\, L_r^\alpha(\xi w)\, e^{-(u+v+w)/2} \tag{40}$$

of perimetric coordinates

$$\begin{aligned}
u &= (-r_1 + r_2 + r_{12}) \\
v &= (\ r_1 - r_2 + r_{12}) \\
w &= 2(r_1 + r_2 - r_{12})\,,
\end{aligned} \tag{41}$$

---

[1] I am grateful to Martin Dunn for emphasising this point to me.

where $\alpha = (D-3)/2$ and ξ can be varied to optimise the variational energy, although as in Pekeris' calculation it was set equal to $\sqrt{-E}$. This basis has the correct singularity structure at both the electron-nucleus and the electron-electron coalescences, but *not* at the 3-particle coalescence. One would accordingly expect somewhat greater convergence difficulties as $D \to 1^+$ due to the increased importance of the 3-particle coalescence, and as Loeser and Herschbach themselves stated [22, p. 3885], "for $D \to \infty$ (the convergence deteriorates) because the charge distribution becomes increasingly focused at fixed values of the coordinates, which makes an adequate representation increasingly difficult to achieve with the chosen basis set".

In their original article [22], Loeser and Herschbach used the variationally computed energies $E(Z, \delta)$, where $\delta = 1/D$, to determine numerically by multiple differencing estimates of the first several coefficients ϵ_n in the $\delta = 1/D$ expansion of the energies for various values of Z:

$$
\begin{aligned}
\epsilon_0 &= \lim_{\delta \to 0} E(\delta) \\[1ex]
\epsilon_1 &= \lim_{\delta \to 0} \frac{E(\delta) - \epsilon_0}{\delta} \\[1ex]
\epsilon_2 &= \lim_{\delta \to 0} \frac{E(\delta) - (\epsilon_0 + \epsilon_1 \delta)}{\delta^2} \\[1ex]
\epsilon_3 &= \lim_{\delta \to 0} \frac{E(\delta) - (\epsilon_0 + \epsilon_1 \delta + \epsilon_2 \delta^2)}{\delta^3} \\[1ex]
\vdots \;\; &= \;\; \vdots \\[1ex]
\epsilon_n &= \lim_{\delta \to 0} \frac{E(\delta) - \sum_{j=0}^{n-1} \epsilon_j \delta^j}{\delta^n}
\end{aligned}
\tag{42}
$$

The degredation in the accuracy of the $E(Z, \delta)$ as $\delta \to 0$ caused the number of significant digits in the estimates of each ϵ_n to decrease rapidly with increasing n.

Shortly thereafter David Goodson and Dudley Herschbach found a way to compute directly the expansion coefficients by a method of moments [23]. This method has the great advantage allowing one to compute the expansion coefficients to much higher order (see Goodson's presentation elsewhere in these proceedings). At one point it became desirable to double check for any possible algorithmic errors in the moments code (which is rather complicated), and at my suggestion John Loeser used Goodson's coefficients ϵ_0, ϵ_1, . . . , ϵ_{n-1} and his own numerically determined $E(Z,\delta)$ in the above formula to determine his own independent estimate of ϵ_n. Through tenth order, when the inaccuracies Loeser's variationally computed $E(Z,\delta)$ began to have a significant effect, the agreement was excellent, and thus indicated that Goodson's code was working correctly.

I should now like to examine the specifics of the electron localisation as $D \to \infty$, and thereby diagnose the cause of the deterioration of the convergence of the $E(Z,D)$ and indicate a cure which will yield *improving* convergence as $D \to \infty$.

As discussed by Dudley Herschbach in his fundamental article on $1/D$ methods [21], for

$$Z \geq Z_c = 1.22791378 \tag{43}$$

the geometry which minimizes the effective potential in the $D \to \infty$ limit consists of the electrons symmetrically located at the apices of an isosceles triangle. The opening angle θ_m obeys

$$\cos(\theta_m) = -\frac{\lambda}{8}\left[\frac{\lambda}{8} + \left(\left(\frac{\lambda}{8}\right)^2 + 2\right)^{1/2}\right], \tag{44}$$

where $\lambda = 1/Z$. It is easy to verify that for $Z \geq Z_c \simeq 1.22...$, $\lambda/8$ is never bigger than about 0.1, and thus $\cos(\theta_m)$ does not differ much from 0 and hence θ_m much from $90°$. In the table below I have listed for several representative values of Z the values of θ_m and the ratios of lengths of the sides of the triangle.

Z	θ_m (deg)	$r_{12}/r_1 = 2\sin(\theta_m/2)$	r_1/r_{12}
1.5	$97.181°$	1.5000	0.6667
2	$95.301°$	1.4781	0.6765
6	$91.713°$	1.4352	0.6968
∞	$90.000°$	1.4142	0.7071

Now as $r_1 \to r_2$ and the geometry of the isosceles triangle is attained in the $D \to \infty$ limit, the perimetric coordinates have the limiting behavior

$$
\begin{aligned}
u &= (-r_1 + r_2 + r_{12}) \to r_{12} \\
v &= (\,r_1 - r_2 + r_{12}) \to r_{12} \\
w &= 2(r_1 + r_2 - r_{12}) \to 2\Big(r_{12}\,\frac{r_1 + r_2}{r_{12}} - 1\Big)\,,
\end{aligned}
\tag{45}
$$

Thus in the $D \to \infty$ limit, the ratios of the perimetric coordinates have the behavior indicated in the table below:

Z	$w_m/u_m = w_m/v_m$
1.5	0.6666
2	0.7060
6	0.7872
∞	0.8284

One can observe that as Z decreases the base r_{12} of the isosceles triangle grows in relation to the other two sides $r_1 = r_2$ and there is an increasing mismatch of the length scales in the Laguerre basis functions with the length scales of the limiting geometry as $D \to \infty$, which is responsible for the deterioration of the convergence of the Rayleigh-Ritz variational calculation as D increases.

Fortunately, this diagnosis suggests an obvious cure: one should use properly (anti-)symmetrised products of generalised Laguerre functions with *three* flexible length scales:

$$
\Big[\Big(L_p^\alpha(\xi_1 u)\, e^{-\xi_1 u/2}\, L_q^\alpha(\xi_2 v)\, e^{-\xi_2 v/2} \pm L_q^\alpha(\xi_2 u)\, e^{-\xi_2 u/2}\, L_p^\alpha(\xi_1 u)\, e^{-\xi_1 u/2}\Big)\Big]
$$

$$
\times\, L_r^\alpha(\xi_3 w)\, e^{-\xi_3 w/2}\,.
\tag{46}
$$

If one were concerned with getting right only the *equilibrium geometry* in the $D \to \infty$ limit, it would suffice to use only *two* length scales, with $\xi_1 = \xi_2 \neq \xi_3$, but allowing ξ_1 to be different from ξ_2 should allow one to get right also the frequency of the *antisymmetric stretch*, and thereby obtain the variational energies *exactly* to $O(1/D)$, with errors of $O((1/D)^2)$. In fact, since as $D \to \infty$

$$
J^{1/2}\left(e^{-(\xi_1 u + \xi_2 v + \xi_3 w)/2} \pm e^{-(\xi_2 u + \xi_1 v + \xi_3 w)/2}\right),
\tag{47}
$$

where J is the Jacobian, tends to a properly (anti-)symmetrised product of Gaussians, which gives *exactly* the correct geometry and frequencies through $O(1/D)$, a *single* basis function should give the right limit and slopes of the ground state energy and the first excited state energy as $1/D \to 0$. Additional properly (anti-)symmetrised products of generalised Laguerre functions, which upon scaling and multiplication by $J^{1/2}$ tend to excited harmonic oscillator functions as $D \to \infty$, will give *exactly* higher-order anharmonic effects. Thus one should be able to obtain results of rapidly *increasing* accuracy as $D \to \infty$.

It would be interesting to test these expectations in a generalisation of the code used by Loeser and Herschbach [22]. However, this discussion suggests another fruitful area for development.

Direct Computation of the Coefficients of the $1/D$-Expansion: An Alternative to the Moment Method

As discussed by David Goodson elsewhere in these proceedings, the moment method has had many successes. It has allowed the computation of the coefficients of the $1/D$ expansion to 35th order and beyond, which have then been used in various summation schemes to obtain energies of the helium atom and of H_2^+ accurate to 8 or more digits. However, the rapid growth of the computing time and the storage requirement with the order n to which the coefficients are computed makes it less likely that the moment method can efficiently be used to compute to significantly higher order the $1/D$-expansion coefficients for He or H_2^+, or to comparably high order the coefficients for larger systems.

The key observation is that the higher-order corrections to the energy, in powers of $1/D$, arise from anharmonic corrections to the normal mode harmonic oscillator motion. Now a given anharmonic correction to the energy, as we all learned long ago when we studied quantum mechanics, can be computed *exactly* from a *finite* number of excited harmonic oscillator functions. This means that a truncated basis which contains properly scaled harmonic oscillator functions can be used to compute *exactly* a finite number of anharmonic corrections. One simply pre-determines to which order one wants to compute the anharmonic corrections, calculates how many excited state harmonic oscillator functions are needed to obtain the anharmonic corrections through that order, and calculates the corrections by standard Rayleigh-Schrödinger perturbation theory.

With this idea alone the storage requirements would still explode exponentially with the number of degrees of freedom. However, we can borrow an idea of Björn Roos [24] for performing 'direct CI' calculations: instead of computing and storing all the matrix elements involving the multi-particle basis functions, we can store only the primitive integrals involving powers of a *single* normal mode coordinate, and powers of the derivatives with respect to that coordinate, sandwiched between the corresponding harmonic oscillator eigenfunctions associated with that normal mode. As is well-known, the matrix of these integrals is sparse. Since at any finite order in the $1/D$ expansion the perturbing operators can be decomposed as finite sums of products of powers of normal mode coordinates and derivatives with respect to them, any matrix element of a perturbing operator can be reduced to a finite sum of products of primitive 1-dimensional integrals. Thus storage requirements should be kept manageable, and thereby allow us to compute to moderately high order the $1/D$-expansion coefficients for the energy levels of small many-electron atoms and molecules of chemical interest.[2]

Summary

I should like to conclude by taking from my discussion in [5] a summary of the key issues, in order of importance, for performing highly accurate Rayleigh-Ritz variational calculations on atoms and molecules:

0. Have many basis functions concentrated in the correct regions of configuration space and momentum space.

1. Get right the analytic structure at the coalescences of single electrons with the nucleus or the nuclei.

2. Get right the analytic structure at coalescences of pairs of electrons.

---

[2]This idea occurred to me during the Dimensional Scaling Workshop itself. I should like to state here that Martin Dunn independently and essentially simultaneously realised the benefits of computing the $1/D$-expansion coefficients perturbatively in a basis of harmonic oscillator eigenfunctions.

3. Get right the analytic structure at coalescences of three or more particles.

.

.

.

∞. Get right the long-distance exponential decay of the wavefunction.

When doing calculations in dimensions D different from 3, the earlier points are weighted more heavily with increasing D.

Acknowledgments

It is a pleasure to thank John Avery, Osvaldo Goscinski, and Dudley Herschbach for organising such a stimulating workshop. I am grateful to Dudley Herschbach and to the many members of his group for their continuing hospitality, which has greatly facilitated my work at Harvard.

I am indebted to Barry Simon for an invitation in June 1988 to present my work on rates of convergence of variational calculations to an audience at Caltech, which included Vince McKoy, who asked a penetrating question which inspired me to find the simple heuristic argument on relating the energy error in a variational calculation to the error in the approximate wavefunction beyond a certain momentum scale.

This work has been supported by my NSF grant PHY-8608155 and by an NSF grant for the Institute for Theoretical Atomic and Molecular Physics at the Harvard-Smithsonian Center for Astrophysics. I am also grateful to the Office of Special Sessions and International Programs at the University of Delaware for helping to provide support for me to attend the Copenhagen workshop.

References

1. C. Schwartz, *Meth. Comp. Phys.* **2**, 241 (1963).
2. B. Klahn and J. D. Morgan III, *J. Chem. Phys.* **81**, 410 (1984).
3. R. N. Hill, *J. Chem. Phys.* **83**, 1173 (1985).

4. D. P. Carroll, H. J. Silverstone, and R. M. Metzger, *J. Chem. Phys.* **71**, 4142 (1979).

5. J. D. Morgan III, "The Analytic Structure of Atomic and Molecular Wavefunctions and its Impact on the Rates of Convergence of Variational Calculations", in M. Defranceschi and J. Delhalle [edd.], *Numerical Determination of the Electronic Structure of Atoms, Diatomic and Polyatomic Molecules*, NATO-ASI Series C **271** (Kluwer, Dordrecht, 1989), pp. 49-84.

6. F. John, *Commun. Pure Applied Math.*

7. T. Kato, *Commun. Pure Appl. Math.* **10**, 151 (1957).

8. V. A. Fock, *Izv. Akad. Nauk SSSR, Ser. Fiz.* **18**, 161 (1954).

9. A. M. Ermolaev, *Vestnik Leningrad. Univ.* **14**, 48 (1958); **16**, 19 (1961); A. M. Ermolaev and G. B. Sochilin, *Dokl. Akad. Nauk SSSR* **155**, 1050 (1964).

10. J. H. Macek, *Phys. Rev.* **160**, 170 (1967).

11. J. Leray, *Actes du 6éme Congres du Groupement des Mathematiciens d'Expression Latine* (Gauthier-Villars, Paris, 1982), p. 179; *Methods of Functional Analysis and Theory of Elliptic Operators* (Università di Napoli, Naples, 1982), p. 165; *Bifurcation Theory, Mechanics and Physics* (Reidel, Dordrecht, 1983), p. 99.

12. J. D. Morgan III, *Theoret. Chim. Acta* **69**, 181 (1986).

13. P. C. Abbott and E. N. Maslen, *J. Phys. A* **20**, 2043 (1987); J. E. Gottschalk, P. C. Abbott, and E. N. Maslen, *ibid.*, **20**, 2077 (1987); J. E. Gottschalk and E. N. Maslen, *ibid.*, **20**, 2781 (1987); K. McIsaac and E. N. Maslen, *Int. J. Quantum Chem.* **31**, 361 (1987).

14. D. E. Freund, B. D. Huxtable, and J. D. Morgan III, *Phys. Rev. A* **29**, 980 (1984).

15. C. L. Pekeris, *Phys. Rev.* **115**, 1216 (1959).

16. K. Frankowski and C. L. Pekeris, *Phys. Rev.* **146**, 46 (1966); **150**, 366 (erratum) (1966).

17. J. D. Baker, D. E. Freund, R. N. Hill, and J. D. Morgan III, *Phys. Rev. A* **41**, 1247 (1990).

18. H. Kleindienst and R. Emrich, *Int. J. Quantum Chem.* **37**, 257 (1990).

19. S. E. Haywood, J. D. Nicholson, and J. D. Morgan III, to be published.

20. K. G. Wilson, to be published.

21. D. R. Herschbach, *J. Chem. Phys.* **84**, 838 (1986).

22. J. G. Loeser and D. R. Herschbach, *J. Chem. Phys.* **84**, 3882 (1986).

23. D. Z. Goodson and D. R. Herschbach, *Phys. Rev. Lett.* **58**, 1628 (1987).

24. B. Roos, *Chem. Phys. Lett.* **15**, 153 (1972).

Chapter 8

TWO-ELECTRON EXCITED STATES

8.1 Dimensional Expansions for Excited States

David Z. Goodson
Department of Chemistry
Harvard University
12 Oxford Street
Cambridge MA 02138, USA
and
  Deborah K. Watson
Department of Physics and Astronomy
University of Oklahoma
Norman, OK 73019-0225, USA

Abstract

*We report calculations of dimensional expansions for the energies of
the three excited S states of helium that correspond to one quantum
in either of the three normal modes of the Langmuir vibrations. Very
accurate energies are obtained for the 1s2s states, which arise from*

D. R. Herschbach et al. (eds.), Dimensional Scaling in Chemical Physics, 359–374.

excitation of the stretching vibrations. Reasonable, but somewhat less accurate, results are found for the energy of the $2p^2$ resonance, which corresponds to excitation in the bending vibration; this is despite an infinite number of curve crossings due to the difference in ordering of the eigenvalues at $D = 3$ and $D \to \infty$. More accurate results are obtained for the three corresponding P^e states, by virtue of the interdimensional degeneracies between $D = 3$ and $D = 5$.

Introduction

One of the major goals in atomic and molecular physics is the accurate calculation of observable properties, such as energies, for many-body systems. There is another goal, however, which is perhaps even more important, and often considerably more difficult—the understanding, in terms of simplified physical models, of *why* a given property has the value it does. Dimensional perturbation theory has the potential of achieving both these goals.

Conventional approaches typically begin with a separable independent-electron approximation and then add configuration interactions to correct for the correlations of the electronic motions. Although accurate numerical results for expectation values can be obtained in this way, these methods do not provide an intuitive picture of the dynamics of electron correlation. The two-electron atom serves as a prototype for highly correlated systems, so during the past two decades a number of studies have attempted to provide a better understanding of the dynamics of this problem. These studies have approached the task in various ways, such as setting up phenomenological classification schemes for the energy spectrum, or searching for approximately separable collective coordinates or for approximate quantum numbers. One unifying idea supported by these studies has been the molecular model of the electron-nucleus-electron system. Originally introduced by Herrick and coworkers [1], this model has been further supported by the hyperspherical coordinate studies of Lin and others [2], the analyses by Berry and coworkers [3] of conditional probability distributions from configuration-interaction wavefunctions, and by the work of Briggs and collaborators [4], who have treated the interelectron distance as a quasiadiabatic parameter. The molecular model has been successful in helping to elucidate the nature of the collective motions of the electrons in doubly excited states, but there remain inconsistencies and unanswered questions.

The large-dimension limit has recently resolved at least some of the difficulties of the molecular model. The molecule-like structure falls out quite naturally from the rigid "bent triatomic" Lewis configuration obtained in the limit $D \to \infty$ [5], and the Langmuir vibrations at finite D can be analyzed in terms of normal modes, which provide a set of approximate quantum numbers [6,7]. These results are obtained directly from the Schrödinger equation, in contrast to the phenomenological basis of some of the earlier studies. When coupled with an analysis of the rotations of the Lewis structure, this approach provides an excellent alternative classification scheme for the doubly-excited spectrum [8]. Furthermore, an analysis [7] of the normal modes offers a simple explanation of the connection between the explicitly molecular approaches of Herrick and of Briggs on the one hand, and the hyperspherical approach, which is rather different in its formulation and basic philosophy.

We will briefly review in this paper some of these qualitative successes of the large-dimension limit. Our main purpose, however, is to demonstrate that the harmonic eigenstates of the large-D limit can be accurately continued to physical three-dimensional solutions using large-order dimensional perturbation theory. We consider the three excited states S corresponding to one quantum of excitation in either of the normal modes of the Langmuir vibrations. We find that the two states that result from the excitation of stretching modes, which are the two lowest-lying excited states at large D, correspond to the two lowest-lying excited states at $D = 3$, and the $1/D$ expansions yield very accurate values for the three-dimensional energies. The accuracy that we obtain is comparable to that obtained for the ground state [9]. We find that the $1/D$ expansion for the state with one quantum in the bending excitation gives a reasonably accurate value for the $2p^2 \, {}^1S$ energy at $D = 3$. This correspondence had been predicted [6] by an analysis of the large-D eigenstates in limit of large Z but it is perhaps surprising to see it borne out by the expansion since at large D this is a low-lying bound state while at $D = 3$ it is a doubly excited resonance embedded in the continuum. Every P^e energy at $D = 3$ is exactly degenerate with the energy of an S state at $D = 5$ [10,11]. By evaluating our $1/D$ expansions at $1/5$ instead of $1/3$ we obtain values for P^e energies that are even more accurate than our results for the corresponding S states.

In Section 2 we set forth the model for excited states of two-electron atoms that is provided by the large-dimension limit and then very briefly describe the method that we used to calculate the expansion coefficients. Further details can be found in Refs. [6] and [7] and in a forthcoming publication [12]. In Section 3 we present our numerical results and discuss implications of this work.

Dimensional Perturbation Theory for Excited States

The asymptotic expansions for the energies have the form

$$E = \delta^2 \sum_{n=0}^{\infty} E_n \delta^n, \tag{1}$$

where $\delta = 1/D$. If we introduce the usual dimensional scalings, then in the limit $\delta \to 0$ all eigenstates of a given total angular momentum collapse to the minimum of the same effective potential. The energy corresponding to this minimum is E_0; hence E_0 in Eq. (1) has the same value for our excited states as for the ground state. E_1 is given by the eigenvalues of the Langmuir vibrations, which are the solutions of the Schrödinger equation within the harmonic approximation to the effective potential.

Let x_1, x_2 and y be dimension scaled displacement coordinates defined by

$$r_1 = r_m(1 + \kappa^{1/2}x_1), \quad r_2 = r_m(1 + \kappa^{1/2}x_2), \quad \theta = \theta_m + \kappa^{1/2}y, \tag{2}$$

where the r_i are the radial distances of the electrons from the nucleus, θ is the included angle, and $\kappa = [(D-2)(D-4)/4]^{-1/2}$. Other definitions for κ would work as well, as long as it is proportional to δ in the large-D limit; this particular choice simplifies the form of the effective potential. In practice, it is more convenient to compute an expansion in powers of κ and then as a final step reexpand it in powers of δ.

The harmonic Schrödinger equation is not separable in terms of the coordinates of Eq. (2). It is nearly separable in terms of symmetry coordinates,

$$x = 2^{-1/2}(x_1 + x_2), \quad \bar{y} = 2^{1/2}y, \quad z = 2^{-1/2}(x_1 - x_2), \qquad (3)$$

except for a small coupling between the symmetric coordinates x and $\bar{y}$. The normal coordinates, which completely separate the harmonic equation, are given in terms of the symmetry coordinates by a transformation of the form

$$q_1 = x \cos \chi + \bar{y} \sin \chi, \quad q_2 = -x \sin \chi + \bar{y} \cos \chi, \quad q_3 = z. \qquad (4)$$

The angle χ is close to zero [7], so q_1 is predominantly a symmetric stretch and q_2 is predominantly a bend. q_3 is exactly an antisymmetric stretch. The eigenvalues in the harmonic approximation are

$$E_1 = 8[(n_s + \tfrac{1}{2})\omega_1 + (n_\theta + \tfrac{1}{2})\omega_2 + (n_a + \tfrac{1}{2})\omega_3] + 6E_0, \qquad (5)$$

where n_s, n_θ and n_a are the quantum numbers corresponding to q_1, q_2 and q_3 respectively and the ω_i are the vibrational frequencies. (The term $6E_0$ and the factor of 8 result from the conversion from a series in κ to a series in δ.) The zeroth-order wavefunction, which is correct to first order in δ, is

$$\Psi_0 \propto h_{n_s}(q_1)h_{n_\theta}(q_2)h_{n_a}(q_3)e^{\frac{1}{2}(\omega_1 q_1^2 + \omega_2 q_2^2 + \omega_3 q_3^2)}, \qquad (6)$$

where the h_n are Hermite polynomials.

It is interesting to note that in the limit of large D the hyperspherical coordinates, R and α, become equivalent to the symmetry coordinates x and z of Eq. (3),

$$R \equiv (r_1^2 + r_2^2)^{1/2} \sim 2^{1/2}r_m + 2^{-1}r_m\kappa^{1/2}x, \qquad (7a)$$

$$\alpha \equiv \arctan(r_1/r_2) \sim \frac{\pi}{4} + 2^{1/2}\kappa^{1/2}z. \qquad (7b)$$

The fundamental assumption of the hyperspherical adiabatic approach to the two-electron atom [2] is that the hyperspherical radius, R, is approximately separable from the other two coordinates. x is approximately equal to the normal coordinate q_1, so at large D the separability of q_1 implies the approximate separability of R. Thus, the symmetry of the Lewis structure appears to be the "hidden symmetry" behind the adiabatic hyperspherical approximation, to the extent that q_1 remains approximately separable at $D = 3$.

Before considering the higher-order terms in the energy expansion, it is important to consider the implications of Eq. (5) for the assignments of the eigenstates. The quantum numbers n_s, n_θ and n_a represent approximate symmetries of the helium atom, which are exact within first-order perturbation theory in δ. At large D these quantum numbers provide a more accurate labeling of the eigenstates than do the traditional hydrogenic labels $1s^2$, $1s2s$, etc., which become exact in the limit $\lambda \to 0$, where λ is the inverse of the nuclear charge, Z. For S states Loeser and Herschbach [6] have established a correspondence between these two sets of quantum numbers by determining the $\lambda \to 0$ limit of the large-D eigenstates. They showed that

$$n_s + n_a \to k_1 + k_2, \quad n_\theta \to l_1 = l_2, \tag{8}$$

where l_i and k_i are the angular momentum quantum numbers and the number of radial nodes, respectively, in either of the two hydrogenic orbitals. The correspondence is completed by the observation that the spin multiplicity of the state is determined by the parity of n_a. This last condition, which holds for any value of λ, is implied by the symmetry of the normal coordinates with respect to interchange of the electrons. q_3 is antisymmetric while q_1 and q_2 are symmetric, according to Eqs. (3) and (4). Thus, an even value of n_a implies a singlet state while an odd value implies a triplet state.

We will use the symbol $|n_a n_s n_\theta\rangle$ to denote the eigenstates of the Langmuir vibrations. The ground state $|000\rangle$ corresponds to $1s^2$ in the hydrogenic labeling, $|100\rangle$ corresponds to the triplet $1s2s$, and $|010\rangle$ to the singlet $1s2s$. The ratios of the frequencies $\omega_a : \omega_s : \omega_\theta$ are $1 : 1.53 : 3.31$; therefore, the triplet is the first excited S state and the singlet is the second, which is in accord with the ordering of the eigenvalues at $D = 3$.

The state with one quantum in the bending vibration, $|001\rangle$, is the seventh excited state of the Langmuir vibrations. It is approximately equal in energy to the states corresponding to $1s4s$ and below the states corresponding to $1s5s$, $1s6s$, etc. However, according to Eqs. (8), the hydrogenic labeling of $|001\rangle$ is $2p^2$. At $D = 3$, the energy of any doubly excited state (both electrons excited) lies well above any singly excited state $1sns$. This implies the existence of level crossings in the energy spectrum as δ goes from 0 to 1/3. Furthermore, the doubly excited S states at $D = 3$ all lie above the $1s$ He$^+$ ion, which implies that they are not true bound states but rather are resonances, which will ionize after a finite lifetime. In the large-D limit all of the eigenvalues lie below the ionization threshold. These observations suggest that the hydrogenic labeling $2p^2$ for $|001\rangle$, which was derived from the small-δ limit, will not continue to be valid at $\delta = 1/3$. What might happen, for example, is that this state could actually become a singly excited state, perhaps $1s4s$, in the manner of an avoided crossing. However, as we will demonstrate, the inclusion of higher-order terms in Eq. (1) does appear to connect the state $|001\rangle$ to a doubly excited state at $\delta = 1/3$.

The higher-order E_n can be calculated from a recursion relation involving the moments of the coordinates. This relation is similar in form to that found for the ground state [13], although some significant modifications are necessary to treat excited states. In the case of the ground state we only needed moments of the form

$$A^p_{jkl} = \langle \Psi_0 | q_1^j q_2^k q_3^l | \Psi_p \rangle, \tag{9}$$

where the functions $\Psi_p(q_1, q_2, q_3)$ are the expansion coefficients of the wavefunction,

$$\Psi = \sum_{n=0}^{\infty} \Psi_n \kappa^{n/2}. \tag{10}$$

For states corresponding to a single excitation in one of the normal modes, we also need moments of the form

$$B^p_{jkl} = \langle \Phi_0 | q_1^j q_2^k q_3^l | \Psi_p \rangle, \tag{11}$$

where Φ_0 is the harmonic eigenfunction of the ground state,

$$\Phi_0 \propto e^{\frac{1}{2}(\omega_1 q_1^2 + \omega_2 q_2^2 + \omega_3 q_3^2)}. \tag{12}$$

The details of this calculation, along with tables of the expansion coefficients, will be presented elsewhere [12].

The resulting energy expansions are divergent, but can be summed using Padé summation. Before summing the expansion, we subtract out the second-order pole in the energy function $E(\delta)$ that is present at $\delta = 1$ for the radially excited states $|100\rangle$ and $|010\rangle$ [14]. The residue of this pole is –2, corresponding to the eigenvalue of the ionized atom in the limit $\delta \to 1$. Thus, we write the expansion in the form

$$E = \frac{-2}{(D-1)^2} + \delta^2 \sum_{n=0}^{\infty} [E_n + 2(n+1)]\delta^n, \tag{13}$$

and apply Padé summation to the sum in Eq. (13) rather than to the original sum in Eq. (1). This procedure improves the convergence of the Padé sequence. The energy for $|001\rangle$ is not expected to be singular at $\delta = 1$. It does have a second-order pole at $\delta = -1$, but explicitly incorporating this pole into the Padé summation does not appear to improve the convergence. Evaluation of the Padé approximants at $\delta = 1/3$ yields S^e states while evaluation at $\delta = 1/5$ yields P^e states [10,11].

Results and Discussion

Our results for energies of helium from Padé summation of the δ expansion are shown in Table 1. We list results from fifteenth-order perturbation theory and, for the states $|010\rangle$ and $|001\rangle$, twentieth-order perturbation theory. The twentieth-order result is not given for $|100\rangle$ on account of roundoff error in the expansion coefficients. The recursion relations from which the coefficients were computed are somewhat unstable to roundoff error, and this appears to be a more severe problem for the ground state and the first excited state than for the two more highly excited states. Our calculations for the expansion coefficients were carried out in double-precision arithmetic. Higher-order results could be obtained by using quadruple-precision arithmetic.

For comparison, the corresponding results for the ground state are also shown. The ground-state calculation [9] was carried out in quadruple-precision arithmetic so roundoff error is not a problem at these orders. (We list the result from nineteenth order rather than twentieth order because the [10/10] approximant for this state is defective [9].) Table 1 also shows results obtained from other calculation methods. For the six *bound* states under consideration the energy

Table 1. Energies from Padé summation of the $1/D$ expansion, compared to values obtained by other methods.

Large-D assignment	Hydrogenic assignment	Energy (a.u.)	Calculation method
$\lvert 000\rangle\, S^e$	$1s^2\ {}^1S^e$	-2.903719	$[7/8]$, $D=3$
		-2.903726	$[9/10]$, $D=3$
		-2.903724^a	Hylleraas-Pekeris
$\lvert 100\rangle\, S^e$	$1s2s\ {}^3S^e$	-2.175226	$[7/8]$, $D=3$
		-2.175229^a	Hylleraas-Pekeris
$\lvert 010\rangle\, S^e$	$1s2s\ {}^1S^e$	-2.144462	$[7/8]$, $D=3$
		-2.145964	$[10/10]$, $D=3$
		-2.145974^a	Hylleraas-Pekeris
$\lvert 001\rangle\, S^e$	$2p^2\ {}^1S^e$ (resonance)	-0.6279	$[7/8]$, $D=3$
		-0.6209	$[10/10]$, $D=3$
		-0.6219^b	Complex-coord. rotation
		-0.6205^c	Close-coupling
		-0.6151^d	Truncated diagonalization
$\lvert 000\rangle\, P^e$	$2p^2\ {}^1P^e$	-0.71050004	$[7/8]$, $D=5$
		-0.71050018	$[9/10]$, $D=5$
		-0.71050016^e	Hylleraas-Pekeris
$\lvert 100\rangle\, P^e$	$2p3p\ {}^1P^e$	-0.5802461	$[7/8]$, $D=5$
		-0.5802465^e	Hylleraas-Pekeris
$\lvert 010\rangle\, P^e$	$2p3p\ {}^3P^e$	-0.5677520	$[7/8]$, $D=5$
		-0.5678124	$[7/8]$, $D=5$
		-0.5678129^e	Hylleraas-Pekeris
$\lvert 001\rangle\, P^e$	$3s3d + 3p^2\ {}^3P^e$ (resonance)	-0.29177	$[7/8]$, $D=5$
		-0.29123	$[10/10]$, $D=5$
		-0.29115^f	Complex-coord. rotation
		-0.29067^d	Truncated diagonalization

^{a}Loeser and Herschbach, Ref. [6]. ^{d}Lipsky, *et al.*, Ref. [18].
^{b}Ho, Ref. [15]. ^{e}Goodson, *et al.*, Ref. [11].
^{c}Oza, Ref. [17]. ^{f}Ho and Callaway, Ref. [16].

eigenvalues from the Hylleraas-Pekeris [6,11] algorithm are certainly accurate within the precision shown, so they can be considered to be the exact results for the purpose of comparison, and they are in excellent agreement with the results from the δ expansion. The energies with the hydrogenic assignments $2p^2$ and $3s3d + 3p^2$ are resonances, and therefore are much more difficult to calculate accurately using standard methods. Table 1 compares our results for the states with large-D designations $|001\rangle$ to $D = 3$ calculations from the literature, for the states with which $|001\rangle$ is predicted to correspond [6]. The literature results were obtained using three different methods: complex-coordinate rotation with a Hylleraas-type wavefunction [15,16], variational close-coupling [17], and truncated diagonalization of a hydrogenic basis set [18]. The first two methods are sophisticated techniques that take into account the fact that these energies correspond to resonances. They are presumably more accurate than the third method, which does not include coupling with the continuum. Our results from fifteenth-order perturbation theory are comparable in accuracy to the truncated diagonalization calculations while the twentieth-order results agree quite well with the complex-rotation calculations. In the case of the S state, we can compare also with the close-coupling method. Our result lies between the close-coupling and complex-rotation results.

It is interesting to examine the apparent large-order behavior of our expansions. Figs. 1 and 2 illustrate the rate of increase of the expansion coefficients. Fig. 1 shows the nth root test for all four of the expansions while Fig. 2 shows the ratio test for $|010\rangle$ and $|001\rangle$. For convergent series these curves will approach the inverse of the radius of convergence. The behavior of $|100\rangle$ is seen to be almost identical to that of the ground state—the nth roots diverge to infinity according to a basically linear rate of increase with a superimposed oscillation. The large-order coefficients alternate in sign except at the cusps, where two neighboring coefficients have the same sign. As discussed in Chapter 7.1, this behavior can be attributed to a branch point in the energy function $E(\delta)$ at $\delta = 0$ that results from a complex-conjugate pair of square-root branch points in the Borel function near the negative real axis of the complex δ plane. The presence of a singularity in $E(\delta)$ at the origin implies that the radius of convergence is zero for these

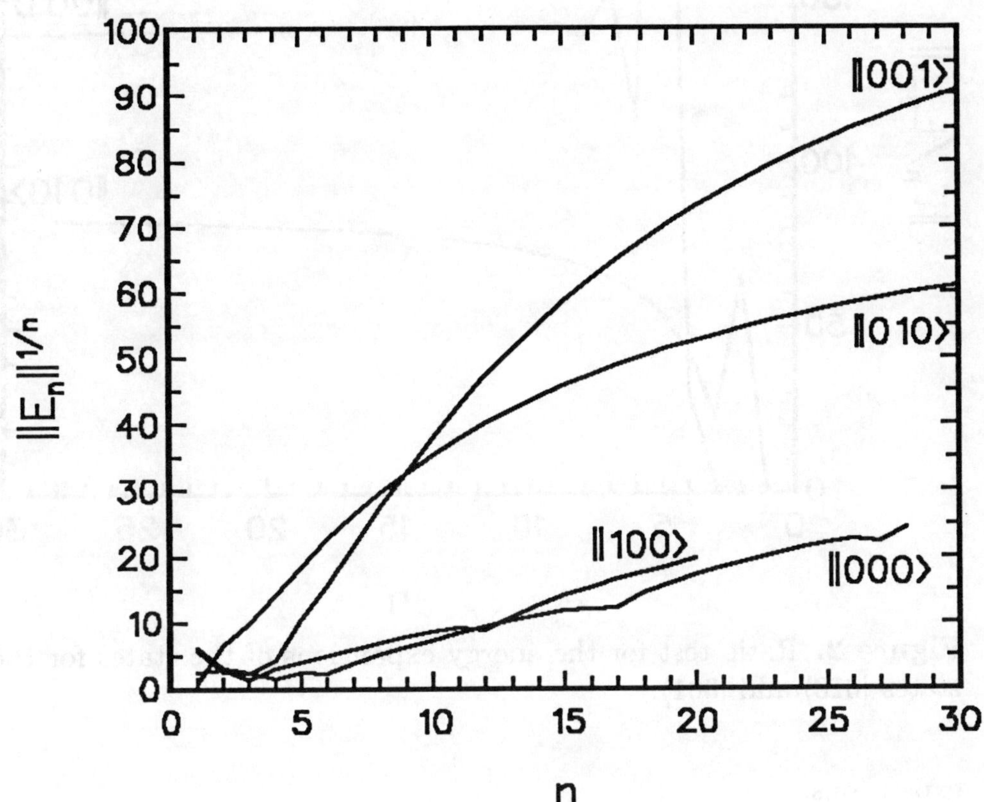

Figure 1. nth-root test for the energy expansions of four eigenstates, labeled $|n_a n_s n_\theta\rangle$ according to the quantum numbers of the large-dimension limit.

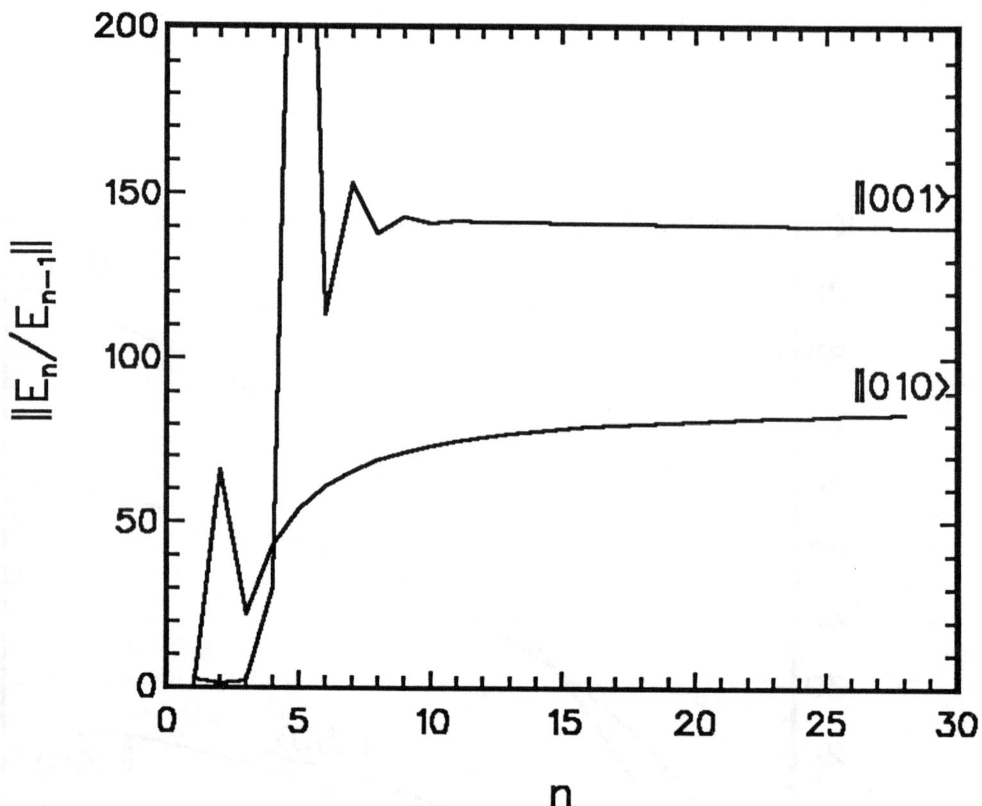

Figure 2. Ratio test for the energy expansions of the states for the states $|010\rangle$ and $|001\rangle$.

expansions.

The behavior of the nth roots for the other two states is rather different. At lower orders the rate of increase is quite steep but at higher orders it appears to be leveling off. Fig. 2 shows the ratio tests for these states. The ratio test here seems to reach its large-order behavior more quickly than the nth root test. Both states appear to have a finite radius of convergence; for $|010\rangle$ it appears to be approximately 0.012 and for $|001\rangle$ it appears to be approximately 0.007. Once the large-order behavior sets in, the expansion coefficients for $|010\rangle$ alternate in sign while those for $|001\rangle$ all have the same sign. This would indicate that the singularity responsible for establishing the radius of convergence lies on the negative real axis in the former case and on the positive real axis in the latter.

Another way to study the singularity structure of $E(\delta)$ is to calculate the zeros of the numerator and denominator polynomials of the Padé approximants. These are shown in Fig. 3 for the [7/8] approximants of all four states. × represents a pole of the approximant while o represents a zero. The series of nearly coincident poles and zeros along the negative real axis, seen in all four panels, is consistent with the presence of a branch point at or near the origin. The first three panels only show singularities at negative dimensions, while the fourth panel in addition shows singular behavior at or near the point $\delta = 0.007$, the singularity implied by the ratio test. This value of δ corresponds to $D \approx 140$. The presence of a singularity between $D \to \infty$ and $D = 3$ would seem to be consistent with the prediction that $|001\rangle$ begins as a bound state but then, as D decreases, at some point becomes a resonance. $D \approx 140$ would then presumably correspond to the point at which the eigenvalue enters into the continuum. Analysis of the singularity structure of the quadratic Padé approximants to the series [12] indicates that the state $|010\rangle$ has a square-root branch point at $\delta = -0.011386$, and $|001\rangle$ has a square-root branch point at $\delta = -0.0178$, while $|000\rangle$ and $|100\rangle$ have rather complicated branch points at $\delta = 0$. The nature of the singularity of $|001\rangle$ at positive δ is not immediately obvious from the quadratic Padé analysis. The apparent finite radius of convergence for $|010\rangle$ and $|001\rangle$ implied by the ratios in Fig. 2 is not inconsistent with the presence of a singularity at $\delta = 0$ similar to those found in the two lower-lying states. It is can be seen from Fig. 1 that the magnitude of the expansion coefficients of the square-root branch points in $|010\rangle$ and $001\rangle$ at nonzero δ is sufficiently large that the slow but steady divergence of $|000\rangle$ and $|100\rangle$ would not become dominant until very high order.

The accuracy of the results in Table 1 from the δ expansions can probably be improved by calculating the expansion coefficients to higher order using quadruple-precision arithmetic. Another way to improve the accuracy would be to replace the Padé summation with a summation method that is better able to model the singularity structure of the energy function, as discussed in Chapter 7.1. Nevertheless, our results shown in the Table appear to be sufficiently accurate to

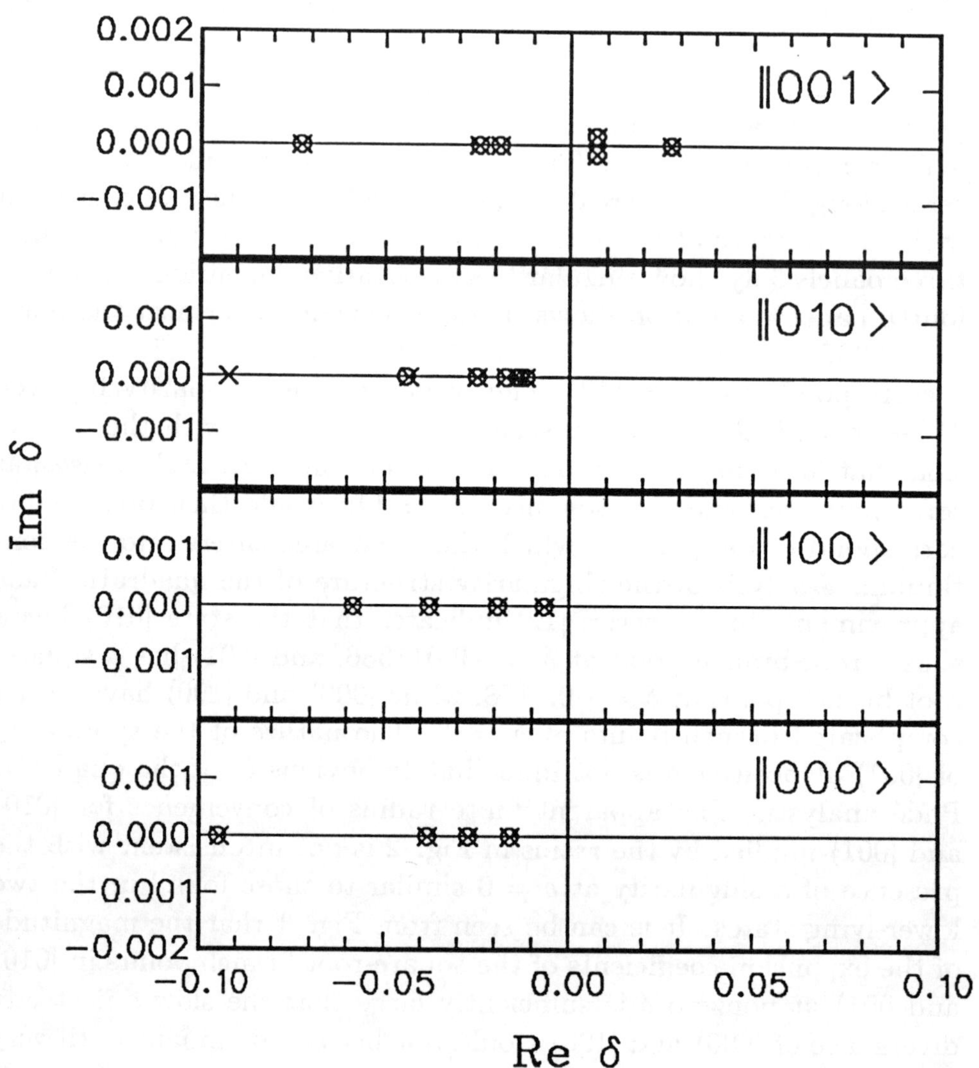

Figure 3. Singularity analysis of the [7/8] Padé approximants of the four energy expansions. × represents a pole of the approximant while ∘ represents a zero. The panels are labeled $|n_a n_s n_\theta\rangle$ according to the quantum numbers of the large-dimension limit.

strongly support the $D = 3$ correspondences of the large-dimension eigenstates that was predicted by Loeser and Herschbach [6]. The correspondence between $|001\rangle$ and $2p^2\ {}^1S$ is particularly remarkable since the eigenvalue must cross an infinite number of states of the form $1sns\ {}^1S$. These states might be expected to mix, resulting in an avoided crossing. Although the $2p^2$ and $1sns$ states share the classification ${}^1S^e$, they differ in their $|n_a n_s n_\theta\rangle$ assignments. The absence of avoided crossings might be an indication that n_a, n_s and n_θ continue to be reasonably accurate quantum numbers as D decreases to 3.

The expansion coefficients for the state $|001\rangle$ are all real numbers, and Padé summation will therefore yield a real number for the energy. Eigenvalues corresponding to resonances are complex numbers, with the imaginary part proportional to the width of the spectral line. If the singularity that we have found at $D \approx 140$ is indeed due to the passing of the eigenvalue into the continuum, then characterizing the functional form of the singularity and explicitly including it in the summation method might yield the imaginary part. Popov and coworkers [19] have shown that by using quadratic Padé approximants to sum the δ expansion for the hydrogen atom in an electric field they could obtain accurate results for the imaginary part of the energy. Quadratic Padé summation of our helium $|001\rangle$ expansion does not give an imaginary part, perhaps because the dimensional singularity in question is more complicated than a simple square-root branch point.

Acknowledgements

We thank John D. Morgan III for helpful discussions. This work was supported in part by a grant from the U. S. Air Force.

References

1. D. R. Herrick, Adv. Chem. Phys. **52**, 1 (1983); and references therein.
2. X. H. Liu, Z. Chen, and C. D. Lin, Phys. Rev. A **44**, 5468 (1991); C. D. Lin, Adv. At. Mol. Phys. **22**, 77 (1986); and references therein.

3. R. S. Berry and J. L. Krause, Adv. Chem. Phys. **70**, 35 (1988); R. S. Berry, Contemp. Phys. **30**, 1 (1989); and references therein; see also Chapter 12.2.

4. J. Feagin and J. S. Briggs, Phys. Rev. A **37**, 4599 (1988); J. M. Rost and J. S. Briggs, Z. Phys. D **5**, 339 (1988); J. M. Rost, R. Gersbacher, K. Richter, J. S. Briggs, and D. Wintgen, J. Phys. B **24**, 2455 (1991); see also Chapter 12.1.

5. D. R. Herschbach, J. Chem. Phys. **84**, 838 (1986).

6. J. G. Loeser and D. R. Herschbach, J. Chem. Phys. **84**, 3882 (1986).

7. D. Z. Goodson and D. R. Herschbach, J. Chem. Phys. **86**, 4997 (1987).

8. D. R. Herschbach, J. G. Loeser, and D. K. Watson, Z. Phys. D **10**, 195 (1988).

9. D. Z. Goodson, M. López-Cabrera, D. R. Herschbach, and J. D. Morgan III, to be published; see also Chapter 7.1.

10. D. R. Herrick, J. Math. Phys. **16**, 281 (1975).

11. D. Z. Goodson, D. K. Watson, J. G. Loeser, and D. R. Herschbach, Phys. Rev. A **44**, 97 (1991).

12. D. Z. Goodson and D. K. Watson, to be submitted to Phys. Rev. A.

13. D. Z. Goodson and D. R. Herschbach, Phys. Rev. Lett. **58**, 1628 (1987).

14. D. J. Doren and D. R. Herschbach, J. Chem. Phys. **87**, 433 (1987); Chem. Phys. Lett. **118**, 115 (1985); see also Chap. 4.1.

15. Y. K. Ho, Phys. Rev. A **34**, 4402 (1986).

16. Y. K. Ho and J. Callaway, J. Phys. B **18**, 3481 (1985).

17. D. H. Oza, Phys. Rev. A **33**, 824 (1986).

18. L. Lipsky, R. Anania, and M. J. Conneely, At. Data Nucl. Data Tables **20**, 127 (1977).

19. V. D. Mur, V. S. Popov, and A. V. Sergeev, Zh. Eksp. Teor. Fiz. **97**, 32 (1990) [Sov. Phys. JETP **70**, 16 (1990)]; and references therein; see also Chapter 6.1.

8.2 Analytic Continuation of Higher Angular Momentum States to D Dimensions and Interdimensional Degeneracies

Martin Dunn and Deborah K. Watson
Department of Physics and Astronomy
University of Oklahoma Norman OK 73019-0225, U.S.A.

Abstract

The difficulty in extending the techniques of dimensional scaling to states other than S-wave, centers on finding a way of factoring out the internal degrees of freedom to produce a tractable set of differential equations in the internal coordinates which admits analytic continuation in D. A solution to this problem for two electrons is discussed. This approach enables one to uncover all of the exact interdimensional degeneracies of the two electron system and is generalizable to more complicated problems.

Introduction

Thus far in the development of dimensional scaling techniques applied to atomic and molecular systems, all of the work has been confined to S-wave states and some P^e states [1] which were obtained by exploiting a known interdimensional degeneracy between S^e states in 5 dimensions and P^e states in 3 dimensions [2,3] to derive results for P-wave states in 3 dimensions. Recently, we have generalized the D-dimensional Schrödinger equation to higher angular momentum states in a way which produces a tractable set of equations and opens up the prospect of extending the previous work to higher symmetries. The major problems in the past have centered on deriving an expansion for the wavefunction which enables one to "factor out" the internal variables from the generalized Euler angles which multiply with increasing D. The most direct way of doing this is to expand the wavefunction

375

D. R. Herschbach et al. (eds.), Dimensional Scaling in Chemical Physics, 375–388.
© 1993 *Kluwer Academic Publishers. Printed in the Netherlands.*

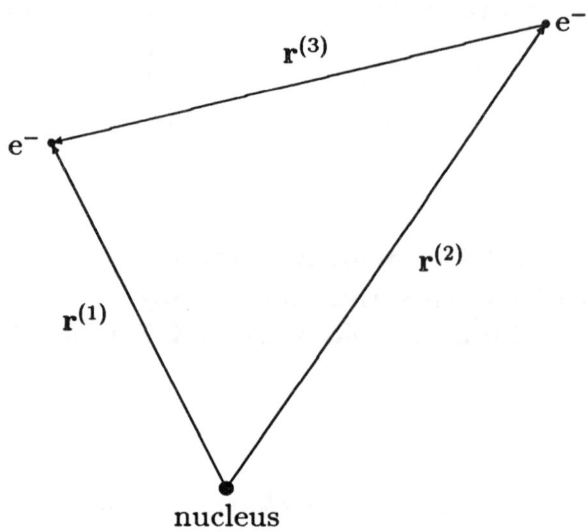

Figure 1. The two electron system.

in terms of the generalized Wigner rotation matrices. However, except for S-wave states, the implementation of this expansion is not as straightforward as might be wished, since the Wigner matrices grow as D is notched up; and so, naively, one also expects the number of expansion coefficients to multiply. If the number of terms proliferates in this fashion, analytically continuing in D, as the $\frac{1}{D}$ expansion requires, would be impossible. It might also be anticipated that the practical task of solving the growing number of coupled differential equations would become increasingly difficult for larger D.

We side-step these issues and derive an expansion for the wave function using the group theoretic method of irreducible tensors [5,6,7] in which each term is a product of two components, one of which is a simple known function of $r^{(1)}$, $r^{(2)}$, $r^{(3)}$, see Figure 1, and the generalized Euler angles while the other component is a function of the internal co-ordinates $r^{(1)}$, $r^{(2)}$ and $r^{(3)}$ only. The number of terms in the expansion is *finite* and *independent* of D. A set of coupled partial differential equations for the terms which are independent of the generalized Euler angles may be derived. These equations admit analytic continuation in D which enables a solution by the $\frac{1}{D}$ expansion to be carried out. Furthermore, the equations clearly show the complete spectrum of interdimensional degeneracies, so extending the work of Herrick and Stillinger, Doren and Herschbach, and Goodson et al. [2,3,1].

In this article we will take the opportunity to derive the expansion for the wavefunction in a heuristic, and we hope accessible, fashion. We refer the reader to references [8] and [9] for a more thorough treatment of the expansion of the wavefunction and further details about the interdimensional degeneracies.

The current work is based on a generalization of Schwartz's [10] expansion for a two electron system to D dimensions and N electrons (although we will not pursue the latter here). Schwartz found that he could expand the wavefunction in the following fashion.

$$\Psi^{L,M}(\mathbf{r}^{(1)}, \mathbf{r}^{(2)}) = \sum_{\{l^{(1)}, l^{(2)}\}} Y^{L,M}_{l^{(1)}, l^{(2)}}(\hat{\mathbf{r}}^{(1)}\hat{\mathbf{r}}^{(2)}) \, f_{l^{(1)}, l^{(2)}}(r^{(1)}, r^{(2)}, r^{(3)}) \quad (1)$$

where

$$Y^{L,M}_{l^{(2)}, l^{(1)}}(\hat{\mathbf{r}}^{(1)}\hat{\mathbf{r}}^{(2)})$$
$$= \sum_{m^{(1)}, m^{(2)}} (l^{(1)}, l^{(2)}; m^{(1)}, m^{(2)}|L, M) \, Y_{l^{(1)}, m^{(1)}}(\hat{\mathbf{r}}^{(1)}) \, Y_{l^{(2)}, m^{(2)}}(\hat{\mathbf{r}}^{(2)})$$

and the sum over the set $\{l^{(1)}, l^{(2)}\}$ is restricted so that $l^{(1)} + l^{(2)} = L$ if the parity $\pi = (-1)^L$ and $l^{(1)} + l^{(2)} = L + 1$ if $\pi = (-1)^{L+1}$. The essential point is that the number of terms in the generalization of equation (1) remains *finite* and *independent* of D.

Cartesian Tensors as Angular Momentum Eigenfunctions

The generalization of Schwartz's result to D dimensions requires the use of angular momentum algebra in D dimensions which can be developed using the method of irreducible Cartesian tensors.

Consider, in three dimensions, a direct product of n $l = 1$ states

$$N_{m_1, m_2, \ldots, m_n} = |1m_1 >^{(1)} |1m_2 >^{(2)} \cdots |1m_n >^{(n)}.$$

This transforms as a nth rank spherical tensor [11,12] , ie

$$N_{m_1,\ldots,m_n} = \sum_{m_1',m_2',\ldots,m_n'} \mathcal{D}^{(1)}_{m_1'm_1} \mathcal{D}^{(1)}_{m_2'm_2} \cdots \mathcal{D}^{(1)}_{m_n'm_n} N'_{m_1',\ldots,m_n'}.$$

which expresses the tensor $N_{m_1,m_2,\ldots,m_n}$ in the unprimed coordinate frame in terms of the same tensor in the primed coordinate frame, $N'_{m_1',\ldots,m_n'}$. $\mathcal{D}^{(1)}_{m'm}$ is the Wigner rotation matrix for $l=1$. These n $l=1$ states may be coupled together to give a state, $|LM\alpha >$ of total angular momentum L and projection M,

$$|LM\alpha >= \sum_{m_1,\ldots,m_n} (111\ldots;m_1m_2\ldots m_n|LM\alpha)N_{m_1,\ldots,m_n} \qquad (2)$$

where α is the coupling scheme. Equation (2) is a unitary change of basis,

$$\sum_{m_1,\ldots,m_n} (111\ldots;m_1m_2\ldots m_n|L'M'\alpha')(LM\alpha|111\ldots;m_1m_2\ldots m_n)$$
$$= \delta_{LL'}\delta_{MM'}\delta_{\alpha\alpha'}$$

$$\sum_{LM\alpha}(LM\alpha|111\ldots;m_1m_2\ldots m_n)(111\ldots;m_1'm_2'\ldots m_n'|LM\alpha)$$
$$= \delta_{m_1m_1'}\delta_{m_2m_2'}\cdots\delta_{m_nm_n'}.$$

Thus one can write

$$N_{m_1,m_2,\ldots,m_n} = \sum_{L\alpha} F^{L\alpha}_{m_1,m_2,\ldots,m_n} \qquad (3)$$

where

$$F^{L\alpha}_{m_1,m_2,\ldots,m_n} = \sum_M (LM\alpha|111\ldots;m_1m_2\ldots m_n)|LM\alpha >. \qquad (4)$$

The spherical tensor $F^{L\alpha}_{m_1,\ldots,m_n}$ is an irreducible spherical tensor of rank n. Irreducible, in this context, may be taken to mean that $F^{L\alpha}_{m_1,\ldots,m_n}$ is an eigenstate of $\mathbf{L}^2$ and that one cannot further decompose $F^{L\alpha}_{m_1,m_2,\ldots,m_n}$ into two components each of which transform as a spherical tensor.

To recapitulate, instead of the usual angular momentum formalism we can use irreducible spherical tensors where we have the correspondence $|LM\alpha > \longrightarrow F^{L\alpha}_{m_1,m_2,\ldots,m_n}$ where we have replaced M by

$m_1, m_2, \ldots, m_n$. For example if we are looking for the spherical tensor of rank 2 which is an eigenstate of $\mathbf{L}^2$ with $L = 2$, from equation (4) we find

$$F^2_{m_1 m_2} = \frac{1}{2}(|1m_1>^{(1)} |1m_2>^{(2)} + |1m_2>^{(1)} |1m_1>^{(2)}$$
$$-\frac{2}{3} g_{m_1 m_2} [\sum_{m_1' m_2'} g_{m_1' m_2'} |1m_1'>^{(1)} |1m_2'>^{(2)}]) \qquad (5)$$

where $g_{m_1 m_2} = (-1)^{m_1} \delta_{m_1, -m_2}$. An $L = 2$ state has five independent components while a general second rank tensor has nine components, however only five of the components of $F^2_{m_1 m_2}$ are independent.

Although we have introduced irreducible spherical tensors, we do not yet have a formalism which admits ready generalization to D dimensions. This can be accomplished by transforming the spherical tensor to a Cartesian tensor. The spherical components of a tensor are related by a unitary transformation to Cartesian components [11,12,13]. For example consider a spherical harmonic of $l = 1$ (a spherical tensor of rank 1) written as a three component vector

$$\begin{pmatrix} Y_1^1 \\ Y_0^1 \\ Y_{-1}^1 \end{pmatrix} = \begin{pmatrix} \left(\frac{3}{4\pi}\right)^{\frac{1}{2}} \left(-\frac{1}{\sqrt{2}} sin\theta \, e^{i\phi}\right) \\ \left(\frac{3}{4\pi}\right)^{\frac{1}{2}} cos\theta \\ \left(\frac{3}{4\pi}\right)^{\frac{1}{2}} \left(\frac{1}{\sqrt{2}} sin\theta \, e^{-i\phi}\right) \end{pmatrix}.$$

One can write this in terms of a Cartesian vector

$$\begin{pmatrix} Y_1^1 \\ Y_0^1 \\ Y_{-1}^1 \end{pmatrix} = \frac{1}{r} \left(\frac{3}{4\pi}\right)^{\frac{1}{2}} \mathbf{U} \begin{pmatrix} x \\ y \\ z \end{pmatrix}$$

where $\mathbf{U}$ is the unitary matrix

$$\mathbf{U} = \frac{1}{\sqrt{2}} \begin{pmatrix} -1 & -i & 0 \\ 0 & 0 & \sqrt{2} \\ 1 & -i & 0 \end{pmatrix}.$$

The normalization factor $\frac{1}{r} \left(\frac{3}{4\pi}\right)^{\frac{1}{2}}$ is irrelevant to the tensor transformation properties, so one can transform the spherical tensor $N_{m_1, \ldots, m_n}$

to a Cartesian tensor $N_{j_1,\ldots,j_n}$ by the following

$$N_{j_1,j_2,\ldots,j_n} = \sum_{m_1,\ldots,m_n} U^{\dagger}_{j_1 m_1} U^{\dagger}_{j_2 m_2} \cdots U^{\dagger}_{j_n m_n} N_{m_1,m_2,\ldots,m_n}$$

$$\propto r^{(1)}_{j_1} r^{(2)}_{j_2} \cdots r^{(n)}_{j_n}$$

Now from equation (3)

$$N_{j_1,j_2,\ldots,j_n} = \sum_{L\alpha} F^{L\alpha}_{j_1,j_2,\ldots,j_n}$$

where

$$F^{L\alpha}_{j_1,j_2,\ldots,j_n} = \sum_{m_1,\ldots,m_n} U^{\dagger}_{j_1 m_1} U^{\dagger}_{j_2 m_2} \cdots U^{\dagger}_{j_n m_n} F^{L\alpha}_{m_1,m_2,\ldots,m_n}. \tag{6}$$

$F^{L\alpha}_{j_1,j_2,\ldots,j_n}$ is an irreducible Cartesian tensor and is an eigenstate of $\mathbf{L}^2$ with total angular momentum L. For example, from equations (5) and (6)

$$F^2_{j_1 j_2} = \sum_{m_1,m_2} U^{\dagger}_{j_1 m_1} U^{\dagger}_{j_2 m_2} F^2_{m_1 m_2}$$

$$\propto r^{(1)}_{j_1} r^{(2)}_{j_2} + r^{(1)}_{j_2} r^{(2)}_{j_1} - \frac{2}{3}\delta_{j_1 j_2} \mathbf{r}^{(1)} . \mathbf{r}^{(2)}. \tag{7}$$

We have succeeded in writing our angular momentum wavefunctions as irreducible Cartesian tensors. However, there is quite a lot of work in coupling the direct product of $l = 1$ states to give states of definite angular momentum, equation (2), then re-expressing them to give an irreducible spherical tensor, equation (4), and finally transforming the tensor to give an irreducible Cartesian tensor, equation (6). One would like to find a way of directly deriving the irreducible Cartesian tensors which generalizes to D dimensions. There is indeed such a method [5,6,7] and we describe it below.

General Method for Generating Irreducible Tensors

A Cartesian tensor $F'_{j_1,j_2,\ldots,j_n}$ of a particular *symmetry type* may be generated from an arbitrary Cartesian tensor $F_{j_1,j_2,\ldots,j_n}$ by a Young Operator $\mathcal{Y}$, ie

$$F'_{j_1,j_2,\ldots,j_n} = \mathcal{Y} F_{j_1,j_2,\ldots,j_n}$$

The symmetry type of F' is denoted by a Young Tableau, for example

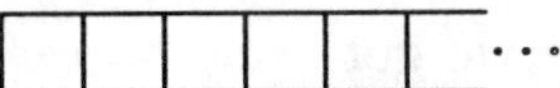

denotes the symmetry type of a 14th rank tensor where there is a box for every index on the tensor. The length of a given row of a tableau cannot be longer than the row above it while the numbers in a Young tableau run from 1 to n and increase along rows from left to right and down columns. In particular

$$\boxed{\ }\boxed{\ }\boxed{\ }\boxed{\ }\boxed{\ } \cdots$$

defines a totally symmetric tensor, one which is invariant under interchange of any pair of indices. For example, the totally symmetric 2nd rank tensor

$$F'_{j_1 j_2} = r^{(1)}_{j_1} r^{(2)}_{j_2} + r^{(1)}_{j_2} r^{(2)}_{j_1}$$

is of symmetry type $\boxed{1}\boxed{2}$. An irreducible Cartesian tensor must satisfy two criteria

1. The tensor must have a definite symmetry.

2. The tensor must be traceless, eg

$$F_{j_1 j_2 j j j_5} = 0,$$

for all pairs of indices. (We will adopt the Einstein summation convention [14] - all repeated indices are summed).

It is readily seen that the 2nd rank Cartesian tensor that we have derived in equation (7) is symmetric and traceless and so $F^2_{j_1 j_2}$ is an irreducible Cartesian tensor of symmetry type $\boxed{1}\boxed{2}$. We have already discussed how to generate a tensor with a specified symmetry type from a given tensor using Young operators, but the question remains how to project out the traceless component of a tensor? The example of an arbitrary third rank tensor [15] will illustrate the method. The tensor $F_{j_1 j_2 j_3}$ may be decomposed into a traceless component $F^0_{j_1 j_2 j_3}$ and other terms involving Kronecker deltas,

$$F^0_{j_1 j_2 j_3} = F_{j_1 j_2 j_3} - H_{j_3}\delta_{j_1 j_2} - K_{j_2}\delta_{j_1 j_3} - L_{j_1}\delta_{j_2 j_3} \tag{8}$$

In general there is a term involving a Kronecker delta for every possible choice of two objects from n objects. Taking the (12)-trace we get

$$F_{jjj_3} = 3H_{j_3} + K_{j_3} + L_{j_3}, \tag{9}$$

taking the (13)-trace we get

$$F_{jj_2 j} = H_{j_2} + 3K_{j_2} + L_{j_2} \tag{10}$$

and taking the (23)-trace we get

$$F_{j_1 jj} = H_{j_1} + K_{j_1} + 3L_{j_1} \tag{11}$$

which can be solved to give

$$H_j = \frac{1}{3^2 + 3 - 2}[(3+1)F_{kkj} - F_{kjk} - F_{jkk}] \tag{12}$$

$$K_j = \frac{1}{3^2 + 3 - 2}[-F_{kkj} + (3+1)F_{kjk} - F_{jkk}] \tag{13}$$

$$L_j = \frac{1}{3^2 + 3 - 2}[-F_{kkj} - F_{kjk} + (3+1)F_{jkk}]. \tag{14}$$

There are in fact only three distinct classes of non-zero traceless tensors with a definite symmetry type, and these are given below together with their more familiar quantum numbers [17]:

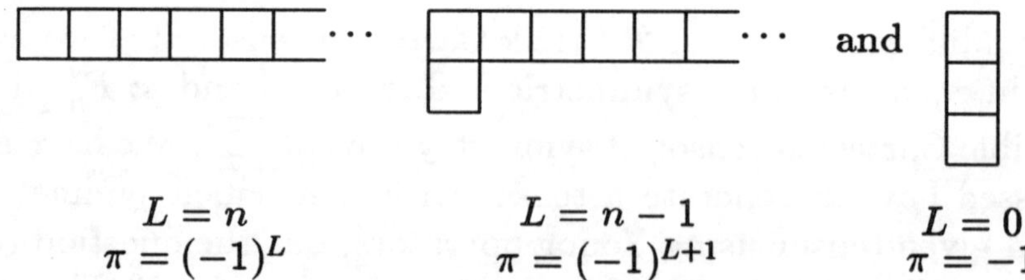

$$L = n \qquad\qquad L = n - 1 \qquad\qquad L = 0$$
$$\pi = (-1)^L \qquad\qquad \pi = (-1)^{L+1} \qquad\qquad \pi = -1$$

where n = the number of boxes in the tableau

      = the rank of the tensor

and M has been replaced by $j_1, j_2, \ldots, j_n$. If $T_{j_1,\ldots,j_n} = r_{j_1} r_{j_2} \cdots r_{j_n}$, T is already of symmetry type

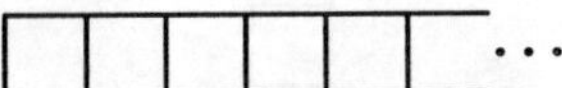

without any need to operate on it with the appropriate Young operator, so extracting the traceless component one obtains the irreducible tensor $T^0_{j_1,\ldots,j_n}$ which is a harmonic polynomial and is proportional to the spherical harmonic Y^n_m expressed as a Cartesian tensor [16].

So far we have only been dealing with a three dimensional space, however, *the above procedure generalizes to D dimensions* [5,6,7].

For example, the second rank irreducible tensor of equation (7) in D dimensions becomes

$$F^2_{j_1 j_2} = r^{(1)}_{j_1} r^{(2)}_{j_2} + r^{(1)}_{j_2} r^{(2)}_{j_1} - \frac{2}{D} \delta_{j_1 j_2} \mathbf{r}^{(1)} . \mathbf{r}^{(2)}.$$

and the $\mathcal{I}$s in equations (9), (10), (11), (12), (13) and (14) become Ds.

This is angular momentum algebra in D dimensions.

There are new states which are non-zero in D dimensions, for example

<table>
<tr><td>1</td><td>2</td><td>4</td><td>5</td><td>7</td><td>14</td></tr>
<tr><td>3</td><td>6</td><td>10</td><td>11</td><td></td><td></td></tr>
<tr><td>8</td><td>9</td><td>13</td><td></td><td></td><td></td></tr>
<tr><td>12</td><td></td><td></td><td></td><td></td><td></td></tr>
</table>

but do not survive as $D \to 3$ [18] (in fact the above traceless tensor is identically zero for $D < 7$), therefore we will always confine our attention to the states

although for the two electron system the

states are identically zero for all D (there are no S^o states in three dimensions).

The Wavefunction in D Dimensions

We will label the symmetry type of the wavefunction by the partition [19] $\Gamma = [\gamma_1, \gamma_2, \gamma_3]$ where γ_1, γ_2 and γ_3 are the number of boxes in the first, second and third rows of the tableau respectively. Since we are only interested in those states which survive in the limit $D \to 3$, $\gamma_2 = 0$ or 1, while $\gamma_3 = 0$ if $\gamma_1 > 1$ and 0 or 1 if $\gamma_1 = \gamma_2 = 1$. Following Schwartz one would expand the wavefunction in a complete set of harmonic polynomials (the tensor equivalent of the spherical harmonics). However, it is much easier if we expand the wavefunction in the monomials

$$T^{\lambda_{(i)}}_{\{J_{\lambda_{(i)}}\}}(\mathbf{r}^{(i)}) = T^{\lambda_{(i)}}_{j_1, j_2, \ldots, j_{\lambda_{(i)}}}(\mathbf{r}^{(i)}) = r^{(i)}_{j_1} r^{(i)}_{j_2} \cdots r^{(i)}_{j_{\lambda_{(i)}}} \qquad (15)$$

where i refers to particle i. The monomials $T^{\lambda_{(i)}}_{\{J_{\lambda_{(i)}}\}}(\mathbf{r}^{(i)})$ form a complete set on the unit hypersphere (a consequence of Weierstrass's Approximation Theorem [20]). Thus the wavefunction, which is required to be an irreducible traceless tensor, may be written in a "Configuration Interaction (or CI)" expansion

$$\Psi^{\Gamma}_{\{I_n\}}(\mathbf{r}^{(1)}, \mathbf{r}^{(2)})$$
$$= \sum_{\lambda_{(1)}, \lambda_{(2)}} W^{\Gamma}_{\lambda_{(1)}, \lambda_{(2)}; \{I_n\}}(\mathbf{r}^{(1)}, \mathbf{r}^{(2)}) \, F_{\lambda_{(1)}, \lambda_{(2)}}(r^{(1)}, r^{(2)})$$

$$(16)$$

where

$$W^{\Gamma}_{\lambda_{(1)}, \lambda_{(2)}; \{I_n\}}(\mathbf{r}^{(1)}, \mathbf{r}^{(2)})$$

$$= \sum_{\{J_{\lambda_{(1)}}\},\{J_{\lambda_{(2)}}\}} C^{\Gamma}_{\{I_n\}\{J_{\lambda_{(1)}}\}\{J_{\lambda_{(2)}}\}} \, T^{\lambda_{(1)}}_{\{J_{\lambda_{(1)}}\}}(\mathbf{r}^{(1)}) \, T^{\lambda_{(2)}}_{\{J_{\lambda_{(2)}}\}}(\mathbf{r}^{(2)}) .$$

$$(17)$$

and the sum in equation (17) is taken over all of the indices on the monomials $T^{\lambda_{(1)}}_{\{J_{\lambda_{(1)}}\}}(\mathbf{r}^{(1)})$ and $T^{\lambda_{(2)}}_{\{J_{\lambda_{(2)}}\}}(\mathbf{r}^{(2)})$. The "Clebsch-Gordon" coefficient

$C^{\Gamma}_{\{I_n\}\{J_{\lambda_{(1)}}\}\{J_{\lambda_{(2)}}\}}$ effects the contraction of $\lambda_{(1)} + \lambda_{(2)} - n$ indices to leave n indices, it then effects the operation of a Young operator on the remaining n indices and the extraction of the traceless component. For example, if the wave function is an irreducible traceless tensor of symmetry type $\boxed{1\,2}$, ie a completely symmetric second rank tensor, and we concentrate on the term in equation (16) for which $\lambda_{(1)} = \lambda_{(2)} = 2$ so that

$$T^{2}_{\{J_2\}}(\mathbf{r}^{(1)}) T^{2}_{\{J_2\}}(\mathbf{r}^{(2)}) = r^{(1)}_{j_1} r^{(1)}_{j_2} r^{(2)}_{j_3} r^{(2)}_{j_4}$$

then

$$C^{[2,0,0]}_{\{I_2\}\{J_2\}\{J_2\}} = [\delta_{i_1 j_1}\delta_{i_2 j_3} + \delta_{i_1 j_3}\delta_{i_2 j_1} - \frac{2}{D}\delta_{i_1 i_2}\delta_{j_1 j_3}]\delta_{j_2 j_4} \qquad (18)$$

where the $\delta_{j_2 j_4}$ contracts $r^{(1)}_{j_2}$ with $r^{(2)}_{j_4}$ to reduce the 4th order monomial $T^{2}_{\{J_2\}}(\mathbf{r}^{(1)})T^{2}_{\{J_2\}}(\mathbf{r}^{(2)})$ to a second order monomial. The "Clebsch-Gordon" coefficient does not contract two indices which would result in the scalar product $\mathbf{r}^{(i)}.\mathbf{r}^{(i)}$, where $i = 1$ or 2, since this could simply be reabsorbed in $F_{\lambda_{(1)},\lambda_{(2)}}(r^{(1)}, r^{(2)})$ making it a term for which $\lambda_{(1)} = \lambda_{(2)} = 1$. The term in square brackets in equation (18) performs the symmetrization of the remaining two indices and the extraction of the traceless component. Thus

$$W^{[2,0,0]}_{2,2;\{I_2\}}(\mathbf{r}^{(1)}, \mathbf{r}^{(2)}) = [r^{(1)}_{i_1} r^{(2)}_{i_2} + r^{(1)}_{i_2} r^{(2)}_{i_1} - \frac{2}{D}\delta_{i_1 i_2}\mathbf{r}^{(1)}.\mathbf{r}^{(2)}] \, \mathbf{r}^{(1)}.\mathbf{r}^{(2)} .$$

We can re-express $\Psi^{\Gamma}_{\{I_n\}}(\mathbf{r}^{(1)}, \mathbf{r}^{(2)})$ by performing all of the contractions first, which gives us

$$\Psi^{\Gamma}_{\{I_n\}}(\mathbf{r}^{(1)}, \mathbf{r}^{(2)})$$

$$= \sum_{\substack{\lambda_{(1)},\lambda_{(2)} \\ (\lambda_{(1)}+\lambda_{(2)}=n)}} W^{\Gamma}_{\lambda_{(1)},\lambda_{(2)};\{I_n\}}(\mathbf{r}^{(1)},\mathbf{r}^{(2)})\, f_{\lambda_{(1)},\lambda_{(2)}}(r^{(1)},r^{(2)},r^{(3)})$$

$$(19)$$

where the $\mathbf{r}^{(1)}.\,\mathbf{r}^{(2)}$ resulting from the contractions have been absorbed into the "CI coefficient" and re-expressed in terms of the inter-electron distance $r^{(3)}$. The sum over $\lambda_{(1)}$ and $\lambda_{(2)}$ is now *finite* since it is subject to the constraint $(\lambda_{(1)} + \lambda_{(2)} = n)$. Equation (19) should be compared with the Schwartz result of equation (1).

Equation (19) is the generalization of the Schwartz result to D dimensions – The expansion contains a finite number of terms which is independent of D.

One can use the expansion for the wavefunction of equation (19) to derive a set of coupled equations for $f_{\lambda_{(1)},\lambda_{(2)}}(r^{(1)},r^{(2)},r^{(3)})$ [9] which are amenable to solution using a variety of methods, notably the $\frac{1}{D}$ expansion. These equations clearly show interdimensional degeneracies between the states in three dimensions associated with the tableaux

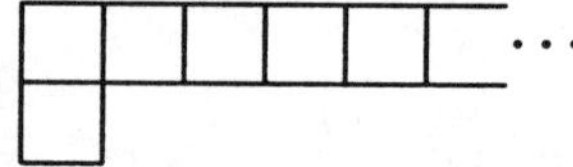

and the states in five dimensions associated with the tableaux

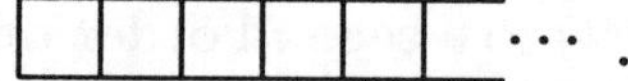

The interdimensional degeneracies are found to be

$D=3$	$D=5$
$^{1,3}P^e$	$^{3,1}S^e$
$^{1,3}D^o$	$^{3,1}P^o$
$^{1,3}F^e$	$^{3,1}D^e$
$\vdots$	$\vdots$

The solution of these equations by dimensional perturbation theory will allow the study of any states of the two electron system, while interdimensional degeneracies will greatly reduce the labor involved. Further, the extension to N electrons [8] opens up the prospect of studying states of any symmetry for more complicated systems [21].

Acknowledgments

We are grateful for the support of this work by the National Science Foundation Grant PHY-9008721. One of the authors, M. Dunn, would like to thank Larry Weaver for helpful discussions regarding the orthogonal group $O(D)$.

References

1. D.Z. Goodson, D.K. Watson, J.G. Loeser and D.R. Herschbach, *Phys. Rev. A* **44**, 97 (1991). D.R. Herrick and F.H. Stillinger *Phys. Rev. A* **11**, 42 (1975).
2. D.R. Herrick, *J. Math. Phys.* **16**, 281 (1975).
3. For another way around this problem see, O. Goscinski and V. Mujica, *Int. J. Quant. Chem.* **29**, 897 (1986).
4. M. Hamermesh, *Group Theory and its Application to Physical Problems* (Dover, New York, 1989).
5. H. Weyl, *The Classical Groups* (Princeton University Press, Princeton, 1939).
6. H. Boerner, *Representations of Groups* (North-Holland, Amsterdam, 1963).
7. M. Dunn and D.K. Watson, *Analytic Continuation of the Wavefunction of Higher Angular Momentum States to D Dimensions*, in preparation.
8. M. Dunn and D.K. Watson, *Analytic Continuation of the Schrödinger Equation for Higher Angular Momentum States to D Dimensions and the Interdimensional Degeneracies of the Two Electron System*, in preparation.
9. C. Schwartz, *Phys. Rev.* **123**, 1700 (1961).
10. J.M. Normand, *A Lie Group: Rotations in Quantum Mechanics* (North-Holland, Amsterdam, 1980).
11. J.M. Normand and J. Raynal, *J. Phys. A* **15**, 1437 (1982).
12. U. Fano and G. Racah, *Irreducible Tesorial Sets* (Academic Press, New York, 1959).
13. D.F. Lawden, *An Introduction to Tensor Calculus and Relativity*, (Methuen, London, 1962), p. 25.
14. This example is taken from Reference [5], p. 393.

15. Z.Y. Wen and J. Avery, *J. Math. Phys.* **26**, 396 (1985); see also J. Avery, *Hyperspherical Harmonics*, (Kluwer Academic, Dordrecht, 1989).

16. One makes the connection by counting the independent components of the irreducible tensor. For the completely symmetric $\Gamma = [\gamma_1, 0, 0]$ states (see the discussion above equation (15) for an explanation of this notation) this is given by equation (18) of Reference [16]. In three dimensions the $\Gamma = [\gamma_1, 1, 0]$ tableau, ie either the two box antisymmetric or the mixed symmetry tableaux, is associate (see Reference [5] p. 396 and Reference [6] p. 164) to the $\Gamma = [\gamma_1, 0, 0]$ tableau, i.e. a completely symmetric tableaux. Thus a second rank antisymmetric or a mixed symmetry tensor has the same dimension as the associate completely symmetric tensor. The three box completely antisymmetric $\Gamma = [1, 1, 1]$ tableau is the associate tableau to the tableau with no boxes and is therefore a scalar. There are more general formulas for the number of independent components of an irreducible tensor in D dimensions, however they are not required to achieve the above identification, see D.E. Littlewood, *The Theory of Group Characters* (Oxford University Press, Oxford, 1950), equation (11.8;6), p. 236.

17. See Reference [6], Theorem (5.7.A), p. 154.

18. See, for example, C.D.H. Chisholm, *Group Theoretical Techniques in Quantum Chemistry* (Academic Press, London, 1976) p. 77.

19. R. Courant and D. Hilbert, *Methods of Mathematical Physics, V. 1* (Interscience, New York, 1953), p. 65. For a simple proof of the one dimensional case see F.D. Murnaghan, *Introduction to Applied Mathematics*, (J. Wiley, New York, 1948), p. 77.

20. For a study of S-wave multi-electron systems see Chapter 3 by Loeser *et al.*

Chapter 9

LARGE-D LIMIT FOR METALIC HYDROGEN

John Loeser
Department of Chemistry
Oregon State University
Corvallis, OR 97331-4003

Abstract

Dimensional scaling is applied to metallic hydrogen by treating the large-D limit of the electronic structure for rigid three-dimensional proton lattices. The uniformly scaled $D \to \infty$ limit localizes the electrons, but approximately preserves $D = 3$ expectation values. The problem is treated for simple, body-centered, and face-centered cubic lattices, and both with and without correlation. Although the absolute errors are much larger than the energy differences associated with structural changes or electron correlation, the small differences are modeled surprisingly well. Correlation effects are especially amenable to dimensional scaling treatment, because solutions can be obtained in the large-D limit without invoking any approximations to the interelectron potentials. Converged correlation energies have been computed for simple cubic lattices for a wide range of densities ($0.1 \leq r_s \leq 10$). Comparison with known results at $D = 3$ reveals that the computed values are too low by a factor of about 2 in the van der Waals regime,

D. R. Herschbach et al. (eds.), Dimensional Scaling in Chemical Physics, 389–427.

and diverge too slowly in the electron gas regime. The fact that the large-D limit can be treated with equal almost equal ease at all densities makes it a potentially useful tool for investigating the effects of correlation on the metallization transition. However, further work will be required if the method is to be used quantitatively.

Introduction

In this chapter we apply the dimensional scaling method to a model solid state problem, namely three-dimensional lattices of hydrogen atoms. In the pressure regimes where such lattices are stable, they are also predicted to be metallic, so we will refer to them by the term metallic hydrogen. The primary goal of this chapter is to outline some basic techniques which will be required for dimensional scaling treatments of electronic structure in extended systems, but we will also look at some specific results for metallic hydrogen.

At normal pressures, of course, hydrogen forms a molecular solid. The stable form is a hexagonal closest packed lattice of H_2 molecules which are essentially freely rotating [1]. This remains the stable form until pressures of the order of one megabar. After this point, a number of phase transitions are expected [2-6], though the picture remains somewhat cloudy [7]. The prevailing opinion appears to be that hydrogen first undergoes a rotational order/disorder transition ($\sim$ 1.5 Mbar), then a molecular metal/insulator transition ($\sim$ 2.5 Mbar), and then at still higher pressure a molecular/atomic transition. It has been suggested that the molecular/atomic transition will actually occur by way of a sequence of phases of lower coordination number [8]. At extremely high pressures (but below those which would destroy the proton lattice [9], it is anticipated that the atomic body-centered cubic phase will be most stable [3].

Metallic hydrogen has been studied extensively for several reasons. First, a symmetric lattice of hydrogen atoms (such as a simple cubic lattice, with one electron and one proton per unit cell) is arguably the simplest realistic solid. Thus it is a natural, though surprisingly difficult, testing ground for theoretical methods. Metallic hydrogen has also been a major goal (some have called it the holy grail [7]) of experimental high-pressure work. There have been reports, starting in 1989, that it has finally been made in diamond anvil cells [5]. However, these experiments are very difficult to perform and interpret, and it should be noted there is still a lot of disagreement concerning the meaning of the experimental observations. Another continuing source of interest in metallic hydrogen has been its importance in planetary physics. The larger planets (especially Jupiter and Saturn) are probably mostly hydrogen, and the pressures in their interiors are measured in tens of Mbar. Thus, metallic hydrogen is probably a major component of these planets [10]. (Because of the temperature, however, the bulk of this may be in the liquid state.) Finally, there is some interest in metallic hydrogen because of the speculation that it might be a room temperature superconductor. Of course, this is of no practical consequence, unless one believes the wild speculation (widely repeated in the popular press [11]) that the superconducting metal might remain stable after the pressure is relieved.

The structure of the chapter is as follows: We begin in Sec. 2 with a qualitative description of the $D \to \infty$ limit of electronic structure in hydrogen atom lattices, using a one-dimensional lattice as an easily pictured example. In Sec. 3 those aspects of the treatment which are of a more generic nature (scaling and Hartree-Fock techniques) are reviewed in the context of simple examples. The lattice problem is treated in the Hartree-Fock approximation in Sec. 4, while Sec. 5 shows how one can undo the Hartree-Fock approximation in a systematic way, thereby bringing electron correlation into the solutions. The results obtained for the correlation energies are presented in Sec. 6. Finally, Sec. 7 outlines a few ways in which the methods described in this chapter might be improved upon.

Qualitative overview

Although protons are light enough that their zero-point motions actually have a significant influence on the thermodynamic properties of metallic hydrogen, we will not consider such motions here. To use the standard terminology, we will use the clamped nucleus approximation. Thus, it is only the electronic degrees of freedom which will be treated by dimensional scaling methods. The asymmetry thereby introduced — that the protons live in a three- dimensional space, while the electrons live in hyperspace — is really just the import for dimensional scaling of the Born-Oppenheimer approximation.

The basic elements of the treatment are the same as in other dimensional scaling treatments of electronic structure problems. In particular, we will focus attention on the probability density $|\Phi|^2 = J|\Psi|^2$, where J is the volume element or Jacobian factor. This probability density is acted upon by two sets of forces: centrifugal forces, which tend to push particles apart, and Coulombic forces, which keep the system bound together. With a suitable scaling of lengths and energies, the primary effect of varying the spatial dimensionality D is simply to alter the degree to which these forces manage to squeeze $|\Phi|^2$. Thus, varying D changes the spread of $|\Phi|^2$, but leaves most expectation values relatively unchanged. The goal is to treat the $D \to \infty$ limit, in which the spread vanishes and $|\Phi|^2$ describes a configuration of point-like particles localized relative to one another, and use this as a model for three- dimensional expectation values.

For multi-particle systems, the electronic structures which arise in the large-dimension limit are hard to picture because they cannot be embedded in three-dimensional space. In fact, the $D \to \infty$ limit for an N-particle system will generally define an $(N-1)$-dimensional structure. This is a consequence of the fact that the volume element vanishes (or equivalently, the centrifugal potential diverges) whenever the structure is not of maximal dimensionality. (It may be useful to think of the extra Cartesian degrees of freedom possessed by the electrons as "replacing" those normally associated with the wavefunction.) In spite of the conceptual challenges they present, the large-D classical geometries are in many ways easier to deal with than many-particle wavefunctions, and useful insights into electronic structure can be obtained from them. For these reasons, we will describe these structures below. In preparation for the problem of three-dimensional lattices, we consider here the simpler example of a uniform one-dimensional chain of hydrogen atoms.

Consider first what happens at large lattice spacing. Recall that for $D \to \infty$ the electron in a hydrogen atom is localized with respect to the nucleus at some distance r (in the sense that $\Delta r/\langle r \rangle \to 0$). Two hydrogen atoms at large internuclear separation $a \gg r$ are therefore like two dipolar diatomics molecules. Due to angular volume element factors, which behave like $\sin^D \theta$ at large D, these dipoles are perpendicular to the internuclear axis, and perpendicular with respect to each other. (The reasons for this can be seen a little more intuitively, perhaps, by noting that for each electron there is one axis defined by its nucleus and the other nucleus, one axis defined by the other nucleus and its electron, and $D-2$ other axes; simple statistics says that in the limit of large D, it will choose with unit probability not to make use of the two special axes.) Now one can use this same reasoning for any pair of atoms, so in the limit of large lattice spacing the electronic structure of the chain of atoms will be such that each electron has a 90° angle with respect to the chain (measured at its nucleus) and a 90° dihedral angle with respect to every other electron. Of course, this structure cannot be pictured in $D=3$ for more than two atoms.

Now consider what happens as the internuclear spacing is reduced. It is convenient to distinguish between a few different regimes on the basis of the relative values of the lattice spacing a and typical electron-nucleus distance r. First, in the van der Waals or low-density regime, where $a \gg r$, the primary effect of decreasing a is an opening of the dihedral angles between nearest neighbors due to the influence of interelectron repulsions. This effect, which is of order $1/a^3$ in the angles and order $1/a^6$ in the electronic energy, is just the $D \to \infty$ limit of the dispersion interaction. Next, in the medium-density or normal metal regime, where $a \sim r$, the dihedral angles are larger, and are important beyond nearest neighbors. However, it is now the shortening of the electron-nucleus distances, rather than the opening of the angles, which has the greatest impact on the energy. (It should be noted that we don't consider here anything but uniform lattices, so there is by construction no molecular regime.) Finally, in the high-density or weak-coupling regime, $a \ll r$, the electrons are strongly delocalized in the lattice (though they remain point-like). Here both angular and radial effects of the electron distribution are strong, but both effects are small compared to the internuclear repulsion.

Dimensional scaling techniques are most useful in connection with the study of many-body effects such as electron correlation. For the hydrogen lattice problem, such effects are embodied primarily in the dihedral angles between the electrons. In fact, the Hartree-Fock (uncorrelated) approximation for the linear chain of atoms can be obtained simply by fixing every dihedral angles at 90°. (This is quite generally the procedure by which one obtains the Hartree-Fock approximation; the reasons for this will be outlined below.) Electron correlation can be introduced by systematically allowing angles to open up, starting from their Hartree-Fock values of 90°. This is essentially the procedure which will be utilized later in treating electron correlation in three-dimensional hydrogen lattices.

Review of scalings and Hartree-Fock procedures

To simplify the discussion below, we consider two aspects of the calculations in the context of simpler problems. First is the matter of scalings. As always, we will utilize units in terms of which the solutions remain finite across the spectrum of dimensionalities. Since we are considering a collection of hydrogen atoms, for which the ground state is characterized by [12]

$$E_{\mathrm{o}} \;=\; -\frac{2}{(D-1)^2}\ \text{hartrees},$$

$$\langle r \rangle_{\mathrm{o}} \;=\; \frac{D(D-1)}{4}\ \text{Bohr radii},$$

$$(1)$$

the simplest set of units which reduces to conventional units at $D=3$, yet finitizes the dimensional limits, is that of $4/(D-1)^2$ hartrees for energies and $D(D-1)/6$ Bohr radii for distances. These are the units that we will use.

As an example, consider a hydrogen atom in a spherical cavity of radius r_o, with the nucleus clamped at the center. (This can be taken as a very primitive model for hydrogen under high pressure). The hamiltonian in unscaled atomic units (hartrees and Bohr radii) may be written succinctly as

$$\mathcal{H}^{(D)} = -\frac{1}{2}\nabla^2 - \frac{1}{r} + \left(r/r_o\right)^\infty.$$

$$(2)$$

Because of the spherical symmetry, the dimensionally generalized radial hamiltonian can be obtained at once by remembering that each extra dimension acts like an extra half unit of angular momentum:

$$\widetilde{\mathcal{H}}^{(D)} = -\frac{1}{2}\frac{\partial^2}{\partial r^2} + \frac{\left(\ell+\frac{D-3}{2}\right)\left(\ell+1+\frac{D-3}{2}\right)}{2r^2} - \frac{1}{r} + \left(r/r_o\right)^{\infty}. \tag{3}$$

(The $\widetilde{}$ indicates that this is a radial hamiltonian, which operates on the radial probability amplitude $\Phi = r^{\beta}\Psi$, where $\beta \equiv \frac{D-1}{2}$; that is, $\widetilde{\mathcal{H}} = r^{\beta}\mathcal{H}\, r^{-\beta}$.)   Introducing scaled units of $4/(D-1)^2$ hartrees and $D(D-1)/6$ Bohr radii, one readily finds the dimensional limit forms to be

$$\begin{aligned}\widetilde{\mathcal{H}}^{(1)} &= -\frac{9}{2}\frac{d^2}{d\rho^2} - 3\delta(\rho) + \left(\rho/\rho_o\right)^{\infty}, \\[2mm] \widetilde{\mathcal{H}}^{(\infty)} &= \frac{9}{8\rho^2} - \frac{3}{2\rho} + \left(\rho/\rho_o\right)^{\infty}.\end{aligned} \tag{4}$$

If desired, the coefficients in these limiting hamiltonians can be given a somewhat more familiar look by rescaling the dynamical variable ρ using $\rho = 3\check{\rho}$ for $\widetilde{\mathcal{H}}^{(1)}$ and $\rho = \frac{3}{2}\hat{\rho}$ for $\widetilde{\mathcal{H}}^{(\infty)}$; this has no effect on the physics, but leads to an apparent scaling of the cavity radius:

$$\begin{aligned}\widetilde{\mathcal{H}}^{(1)} &= -\frac{1}{2}\frac{d^2}{d\check{\rho}^2} - \delta(\check{\rho}) + \left(\check{\rho}/\tfrac{1}{3}\rho_o\right)^{\infty}, \\[2mm] \widetilde{\mathcal{H}}^{(\infty)} &= \frac{1}{2\hat{\rho}^2} - \frac{1}{\hat{\rho}} + \left(\hat{\rho}/\tfrac{2}{3}\rho_o\right)^{\infty}.\end{aligned} \tag{5}$$

Because of the uniform procedure used to derive the limiting forms, the solutions obtained from them can be expected to bracket the $D=3$ solutions. In fact, simple dimensional interpolation (linear interpolation in $1/D$ between dimensional limits evaluated in the above scaling) allows one to obtain rough estimates of $D=3$ energies for this and a variety of other simple systems, as shown in Fig. 1. It is important to note that it is only because of the use of a *uniform* scaling that this is true. For example, if one used units of $(D-1)/2$ Bohr radii for the $D \to 1$ limit and $(D-1)^2/4$ Bohr radii for the $D \to \infty$ limit [12], one would obtain Eq. (5) without the factors of $\frac{1}{3}$ and $\frac{2}{3}$ in front of the cavity radius, and the solutions obtained from these hamiltonians would not be useful for quantitative calculations.

It is also useful to consider the implementation of Hartree-Fock theory in the context of a simple example. Here we use the helium

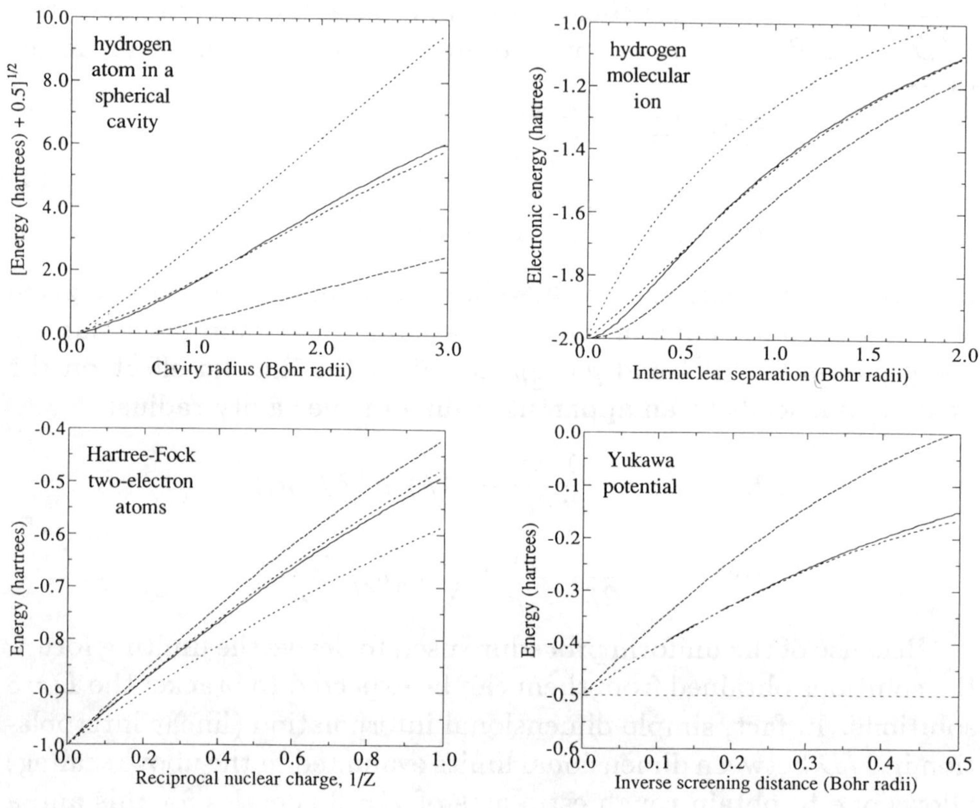

Figure 1. Energies for a hydrogen atom in a spherical cavity and for several other representative problems. Uniform scalings were used for all four problems. Energies plotted are dimensional limit values $E^{(1)}$ ($\cdots$) and $E^{(\infty)}$ ($----$), dimensionally interpolated values $\frac{1}{3}E^{(1)} + \frac{2}{3}E^{(\infty)}$ ($---$), and reference values $E^{(3)}$ ($\underline{\quad\quad}$).

atom as an example. Since the HF wavefunction depends only on the electron-nucleus distances ρ_1 and ρ_2, the probability distribution with respect to the interelectron angle θ_{12} will be determined purely by volume element factors. As pointed out above, this means that the angle will become fixed at 90° in the $D \to \infty$ limit. (Consider two randomly chosen D-dimensional unit vectors. For each, there are D independent directions, only one which is parallel to the other vector. The expectation value for the parallel component will therefore decrease in magnitude as the dimension is increased, and one can say with certainty that randomly chosen vectors will be orthogonal in the $D \to \infty$ limit [13].) Therefore the $D \to \infty$ limit of HF-helium can be obtained by pre-assigning the interelectron angle to its known limiting value of 90°, and minimizing the resulting hamiltonian,

$$\widetilde{\mathcal{H}} = \frac{9}{8\rho_1^2} + \frac{9}{8\rho_2^2} - \frac{3Z}{2\rho_1} - \frac{3Z}{2\rho_2} + \frac{3}{2\sqrt{\rho_1{}^2 + \rho_2{}^2}}. \tag{6}$$

This gives $\rho_1 = \rho_2 = \frac{3}{2}/(Z - \frac{1}{\sqrt{8}})$ (in Bohr radii) and $E_{\mathrm{HF}} = -(Z - \frac{1}{\sqrt{8}})^2$ (in hartrees).

Assuming that the coordinate system has been chosen in such a way that the angles correspond precisely to the degrees of freedom over which the Hartree-Fock approximation averages, the procedure of pre-assigning angles to 90° in order to obtain the HF approximation in the $D \to \infty$ limit is quite general. For example, for the many-electron atom one can set all interelectron angles measured at the nucleus to 90°, and for chains of hydrogen atom one can set all dihedral angles measured about the chain axis to 90°. The equivalent procedure for three-dimensional lattices will be discussed below.

The $D \to \infty$ solution to the HF-helium problem has $\rho_1 = \rho_2$. If this were assumed from the beginning, then it is clear that the problem could also be solved by symmetrizing the hamiltonian and treating the contribution from either electron, namely

$$\widetilde{\mathcal{H}} = \frac{9}{8\rho^2} - \frac{3Z}{2\rho} + \frac{3}{4\sqrt{2}\rho^2}. \tag{7}$$

Upon minimization, this gives the energy per electron, $-\frac{1}{2}(Z - \frac{1}{\sqrt{8}})^2$. A symmetrized hamiltonian per electron of this sort is what will be used to treat hydrogen lattices in the following section. It should be noted that Eq. (7) is not the usual one-electron Hartree-Fock operator. For example, for helium the $D \to \infty$ transcription of the usual unrestricted Hartree-Fock method would involve the iterative minimization (to self-consistency) of the one-electron operators

$$\widetilde{\mathcal{H}}_1 = \frac{9}{8\rho_1{}^2} - \frac{3Z}{2\rho_1} + J_{12}, \qquad \widetilde{\mathcal{H}}_2 = \frac{9}{8\rho_2{}^2} - \frac{3Z}{2\rho_2} + J_{21}, \qquad (8)$$

where

$$J_{12} = J_{21} = \frac{3}{2\sqrt{\rho_1{}^2 + \rho_2{}^2}}. \qquad (9)$$

From the one-electron operators one obtains the orbital energies $E_1 = E_2 = -\frac{1}{2}(Z - \frac{1}{\sqrt{8}})(Z - \frac{3}{\sqrt{8}})$. Since both of these include the effect of the single interelectron repulsion or Coulomb integral $J_{12} = \frac{1}{\sqrt{2}}(Z - \frac{1}{\sqrt{8}})$, the total Hartree-Fock energy is given by $E_{\mathrm{HF}} = E_1 + E_2 - J_{12}$. Of course, the final result is the same as before.

It should be noted that when lattices are treated in an analogous way by setting all dihedral angles to 90°, the resulting uncorrelated reference state is not strictly speaking the Hartree-Fock approximation. One can see this from the fact that the lattice dissociates correctly to hydrogen atoms. (It appears that the approximation is more closely related to an extended valence bond treatment, and its error to the "dynamic" correlation energy.) However, we will continue to refer to the natural $D \to \infty$ uncorrelated reference state as Hartree-Fock, and to its error as correlation energy.

Cubic hydrogen lattices: HF approximation

The hamiltonian for one electron in a lattice of hydrogen atoms with clamped nuclei can be written [14]

$$\mathcal{H} = -\tfrac{1}{2}\nabla^2 - \frac{1}{|\mathbf{r}|} + \tfrac{1}{2}\sum\sum\sum_{l,m,n \in \mathcal{L}'} \left(\frac{1}{|\mathbf{R}_{lmn}|} - \frac{1}{|\mathbf{r} - \mathbf{R}_{lmn}|} - \frac{1}{|\mathbf{r}_{lmn}|} + \frac{1}{|\mathbf{r} - \mathbf{r}_{lmn}|} \right).$$

$$(10)$$

Here $\mathcal{L}$ designates a three-dimensional lattice, while $\mathcal{L}'$ designates the lattice minus one site, which is taken to be $(0, 0, 0)$. For simplicity, we will consider only lattices of cubic symmetry, so the lattice is characterized by a single lattice constant a, in terms of which the nuclear positions are given by $\mathbf{R}_{\ell mn} = (\ell a, ma, na)$ for some set of integer triples ℓ, m, n. (In particular, the set of all integer triples gives the SC lattice, the set with even $\ell + m + n$ gives the FCC lattice, and the set with ℓ, m, n all even or all odd gives the BCC lattice.) One electron is "assigned" to each nucleus, so their positions $\mathbf{r}_{\ell mn}$ are labeled analogously.

We now proceed to obtain the $D \to \infty$ limit of this one-electron hamiltonian. It is first necessary to choose an appropriate coordinate system. We will describe the position of an electron by means of its perpendicular distance from the three-dimensional nuclear lattice, by the point in the lattice on which the perpendicular stands, and by the dihedral angles of this perpendicular with respect to the others. (The geometry is of course not easy to picture. It may help to think by analogy with the somewhat more accessible linear chain of atoms; the corresponding coordinates here would be the distance of an electron from the chain, the point along the chain above that it sits above, and the set of dihedral angles measured by looking down the chain.) As described in the previous section, for the Hartree-Fock approximation we can simply fix all of the dihedral angles at 90°. Again, this is due to absence of any explicit dependence on the interelectron coordinates (and consequently on the dihedral angles) in the Hartree-Fock wavefunction. Later, we will systematically allow for correlation effects through the opening up of the dihedral angles between the electrons.

Finally, in order to proceed quickly, we will also make a rather strong symmetry assumption, namely that at the minimum of the $D \to \infty$ Hartree-Fock hamiltonian, each electron will be positioned at the same distance from the three- dimensional nuclear lattice, and directly "above" one of the protons in that lattice. This is the three-dimensional equivalent (in the HF approximation) of the geometry described for the linear chain in Sec. 2. It amounts to assuming that the $D \to \infty$ solution will possess the full symmetry of the hamiltonian (though the assumed geometry is not the *only* one having this symmetry). We make the assumption because it greatly facilitates solution, and because there are several indications or suggestions that it will turn out to be correct. The grounds include an inability to locate any alternative minima of lower energy, and the observation that the $D \to \infty$ limit gives rise to highly or fully symmetric solutions in a variety of contexts [15,16]. However, so far only a few of the possible symmetry-breaking modes (such as uniform displacements of the electrons parallel to the nuclear lattice) and alternative symmetric geometries (such as that in which the electrons lie above the centers of the cells defined by the nuclei) have been considered, so this symmetry assumption definitely needs further testing.

We can now write down the Hartree-Fock one-electron hamiltonian in the $D \to \infty$ limit. As usual, we will write the hamiltonian as one for the probability amplitude, and will remove the dominant dimension-dependence of the solutions through use of appropriately scaled units. As discussed in the previous section, this means that energies will be in units of $4/(D-1)^2$ hartrees, and distances in units of $D(D-1)/6$ Bohr radii. The symmetry assumption allows us to equate all electron-nucleus distances, and to constrain the electrons to positions directly above the nuclei, prior to minimization of the hamiltonian. Also, the Hartree-Fock approximation allows dihedral angles to be fixed at 90°. With these scalings and simplifications, the $D \to \infty$ limit hamiltonian can be written

$$\widetilde{\mathcal{H}} = \frac{9}{8\rho^2} - \frac{3}{2\rho} + \frac{3}{2}W \tag{11}$$

where, with $\sigma^2 \equiv \ell^2 + m^2 + n^2$,

$$W = \tfrac{1}{2}\sum\sum\sum_{\ell,m,n \in \mathcal{L}'} \left(\frac{1}{\sqrt{\sigma^2\alpha^2}} - \frac{2}{\sqrt{\sigma^2\alpha^2 + \rho^2}} + \frac{1}{\sqrt{\sigma^2\alpha^2 + 2\rho^2}} \right). \tag{12}$$

For any specified lattice type $\mathcal{L}$ and (scaled) lattice constant α, the minimum of Eq. (11) with respect to ρ gives the energy per electron.

The triple sum W is a kind of Madelung sum, though of course both its reference to fixed electron positions and its implicitly high-dimensional character make it a somewhat unusual Madelung sum. After pulling out a factor of $1/\alpha$, it becomes a function of the single variable ρ/α (the ratio of orbit radius to lattice spacing, which can be used to characterize the different density regimes). However, for convenience we will continue to view W as a function of two variables. These may be taken to be ρ and either α or r_s (the standard solid state parameter, defined as the radius in a_o of a sphere which contains on average one electron). For the three cubic lattices which will be considered here, r_s and α are related by

$$\tfrac{4}{3}\pi r_s{}^3 = \begin{cases} \alpha^3 & \text{for SC} \\ 2\alpha^3 & \text{for FCC} \\ 4\alpha^3 & \text{for BCC.} \end{cases} \tag{13}$$

Using r_s shifts attention from the nuclear lattice to the electrons, and is especially appropriate at very high density.

It turns out that an excellent approximation to the Madelung sums can be obtained by means of a suitable interpolation between its low- and high-density limiting forms, both of which are easily evaluated (Table 1). Consider first the low-density limit. Here $\rho \ll \alpha$, and one can expand the square roots in powers of the small quantity $\rho/\sigma\alpha$ and consider just the leading-order contribution (which amounts to a sum of dipole-dipole interactions), namely

$$\begin{aligned}
\lim_{\alpha \to \infty} W &= \tfrac{1}{2}\sum\sum\sum_{\ell,m,n \in \mathcal{L}'} \frac{1}{\sigma\alpha} \frac{3}{4} \left(\frac{\rho}{\sigma\alpha}\right)^4 \\
&= \frac{3\rho^4}{8\alpha^5} \sum\sum\sum_{\ell,m,n \in \mathcal{L}'} \frac{1}{(\ell^2 + m^2 + n^2)^{5/2}} \\
&= \frac{3\rho^4}{8\alpha^5} \begin{cases} 10.3775 & \text{for SC} \\ 2.9995 & \text{for FCC} \\ 0.94676 & \text{for BCC.} \end{cases}
\end{aligned} \tag{14}$$

The sums over integers are taken from the tables of Born and Misra [17].

For the high-density limit, where $\rho \gg \alpha$, the sum can be replaced by an integral. Since the volume per atom is $\tfrac{4}{3}\pi r_s{}^3$, one has

$$\begin{aligned}
\lim_{\alpha \to 0} W &= \tfrac{1}{2}\int_0^\infty \left(\frac{1}{\sqrt{r^2}} - \frac{2}{\sqrt{r^2 + \rho^2}} + \frac{1}{\sqrt{r^2 + 2\rho^2}}\right) \frac{4\pi r^2\, dr}{\tfrac{4}{3}\pi r_s{}^3} \\
&= \frac{3\rho^2}{2r_s{}^3}\int_0^\infty \left(\frac{1}{\sqrt{\eta}} - \frac{2}{\sqrt{\eta+1}} + \frac{1}{\sqrt{\eta+2}}\right)\sqrt{\eta}\, d\eta \\
&= 0.51986\, \frac{\rho^2}{r_s{}^3}.
\end{aligned} \tag{15}$$

Table 1. Low- and high-density limits for Coulomb sums W. Limits are given in terms of both α (the lattice constant) and r_s (the radius of a sphere containing on average one electron).

limit	SC	FCC	BCC
low density	$0.35753\dfrac{\rho^4}{r_s{}^5}$ $3.89157\dfrac{\rho^4}{\alpha^5}$	$0.32808\dfrac{\rho^4}{r_s{}^5}$ $1.12480\dfrac{\rho^4}{\alpha^5}$	$0.32877\dfrac{\rho^4}{r_s{}^5}$ $0.35503\dfrac{\rho^4}{\alpha^5}$
high density	$0.51986\dfrac{\rho^2}{r_s{}^3}$ $2.17759\dfrac{\rho^2}{\alpha^3}$	$0.51986\dfrac{\rho^2}{r_s{}^3}$ $1.08879\dfrac{\rho^2}{\alpha^3}$	$0.51986\dfrac{\rho^2}{r_s{}^3}$ $0.32877\dfrac{\rho^2}{\alpha^3}$

Note that the high-density limit is independent of lattice type when expressed in terms of r_s.

From the low- and high-density limits, one can obtain an approximate sum at any density by means of a simple harmonic interpolant. That is,

$$W \approx \left(\frac{1}{\lim\limits_{\alpha \to \infty} W} + \frac{1}{\lim\limits_{\alpha \to 0} W} \right)^{-1} \tag{16}$$

(in either the α or r_s representation). For example, working in terms of α, and using the limits from Table 1, one finds that for the simple cubic lattice

$$W \approx \left(\frac{\alpha^5}{3.89157\rho^4} + \frac{\alpha^3}{2.17759\rho^2} \right)^{-1}. \tag{17}$$

Comparing this approximation with the actual lattice sums (as determined by direct summation), one finds that it is accurate to within $\frac{1}{2}\%$ for all densities. Since this error is smaller than that introduced by the dimensional scaling approximation, we will use the approximate form for the Coulomb sum in order to simplify calculations. It should be noted that these lattice sums will be corrected systematically when electron correlation is introduced in Sec. 5.

The electronic energy per atom is given, for any lattice $\mathcal{L}$ and any lattice constant α (or r_s), by the minimum with respect to ρ of Eq. (11). The energies obtained for the simple cubic lattice are compared in Fig. 2 with reference values obtained from density functional theory [18]. The behavior is seen to be only qualitatively correct. Although several other factors may be contributing to the errors in the dimensional scaling values (including, for example, the assumption of unbroken symmetry and the approximation to the Madelung-like sum), probably the main difficulty is that only the $D \to \infty$ limit has been used so far. As Fig. 1 makes clear, it is only with the help of actual results from low-D or order-$1/D$ calculations that one can hope to obtain accurate results at $D = 3$.

The most serious errors are at small r_s. In fact, the asymptotic behavior as $r_s \to 0$ is not even qualitatively correct. (The total energy diverges as $1.873\,r_s^{-3/2}$, rather than $2.210\,r_s^{-2}$.) The problem is that the $D \to \infty$ limit is insufficient to describe the band structure. In the

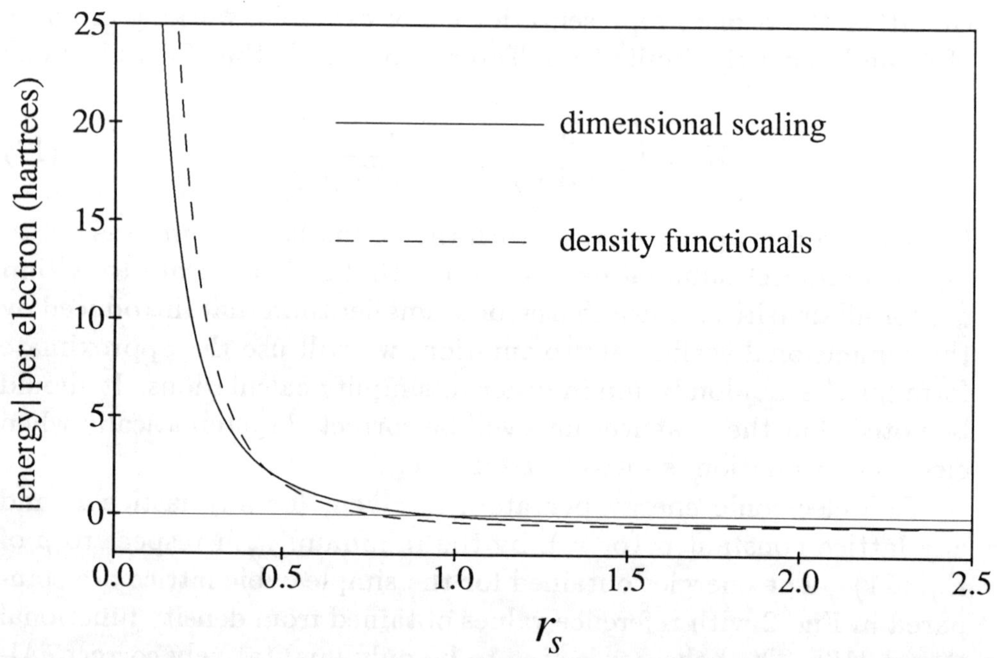

Figure 2. Total energies for simple cubic hydrogen lattices from dimensional scaling and from density functional theory [18].

high-density regime it will apparently be necessary to go beyond the $D \to \infty$ limit approximation to obtain satisfactory results. A possible means for doing this without explicitly calculating any higher-order effects will be described in the last section.

In spite of significant systematic errors, the $D \to \infty$ limit solutions still contain much useful information. For example, one obtains quite reasonable values for the relative energies of the three cubic lattice types. It isn't hard to show from the approximate forms for the Madelung-like sums W that the order of stability is predicted to be FCC > BCC > SC, but with FCC and BCC much closer in energy, at any pressure. At 2.5 Mbar, which is where recent experiments suggest hydrogen metallization takes place [5], the SC – BCC energy difference is predicted to be 0.08 eV per atom, while the BCC – FCC energy difference is about 0.002 eV per atom. (The pressure in Mbar is related to the energy in hartrees by $P = 23.7 \, r_s^{-2} \, \partial E / \partial r_s$.) These values are in good agreement with quantum Monte Carlo calculations [4]. We now turn to the facet of the calculations where the $D \to \infty$ limit solutions appear to have the most to offer, namely electron correlation.

Introducing correlation

The primary effect of electron correlation in the $D \to \infty$ limit is to open up the dihedral angles from their Hartree-Fock values of exactly 90°. Angles in the correlated solution are determined by the balance between centrifugal effects (which always favor 90°) and interelectron repulsions (which always favor 180°). The balance turns out to be achieved at angles quite close to 90°. Not surprisingly, the largest effects, of the order of a few degrees, are found at moderately high (but not very high) densities, and for electrons that are close to one another.

We will calculate the effects of correlation in a systematic way. We do so by treating the interactions of each electron with larger and larger clusters of other electrons. Except at quite high density, one doesn't have to go very far to achieve convergence, or at least to get to a point from which one can extrapolate reliably. For the simple cubic lattice, for example, we will treat correlations out to sixth nearest neighbors (80 electrons), and then extrapolate.

We take the uncorrelated or Hartree-Fock hamiltonian as our point of reference. Since correlation effects are determined by a balance of centrifugal and Coulombic angular forces, it is clear that angular terms need to be inserted in both the centrifugal and Coulombic terms of the hamiltonian. The angular parts of the centrifugal potential were just what we used, implicitly, when we argued that electrons will assume positions at 90° relative to one another when interelectron repulsions are absent or averaged over. It requires a fair amount of work to obtain the explicit form of the angular centrifugal potential. Below, we will first treat the angular aspects of the interelectron repulsions (which are easy), and then present the long derivation of the centrifugal terms.

In treating the angular aspects of the problem, we will again make the assumption that symmetry is not broken. Specifically, we will assume that dihedral angles which are equivalent in the hamiltonian will turn out to be equal when the $D \to \infty$ hamiltonian is minimized. In a simple cubic lattice, for example, we assume that the dihedral angles between a reference electron and the 6 nearest neighbor electrons will all be equal, that the dihedral angles with respect to the 12 next nearest neighbor electrons will all be equal, and so forth. We will also continue to assume that the electrons are all equidistant from the nuclear lattice, and directly above the nuclei. (However, the radial distance is not assumed to be the same as in the uncorrelated problem.) As mentioned above, there is reason to believe that this strong symmetry assumption will turn out to be correct, but it has not yet been tested in any systematic way.

The symmetry assumptions mean that a single variable $\theta_{\ell m n}$ with $\ell \geq m \geq n \geq 0$ can be used for a complete set of equivalent dihedral angles. Actually, it is simpler to work directly with the cosines $\gamma_{\ell m n} = \cos \theta_{\ell m n}$, and we will use these as the fundamental variables throughout. Each $\gamma_{\ell m n}$ is associated with a distance $\sigma \alpha$, where again $\sigma^2 = \ell^2 + m^2 + n^2$, and with a multiplicity or symmetry number [19]. (For the SC lattice, for example, the cosine of the dihedral angle between the reference electron and its nearest neighbor electrons is γ_{100}, the corresponding internuclear distance is α, and the multiplicity is 6.) These quantities are used repeatedly, and for convenience are

Table 2. Canonical indices for sets of equivalent lattice sites, one-character abbreviations (for Table 3), squared distances (in units of α^2), and multiplicities for simple, face-centered, and body- centered cubic lattices.

standard index set	abbre- viation	squared distance	number of positions		
			SC	FCC	BCC
000	0	0	1	1	1
100	1	1	6	0	0
110	2	2	12	12	0
111	3	3	8	0	8
200	4	4	6	6	6
210	5	5	24	0	0
211	6	6	24	24	0
220	8	8	12	12	12
221	9	9	24	0	0
300	I	9	6	0	0
310	J	10	24	24	0
311	K	11	24	0	24
222	L	12	8	8	8

summarized in Table 2. Also listed in the table are single-character abbreviations (a number or letter corresponding to σ^2), which will also be useful below.

Consider first the relatively straightforward insertion of angle- dependence into the Coulomb terms. The Madelung-like sums used in the Hartree-Fock approximation, Eq. (12), were constructed by assuming that all dihedral angles were fixed at 90°. Now we wish to undo this assumption. Suppose we wish to treat correlation between the electron at the reference site $(0,0,0)$ and the electron at some other site (ℓ, m, n). This means that the dihedral angles must be permitted to open up from 90°. The Madelung-like sum contained somewhere within it (perhaps implicitly, if the sum has been approximated) a term $\frac{1}{2}[\sigma^2\alpha^2 + 2\rho^2]^{-1/2}$. (Recall that half of an interelectron repulsion is assigned to each of two interacting electrons.) We need to take this out and replace it with a term which no longer assumes a 90° dihedral angle, namely $\frac{1}{2}[\sigma^2\alpha^2 + 2\rho^2(1-\gamma_{\ell mn})]^{-1/2}$. That is, we need to make a correction

$$\Delta W(\rho, \sigma\alpha, \gamma_{\ell mn}) = \frac{1}{2}\left(\frac{1}{\sqrt{\sigma^2\alpha^2 + 2\rho^2(1-\gamma_{\ell mn})}} - \frac{1}{\sqrt{\sigma^2\alpha^2 + 2\rho^2}}\right) \quad (18)$$

to the Coulomb sum. In the SC lattice, for example, the corrected Coulomb sum necessary to treat correlation through third-nearest neighbors (the first cubic shell around the reference site) would be

$$W^{(3)} = \left(\frac{\alpha^5}{3.89157\rho^4} + \frac{\alpha^3}{2.17759\rho^2}\right)^{-1} \quad (19)$$

$$+ 6\,\Delta W(\rho, \alpha, \gamma_{100}) + 12\,\Delta W(\rho, \sqrt{2}\alpha, \gamma_{110}) + 8\,\Delta W(\rho, \sqrt{3}\alpha, \gamma_{111}).$$

This is all that needs to be done to the Coulomb part of the $D\to\infty$ hamiltonian.

Now we come to the second and harder place in the hamiltonian where the dihedral angles come into play, namely the centrifugal terms. The angular centrifugal terms quantify the tendency toward orthogonality that was described and used above. (One can consider that these terms were actually present in the Hartree-Fock hamiltonians, but could be ignored because Hartree- Fock averaging of the interelectron repulsions destroyed any tendency toward nonorthogonality, thereby allowing all angles to be fixed at 90° prior to minimization.) Because of the choice of coordinate system and the symmetry assumption, the basic structure of the angular centrifugal potential is similar to that which arose in the treatment of many-electron atoms (Chapter 3). In particular, for each electron it takes the form of a ratio of Gramian determinants. For the lattice problem, the determinants are infinite in size and have a more complex internal structure, but the geometrical interpretation remains largely the same as that encountered in atoms. We therefore review the many-electron atom centrifugal potential briefly.

For each electron in a many-electron atom, the centrifugal potential was proportional to an inverse squared distance, namely the distance of the electron from the subspace defined by the other $N-1$ electrons and the nucleus. In the units of this chapter, namely $4/(D-1)^2$

hartrees for energies and $D(D-1)/6$ Bohr radii for distances (which are different from those of Chapter 3), the centrifugal potential for electron i is in fact $9/(8\pi_i{}^2)$, where π_i is the perpendicular distance. Gramian determinants arise when one rewrites π_i in terms of interparticle distances and angles. Specifically, if ρ_i denotes the distance of the ith electron from the nucleus, and γ_{jk} denotes the cosine of the angle between the jth and kth electrons measured at the nucleus, then $\pi_i = \rho_i(\Gamma/\Gamma^{(i)})^{1/2}$. Here $\Gamma = |\gamma_{jk}|$ is the Gramian determinant for all electrons in the atom, while $\Gamma^{(i)}$ is that for all but the ith electron (the determinant obtained by striking out row i and column i from the all-electron determinant).

To see the relevance of this to the lattice problem, note that when any finite piece of the lattice is projected sequentially down all three nuclear axes (thereby superimposing all of the nuclei and creating a big "atom"), the dihedral angles between the electrons map into simple interelectron angles. Therefore the angular centrifugal potential for the finite lattice is identical in form to that for the corresponding atom. (Note that it is only because we are working in the $D \to \infty$ limit, and under the assumption of unbroken symmetry, that motions within the three-dimensional subspace of the nuclear lattice can be ignored to leading order in $1/D$.) Of course, this only refers to the form of the potential — clearly dihedral angles associated with different distances in the lattice are not equivalent, as they are after the projection. From this mapping, one can see that the $D \to \infty$ limit hamiltonian for any finite piece of the hydrogen lattice can be obtained by replacing the simple distance ρ in the one-electron Hartree-Fock centrifugal potential by the corresponding perpendicular distance, $\rho(\Gamma/\Gamma')^{1/2}$. Here the prime is used to denote the Gramian determinant for all electrons except the one being treated.

It is difficult to picture the structure of the lattice Gramian determinants, since each row or column is labeled by a triple of indices (ℓ, m, n), each of which ranges from $-\infty$ to ∞. Although it will be shown shortly that we can treat the Gramian determinant ratios without worrying about their structure, it is still useful to consider an explicit example, in order to see in concrete terms the workings of the angular centrifugal potentials. For this purpose, we consider again the linear chain of atoms. We start by considering a finite chain with periodic boundary conditions and nearest neighbor correlations only. That is, we consider a finite repeated chain in which dihedral angles between nearest neighbor electrons are allowed to open, while all others are kept fixed at their Hartree-Fock values of 90°.

If the atoms of a linear chain of N atoms are labeled in order, then the matrix for the complete Gramian determinant Γ will have a diagonal of ones, a constant value in all positions immediately adjacent to the diagonal (the cosines of the nearest neighbor dihedral angles, which by symmetry are all equal), and zeros elsewhere (since we are assuming that all other angles are fixed at 90°). Note that for the chain with periodic boundary conditions (a ring), the upper-right and lower-left corner positions are also considered adjacent to the diagonal. Call the nearest neighbor cosine β, and temporarily replace the ones on the diagonal by the variable α. Then the Gramian takes the form of the Hückel determinant for a cyclic conjugated polyene [20], $C_N H_{2N}$:

$$\Gamma = \begin{vmatrix} \alpha & \beta & 0 & \cdot & 0 & 0 & \beta \\ \beta & \alpha & \beta & \cdot & 0 & 0 & 0 \\ 0 & \beta & \alpha & \cdot & 0 & 0 & 0 \\ \cdot & \cdot & \cdot & & \cdot & \cdot & \cdot \\ 0 & 0 & 0 & \cdot & \alpha & \beta & 0 \\ 0 & 0 & 0 & \cdot & \beta & \alpha & \beta \\ \beta & 0 & 0 & \cdot & 0 & \beta & \alpha \end{vmatrix} = \prod_{i=1}^{N} \left(\alpha - 2\beta \cos\tfrac{2\pi i}{N} \right). \qquad (20)$$

The Gramian determinant Γ' is obtained by removing one row and one column from the matrix for Γ. All atoms in the ring (that is, the chain with periodic boundary conditions) were equivalent, but it is simplest to eliminate the atom labeled 1 or N. This gives the Hückel determinant for a linear conjugated polyene [20], $C_{N-1} H_{2N}$:

$$\Gamma' = \begin{vmatrix} \alpha & \beta & 0 & \cdot & 0 & 0 \\ \beta & \alpha & \beta & \cdot & 0 & 0 \\ 0 & \beta & \alpha & \cdot & 0 & 0 \\ \cdot & \cdot & \cdot & & \cdot & \cdot \\ 0 & 0 & 0 & \cdot & \alpha & \beta \\ 0 & 0 & 0 & \cdot & \beta & \alpha \end{vmatrix} = \prod_{i=1}^{N-1} \left(\alpha - 2\beta \cos\tfrac{\pi i}{N} \right). \qquad (21)$$

Alternatively, and much more usefully, we can express Γ' in terms of Γ as [21]

$$\Gamma' = \frac{1}{N}\frac{\partial \Gamma}{\partial \alpha}. \tag{22}$$

Thus, the ratio of Gramian determinants is

$$
\begin{aligned}
\frac{\Gamma'}{\Gamma} &= \frac{1}{N}\frac{\partial \log \Gamma}{\partial \alpha} \\
&= \frac{1}{N}\frac{\partial}{\partial \alpha} \log \prod_{i=1}^{N}\left(\alpha - 2\beta \cos\tfrac{2\pi i}{N}\right) \\
&= \frac{1}{N}\sum_{i=1}^{N}\frac{1}{\alpha - 2\beta \cos\frac{2\pi i}{N}}.
\end{aligned}
\tag{23}
$$

Finally, setting $\alpha = 1$ and $\beta = \cos\theta$, where θ is the nearest neighbor dihedral angle, and letting $N \to \infty$, we obtain for the centrifugal potential

$$
\begin{aligned}
\frac{9}{8\rho^2}\frac{\Gamma'}{\Gamma} &= \frac{9}{8\rho^2}\lim_{N\to\infty}\frac{1}{N}\sum_{i=1}^{N}\frac{1}{1 - 2\cos\theta \cos\frac{2\pi i}{N}} \\
&= \frac{9}{8\rho^2}\frac{1}{2\pi}\int_0^{2\pi}\frac{d\xi}{1 - 2\cos\theta \cos\xi} \\
&= \frac{9}{8\rho^2}\frac{1}{\sqrt{1 - 4\cos^2\theta}}.
\end{aligned}
\tag{24}
$$

For any ρ, this generalized centrifugal potential has its minimum at $\theta = 90°$ (so that it favors orthogonality) and diverges when $\theta = 60°$ or $120°$. The divergence signals the fact that the electronic coordinates for the reference electron have become linearly dependent on those of the other electron [22] (a generalization of the observation that three vectors separated pairwise by $120°$ are linearly dependent).

The generalization of the above to include interactions beyond nearest neighbors is straightforward. If β_1, β_2, ... denote the cosines of the dihedral angles for nearest neighbor, next nearest neighbor, ...

interactions, then from [23]

$$
\Gamma \;=\;
\begin{vmatrix}
\alpha & \beta_1 & \beta_2 & \cdot & \beta_3 & \beta_2 & \beta_1 \\
\beta_1 & \alpha & \beta_1 & \cdot & \beta_4 & \beta_3 & \beta_2 \\
\beta_2 & \beta_1 & \alpha & \cdot & \beta_5 & \beta_4 & \beta_3 \\
\cdot & \cdot & \cdot & \cdot & \cdot & \cdot & \cdot \\
\beta_3 & \beta_4 & \beta_5 & \cdot & \alpha & \beta_1 & \beta_2 \\
\beta_2 & \beta_3 & \beta_4 & \cdot & \beta_1 & \alpha & \beta_1 \\
\beta_1 & \beta_2 & \beta_3 & \cdot & \beta_2 & \beta_1 & \alpha
\end{vmatrix}
$$

$$
=\; \prod_{i=1}^{N} \left(\alpha - 2\beta_1 \cos\tfrac{2\pi i}{N} - 2\beta_2 \cos\tfrac{4\pi i}{N} - \ldots \right)
\tag{25}
$$

one readily finds that the above generalizes to

$$
\frac{9}{8\rho^2}\frac{\Gamma'}{\Gamma} \;=\; \frac{9}{8\rho^2}\frac{1}{2\pi}\int_0^{2\pi} \frac{d\xi}{1 - 2\beta_1 \cos\xi - 2\beta_2 \cos 2\xi - \ldots}.
\tag{26}
$$

This integral, like those for three-dimensional lattices considered below, cannot be done in closed form. However, as pointed out above, all of the dihedral angle cosines turn out to be small (especially those at longer range), so the integrand can be expanded in the β's and integrated term by term.

The generalization to three-dimensional lattices may be accomplished most easily by using some results from the theory of graph spectra [21]. The spectrum of a graph is the set of eigenvalues of its adjacency matrix, which for an undirected graph is the symmetric matrix in which each entry is just the numbers of edges connecting the two vertices specified by its row and column. For example, the spectrum of a simple cycle graph C_N (consisting of N vertices and N edges) is $\{2\cos\frac{2\pi i}{N}\,|\,i=1,\ldots,N\}$, while that for the open chain (or path) P_{N-1} obtained by removing one vertex and its adjoining edges from C_N is $\{2\cos\frac{\pi i}{N}\,|\,i=1,\ldots,N-1\}$. Eqs. (20) through (24) follow from these results. Similarly, Eqs. (25) and (26) follow from the spectra of the graphs $C_{N/K}$ in which N lines connect Kth nearest neighbors in a cycle of N vertices, namely $\{2\cos\frac{K\pi i}{N}\,|\,i=1,\ldots,N\}$.

Only a few results and concepts from graph spectral theory are needed in order to generalize to three-dimensional lattices. First it is necessary to know how to express certain graphs and their spectra in terms of simpler graphs and spectra. By way of example, an $M \times N$ square lattice graph with periodic boundary conditions (equivalently, the grid generated by sets of big and small circles on a torus) can be called the sum of the two cycles C_M and C_N, and its spectrum is just the set of sums of eigenvalues for those two cycles, $\{2\cos\frac{2\pi i}{M} + 2\cos\frac{2\pi j}{N}\}$. Similarly, the graph obtained from exactly the same set of vertices, but with the diagonals rather than the sides of the squares as its edges, is the product of the cycles C_M and C_N, and its spectrum is the set of eigenvalue products, $\{2\cos\frac{2\pi i}{M} \times 2\cos\frac{2\pi j}{N}\}$. As one might guess, these sum and product spectra are what one needs in order to treat nearest neighbor and next nearest neighbor correlations in a square planar lattice.

The second useful general result relates the spectrum of a graph $\mathcal{G}$ in which all vertices are equivalent to that of the subgraph $\mathcal{G}'$ in which any one of those vertices and its associated edges have been removed. The relationship is given most simply in terms of the characteristic polynomials (whose zeros are the eigenvalues) of the two graphs. If $|\mathcal{G}|$ denotes the number of vertices in $\mathcal{G}$, then the characteristic polynomials are related by $\frac{\partial}{\partial\lambda}P_{\mathcal{G}}(\lambda) = |\mathcal{G}|P_{\mathcal{G}'}(\lambda)$. Use of this result allows one to generalize Eq. (23).

Consider now a three-dimensional lattice in which correlations within some region $\mathcal{R}$ about the reference point are considered. For concreteness, consider a finite simple cubic lattice of size $N \times N \times N$ with periodic boundary conditions. The above results can be combined together most readily by defining a generalized eigenvalue in which the spectral contributions are weighted by the corresponding dihedral angle cosines $\gamma_{\ell m n}$,

$$\lambda_{ijk} = \sum_{\ell,m,n \in \mathcal{R}} \sum \sum \gamma_{\ell m n} \cos\frac{2\pi i\ell}{N} \cos\frac{2\pi jm}{N} \cos\frac{2\pi kn}{N}. \tag{27}$$

(This may be viewed as the eigenvalue for a generalized graph in which vertices can be connected by fractional numbers of edges, as given by the $\gamma_{\ell m n}$.) Note that in this expression, the triple sum includes negative as well as positive values of ℓ, m, n, and thereby generates the factors of 2 in front of the cosines that seem at first to be missing.

The ratio of Gramian determinants needed for the angular centrifugal potential may now be computed:

$$\frac{\Gamma'}{\Gamma} = \frac{P_{\mathcal{L}'}(1)}{P_{\mathcal{L}}(1)}$$

$$= \frac{1}{N^3}\frac{\partial}{\partial\lambda}\log P_{\mathcal{L}}(\lambda)\Big|_{\lambda=1}$$

$$= \frac{1}{N^3}\frac{\partial}{\partial\lambda}\log\prod_{i=1}^{N}\prod_{j=1}^{N}\prod_{k=1}^{N}(\lambda-\lambda_{ijk})\Big|_{\lambda=1}$$

$$= \frac{1}{N^3}\sum_{i=1}^{N}\sum_{j=1}^{N}\sum_{k=1}^{N}\frac{1}{1-\lambda_{ijk}}.$$

(28)

Finally, taking the limit $N \to \infty$ gives the angular centrifugal potential for the infinite lattice

$$\frac{\Gamma'}{\Gamma} = \frac{1}{(2\pi)^3}\int_0^{2\pi}\int_0^{2\pi}\int_0^{2\pi}\frac{d\xi\,d\eta\,d\zeta}{\tilde{\gamma}(\xi,\eta,\zeta)},$$

(29)

where

$$\tilde{\gamma}(\xi,\eta,\zeta) = \sum_{\ell,m,n=-\infty}^{\infty}\gamma_{\ell mn}\cos\ell\xi\,\cos m\eta\,\cos n\zeta.$$

(30)

Since this general integral cannot be done in closed form, the most effective way to use it is to expand the integrand as a Taylor series in the small quantities $\gamma_{\ell mn}$, and integrate term by term. Once cosines of multiple angles have been removed using the Chebyshev polynomial expansion

$$\cos n\theta = T_n(\cos\theta),$$

(31)

this can be done using the single integral

$$\frac{1}{2\pi}\int_0^{2\pi}\cos^n\varphi\,d\varphi = \begin{cases} 0 & n \text{ odd,} \\ (n-1)!!/n!! & n \text{ even.} \end{cases}$$

(32)

For example, for the simple cubic lattice with correlation treated to third-nearest neighbors (the first cubic shell around the reference site), one has

$$\tilde{\gamma}(\xi,\eta,\zeta) = 1 + 2\gamma_{100}(\cos\xi + \cos\eta + \cos\zeta)$$

$$+ 4\gamma_{110}(\cos\xi\,\cos\eta + \cos\eta\,\cos\zeta + \cos\zeta\,\cos\xi)$$

$$+ 8\gamma_{111}\cos\xi\,\cos\eta\,\cos\zeta. \tag{33}$$

Expanding this as a Taylor series in the $\gamma_{\ell mn}$ and performing the integrals, one obtains

$$\frac{\Gamma'}{\Gamma} \approx 1 + 6\gamma_{100}{}^2 + 12\gamma_{110}{}^2 + 8\gamma_{111}{}^2 - 72\gamma_{100}{}^2\gamma_{110} - 48\gamma_{110}{}^3 + \dots. \tag{34}$$

Because of the large number of terms, it is convenient to introduce a shorthand. If only interactions at fairly short range are being considered, then it is convenient to replace each triple of indices ℓmn by the corresponding squared distance $\sigma^2 = \ell^2 + m^2 + n^2$, and each product of cosines $\gamma_{\ell_1 m_1 n_1}\gamma_{\ell_2 m_2 n_2}\dots$ by a symbol $\Lambda_{\sigma_1{}^2\sigma_2{}^2\dots}$. (This notation eventually becomes ambiguous due to the fact that more than one integer triple ℓmn can give rise to the same σ^2; this first happens, however, only at $\sigma^2 = 9$ for the SC lattice, 18 for the FCC lattice, and 27 for the BCC lattice.) With this notation, the expansion through third-nearest neighbors, and to fifth order in the dihedral angles, becomes

$$\left(\frac{\Gamma'}{\Gamma}\right)^{(3)} = 1$$

$$+ (\quad 6\,\Lambda_{11} \quad + \quad 12\,\Lambda_{22} \quad + \quad 8\,\Lambda_{33}\quad)$$

$$- (\quad 72\,\Lambda_{112} \quad + \quad 48\,\Lambda_{222} \quad + \quad 144\,\Lambda_{123}\quad)$$

$$+ (\quad 90\,\Lambda_{1111} \quad + \quad 1152\,\Lambda_{1122} \quad + \quad 540\,\Lambda_{2222} \quad + \quad 192\,\Lambda_{1113}$$

$$+ \quad 1728\,\Lambda_{1223} \quad + \quad 432\,\Lambda_{1133} \quad + \quad 1296\,\Lambda_{2233} \quad + \quad 216\,\Lambda_{3333}\quad)$$

$$- (\quad 2880\,\Lambda_{11112} \quad + \quad 15120\,\Lambda_{11222} \quad + \quad 4320\,\Lambda_{22222} \quad + \quad 10080\,\Lambda_{11123}$$

$$+ \quad 30240\,\Lambda_{12223} \quad + \quad 12960\,\Lambda_{11233} \quad + \quad 12960\,\Lambda_{22233} \quad + \quad 12960\,\Lambda_{12333}\quad)$$

$$+ \quad \dots$$

$$\tag{35}$$

The coefficient of each term is, up to its sign, just the number of distinct paths beginning and ending at the reference site, and composed of segments specified by the indices [24]. For example, the term $-72\Lambda_{112}$ term indicates that there are 72 three-segment round-trip paths from the reference site consisting of legs of lengths 1, 1, and $\sqrt{2}$ (in any order). One sees immediately from this that the leading terms in the expansion are quadratic and diagonal, with coefficients

Table 3. Cubic terms in centrifugal potential through twelfth nearest neighbors; see Eq. (34). Table 2 gives key for single-character lattice site abbreviations (as well as quadratic terms in expansion).

term	coef.	term	coef.	term	coef.	term	coef.	term	coef.
112	-72	556	-288	369	-144	66J	-288	69K	-432
222	-48	266	-144	569	-288	28J	-144	2IK	-144
123	-144	466	-144	189	-144	68J	-288	1JK	-288
114	-18	666	-48	589	-288	39J	-288	5JK	-288
224	-72	228	-36	299	-144	59J	-288	9JK	-288
334	-72	338	-72	499	-72	99J	-288	4KK	-144
125	-144	448	-72	14I	-36	1IJ	-144	8KK	-216
235	-288	158	-144	25I	-144	2JJ	-144	33L	-24
145	-144	558	-144	36I	-144	4JJ	-72	55L	-144
255	-216	268	-288	58I	-144	8JJ	-72	26L	-144
455	-72	668	-72	69I	-288	34K	-144	48L	-144
226	-144	888	-48	24J	-144	25K	-288	19L	-144
136	-144	239	-144	15J	-144	16K	-144	9IL	-144
246	-144	259	-288	35J	-288	56K	-576	6JL	-288
156	-288	459	-288	55J	-144	38K	-288	3KL	-144
356	-288	169	-288	26J	-288	29K	-288	KKL	-144

that simply count equivalent lattice positions (since there is just one two-segment round-trip path from the reference site to each position in the lattice). Thus, the leading-order terms out to twelfth-nearest neighbors are given by Table 2. The cubic terms are given in Table 3. (Quartic and higher terms can also be computed by the above algorithm, but because of the small values assumed by the $\gamma_{\ell mn}$, these generate relatively small corrections.) Note that the expansions for the FCC and BCC lattices may be obtained by simply dropping those terms which contain irrelevant dihedral angles.

Finally, to treat the correlated lattice problem, it is only necessary to use a one-electron hamiltonian in which dihedral angles have been introduced into both the centrifugal and Coulombic terms. For example, for the simple cubic lattice including electron correlation out to third nearest neighbors, one would use in place of Eq. (11)

$$\widetilde{\mathcal{H}} = \frac{9}{8\rho^2}\left(\frac{\Gamma'}{\Gamma}\right)^{(3)} - \frac{3}{2\rho} + \frac{3}{2}W^{(3)}, \tag{36}$$

with the Gramian ratio given by Eq. (35) and the Madelung sum given by Eq. (19). (The centrifugal potential in Eq. (35) extends only to fifth order in the cosines, but because the angles wind up quite close to 90°, this should be more than sufficient.) Minimization of this potential with respect to the free variables ρ, γ_{100}, γ_{110}, and γ_{111} gives the total energy per electron or atom, including correlation effects out to third-nearest neighbors.

Correlation energy results

In this section we discuss the results obtained for the correlation energies of simple cubic hydrogen lattices. Correlation energies were computed as the differences between correlated and uncorrelated (Hartree-Fock) total energies. For reasons outlined more fully in Chapter 3, we expect the small energy difference associated with electron correlation to be modeled relatively well, in spite of the rather large absolute errors in the total energies. Roughly speaking, this expectation is based on the idea that the complementary characters of the dimensional scaling and Hartree- Fock approximations will result in relatively little "interference" between them.

Probably the most significant source of interference is the absence of exchange effects in the $D \to \infty$ limit. (Exchange enters at first order in the $1/D$ expansion.) Exchange reduces the extent of true correlation between parallel spin electrons, so its absence would be expected to result in some overestimation of correlation energies. In both the high- and low- density limits, however, this effect becomes negligible (because the radius of the Fermi or exchange hole vanishes in comparison with the Coulomb screening distance at high density [25], while the exponential fall-off of the exchange interaction with distance renders exchange effects negligible in comparison with correlation in the low-density limit [26,27]), which suggests that the problem may not be that severe.

The results of correlation energy calculations for the simple cubic lattice are shown in Fig. 3. Correlation was treated by starting with the Hartree-Fock approximation, and then introducing correlation effects in larger and larger regions about the reference site. Specifically, for each density, correlation was treated in clusters consisting of the reference site together with all sites out to nth-nearest neighbors, for $n = 1, \ldots, 6$. From Table 2, the total numbers of electrons correlating with the reference electron in these calculations were therefore 6, 18, 26, 32, 56, and 80. For each calculation, all terms in the centrifugal potential through quartic were included, and all independent dihedral angles were optimized. (The possibility of symmetry breaking was not explored.) The six correlation energies obtained for the successively larger clusters were then extrapolated to infinite cluster size. In Fig. 3, the three dotted curves show the values obtained by treating groups through first-, third-, and sixth nearest neighbors, while the solid line shows the extrapolated values. (The extrapolations were obtained by plotting the finite-cluster correlation energies as a function the inverses of the cluster sizes cited above, fitting a straight line to the values for the three largest clusters, and using the intercept as the infinite-cluster value.)

The correlation energy of a lattice of hydrogen atoms should be described at sufficiently high density by the electron gas model, and at sufficiently low density by a sum of van der Waals (dispersion) energies. These limiting behaviors are shown by dashed lines in Fig. 3. The electron gas results [30,31] are for the paramagnetic form, which

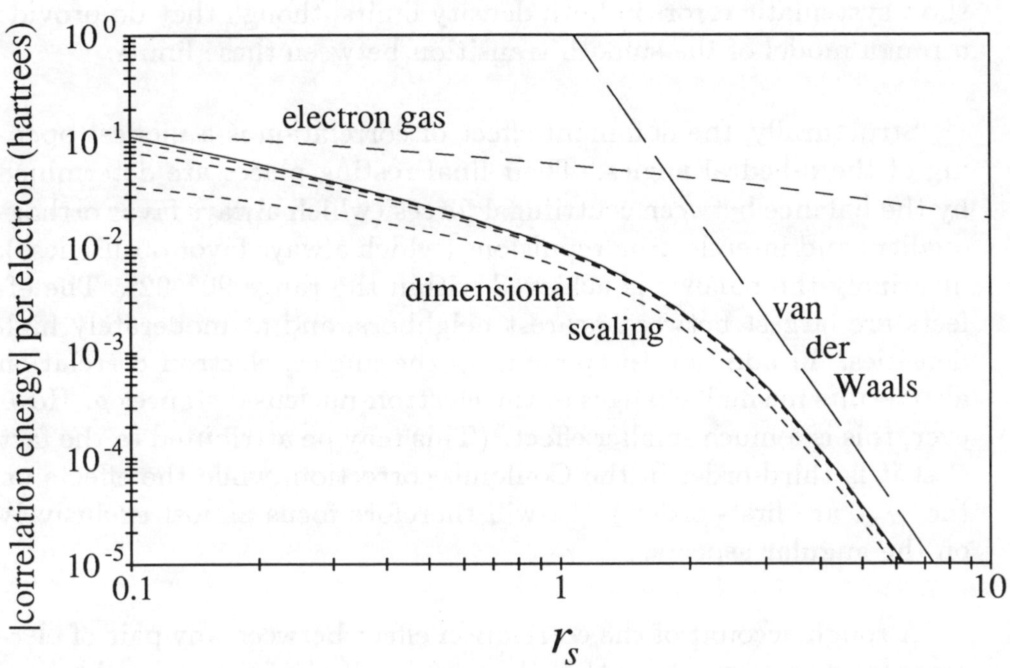

Figure 3. Correlation energies in simple cubic hydrogen lattices. Dimensional scaling values were calculated with correlation through first-, third-, and sixth-nearest neighbors (dotted lines) and by extrapolation of values with correlation through fourth-, fifth-, and sixth-nearest neighbors (solid line). Reference values in the high- and low-density limits are given respectively by electron gas calculations [30,31] and by a sum of van der Waals interactions [17,29].

in the presence of correlation is probably the ground state at all densities [32]. (It is certainly the ground state wherever the electron gas model is appropriate.) The van der Waals result for the SC lattice is just $-\frac{1}{2}(8.4019)(6.49903)/(\frac{4}{3}\pi r_s{}^3)^2$ hartrees, as described below. There are no reliable correlation energies currently available for three-dimensional hydrogen atom lattices at intermediate densities. It can be seen that the correlation energies obtained by dimensional scaling show systematic errors in both density limits, though they do provide a rough model of the smooth transition between these limits.

Structurally, the dominant effect of correlation is a modest opening of the dihedral angles. Their final resting places are determined by the balance between centrifugal forces (which always favor orthogonality) and interelectron repulsions (which always favor obtuseness); invariably the balance is achieved within the range 90°–92°. The effects are largest between nearest neighbors, and at moderately high densities. In addition to opening up the angles, electron correlation also results in small changes to the electron-nucleus distances ρ. However, this is a much smaller effect. (This may be attributed to the fact that it is third-order in the Coulomb correction, while the effects on the $\gamma_{\ell m n}$ are first- order.) We will therefore focus almost exclusively on the angular aspects.

A rough account of the correlation effect between any pair of electrons (and an exact account in the low-density limit) may be obtained by expanding the correlated, angle-dependent hamiltonian in powers of the $\gamma_{\ell m n}$, and considering the leading-order terms. One finds that the lowest- order contribution from the centrifugal terms is quadratic, while that from the Coulomb terms is linear. Explicitly, for the correlation between two electrons separated by a distance α, these lowest-order corrections to the HF hamiltonian read

$$\Delta\widetilde{\mathcal{H}}(\alpha) = \frac{9}{8\rho^2}\,\gamma^2 + \frac{3\rho^2}{4(\alpha^2 + 2\rho^2)^{3/2}}\,\gamma. \qquad (37)$$

The minimum with respect to γ is given by

$$\gamma(\alpha) = -\frac{\rho^4}{3(\alpha^2 + 2\rho^2)^{3/2}},$$

$$\Delta E(\alpha) = -\frac{\rho^6}{8(\alpha^2 + 2\rho^2)^3}. \tag{38}$$

Here we have assumed that ρ can be held fixed at its Hartree-Fock value, given by the minimum of the Hartree-Fock hamiltonian, Eq.(11).

Eq. (38) can be used to evaluate the low-density limit of a pair correlation exactly, since there the $\gamma_{\ell m n}$ all tend to zero, and a leading-order treatment is sufficient. In this limit we have $\rho = \frac{3}{2}$ and $\alpha \gg \rho$, so

$$\lim_{\alpha \to \infty} \gamma(\alpha) = -\frac{27}{16\alpha^3},$$

$$\lim_{\alpha \to \infty} \Delta E(\alpha) = -\frac{729}{512\alpha^6}. \tag{39}$$

This may be compared to half of the known dispersion interaction [29] between two hydrogen atoms (which is attributable to correlation), namely $\frac{1}{2}(-6.49903/\alpha^6)$. (Again, the factor of $\frac{1}{2}$ is necessary because half of the total correlation energy is assigned to each of the two atoms.) Thus, the dimensional scaling treatment underestimates the correlation energy between any pair of atoms by 56% in the low-density limit.

Now consider what happens in a three-dimensional lattice. In the low-density limit, coupling terms between different pair correlations can be ignored, since they are always of higher order than the terms in Eq. (37). Therefore the total correlation energy can be obtained as a simple lattice sum [17]

$$\Delta E = -\frac{729}{512\alpha^6} \sum_{\ell, m, n \in \mathcal{L}'} \sum \sum \frac{1}{(\ell^2 + m^2 + n^2)^3}$$

$$= -\frac{729}{512\alpha^6} \begin{cases} 8.4019 & \text{for SC} \\ 1.8067 & \text{for FCC} \\ 0.4538 & \text{for BCC}. \end{cases} \tag{40}$$

The same Madelung factors apply to the exact correlation energies, so for any lattice the low-density limit value obtained by dimensional scaling will again be low by 56%. It should be noted that it is only because of the high dimensionality of the electronic configuration space that the pair correlations decouple, so that they can simply be added together. (Geometrically, the dihedral angles define orthogonal planes, so that the infinitessimal openings from 90° which take place in the low- density limit can be treated independently.) With three-dimensional classical models, it is impossible to reproduce the scalar additivity of dispersion interactions [33].

Next consider the high-density limit. No analysis of the asymptotic behavior has been attempted in this limit, but the general features of electron correlation in this limit may be described as follows: First, correlation effects become increasingly long-range relative to the lattice spacing, though increasingly short-range in absolute units. This is simply because ρ grows compared to α (so that larger and larger numbers of lattice sites enter the purview of any given electron), though it decreases in absolute units. (In fact, $\rho \sim 1.096 r_s^{3/4}$ in the high- density limit, according to the model; it appears, however, that the incorporation of order-$1/D$ effects would change this relationship, possibly to $\rho \sim r_s^{1/2}$.) One can see the need for treating interactions with electrons at more distant lattice sites in Fig. 3, where the three dotted lines give the results through first-, third-, and sixth- nearest neighbors. Second, correlation effects become increasingly many- body in character at high density. This is because the correlation angles become larger, so that the higher-order terms in the centrifugal expansions (which couple the interactions of different electron pairs) assume more and more importance. The effect of these terms is always a moderating one. (This can be seen in Eq. (35): since all of the $\gamma_{\ell m n}$ are negative, each term in the centrifugal expansion contributes positively.) In particular, the independent electron pair approximation always overestimates the correlation energy, and this problem becomes increasingly severe as r_s decreases.

Fig. 3 suggests that the present treatment is insufficient to describe the high-density limit. It is not known just what the asymptotic behavior is, but it apparent that it will not agree with the known behavior ($0.0311 \ln r_s - 0.047$, in hartrees [34]. This problem is another consequence of the inadequate representation of excitation effects in the $D \to \infty$ limit mentioned above. (Recall that the total energies diverged as $r_s^{-3/2}$ rather than r_s^{-2}.)

Finally, there is the important domain of intermediate density, which is really the only experimentally accessible one: lattices with $r_s \gtrsim 1.3$ are unstable relative to the molecular solid, while those with $r_s \lesssim 1.0$ are likely to remain experimentally inaccessible in the foreseeable future. In this domain, neither the van der Waals nor the electron gas model is appropriate. There are no calculations in this regime to which comparison can be made, although results from calculations on one-dimensional chains of hydrogen atoms [35] suggest that the dimensional scaling values may be somewhat low here.

Prospects

Although the model outlined in this chapter requires some new techniques and some geometric imagination, it turns out to be both conceptually and computationally quite simple. The Hartree-Fock version barely requires the use of a computer, and even the correlated model presents no great challenge. The price for this simplicity is that the model is in many respects still quite crude. Some important aspects of the electronic structure, such as exchange effects, are not modeled at all. In this section we consider how one can improve upon the present model.

First, one can combine the dimensional scaling treatment with others having complementary strengths. In general, dimensional scaling should complement band theory calculations, since the former incorporates many-body effects, and clearly lies at the localized end of the localized/itinerant continuum. The simplest procedure for combining methods would be to correct a band theory calculation with correlation energies computed by dimensional scaling. This approach will break down to the extent that those effects which are not adequately treated by dimensional scaling (especially exchange) alter the degree of electron correlation.

A second possibility would be to improve systematically upon the $D \to \infty$ model within the dimensional scaling framework. This might be done by considering the electronic behavior at low D (specifically $D = 1$), and using a dimensional interpolation strategy. However, it seems more likely that progress could be made through consideration of the first-order term in the $1/D$ expansion (the $D \to \infty$ limit being nothing but the zeroth-order term). This term arises from the normal mode vibrations of the electronic lattice. In fact, the electronic band structure would be treated in a manner analogous to that ordinarily used for the phonon spectrum. The fact that the electrons live in a higher-dimensional space would of course render the calculations somewhat different (and probably in some respects simpler) than for the nuclear normal mode problem.

It should be noted that the symmetry assumption utilized in this chapter has hidden some electronic degrees of freedom which must be treated at order $1/D$. Specifically, the treatment of this chapter refers explicitly to the electronic distances above the nuclear lattice and to the dihedral angles between the electrons, but not to the electronic coordinates within the three-dimensional space of the nuclear lattice. This was simply because of the assumption that at the global minimum of the $D \to \infty$ hamiltonian, each electron would occupy a position directly "above" one of the nuclei. Motions of the electrons within the space of the nuclear lattice are of course central, and would need to be treated at order $1/D$; at $D = 3$, the zero-frequency modes correspond to electronic conduction.

A third and very promising route toward improvement of the model would be to meld the two approaches just outlined, as follows: A low-order $1/D$ expansion can generally be rendered significantly more accurate through an appropriate re-expansion. For example, one can re-expand in such a way as to make the first-order term vanish [36]. This is in general the optimal choice if one wants to use the zeroth-order solution as a model. As shown in Chapter 3, the new zeroth-order hamiltonian (called there a subhamiltonian) has "boosted" centrifugal terms, where the boosts simulate the effects of quantal excitation. In principle, determining the appropriate re-expansion is tantamount to solving the first-order problem. However, it may be possible to use results obtained from standard methods to determine the appropriate shifts empirically. For example, if one boosts the centrifugal potential in Eq. (11) by writing

$$\widetilde{\mathcal{H}} = \frac{9}{8\rho^2}\left(1 + \frac{B}{r_s}\right) - \frac{3}{2\rho} + \frac{3}{2}W, \qquad (41)$$

then the total energy will have qualitatively correct behavior at high density (it will diverge as r_s^{-2}). B might be chosen to be that probability distribution which reproduces the energy spectrum of the high-density electron gas.

No systematic investigations using calculated or simulated higher-order terms in $1/D$ have yet been attempted. It is anticipated, however, that their use would significantly improve the overall description; clearly they will bring in qualitative features which are absent in the $D \to \infty$ limit, and if convergence problems are not overwhelming, they should be quantitatively useful as well. Some small indication that the effects will be of the right magnitude to generate a reasonable band structure is provided by the following simple calculation of the metal-insulator transition density in the atomic hydrogen lattice. Herzfeld [37,38] argued that the metal-insulator transition would occur roughly at the point where the molar refractivity ($\frac{4}{3}N_A\pi\alpha$, where α is the atomic polarizability) equals the molar volume ($\frac{4}{3}N_A\pi r_s^3$). In the $D \to \infty$ limit, one finds $\alpha = \frac{9}{4}$. (This is obtained by minimizing the $D \to \infty$ limit of the uniformly scaled hamiltonian, namely

$$\widetilde{\mathcal{H}} = \frac{9}{8\rho^2}\csc^2\theta - \frac{3}{2\rho} + \mathcal{E}\,\rho\cos\theta, \tag{42}$$

for which the minimum is $E = -\frac{1}{2} - \frac{9}{8}\mathcal{E}^2$.) Thus, the metal-insulator transition is predicted to occur at $r_s = (9/4)^{1/3} \approx 1.310$, which is quite close to the values which have been suggested before [2,6,39].

Acknowledgments

This material is based upon work supported by the National Science Foundation under Grant. No. CHE-9007620 and by the Donors of The Petroleum Research Fund administered by the American Chemical Society.

References

1. I. F. Silvera, in A. Polian, P. Loubeyre, and N. Boccara (eds.), *Simple Molecular Systems at Very High Density* (Plenum, NY, 1989), p. 33.

2. B. I. Min, H. J. F. Jansen, and A. J. Freeman, *Phys. Rev. B* **33**, 6383 (1986).

3. T. W. Barbee III, A. García, M. L. Cohen, and J. L. Martins, *Phys. Rev. Lett.* **62**, 1150 (1989).

4. D. M. Ceperley and B. J. Alder, *Phys. Rev. B* **36**, 2092 (1987).

5. H. K. Mao and R. J. Hemley, *Science* **244**, 1462 (1989).

6. J. van Straaten and I. F. Silvera, *Phys. Rev. B* **37**, 1989 (1988).

7. F. Golden, *Mosaic* **22**, 22 (1991).

8. I. V. Brovman, Yu. Kagan, and A. Kholas, *Sov. Phys. JETP* **34**, 1300 (1972).

9. T. Tsuneto, in T. Matsubara (ed.), *The Structure and Properties of Matter* (Springer, Berlin, 1982), p. 59.

10. T. Hey and P. Walters, *The Quantum Universe* (Cambridge Univ., Cambridge, 1987), p. 96.

11. *Scientific American* **261**(5), 26 (Nov. 1989).

12. D. R. Herschbach, *J. Chem. Phys.* **84**, 838 (1986).

13. Z. A. Melzak, *Mathematical Ideas, Modeling, and Applications* (Wiley, NY, 1976), p. 86.

14. S. Raimes, *The Wave Mechanics of Electrons in Metals* (North-Holland, Amsterdam, 1961), p. 135.

15. J. G. Loeser, *J. Chem. Phys.* **86**, 5635 (1987).

16. J. G. Loeser, Z. Zhen, S. Kais, and D. R. Herschbach, *J. Chem. Phys.* **95**, 4525 (1991).

17. M. Born and R. D. Misra, *Proc. Camb. Phil. Soc.* **36**, 466 (1940).

18. B. I. Min, H. J. F. Jansen, and A. J. Freeman, *Phys. Rev. B* **30**, 5076 (1984).

19. J. H. Conway and N. J. A. Sloane, *Sphere Packings, Lattices, and Groups* (Springer, NY, 1988), pp. 107–116.

20. I. S. Dmitriev, *Molecules Without Chemical Bonds* (Mir, Moscow, 1981), pp. 24–26.

21. D. M. Cvetković, M. Doob, and H. Sachs, *Spectra of Graphs* (Academic, NY, 1979), pp. 59–72.

22. H. S. M. Coxeter, *Regular Polytopes* (Dover, NY, 1973), pp. 173–193.

23. T. Muir, *A Treatise on the Theory of Determinants* (Dover, NY, 1960), p. 442.

24. C. Domb, *Adv. Phys.* **9**, 245 (1960).

25. S. Raimes, *The Wave Mechanics of Electrons in Metals* (North-Holland, Amsterdam, 1961), pp. 150 and 293.

26. H. Margenau and N. R. Kestner, *Theory of Intermolecular Forces* (Pergamon, Oxford, 1969), p. 317.

27. W. J. Carr and M. Ashkin, *J. Chem. Phys.* **42**, 2796 (1965).

28. S. Raimes, *Many-Electron Theory* (North- Holland, Amsterdam, 1972), pp. 64 and 176.

29. L. Pauling and J. Y. Beach, *Phys. Rev.* **47**, 686 (1935).

30. D. M. Ceperley and B. J. Alder, *Phys. Rev. Lett.* **45**, 566 (1980).

31. S. H. Vosko, L. Wilk, and M. Nusair, *Can. J. Phys.* **58**, 1200 (1980).

32. R. Brout, *Phase Transitions* (Benjamin, New York, 1965), p. 124.

33. W. E. Thirring, in C. W. Kilmister (ed.), *Schrödinger* (Cambridge, Cambridge, 1987), p. 65.

34. M. Gell-Mann and K. A. Brueckner, *Phys. Rev.* **106**, 364 (1957).

35. S. Suhai and J. Ladik, *J. Phys. C* **15**, 4327 (1982).

36. U. Sukhatme and T. Imbo, *Phys. Rev. D* **28**, 418 (1983).

37. K. F. Herzfeld, *Phys. Rev.* **29**, 701 (1927); *J. Chem. Phys.* **44**, 429 (1966).

38. P. P. Edwards and M. J. Sienko, *Acc. Chem. Res.* **15**, 87 (1982).

39. M. Ross, *J. Chem. Phys.* **56**, 4651 (1972) and references therein.

Chapter 10

D-INTERPOLATION OF VIRIAL COEFFICIENTS

Zheng Zhen[1] and John Loeser
Department of Chemistry
Oregon State University
 Corvallis, OR 97331-4003

Abstract

Dimensional interpolation is used to approximate the configuration space integrals required in the computation of higher-order hard sphere virial coefficients. Simple analytic results can be obtained at $D = 0$, $D = 1$, and $D \to \infty$; the smooth and generic dimension-dependence of the integrals enables one to interpolate reasonably accurate $D = 3$ values (rms error 1%) from the dimensional limit results. The interpolated integrals can be used either on their own, or in conjunction with an integral equation approximation which sums some subset of the required integrals exactly (such as the hypernetted-chain or Percus-Yevick methods); the combination methods are invariably better than either dimensional interpolation or integral equations alone. Interpolation-corrected Percus-Yevick values can be computed quite easily at arbitrary order; however, errors in higher-order values are

[1]current address: Department of Radiation Oncology, Division of Medical Physics, University of Pittsburgh, Pittsburgh, PA 15213

D. R. Herschbach et al. (eds.), Dimensional Scaling in Chemical Physics, 429–458.
© 1993 *Kluwer Academic Publishers. Printed in the Netherlands.*

probably still quite large at the current level of calculation. Geometric perspectives suggested by the dimensional scaling treatment are stressed throughout.

Introduction

The hard-sphere fluid is undoubtedly the most familiar model for dense gases and liquids. At least for relatively spherical molecules, it provides a reasonable representation of the short-range interactions, and therefore serves to describe the packing problem encountered at high density. The model has been studied extensively, but remains a challenge. Indeed, our present knowledge of the behavior of hard sphere fluids comes from an interesting mix of theoretical calculations (at low densities, for example for the virial expansion [1]), computer simulations (at densities extending through the fluid-solid transition [2]), and experiments using small plastic or metal spheres (at densities such as that of random closest packing [3]).

It is most convenient to treat the equation of state in dimensionless form. This may be done by writing the compressibility factor ($z \equiv pV/NkT$) as a function of the packing fraction ($\eta \equiv Nv_0/V$, where v_0 is the volume of a sphere). Consider first a hard sphere fluid in $D=1$. A one-dimensional sphere is a line segment, so this is nothing but a fluid of hard rods interacting along a line. Clearly the fluid should approach ideality ($z \rightarrow 1$) at very low density ($\eta \rightarrow 0$), and should become incompressible ($z \rightarrow \infty$) at closest packing ($\eta \rightarrow 1$). In fact, the equation of state is known exactly [4], and is just

$$z = \frac{1}{1-\eta}. \tag{1}$$

As shown in Fig. 1, hard sphere fluids in $D>1$ exhibit a fluid-solid phase transition. (Because there are no attractive forces, however, there is no liquid-vapor transition). In the fluid region, the equation of state can be expanded as a Taylor series in the packing fraction,

$$z = 1 + b_2\eta + b_3\eta^2 + \dots . \tag{2}$$

For $D=3$, for example,

$$z = 1 + 4\eta + 10\eta^2 + 18.36\eta^3 + 28.22\eta^4 + \dots . \tag{3}$$

At least at low order, this expansion is reproduced to good accuracy by the well-known Carnahan-Starling equation [6], $z = (1+\eta+\eta^2-\eta^3)/(1-\eta)^3$, which yields $b_n = n^2+n-2$ for $n > 1$.) The expansion in terms of the packing fraction η is related to the more familiar virial expansion in terms of the particle density ρ,

$$z = 1 + B_2\rho + B_3\rho^2 + \cdots, \tag{4}$$

by

$$\eta = \frac{Nv_0}{V} = \rho v_0, \tag{5}$$

where v_0 is the volume of a sphere; in D-dimensional space, the sphere volume is given in terms of the sphere diameter σ by

$$v_0 = \pi^{D/2}(\tfrac{1}{2}\sigma)^D/\Gamma(\tfrac{D}{2}+1).$$

Thus, the two sets of virial coefficients are related by $B_n = b_n(v_0)^{n-1}$. Finally, a third set of coefficients is that of the reduced values,

$$\tilde{B}_n = \frac{b_n}{(b_2)^{n-1}} = \frac{B_n}{(B_2)^{n-1}}, \tag{6}$$

in terms of which the virial coefficients are often reported. (These can be unreduced by using $b_2 = 2^{D-1}$ or $B_2 = 2^{D-1}v_0$.) Values for the $\tilde{B}_n$ for hard rods, disks, and spheres are given in the first section of Table 1.

At high density, hard spheres with $D > 1$ have a solid phase which terminates in the close-packed crystal. The equation of state for this phase has a first-order pole at the crystal packing fraction η_0. (For hard disks and spheres, the close-packed crystals are those of hexagonal closest packing and cubic closest packing; the packing fractions η_0 are respectively $\frac{\pi}{2\sqrt{3}} \approx 0.90690$ and $\frac{\pi}{3\sqrt{2}} \approx 0.74048$.) The equation of state is usually expressed as a series in the relative free volume

$$\alpha = \frac{V}{V_0} - 1 = \frac{\eta_0}{\eta} - 1, \tag{7}$$

in terms of which it reads

$$z = C_{-1}\alpha^{-1} + C_0 + C_1\alpha + \cdots. \tag{8}$$

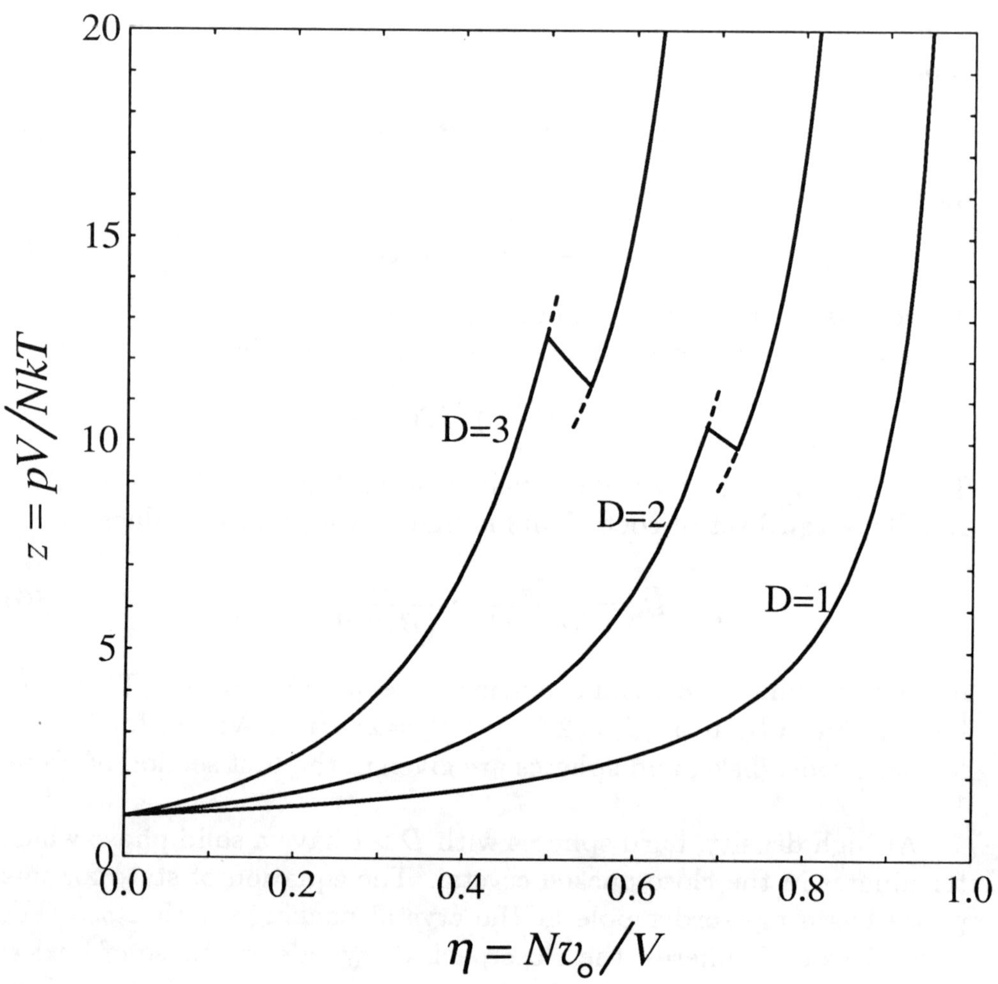

Figure 1. Equations of state for hard rod, hard disk, and hard sphere fluids [2,4,5,6].

Table 1. Parameters for equations of state of hard rod, hard disk, and hard sphere fluids [1,2,4,7,8].

		$D=1$	$D=2$	$D=3$
fluid branch	$\tilde{B}_2$	1	1	1
	$\tilde{B}_3$	1	0.78200	0.625
	$\tilde{B}_4$	1	0.53223	0.28695
	$\tilde{B}_5$	1	0.33356	0.11025
	$\tilde{B}_6$	1	0.19893	0.0389
	$\tilde{B}_7$	1	0.1148	0.013
tie line	X	—	7.104	6.239
	η_f	—	0.683	0.495
	η_m	—	0.720	0.548
solid branch	η_o	1	0.90690	0.74048
	C_{-1}	1	2	3
	C_0	1	1.90	2.566
	C_1	0	0.67	0.55
	C_2	0	1.5	−1.19
	C_3	0	?	5.95

For hard spheres, for example [2],

$$z = 3\alpha^{-1} + 2.566 + 0.55\alpha - 1.19\alpha^2 + 5.95\alpha^3 + \ldots \qquad (9)$$

The closest packing fractions η_o and known coefficients C_n are also listed in Table 1.

Finally, connecting the fluid and solid branches is an equilibrium tie line, given by

$$z = X/\eta \qquad\qquad (\eta_f < \eta < \eta_m), \qquad (10)$$

where η_f and η_m are the packing fractions at the onsets of freezing and melting. For the hard-sphere fluid, the tie line is given by [2]

$$z = 6.239/\eta \qquad\qquad (0.495 < \eta < 0.548). \qquad (11)$$

Again, data are collected in Table 1.

It is only for the fluid phase that theoretical methods are well-developed. Here, the Mayer cluster integral approach [9] provides an exact prescription for computing the virial coefficients, and a variety of integral equation methods [10] (Born-Green-Yvon, Percus-Yevick, hypernetted chain, etc.) provide approximate solutions which become exact at low density. Neither approach can be taken to high density, however, due to the computational complexity of the Mayer approach, and to the inherent errors of the integral equation approximations. For the high-density fluid and for the solid, Monte Carlo simulations or experiments remain the best source of information for the equation of state. (Indeed, all of the data for the tie line and for the solid branch in Table 1, except the location and residue of the pole at closest packing, η_o and C_{-1}, were obtained from numerical fits to Monte Carlo data [2].)

In this chapter, dimensional scaling methods will be used to study the hard sphere equation of state. We do not seek here to treat the subject completely anew, but rather to use dimensionality-based methods to compute approximately the quantities that are needed within the usual formalism. We begin by reviewing in Secs. 2 and 3 those aspects of this formalism to which dimensional scaling will be applied. Then in Sec. 4 we show how the required quantities (cluster integrals) can be treated exactly in the low-D and high-D limits.

These results are used in Sec. 5 to generate approximate values at $D=3$. Finally, routes for improving upon and extending the methods of this chapter are discussed in Sec. 6. Further details and related aspects are developed in Ref. [11].

Mayer expansions

In the Mayer approach, each virial coefficient is represented exactly as a sum of integrals [10]. We will write the sums in the form

$$\tilde{B}_n^{(D)} = \sum_k C(n_k)\, n_k^{(D)}, \tag{12}$$

where the n_k are the integrals (called cluster integrals), and the $C(n_k)$ are combinatorial coefficients. Roughly speaking, each cluster integral corresponds to a different kind of irreducible interaction among n particles, and is conveniently represented by a diagram in which the interaction between a pair of particles is denoted by an edge connecting the corresponding vertices. The diagrams for all integrals contributing to the virial coefficients through $\tilde{B}_7$ are available [13]. Those contributing to $\tilde{B}_3$, $\tilde{B}_4$ and $\tilde{B}_5$ are shown here along with their standard n_k labels:

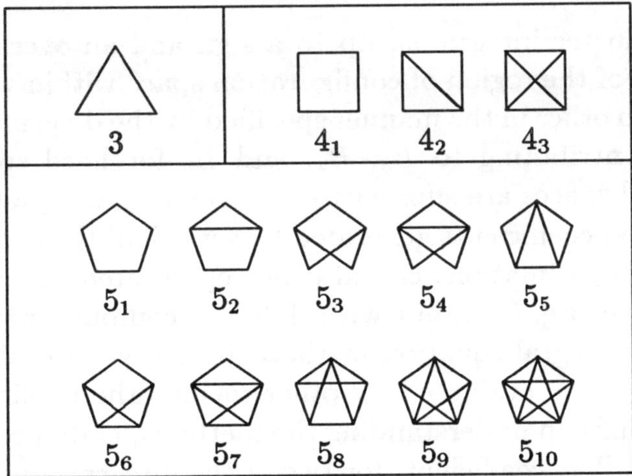

The derivation of the cluster expansions is given in many places. In very brief summary [12], the expansion is obtained by starting with the expression for the compressibility factor z in terms of the

N-particle configuration integral Z_N, namely $z = d\log Z_N/d\log V^N$ where $Z_N = \int e^{-\beta U_N} d^N\mathbf{r}$; then defining the Mayer f-functions by $f_{ij} = e^{-\beta u_{ij}} - 1$ and expanding the integrand of Z_N in terms of these,

$$\exp(-\beta U_N) = \exp\left(-\beta \sum_{i<j} u_{ij}\right) = \prod_{i<j}(1 + f_{ij}) = 1 + \sum_{i<j} f_{ij} + \dots ; \quad (13)$$

and finally going through quite a bit of algebra to obtain the expansions for the virial coefficients. In the end one obtains the expansions in Eq. (12), with

$$n_k^{(D)} = \frac{\dfrac{1-n}{n!} \displaystyle\int d\mathbf{r}_1\, d\mathbf{r}_2 \cdots d\mathbf{r}_{n-1} \prod_{lm \in n_k} f_{lm}}{(B_2)^{n-1}}. \quad (14)$$

For each integral the product of f-functions is over all pairs represented by lines in the integral's diagram.

For hard sphere fluids, the Mayer integrals take an especially simple form, since then

$$f_{ij} = \exp(-\beta u_{ij}) - 1 = \begin{cases} -1 & \text{if } |\mathbf{r}_i - \mathbf{r}_j| < \sigma \\ 0 & \text{if } |\mathbf{r}_i - \mathbf{r}_j| > \sigma, \end{cases} \quad (15)$$

and each cluster integral is, up to a sign and an overall factor, just the volume of the region of configuration space within which particles overlap each other in the manner specified by the diagram. The cluster integrals contributing to $\tilde{B}_3$, $\tilde{B}_4$, and $\tilde{B}_5$ for hard spheres in low-dimensional spaces are summarized in Table 2, and plotted in Fig. 2.

The virial coefficients generated by several of the common integral equation approximations can also be represented as sums of cluster integrals as in Eq. (12), but with different combinatorial factors. Although the integral equation methods in many cases allow the virial coefficients to be evaluated by other means, the explicit expansions are very helpful in understanding the methods, and in correcting their deficiencies. The coefficients for two of the integral equation methods (the compressibility equation forms of the Percus-Yevick and hypernetted chain approximations, designated $\mathrm{PY_c}$ and $\mathrm{HNC_c}$) are listed in Table 2 along with the exact coefficients.

Table 2. Combinatorial factors [10] and integrals [13-16] for Mayer expansions of the virial coefficients $\tilde{B}_3$, $\tilde{B}_4$, and $\tilde{B}_5$ at low D.

n_k	$C(n_k)$			$n_k^{(D)}$			
	exact	PY_c	HNC_c	$D=0$	$D=1$	$D=2$	$D=3$
3	1	1	1	$\frac{4}{3}$	1	0.78200	0.625
4_1	3	2	3	-1	$-\frac{2}{3}$	-0.45962	-0.32381
4_2	6	4	5	1	$\frac{7}{12}$	0.36424	0.23612
4_3	1	0	0	-1	$-\frac{1}{2}$	-0.27433	-0.15836
5_1	12	6	12	$\frac{8}{15}$	$\frac{23}{72}$	0.19910	0.12695
5_2	60	30	48	$-\frac{8}{15}$	$-\frac{49}{180}$	-0.14899	-0.08484
5_3	10	0	7	$-\frac{8}{15}$	$-\frac{4}{15}$	-0.14156	-0.07810
5_4	10	0	7	$\frac{8}{15}$	$\frac{1}{4}$	0.12757	0.06835
5_5	60	30	42	$\frac{8}{15}$	$\frac{29}{120}$	0.12061	0.06349
5_6	30	0	0	$\frac{8}{15}$	$\frac{41}{180}$	0.10676	0.05279
5_7	30	0	0	$-\frac{8}{15}$	$-\frac{19}{90}$	-0.09369	-0.04433
5_8	15	0	0	$-\frac{8}{15}$	$-\frac{1}{5}$	-0.08432	-0.03798
5_9	10	0	0	$\frac{8}{15}$	$\frac{11}{60}$	0.07202	0.03049
5_{10}	1	0	0	$-\frac{8}{15}$	$-\frac{1}{6}$	-0.06030	-0.02371

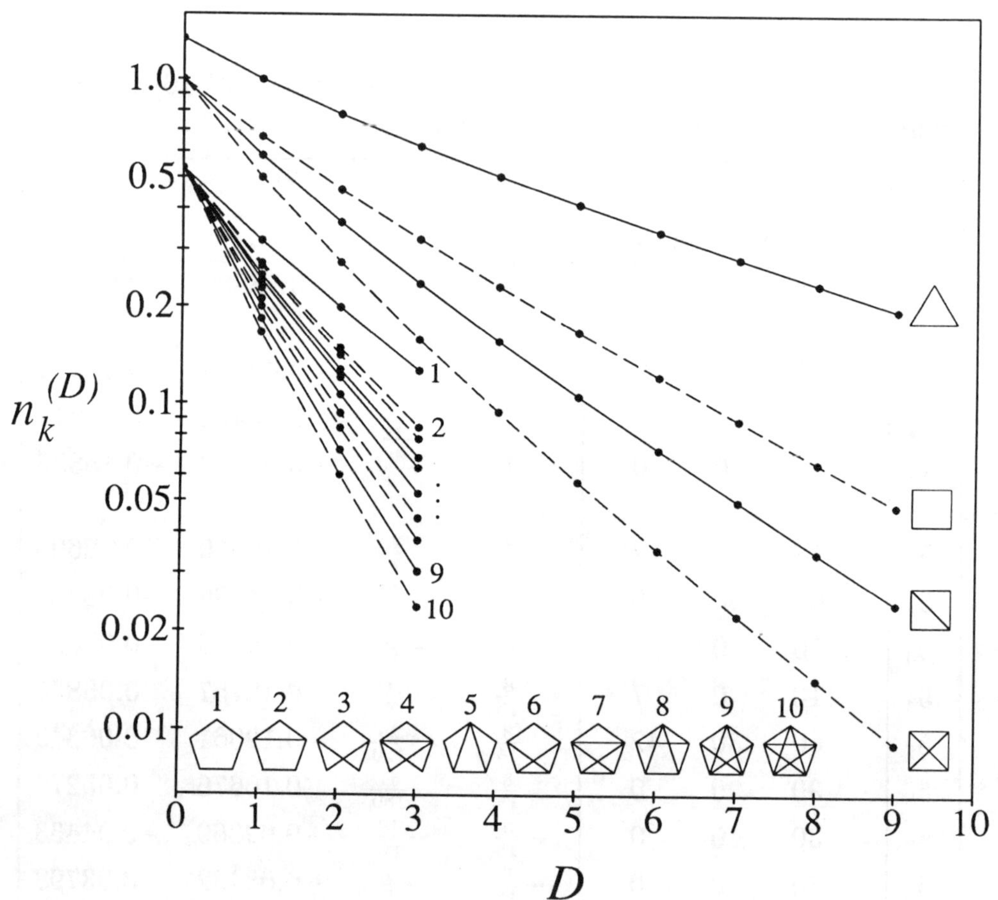

Figure 2. Dimension-dependence of the cluster integrals $n_k^{(D)}$ contributing to the Mayer expansions for the virial coefficients $\widetilde{B}_3$, $\widetilde{B}_4$, and $\widetilde{B}_5$ [13,14,17]. Dashed lines indicate negative values.

As an example, consider $\tilde{B}_4$ for hard spheres. Using the combinatorial factors and numerical integrals in Table 2, the virial coefficient and the two integral equation approximations to it are

$$
\begin{aligned}
\tilde{B}_4^{(3)} &= 3(4_1) + 6(4_2) + (4_3) &&= 0.28695 \\
\tilde{B}_4^{(3)}(\text{PY}_c) &= 2(4_1) + 4(4_2) &&= 0.29688 \\
\tilde{B}_4^{(3)}(\text{HNC}_c) &= 3(4_1) + 5(4_2) &&= 0.20919.
\end{aligned}
\tag{16}
$$

Geometrically, 4_3 ($\boxtimes$) is proportional to the volume of the region of configuration space defined by the criterion that each of four spheres must overlap each of the other spheres, while 4_2 ($\boxslash$) and 4_1 ($\square$) correspond to the successively larger regions in which one or two of these overlap criteria have been removed. Note that the signs of the integrals alternate with the number edges in the diagram (f-functions in the integrand), and that as a consequence there is a fairly high degree of cancellation between positive and negative contributions in the above sums. This cancellation problem becomes progressively worse as one goes to higher order in the virial series.

Although the procedure above sounds straightforward, the cluster integral computations quickly become very demanding. This is because each increase in order entails an increase in the total number of integrals required [18], in the dimensionality of the individual integrals, and in the accuracy to which each integral must be evaluated (due to the cancellation problem), as shown in the following table:

	number of integrals required	dimensionality of each integral	extra digits required
$\tilde{B}_3$	1	3	0
$\tilde{B}_4$	3	6	1
$\tilde{B}_5$	10	9	2
$\tilde{B}_6$	56	12	3
$\tilde{B}_7$	468	15	4
$\tilde{B}_8$	7123	18	5
$\tilde{B}_9$	194066	21	6

(Asymptotically, the number of Mayer integrals required for the calculation of $\tilde{B}_n$ grows like $2^{n(n-1)/2}/n!$ [19]. The dimensionality of the

$\tilde{B}_n$ integrals in $D=3$ is $3(n-2)$, since by Eq. (14) one particle is not integrated over, and an overall orientation can be factored out. The numbers in the last column refer to the approximate number of extra significant digits which must be kept in the calculation of the individual integrals n_k, beyond the number required for $\tilde{B}_n$ itself; these numbers were obtained by comparing the contributions of typical diagrams, as computed by the methods described in this chapter, with the final total virial coefficient.)

Ree-Hoover reformulation

Two of the problems mentioned in connection with the Mayer expansion are at least partially removed by means of a reformulation due to Ree and Hoover [20]. This expansion makes use of the modified (or Ree-Hoover) integrals, defined by

$$\overline{n}_k^{(D)} = \frac{\dfrac{1-n}{n!} \displaystyle\int d\mathbf{r}_1 \, d\mathbf{r}_2 \cdots d\mathbf{r}_{n-1} \prod_{lm \in n_k} f_{lm} \prod_{l'm' \notin n_k} \tilde{f}_{l'm'}}{(B_2)^{n-1}}, \tag{17}$$

where $\tilde{f}_{ij} \equiv 1 + f_{ij}$. Note that $\overline{n}_k$ is obtained from n_k by inserting an $\tilde{f}$ into the integrand between every pair of particles not connected by an f. Each $\overline{n}_k$ can be expressed as a linear combination of n_k integrals, and vice versa (since $\tilde{f} \equiv 1 + f$), so the Mayer expansions in Eq. (12) are readily recast as

$$\tilde{B}_n^{(D)} = \sum_k C(\overline{n}_k) \, \overline{n}_k^{(D)}. \tag{18}$$

The Ree-Hoover coefficients and integrals needed for the hard sphere $\tilde{B}_3$, $\tilde{B}_4$, and $\tilde{B}_5$ are summarized in Table 3.

From a numerical standpoint the Ree-Hoover expansion is far better to work with than the Mayer expansion. There are two reasons for this. First, the inserted $\tilde{f}$ functions act as additional constraints which render each integrand nonzero only over a much more limited region of configuration space. (It isn't hard to see that each Ree-Hoover integral corresponds to a different region of configuration space, and that some of those regions will be very small. For hard rods, disks,

Table 3. Combinatorial factors and integrals for Ree- Hoover expansions of the virial coefficients $\tilde{B}_3$, $\tilde{B}_4$, and $\tilde{B}_5$ at low D.

$\overline{n}_k$	$C(\overline{n}_k)$			$\overline{n}_k^{(D)}$			
	exact	PY$_c$	HNC$_c$	$D=0$	$D=1$	$D=2$	$D=3$
$\overline{3}$	1	1	1	$\frac{4}{3}$	1	0.78200	0.625
$\overline{4}_1$	3	2	3	0	0	-0.00548	-0.00992
$\overline{4}_2$	0	0	-1	0	$\frac{1}{12}$	0.08991	0.07776
$\overline{4}_3$	-2	-2	-2	-1	$-\frac{1}{2}$	-0.27433	-0.15836
$\overline{5}_1$	12	6	12	0	0	0.00070	0.00137
$\overline{5}_2$	0	0	-12	0	0	-0.00053	-0.00093
$\overline{5}_3$	10	0	7	0	0	0	-0.00002
$\overline{5}_4$	0	0	0	0	0	0.00223	0.00314
$\overline{5}_5$	0	0	6	0	$\frac{1}{360}$	0.00465	0.00463
$\overline{5}_6$	-60	-30	-57	0	0	0.00017	0.00028
$\overline{5}_7$	0	0	0	0	$-\frac{1}{90}$	-0.00996	-0.00705
$\overline{5}_8$	45	30	63	0	0	-0.00059	-0.00070
$\overline{5}_9$	0	0	-13	0	$\frac{1}{60}$	0.01171	0.00678
$\overline{5}_{10}$	-6	-6	-6	$-\frac{8}{15}$	$-\frac{1}{6}$	-0.06030	-0.02371

and spheres, many of the Ree- Hoover integrals either vanish identically or are so small that they can be neglected [20].) Second, it turns out that many of the remaining integrals don't have to be evaluated because their combinatorial factors $C(\overline{n}_k)$ vanish.

Partially as a consequence of these two simplifying characteristics, the Ree-Hoover sums lend themselves well to partial summation. The leading-order, dominant terms turn out to be those which contain the most f (and therefore fewest $\tilde{f}$) functions. Ordinarily these are placed at the end of the list of $\overline{n}_k$ integrals, so they may conveniently be referred to as $\overline{n}_{\max}$, $\overline{n}_{\max-1}$, etc. Thus, $\overline{n}_{\max}$ is the integral with no $\tilde{f}$, $\overline{n}_{\max-1}$ is the one with a single $\tilde{f}$, $\overline{n}_{\max-2}$ is the one with a pair $\tilde{f}$ functions that don't share an index, and $\overline{n}_{\max-3}$ is the one with a pair that do share an index. Geometrically, these integrals measure the regions of the n-particle configuration space within which there are very few nonoverlapping pairs of particles. If we measure all particle positions with respect to the center of mass, they may be viewed as the centermost regions. Thus, at the very center is $\overline{n}_{\max}$, which describes the region where every particle overlaps every other particle. Immediately across the bounding surfaces of this region are $\binom{n}{2}$ equivalent $\overline{n}_{\max-1}$ regions (each of which may consist of more than one part). After this come the $3\binom{n}{4}$ regions corresponding to $\overline{n}_{\max-2}$ and the $3\binom{n}{3}$ regions $\overline{n}_{\max-3}$.

To make this a little more concrete, consider the integrals $\overline{4}_k^{(1)}$, which measure regions of configuration space for 4 hard rods. The space is 4-dimensional, but the degree of freedom associated with the center of mass can be eliminated in order to render the space visualizable. The central region is shown in Fig. 3. In this center-of-mass projection, the 4 particle positions x_i are measured along the 4 C_3 axes of the cube, while the 6 possible coincidences $x_i = x_j$ define the 6 σ_d planes Thus, each pairwise overlap eliminates one of the six diagonal slabs from the configuration space. Now consider the integrals $\overline{n}_{\max-0,1,2,3}$, or $\overline{4}_3$, $\overline{4}_2$, $\overline{4}_1$, $\overline{4}_0$. (We extend the usual list of integrals for $n = 4$ in considering the last of these.) One of these integrals, namely $\overline{4}_1$, is identically zero for hard rods — there is no way to place four rods such that the pairs 12, 23, 34, and 41 all overlap, while 13 and 24 do not. The remaining three integrals generate the regions shown to the right in Fig. 3; these are the regions within which precisely 6, 5, or 4 of the slabs overlap each other.

For general n, the exact combinatorial coefficients for the four integrals just described are

$$
\begin{aligned}
C(\overline{n}_{\max}) &= -(-)^{n(n-1)/2}(n-2)! \\
C(\overline{n}_{\max-1}) &= 0 \\
C(\overline{n}_{\max-2}) &= \tfrac{3}{2}\binom{n}{4}(-)^{n(n-1)/2}(n-2)! \\
C(\overline{n}_{\max-3}) &= 0,
\end{aligned}
\tag{19}
$$

and the first two Ree-Hoover partial sums are simply

$$
\begin{aligned}
\tilde{B}_n(1) &= C(\overline{n}_{\max})\,\overline{n}_{\max} \\
\tilde{B}_n(2) &= C(\overline{n}_{\max})\,\overline{n}_{\max} + C(\overline{n}_{\max-2})\,\overline{n}_{\max-2}.
\end{aligned}
\tag{20}
$$

These partial sums can also be expressed in terms of Mayer integrals by using

$$
\begin{aligned}
\overline{n}_{\max} &= n_{\max} \\
\overline{n}_{\max-2} &= n_{\max} + 2n_{\max-1} + n_{\max-2}.
\end{aligned}
\tag{21}
$$

(It should be noted, however, that the resulting expansions in terms of Mayer integrals are *not* truncations of the Mayer expansions.)

The Ree-Hoover series is really not understood very well, and as a consequence it is difficult to estimate the errors associated with its partial sums. However, the kind of accuracy to be expected may be judged to some extent from the following table, which shows the breakdown of the known virial coefficients for hard disks and hard spheres into contributions from the first, second, and remaining terms of the Ree Hoover series [21]:

	$D = 2$			$D = 3$		
	$\overline{n}_{\max}$	$\overline{n}_{\max-2}$	other	$\overline{n}_{\max}$	$\overline{n}_{\max-2}$	other
$\tilde{B}_3$	100%	0	0	100%	0	0
$\tilde{B}_4$	103%	-3%	0	110%	-10%	0
$\tilde{B}_5$	108%	-8%	$-\tfrac{1}{2}\%$	129%	-29%	$-\tfrac{1}{2}\%$
$\tilde{B}_6$	115%	-14%	-1%	152%	-55%	3%
$\tilde{B}_7$	123%	-20%	-3%	169%	-83%	14%

The Ree-Hoover reformulation is also very useful as a tool for understanding and for correcting the integral equation treatments.

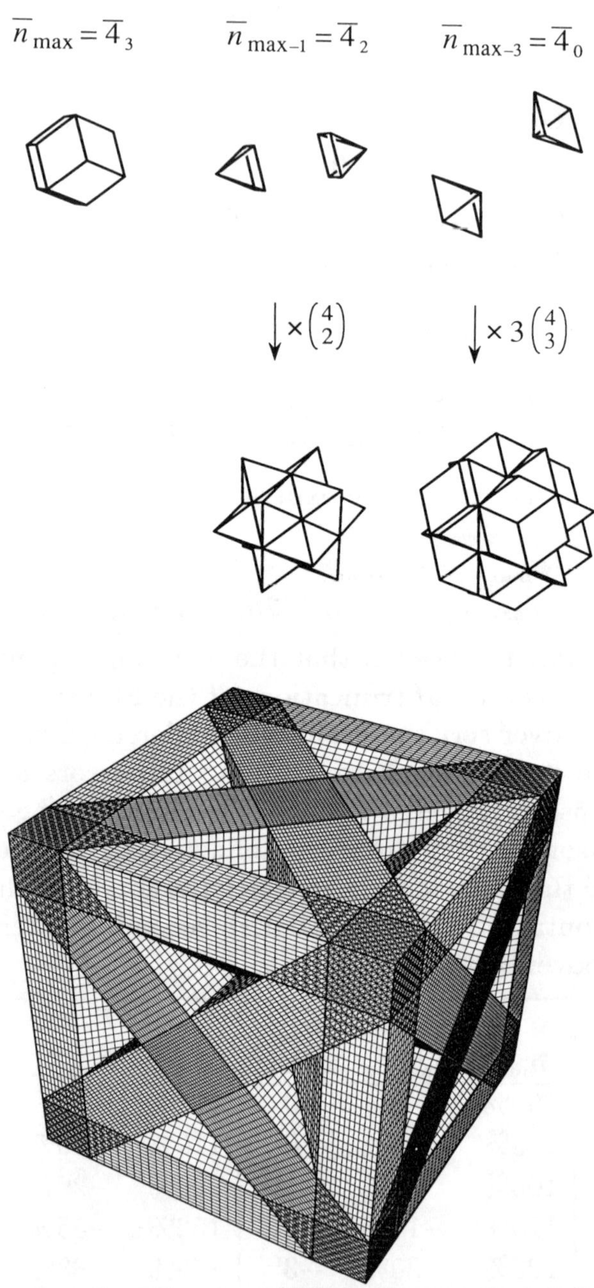

Figure 3. Central regions of configuration space for 4 hard rods on a line, rendered three-dimensional by projecting out the center of mass. Only open portions involve no particle overlaps and contribute to the configuration integral. Centermost regions and corresponding Ree-Hoover integrals are shown above.

As Table 3 illustrates, both the PY_c and HNC_c approximations give the correct zeroth-order contribution $C(\overline{n}_{\max})\overline{n}_{\max}$. The leading-order and dominant error in the HNC_c approximation arises from the non-zero coefficient which it assigns to $\overline{n}_{\max-1}$. One way to correct the approximation would be to take this contribution back out. For the PY_c approximation, the leading-order contribution to the error arises from an incorrect $\overline{n}_{\max-2}$ coefficient. In fact, to this order in the Ree-Hoover series, PY_c gives precisely

$$\tilde{B}_n(PY_c, 2) = C(\overline{n}_{\max})\overline{n}_{\max} + \tfrac{2}{3}C(\overline{n}_{\max-2})\overline{n}_{\max-2}, \tag{22}$$

so the leading order correction is

$$\tilde{B}_n(\overline{PY_c}, 2) = \tfrac{1}{3}C(\overline{n}_{\max-2})\overline{n}_{\max-2}. \tag{23}$$

As an example of the Ree-Hoover series, we again consider $\tilde{B}_4$ for hard spheres. The exact, PY_c, and HNC_c expressions for the virial coefficient are

$$
\begin{aligned}
\tilde{B}_4^{(3)} &= -2(\overline{4}_3) &&+ 3(\overline{4}_1) &&= 0.28695 \\
\tilde{B}_4^{(3)}(PY_c) &= -2(\overline{4}_3) &&+ 2(\overline{4}_1) &&= 0.29688 \\
\tilde{B}_4^{(3)}(HNC_c) &= -2(\overline{4}_3) - 1(\overline{4}_2) &&+ 3(\overline{4}_1) &&= 0.20919,
\end{aligned}
\tag{24}
$$

in agreement with the Mayer series results. In this case, the first partial sum gives 110% of the exact result, while the second partial sum and the corrected Percus-Yevick approximation are exact.

Dimensional limits

The problem of computing a given virial coefficient has been broken down into one of determining the form of the expansion, including the combinatorial factors $C(n_k)$ or $C(\overline{n}_k)$, and one of determining the values of the integrals n_k or $\overline{n}_k$. It is the latter problem which is invariably the most difficult, and it is for this that we will utilize dimensional methods. We consider first the Mayer integrals. Values for the integrals $n_k^{(D)}$ with $n \leq 5$ which are available from the literature [13,14,17] were plotted as a function of D in Fig. 2; the values for $D \leq 3$ were also given in Table 2. One sees that each integral decreases in roughly geometric fashion with D.

Not surprisingly, it turns out to be easiest to obtain results at low D and at high D. Specifically, one can obtain results at $D = 0$ (the trivial problem of "hard points"), at $D = 1$ (the hard rod problem), and in the $D \to \infty$ limit. As in other applications of dimensional scaling presented in this book, it is helpful to deal with quantities which remain finite across the spectrum of dimensionalities. For the cluster integrals, such a quantity is provided by the interdimensional ratio,

$$\rho(n_k; D) \equiv \frac{n_k^{(D-1)}}{n_k^{(D)}}. \tag{25}$$

These ratios measure the slopes of the lines plotted in Fig. 2. By treating these ratios in the low-D and high-D regimes exactly, we will be able to approximate their behavior at intermediate D.

It is relatively straightforward to obtain the low-D results. For $D = 0$ there are no integrals, and one readily finds

$$n_k^{(0)} = -(-)^{[n_k]} \frac{2^{n-1}}{n(n-2)!}, \tag{26}$$

where $[n_k]$ denotes the number of f functions in the integral (or edges in the diagram). In particular, for the last three integrals at any n, which were the ones needed for the first two Ree-Hoover partial sums,

$$n_{\text{max}-\mu}^{(0)} = -(-)^{n(n-1)/2-\mu} \frac{2^{n-1}}{n(n-2)!} \qquad (\mu = 0, 1, 2). \tag{27}$$

For $D = 1$, results are available in the literature up to $\tilde{B}_7$, and further results could be computed by the same methods [13]. For the last three integrals, general expressions can be found:

$$n_{\text{max}-\mu}^{(1)} = -\frac{(-)^{n(n-1)/2-\mu}}{(n-2)!} \left[1 + \frac{2\mu}{n(n-1)} \right] \qquad (\mu = 0, 1, 2). \tag{28}$$

Now consider the $D \to \infty$ limit. This problem is conceptually harder, but computationally simpler, than the low-D problems. Roughly speaking, the reason that the analysis becomes simple in the $D \to \infty$ limit is that the integrals come to be dominated by ever smaller (in a relative sense) regions of configuration space as D is increased. This is due to the fact that there is only a bounded region of configuration space within which the integrand of any Mayer or Ree-Hoover integral is non-zero, while volume element factors weight ever more heavily those regions in which particles are as far from one another as possible. The net result is that the integrand gets pushed up against the outermost boundary of the region of integration. This is quite analogous to the situation encountered in electronic structure, where Jacobian factors (which were multinomials in the interparticle distances) played off against wavefunctions tails (which were asymptotically exponentials in those distances), again resulting in localization.

We start by rewriting Eq. (14) with the denominator expanded as one big integral, in order to give it the same form as the numerator:

$$n_k^{(D)} = \frac{\dfrac{1-n}{n!} \displaystyle\int d\mathbf{r}_1 \, d\mathbf{r}_2 \cdots d\mathbf{r}_{n-1} \prod_{lm \in n_k} f_{lm}}{\left(-\frac{1}{2}\right)^{n-1} \displaystyle\int d\mathbf{r}_1 \, d\mathbf{r}_2 \cdots d\mathbf{r}_{n-1} \prod_{l \neq n} f_{ln}}. \tag{29}$$

Next, motivated by the ideas discussed in the preceding paragraph, we recast this expression in terms of the interparticle coordinates r_{ij}. For sufficiently large D, these $\frac{1}{2}n(n-1)$ coordinates are independent. Since integrals involving the overall orientation of the configuration of particles (generalized Euler angles) are clearly the same in the numerator and denominator, these can be cancelled out, leaving

$$n_k^{(D)} = \frac{\dfrac{1-n}{n!} \displaystyle\int \prod_{i<j} dr_{ij} \; J_n^{(D)} \prod_{lm \in n_k} f_{lm}}{\left(-\frac{1}{2}\right)^{n-1} \displaystyle\int \prod_{i<j} dr_{ij} \; J_n^{(D)} \prod_{l \neq n} f_{ln}}. \tag{30}$$

Here $J_n^{(D)}$ is the Jacobian for the transformation, which is given (up to a constant which again cancels) by [22]

$$J_n^{(D)} = \left(\prod_{i<j} r_{ij}\right)
\begin{vmatrix}
0 & 1 & 1 & 1 & \cdots & 1 \\
1 & 0 & r_{12}{}^2 & r_{13}{}^2 & \cdots & r_{1n}{}^2 \\
1 & r_{12}{}^2 & 0 & r_{23}{}^2 & \cdots & r_{2n}{}^2 \\
1 & r_{13}{}^2 & r_{23}{}^2 & 0 & \cdots & r_{3n}{}^2 \\
\vdots & \vdots & \vdots & \vdots & \ddots & \vdots \\
1 & r_{1n}{}^2 & r_{2n}{}^2 & r_{3n}{}^2 & \cdots & 0
\end{vmatrix}^{\frac{1}{2}(D-n)} \tag{31}$$

The determinant within this Jacobian has a simple geometric interpretation [23], since it is proportional to the squared volume of the simplex (generalized triangle or tetrahedron) defined by the $\frac{1}{2}n(n-1)$ distances r_{ij}. Alternatively, it is proportional to the squared volume of any one of the n parallelotopes (generalized parallelograms or parallelepipeds) defined by the edges incident with any given vertex of the simplex. It doesn't matter whether we use the simplex or parallelotope volume, but it turns out to be simpler to use the latter. If we denote this volume simply by V, then the expression for n_k becomes

$$
n_k^{(D)} = \frac{\dfrac{1-n}{n!} \displaystyle\int \prod_{i<j} d(r_{ij}{}^2) \; V^{D-n} \prod_{lm \in n_k} f_{lm}}{\left(-\dfrac{1}{2}\right)^{n-1} \displaystyle\int \prod_{i<j} d(r_{ij}{}^2) \; V^{D-n} \prod_{l \neq n} f_{ln}} \; .
\tag{32}
$$

Now consider what happens as D gets large. The simplification here arises from the fact that both numerator and denominator integrands develop ever sharper maxima as D increases, so that in the $D \to \infty$ limit only those configurations which define the maxima need to be considered. In each case, the maximum arises because the V^{D-n} factor gives ever greater weight to those configurations which enclose large volumes in configuration space, while the $\prod f$ factor (which is everywhere either ± 1 or 0) confines the integration to a bounded region of configuration space. Loosely speaking, V^{D-n} pushes the integrand up against the walls defined by $\prod f$, and the integrand as a whole develops a very sharp maximum.

As the above expressions for n_k make clear, the reduced virial coefficients are dimensionless, and therefore independent of the sphere diameter σ. To simplify the following discussion, we will therefore set $\sigma = 1$. With this choice, the product of f-functions in the denominator is nonzero in the region where $r_{ln} \leq 1$ for all $l \neq n$. The maximum volume assumed by a parallelotope within this region is just 1. (If the edges incident with vertex n are used to define the parallelotope, it is just a unit cube; for other choices of reference vertex the parallelotope is skewed, but still has unit volume.) For the numerator, the product of f- functions is nonzero when $r_{lm} \leq 1$ for all $lm \in n_k$. The maximum parallelotope volume within this region can be determined simply by maximizing the expression for the parallelotope volume, subject to the constraints imposed by the f-functions. Let this volume be denoted $V_{\mathrm{max}}(n_k)$.

Finally, in order to evaluate $\rho(n_k; \infty)$, consider what happens to n_k when D is incremented by 1 at very large D. In both numerator and denominator, the power of V is incremented by 1. However, the maximum values assumed by V, namely $V_{\mathrm{max}}(n_k)$ and 1, remain unchanged. At large D, the net effect is to multiply the integral as whole by $V_{\mathrm{max}}(n_k)$. Thus,

$$
\begin{aligned}
\rho(n_k; \infty) &= \lim_{D \to \infty} \frac{n_k^{(D-1)}}{n_k^{(D)}} \\
&= \frac{1}{V_{\mathrm{max}}(n_k)}.
\end{aligned}
\tag{33}
$$

Table 4 gives the ratios $\rho(n_k; \infty)$, as well as the interparticle distances which yield V_{max}, for all integrals contributing to $\tilde{B}_3$, $\tilde{B}_4$, and $\tilde{B}_5$.

Dimensional Interpolation

We will now use dimensional interpolation to proceed from the known $n_k^{(D)}$ results, namely those at low n, at low D, and at high D, to the desired results at higher n and at $D=2$ or $D=3$. The interpolation will actually be performed on the cluster integral ratios $\rho(n_k; D)$, since these quantities had finite but non-zero dimensional limits. Given interpolated ratios, one can simply step up from the known integrals for hard points or rods to those for hard disks or spheres (or even higher-dimensional fluids), since

$$
\begin{aligned}
n_k^{(D)} &= n_k^{(0)} \frac{n_k^{(1)}}{n_k^{(0)}} \times \cdots \times \frac{n_k^{(D)}}{n_k^{(D-1)}} \\
&= n_k^{(0)} \frac{1}{\rho(n_k; 1)} \times \cdots \times \frac{1}{\rho(n_k; D)}.
\end{aligned}
\tag{34}
$$

Several procedures can be used to perform the interpolation, but the one that we will use here is to assume that each $\rho(n_k; D)$ varies

Table 4. Cluster integral ratios in the $D{\to}\infty$ limit and the volume-maximizing geometries from which these are derived. Values listed are all squared, and distances are in units of sphere diameter. Only distances unconstrained by f-functions are listed; all others are unity. Labeling of spheres is based on the standard diagrams in Sec. 2, and is clockwise starting at the top.

n_k	$[\rho(n_k;\infty)]^2$	Geometric parameters
3	$4/3$	
4_1	$27/16$	$r_{13}^2 = r_{24}^2 = 4/3$
4_2	$16/9$	$r_{13}^2 = 3/2$
4_3	2	
5_1	$256/125$	$r_{13}^2 = r_{14}^2 = r_{24}^2 = r_{25}^2 = r_{35}^2 = 3/2$
5_2	$9/4$	$\begin{cases} r_{13}^2 = r_{14}^2 = 5/3 \\ r_{24}^2 = r_{35}^2 = 4/3 \end{cases}$
5_3	$64/27$	$\begin{cases} r_{13}^2 = r_{14}^2 = r_{34}^2 = 3/2 \\ r_{25}^2 = 1 \end{cases}$
5_4	$64/27$	$r_{13}^2 = r_{14}^2 = r_{34}^2 = 3/2$
5_5	$64/27$	$\begin{cases} r_{24}^2 = r_{35}^2 = 3/2 \\ r_{25}^2 = 7/4 \end{cases}$
5_6	$(26\sqrt{13} - 70)/9$	$\begin{cases} r_{13}^2 = r_{14}^2 = (5 + \sqrt{13})/6 \\ r_{25}^2 = (7 - \sqrt{13})/3 \end{cases}$
5_7	$8/3$	$r_{13}^2 = r_{14}^2 = 3/2$
5_8	$(8\sqrt{3} + 12)/9$	$r_{24}^2 = r_{35}^2 = 3 - \sqrt{3}$
5_9	3	$r_{25}^2 = 4/3$
5_{10}	$16/5$	

between its low- and high-D limiting values with the same dimension-dependence as $\rho(3; D)$ (which is known analytically for any D). Specifically, we will approximate $\rho(n_k; D)$ as

$$\rho(n_k; D) \approx \rho(n_k; 1)\xi_D + \rho(n_k; \infty)(1 - \xi_D), \tag{35}$$

where ξ_D is defined so as to make this equation exact for $\rho(3; D)$. In particular, using

$$
\begin{aligned}
\rho(3,1) &= \tfrac{4}{3} & \rho(3,2) &= \left[\tfrac{4}{3} - \tfrac{\sqrt{3}}{\pi}\right]^{-1} \\
\rho(3,\infty) &= \tfrac{2}{\sqrt{3}} & \rho(3,3) &= \tfrac{8}{5}\left[\tfrac{4}{3} - \tfrac{\sqrt{3}}{\pi}\right]
\end{aligned}
\tag{36}
$$

we obtain the interpolation parameters needed for hard disks and hard spheres, $\xi_2 \approx 0.69452$ and $\xi_3 \approx 0.54025$.

Hard sphere Mayer integrals computed by this interpolation procedure have errors of the order of 1% wherever comparisons can be made. (For the 13 integrals contributing to $\tilde{B}_4$ and $\tilde{B}_5$, the r.m.s. percentage errors are 0.5% for hard disks and 1.0% for hard spheres.) In general it is the more complex integrals which are treated more accurately. Apparently the cluster integral ratio for the triangle diagram, $\rho(3; D)$, which has been used as a basis for the interpolation, serves as a better model for the dimension-dependence of the complex diagrams than for that of the simple ones. This is perhaps explained by the observation that the more complex diagrams are the more fully triangulated ones. (Indeed, for the 13 $\tilde{B}_4$ and $\tilde{B}_5$ integrals, one encounters the largest errors for precisely those diagrams which contain no triangles, namely 4_1, 5_1, and 5_3.)

For $n \leq 7$, all of the low-D limiting ratios $\rho(n_k; 1)$ are available from the literature [13], while the high-D ratios are easy to compute; values through $n = 5$ for the two limits can be obtained from Tables II and IV. From these, it is straightforward to interpolate all of the integrals n_k for hard disks or spheres. The interpolated values can be used in two ways. First, one can simply plug them into Eq. (12) to obtain the virial coefficients. Alternatively, and preferably, one can start with an integral equation approximation which sums some subset of the integrals exactly, and then use the dimensionally interpolated values only for the remainder. Since the HNC$_c$ approximation sums a larger class of Mayer integrals than the PY_c approximation [10], it is clearly preferable to use the former. In the following table, the exact hard sphere virial coefficients $\tilde{B}_4$, $\tilde{B}_5$, and $\tilde{B}_6$ are compared with those computed by dimensional interpolation, by the HNC_c approximation [24], and by HNC_c with DI correction. For all three coefficients, the $HNC_c + DI$ approximation is the best.

	$\tilde{B}_4^{(3)}$	$\tilde{B}_5^{(3)}$	$\tilde{B}_6^{(3)}$
exact	0.28695	0.11025	0.0389
DI	0.27680	0.08014	−0.0053
HNC_c	0.20919	0.04927	0.0281
$HNC_c + DI$	0.28773	0.11256	0.0357

It would probably be a waste of time to try to apply the procedure just described at higher order. This is because the errors associated with dimensional interpolation for any given integral (of the order of 1% at $D = 3$) would render the final sum, which is quite small compared to the contribution of any given integral, unreliable. Indeed, for $\tilde{B}_6$, the final sum is only about $\frac{1}{3500}$ of the sum of absolute values, and the dimensionally interpolated result does not even have the right sign.

In order to proceed to higher order, we therefore turn to the Ree-Hoover reformulation, because here we can consider partial sums of only a few terms, and thereby avoid the problem just discussed. Of course, the use of partial sums introduces a new error, namely that of the truncation. We will consider here only the first two partial sums, as given by Eq. (21). The dimensional limit results needed for these sums are the following:

$$\rho(n_{\max}; 1) = \frac{2^{n-1}}{n}$$

$$\rho(n_{\max-1}; 1) = \frac{2^{n-1}}{n}\left[1 + \frac{2}{n(n-1)}\right]^{-1}$$

$$\rho(n_{\max-2}; 1) = \frac{2^{n-1}}{n}\left[1 + \frac{4}{n(n-1)}\right]^{-1}$$

$$\rho(n_{\max}; \infty) = \left[\frac{2^{n-1}}{n}\right]^{1/2}$$

$$\rho(n_{\max-1}; \infty) = \left[\frac{2^{n-1}(n-2)}{(n-1)^2}\right]^{1/2}$$

$$\rho(n_{\max-2}; \infty) = \left[\frac{2^n(n-4)^3}{(n^2-6n-1)^2 + (n-3)(n^2-6n+17)^{3/2} - 108}\right]^{1/2} \tag{37}$$

From these, one can interpolate the three Mayer integrals $n_{\max-0,1,2}$, and then from these obtain the two required Ree-Hoover integrals $\overline{n}_{\max-0,2}$.

Once the Ree-Hoover integrals have been computed, there is again a choice. First, one can simply construct the Ree-Hoover partial sums using the interpolated integrals. Second, one can start with an integral equation approximation, and correct it. When one uses the Ree-Hoover expansion or goes to higher order, it is the PY_c approximation which is most useful. This is first of all because higher-order HNC_c virial coefficients are unavailable, whereas the analytic $D = 3$ equation of state in the PY_c approximation, namely $z = (1+\eta+\eta^2)/(1-\eta)^3$, yields explicit virial coefficients for all n:

$$\tilde{B}_n^{(3)}(PY_c) = \frac{3n^2 - 3n + 2}{2^{2n-1}} . \tag{38}$$

Even if higher-order HNC_c virial coefficients were available, however, they would probably not be as useful as the PY_c coefficients, since the Ree-Hoover expansions for the latter more closely match the exact expansions at low order.

The following table compares hard sphere virial coefficients through $\tilde{B}_9$ obtained in three different ways: by using dimensionally interpolated integrals to compute the second Ree-Hoover partial sum,

Eq. (21); by using the PY_c approximation, Eq. (38); and by starting with the PY_c approximation and then correcting it using a dimensionally interpolated value for the leading-order correction, Eq. (23). Where comparisons with exact values can be made, the last of these procedures is the best, though the errors are still somewhat large.

	$\widetilde{B}_4^{(3)}$	$\widetilde{B}_5^{(3)}$	$\widetilde{B}_6^{(3)}$	$\widetilde{B}_7^{(3)}$	$\widetilde{B}_8^{(3)}$	$\widetilde{B}_9^{(3)}$
exact	0.28695	0.11025	0.0389	0.013	?	?
DI(2)	0.27680	0.10219	0.03348	0.00993	0.00268	0.00066
PY_c	0.29688	0.12109	0.04492	0.01562	0.00519	0.00166
PY_c+DI(2)	0.28338	0.10797	0.03691	0.01169	0.00350	0.00101

The errors in the DI and $PY_c + DI(2)$ values have two sources: those associated with truncating the Ree-Hoover series, and those associated with dimensional interpolation. In general, one can expect the truncation errors to increase with n [20]. For example, the errors caused by truncating the Ree-Hoover series for $\widetilde{B}_4$ through $\widetilde{B}_7$ after $\overline{n}_{\mathrm{max}-2}$ are respectively 0%, $+\frac{1}{2}$%, -3%, and -14%. On the other hand, for any given level of truncation, the error generated by dimensional interpolation of the integrals can be expected to decrease with n. This is because contributing diagrams become more and more complex (specifically, more fully triangulated) as n increases, and as pointed out above this actually improves the accuracy of the interpolations. For example, using the Ree-Hoover Monte Carlo calculations [21] as a reference, we find that the interpolation errors in $\overline{n}_{\mathrm{max}-2}$ for $\widetilde{B}_4$ through $\widetilde{B}_7$ are respectively 36%, 24%, 13%, and 3%. Overall, in the $PY_c + DI(2)$ approximation the interpolation errors dominate for $n < 6$ and the truncation errors for $n > 6$; the two sources contribute about equally (and additively) to the error in $\widetilde{B}_6$ itself.

Improvements and extensions

The dimensional interpolation strategy provides an alternative to Monte Carlo calculations for the evaluation of cluster integrals. Compared to the Monte Carlo technique, interpolation has several disadvantages. In particular, the errors associated with the method are difficult to estimate, and cannot be reduced through further calculations. On the other hand, interpolation is simple enough that it can be carried to higher order very easily; for many integrals results of arbitrary order can be obtained analytically.

From a computational standpoint, the results given above make it clear that dimensional interpolation is best employed as an adjunct to a method which sums some subset of the integrals exactly (such as the Percus-Yevick or hypernetted chain approximation). Interpolation is then used only for the integrals which have been left out, and the interpolation errors are correspondingly limited. What makes this procedure especially effective is that dimensional interpolation turns out to be most accurate (and even easiest) for the more complex Mayer integrals, which are generally speaking the ones which the integral equation approximations like HNC and PY omit. By itself, dimensional interpolation yields virial coefficients which are fairly comparable in accuracy to those provided by these more familiar methods. On the other hand, the best of the combined approximations ($HNC_c + DI$ in the Mayer formulation, or $PY_c + DI(2)$ in the truncated Ree-Hoover formulation) are uniformly better than either approximation individually.

This situation is quite analogous to that encountered in electronic structure applications of dimensional scaling. Here, too, the dimensional approach could be used on its own as a conceptual tool or semi-quantitative model, or it could be used quantitatively as a corrective for more familiar approximation methods. In particular, dimensional scaling could be used to treat the many- body effects omitted from self-consistent field calculations, in much the same way that it has been used here to treat the cluster integrals omitted from the HNC and PY calculations.

Even when used in conjunction with integral equation approximations, dimensional interpolation appears not to be accurate enough to obtain reliable higher-order virial coefficients. However, there are several ways in which the calculations described in this chapter might be improved. First, since the primary error in the higher-order virial coefficients is probably that generated by truncation of the Ree-Hoover series, one could attempt to carry this series further. Actually, this might be a somewhat problematic venture. For example, from Monte Carlo calculations [21] we know that for $\tilde{B}_6$ and $\tilde{B}_7$, the seemingly natural third partial sum of the Ree-Hoover series (which includes a total of five terms) turns out to be worse than the second. However, the partial sums needed to correct the Percus-Yevick approximation may show better behavior. Also, the Ree-Hoover series has really not been studied that extensively, and there may be other ways to proceed to higher order.

If the Ree-Hoover series is carried further, then the problems associated with the limited accuracy of interpolated integrals will become more severe, because as one goes further in the series the required Ree-Hoover integrals become sums of ever larger numbers of Mayer integrals. Thus, another area that needs to be investigated is that of increasing the accuracy of the interpolated values. This could probably be done by means of a more detailed characterization of the dimension-dependence of the cluster integrals. For example, we have noted that there are systematic overestimation errors in the dimensionally interpolated values for the integrals with relatively "open" diagrams (in particular the triangle-lacking diagrams 4_1, 5_1, and 5_3). These diagrams apparently have a somewhat different behavior with respect to D, and might be better modeled using $\rho(4_1; D)$ (for which an analytic expression is also available [25]).

Perhaps the most promising avenue to improved accuracy is that of interpolating the Ree-Hoover integrals directly, instead of interpolating the Mayer integrals and then combining them. In general, one will have to deal with a slightly more complex dimension-dependence in the $\overline{n}_k$ integrals. Thus, except for $\overline{n}_{\max}$, which was identical to $n_{\max}$, each Ree-Hoover integral is not monotonic in D (since it is identically zero at sufficiently low D, and approaches zero asymptotically at high D). However, it appears that the Ree-Hoover integrals invariably show smooth and relatively generic dimension-dependence.

Finally, we note that it is also possible to extend the methods of this chapter to study other aspects of the hard sphere problem. For example, by essentially the same approach one can treat the radial distribution function $g(r)$. (Like the equation of state, this can be expanded in powers of the density, and then the r-dependent coefficients expanded in terms of r-dependent Mayer or Ree-Hoover integrals.) There are again several options, depending on whether one uses dimensional scaling in isolation or in conjunction with an integral equation approximation, and whether one uses the Mayer or truncated Ree-Hoover formulation. The results parallel quite closely those for virial coefficients. On its own, dimensional interpolation gives results which are roughly comparable in accuracy to the integral equation approximations. However, when used as a corrective for those approximations, it generates significantly more accurate results.

It is also possible, and in many ways preferable, to work with the direct correlation function $c(r)$, from which $g(r)$ can be obtained by a Fourier transform. A simple analytic expression for $c(r)$ is available in the Percus-Yevick approximation, and it appears appropriate to use dimensional interpolation to correct some of the deficiencies of this approximation (in particular, its lack of a long-range tail). This can be done at low order in the Ree-Hoover series quite easily. Curiously, when the leading- order corrections are summed to infinite order in the density, the resulting $c(r)$ becomes singular at packing fractions ($\frac{1}{\sqrt{2}}$ for disks and $\frac{1}{2}$ for spheres) which are quite close to those where freezing begins (0.683 for disks and 0.495 for spheres).

Acknowledgments

We are very grateful to Dudley Herschbach and Sabre Kais for initiating the application of dimensional scaling to hard sphere fluids, and to them and Gerardo Beltrame for sharing their hopes, insights, and discoveries with us. This material is based upon work supported by the National Science Foundation under Grant. No. CHE-9007620 and by the Donors of The Petroleum Research Fund administered by the American Chemical Society.

References

1. J. E. Kilpatrick, *Adv. Chem. Phys.* **20**, 39 (1971) and references therein.
2. D. A. Young and B. J. Alder, *J. Chem. Phys.* **70**, 473 (1979) and references therein.
3. J. G. Berryman, *Phys. Rev. A* **27**, 1053 (1983) and references therein.
4. L. Tonks, *Phys. Rev.* **50**, 955 (1936).

5. M. Baus and J. L. Colot, *Phys. Rev. A* **36**, 3912 (1987).

6. N. F. Carnahan and K. E. Starling, *J. Chem. Phys.* **51**, 635 (1969).

7. K. W. Kratky, *J. Chem. Phys.* **69**, 2251 (1978).

8. K. W. Kratky, *Physica* **87A**, 584 (1977).

9. J. E. Mayer and M. G. Mayer, *Statistical Mechanics*, 2nd ed. (Wiley, New York, 1977), Chapter 8.

10. C. A. Croxton, *Liquid State Physics* (Cambridge, Cambridge, 1974), Chapters 2 and 3.

11. J. G. Loeser, Z. Zhen, S. Kais, and D. R. Herschbach, *J. Chem. Phys.* **95**, 4525 (1991).

12. R. P. Feynman, *Statistical Mechanics* (Benjamin, Reading, MA, 1972), Chapter 4.

13. W. G. Hoover and A. G. DeRocco, *J. Chem. Phys.* **36**, 3141 (1962).

14. K. W. Kratky, *Physica* **85A**, 607 (1976); **87A**, 584 (1977).

15. G. S. Rushbrooke and P. Hutchinson, *Physica* **27**, 647 (1961).

16. S. Kim and D. Henderson, *Phys. Lett.* **27A**, 378 (1968).

17. F. H. Ree and W. G. Hoover, *J. Chem. Phys.* **40**, 2048 (1964).

18. F. Harary and E. M. Palmer, *Graphical Enumeration* (Academic, New York, 1973), p. 188.

19. R. J. Riddell and G. E. Uhlenbeck, *J. Chem. Phys.* **21**, 2056 (1953).

20. F. H. Ree and W. G. Hoover, *J. Chem. Phys.* **41**, 1635 (1964).

21. F. H. Ree and W. G. Hoover, *J. Chem. Phys.* **40**, 939 (1964); **46**, 4181 (1967).

22. L. Kronecker, *Journal für die Reine und Angewandte Mathematik* **72**, 152 (1870).

23. M. G. Kendall, *A Course in the Geometry of n Dimensions* (Charles Griffin, London, 1961), p. 38.

24. P. Hutchinson and G. S. Rushbrooke, *Physica* **29**, 675 (1963).

25. M. Luban and M. Baram, *J. Chem. Phys.* **76**, 3233 (1982).

Part III
RELATED METHODS

Chapter 11
NONSEPARABLE DYNAMICS

Ugo Fano
Department of Physics and
James Franck Institute
University of Chicago
5640 South Ellis Avenue
Chicago, Illinois 60637, USA

Abstract

Regularities of collisions and spectra are generally classified successfully by approximate constants of the motion in spite of their nonseparable dynamics. These "constants" belong to channels of excitation that point toward alternative fragmentations of the system but become identifiable only at sufficient distance from the system's center of mass. It is shown how the nonseparability of channels originates in the critical transition from centrifugal dominance at short ranges to the independence of fragmentation channels.

Introduction

Observed regularities of diverse collisions and spectra have been classified by parameters that represent approximate constants of the motion

D. R. Herschbach et al. (eds.), Dimensional Scaling in Chemical Physics, 461–470.

for processes whose dynamics are clearly nonseparable. Prototype examples are the (diamagnetic) Zeeman effect of Rydberg   spectra and the double excitation and ionization of helium [1], together with the vibrational spectrum of H_3^+ [2].

I shall indicate here how these regularities fit into an elementary analysis of the Schrödinger equation for aggregates of electrons and nuclei, without recourse to modeling or to perturbative procedures. The equation for an N particle aggregate will be cast in hyperspherical coordinates [3]. The $3N$ position coordinates of the particles reduce to $3(N-1)$ after separation of the center of mass, and to $3(N-2)$ after further separation of the Euler angles in an inertial body frame. Each of these coordinate sets, complemented by spin coordinates, identifies one configuration of the aggregate in physical space.

Hyperspherical coordinates are polar coordinates in $3(N-2)$ dimensions, consisting of one radius of inertia R and a set $\{\omega\}$ of $3(N-2)-1$ angles [3]. Different sets of Jacobi coordinates of the aggregate, appropriate ,e.g., to its alternative break-ups, correspond to alternative sets $\{\omega\}$. Transformations among these sets, at constant R, are provided by the standard algebra of the $O[3(N-2)]$ group. To each set of angles $\{\omega\}$ corresponds a *complete set* $\{Y_\alpha\}$ of hyperspherical harmonics, labeled by sets of quantum numbers $\{\alpha\}$ conjugate to the angles $\{\alpha\}$ [4].

A fundamental shift of emphasis results from expanding the Hamiltonian into harmonics $\{Y_\alpha\}$. The expansion focuses attention onto quantum numbers $\{\alpha\}$ generally accessible to observation, in place of angles $\{\omega\}$ that may seem more intuitive, but escape direct observation. It thus replaces all but one space coordinate by *collective* coordinates apt to describe the static or evolving features of any aggregate of interest.

The aggregate's Hamiltonian is thus cast in the form [3]

$$H = -\frac{1}{2M}\left[-\frac{d^2}{dR^2} + \frac{[\lambda+(3N-5)/2]^2 - 1/4}{R^2}\right] + \frac{(\alpha|C|\alpha')}{R} \tag{1}$$

whose second and third terms are viewed as generalizations of the corresponding centrifugal and Coulomb terms of the H atom prototype. The *grand angular momentum* quantum number λ, an analogue of the orbital angular momentum quantum number l for a single particle, is generally included in the set $\{\alpha\}$. Equation (1) is cast in atomic units and M represents the aggregate's mass. (Note how the numerator of the centrifugal term reduces to $\lambda(\lambda + 1)$ for $N = 2$, and how the dimensionality index $(3N - 5)/2$ is added to λ as it is to l in the dimensional scaling formulas of this volume.) The Coulomb interaction $(\alpha|C|\alpha')/R$ results from the harmonic expansion of all electrostatic interactions within the aggregate [5]. The dynamics of interest stems mainly from the elements of $(\alpha|C|\alpha')$ off diagonal in α. Spin interactions and Coriolis terms have been omitted in equation (1) for simplicity.

Shifting Dominance of Centrifugal and Coulomb Energies

The structural similarity of equation (1) to the H atom Hamiltonian implies a corresponding analogy of the dynamics of all atomic systems. Recall how the size and geometry of hydrogen states reflect the balance of their kinetic and Coulomb energies. The rotational component of the kinetic energy acts as a centrifugal potential and may vanish in the H atom. This component is instead boosted greatly in larger aggregates by the correlations among the $\{\omega\}$ variables embodied in the N-dependence of the centrifugal term of (1). The radial kinetic energy takes thus a subsidiary role.

The centrifugal term of (1) predominates over the Coulomb term for sufficiently small radii of inertia R. Under this condition each radial component $f_\alpha(R)$ of a wavefunction

$$\psi(R, \omega) = \sum_\alpha f_\alpha(R) Y_\alpha(\omega) \tag{2}$$

vanishes as $R \to 0$ as $R^{\lambda + (3N-4)/2}$, independently of the Coulomb fields. Within the domain of the Hamiltonian's variables we thus identify a *first range* from which the centrifugal barrier *excludes* significant phenomena. The boundaries of this range depend on the matrix $(\alpha|C|\alpha')$ of each system; they will be described below for a simple example.

A *second range* is envisaged next, at larger R values, within which the centrifugal and Coulomb energies are comparable. In this *tran-*

sition range the offdiagonal elements of $(\alpha|C|\alpha')$ interlink the radial components $f_\alpha(R)$ strongly thus generating effects that prove critical to the breakdown of separability. This transition from centrifugal to Coulomb dominance proceeds rather rapidly as R increases by half an order of magnitude at constant values of the other parameters. Only a limited subset of the radial functions $\{f_\alpha(R)\}$ may thus be coupled strongly at any one value of R.

The Schrödinger equation with the Hamiltonian (1) may be conveniently integrated through this transition range by a long-established phase-amplitude transformation [6], never applied in the past, leading to a system of first-order differential equations. This system, readily accessible to numerical solution, may also afford a physical picture of the evolving orthogonal eigenvectors of a reaction matrix $K_{\alpha\alpha'}(R)$. The influence of the Coulomb field $(\alpha|C|\alpha')/R$ accumulated over the range $0 \leq R' \leq R$ is embodied in this matrix. The initial application of this procedure to a simple prototype is being reported elsewhere [7], and is outlined below.

Dominance of the Coulomb over the centrifugal energy emerges at sufficiently large values of R. In this *third range*, the eigenvectors of the $(\alpha|C|\alpha')$ matrix would identify components of $\psi(R,\omega)$ that would *propagate independently* as R increases, if the if the centrifugal field were altogether negligible. The propagation would then be represented by hydrogenic functions for the field of a charge equal to the relevant eigenvalue of $(\alpha|C|\alpha')$. The residual influence of the centrifugal field will indeed distort these channels from independent propagation but should not impair their inherent separability.

The origin of the observed regularities in collisions and spectra appears thus to lie in the hydrogenlike structure of the Hamiltonian (1), where the manifold of interactions within an aggregate condense in a single matrix $(\alpha|C|\alpha')$. The separability that prevails at large R pertains to the diagonal form of this matrix, but is thwarted at shorter ranges where the balance of Coulomb and centrifugal energies shifts rapidly. The matrix $(\alpha|C|\alpha')$ has infinite dimension, in principle, but its relevant portion is finite and hence amenable to practical diagonalization, as we shall discuss next.

The Finite Subset of Relevant Channels

In the expansion (2) of a wavefunction, each of the terms $f_\alpha(R)Y_\alpha(\omega)$ may be viewed as identifying a *channel* of propagation of the radial function $f_\alpha(R)$ from $R = 0$ toward $R = \infty$. the set of quantum numbers α represents a specific partition of the available *grand angular momentum* λ into several components, a partition that attains full manifestation at the $R \to \infty$ limit where it also governs the partition of energy among fragments of the aggregate. The λ quantum number is, in essence, the sum of all other elements of the set α (much as the index l of a harmonic $Y_{lm}(\theta, \phi)$ is the sum of the number of nodes along the ϕ and θ coordinates). The set of channels with equal λ value is accordingly finite.

The λ value itself ranges to infinity. The subset $\{\lambda\}$ *relevant* at any given value of R is, however, finite as the quadratic rise of the centrifugal barrier with increasing λ places large values of λ in the *first range*, from which significant phenomena are excluded. The entire subset of α channels relevant at any given R is accordingly also *finite*.

This result has presumably a most important consequence for the eigenvectors of the matrix $(\alpha|C|\alpha')$. It has been noted recently for a class of finite atomic matrices that the eigenvectors corresponding to higher and lower subsets of eigenvalues localize sharply in the space of their indices [8]. So do their transforms in physical space. A prototype example of this phenomenon is seen in the stability of asymmetric rotor states spinning about their axes of either highest or lowest inertia. A more immediately relevant example will be introduced below.

The progressive expansion of the set of relevant channels as R increases also complements, indeed rectifies, the current treatments of multichannel atomic and molecular processes [9]. These treatments center on the excitation of one (or a very few) electrons, or other degrees of freedom, out of a ground state core. Wavefunctions representing such alternative excitations in different channels are represented semianalytically in the space surrounding the core radius r_0, but are interrelated through their respective boundary conditions at r_0. The number of relevant channels is chosen judiciously for each application; it would increase substantially upon the inclusion of inner-shell excitations. The present hyperspherical approach replaces this ad-hoc selection of channels with the systematic enlargement of the relevant channel set at increasing values of R. (It also replaces the usual opaque variational procedures that determine boundary values at a fixed r_0 by the semianalytic calculation of reaction matrices $K_{\alpha\alpha'}(R)$ at successive values of R, as indicated above.)

An Illustrative Example: Rydberg Diamagnetism

Key elements of our study of the Hamiltonian (1) reappear in the much simpler setting of a single electron subject to the combination of the Coulomb field of a unit positive charge at the origin and of a uniform magnetic field parallel to the z axis [1]. The Larmour precession energy generated by the field is represented by the diamagnetic potential $\omega_c^2 r^2 \sin^2 \theta/8$ in terms of the cyclotron frequency ω_c. The electron's Hamiltonian, analogous to (1), is

$$H = \frac{1}{2}\left[-\frac{d^2}{dr^2} + \frac{l(l+1)}{r^2}\right] - \frac{1}{r} + \frac{1}{8}\omega_c^2 r^2(l|\sin^2\theta|l') \qquad (3)$$

The functions of the Coulomb term of the Hamiltonian (1) are subdivided here among the last two terms of (3). The Coulomb term $-1/r$ becomes dominant as the centrifugal term declines, whereas the nondiagonal matrix elements $(l|\sin^2\theta|l')$ interlink the radial functions $f_l(r)$ of different Coulomb channels. The extremely small value of the coefficient ω_c^2 for laboratory fields prevents this coupling from competing with the centrifugal field except at large radial distances where the coefficient product $\omega_c^2 r^2$ becomes very significant.

The shifting dominance of centrifugal and Coulomb energies is thus substantially unaffected in equation (3) by the diamagnetic term. This dominance can accordingly be illustrated through the behavior of Coulomb wavefunctions which we cast in the Milne form [10]:

$$f_l(r) = \left[\frac{2}{\pi}\frac{d\phi_l}{dr}\right]^{1/2}\sin\phi_l(r) \qquad (4)$$

Figure (1) displays the growth of the phase functions $\phi_l(r)$ for a set of zero-energy Coulomb functions with even values of the orbital number l. Dominance of the centrifugal barrier at small values of r is manifested for each $l \neq 0$, by zero-growth of $\phi_l(r)$. The range of the (l,r) diagram in Figure (1) where each $\phi_l(r)$ remains flat corresponds thus to the *first range* introduced above.

Phase growth of $\phi_l(r)$ sets in as the centrifugal dominance declines and r approaches the value $l(l+1)/2$ at which the combined energy $l(l+1)/2r^2 - 1/r$ vanishes. The rate of growth increases further, peaking at $r = l(l+1)$ where the *second range* may be said to terminate. Thereafter each $\phi_l(r)$ keeps growing on a course nearly parallel, but slowly converging, to that of the zero-l phase $\phi_0(r)$. In this *third range* the Coulomb functions with different l propagate with comparable wavelengths.

Consider now the influence of the diamagnetic energy term in equation (3). This term commutes with the Coulomb energy thus becoming effective, regardless of its small value, as soon as the centrifugal dominance subsides. Its influence has been evaluated through calculation of the reaction matrix $K_{ll'}(r)$ for $r \leq 200$ a.u. and for all l components in Figure 2 which emerge from centrifugal dominance in this range [7]. This procedure could extend to larger r values but its results would not reflect the separation of variables to be afforded locally, i.e., within the *third range* to the right of the boundary in Figure (1).

The calculation casts eigenfunctions of the Hamiltonian (3) as superpositions of Coulomb field functions with alternative phases,

$$\sum_l \left[a_l(r) \sin \phi_l(r) + b_l(r) \cos \phi_l(r) \right] \left[\frac{2}{\pi} \frac{d\phi_l}{dr} \right]^{1/2} Y_{lm}(\theta, \phi) \qquad (5)$$

The Schrödinger equation reduces then to a system of first order equations in the coefficients $\{a_l(r), b_l(r)\}$ whose evolution can be displayed as functions of r. This set of nonzero coefficients expands, of course, as the centrifugal barrier subsides for successive l values.

Solutions $\{a_l(r), b_l(r)\}$ for $r \leq 200$ a.u. have served to construct the eigenvectors and eigenvalues of $K_{ll'}(r)$ in the transition range. Figure (1) indicates that several l-components of (5) lie in the *third range* where separability should operate. It should thus be possible to recast the solutions of (5) into products of two functions that propagate along and across wavefronts $\phi(r,l) = constant$, respectively [7].

This result has not yet been achieved in the desired form, but the localization of eigenfunctions at the lowest and highest values of $\sin^2\theta$ has been demonstrated in a different setting [11].

Construction of eigenfunctions in the *third range* and of their *connection* with the $K_{ll'}(r)$ matrix in the *transition range* will be required to provide a transparent interpretation of the observed Zeeman effect on Rydberg spectra.

Discussion

The analysis of the hyperspherical Hamiltonian (1) conducted above has brought out several simplifying features common to all atomic and molecular problems:

a) Energy eigenfunctions can be conveniently represented at sufficiently low R values in terms of eigenfunctions of the squared grand angular momentum, which propagate independently in this range.

b) At sufficiently large R, the eigenvectors of the Coulomb matrix $(\alpha|C|\alpha')$ should perform the same function.

c) Nonseparability stems mainly from the transition between the ranges of R where a) and b) hold. (Additional contributions to nonseparability, stemming from boundary conditions at $R \to \infty$, are treated in Reference [3].)

d) The transition range occurs over a modest range of R centered about a value $\sim [\lambda + (3N - 5)/2]^2$.

e) The evolution of eigenfunctions through the transition range can be traced numerically through the eigenvectors of the reaction matrix $K_{\alpha\alpha'}(R)$.

f) The eigenvectors of $(\alpha|C|\alpha')$, which manifest themselves in high excitation spectra and in collisions, are thus coupled by parameters of the transition range whose identity and evaluation remain undeveloped.

Elaboration and exploration of these features will require extended effort. Its results might eventually replace current technologies, which are rooted in the independent particle approximations appropriate to simple spectra.

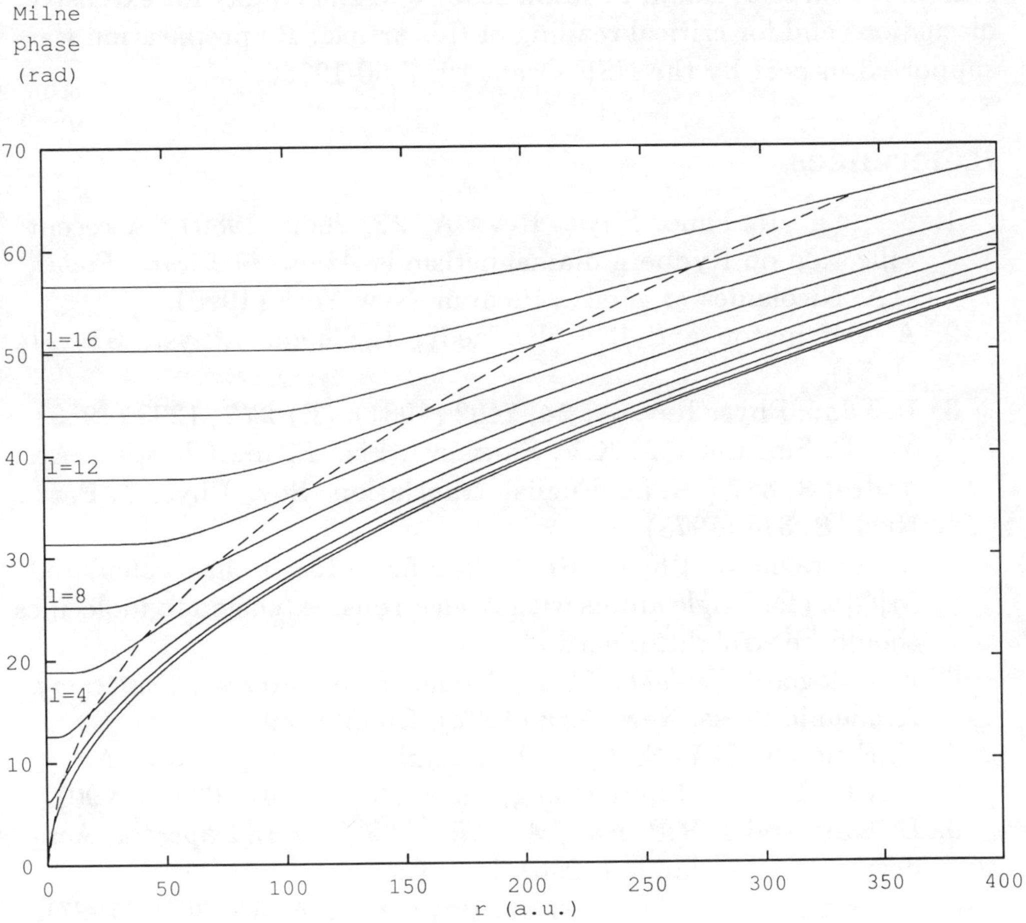

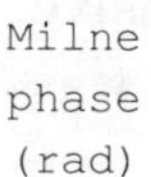

Figure 1: Growth of the Milne phase $\phi_l(r)$ of H atom wavefunctions vs. increasing radial distances, at the ionization threshold energy. Note the different behavior of each curve with $l \neq 0$ in the three ranges of r: a) the flat portion; b) the rising, upward concave, portion; c) the rising, but upward convex, portion to the right of the line $r = l(l+1)$. (Courtesy, E.Y. Sidky)

Acknowledgement

I am indebted to J. Bohn, F. Robicheaux and E.Y. Sidky for extensive discussions and for critical reading of this article. Its preparation was supported in part by the NSF Grant PHY 90-19966.

References

1. See, e.g. U. Fano, Phys. Rev. A, **22**, 2660 (1980). A recent collection on Rydberg diamagnetism is *Atoms in Strong Fields*, C.A. Nicolaides *et al* eds., Plenum, New York (1990).
2. A. Carrington and R.A. Kennedy, J. Chem. Phys. **81**, 91 (1984).
3. U. Fano, Phys. Rev. A, **24**, 2402 (1981); (E) **277**, 1208 (1983).
4. Yu. F. Smirnov and K.V. Shitikova, Fiz. Elem. Chastits. At. Yadra, **8**, 847 (1977). English translation: Sov. Phys. J. Part. Nucl. **8**, 344 (1978).
5. M. Cavagnero, Phys. Rev. **36**, 523 (1987), has calculated $(\alpha|C|\alpha')$ for single atoms with N electrons; extension to molecules should be straightforward.
6. F. Calogero, *Variable Phase Approach to Potential Scattering*, Academic Press, New York (1967), Chapter 19.
7. U. Fano and E.Y. Sidky, to be submitted to Phys. Rev. A.
8. A.R.P. Rau and Lijun Zhang, Phys. Rev. A, **42**, 6342 (1990).
9. U. Fano and A.R.P. Rau, *Atomic Collisions and Spectra*, Academic Press, Orlando (1986), Part C.
10. See, e.g., F. Robicheaux *et al.*, Phys. Rev. A, **35**, 3619 (1987).
11. U. Fano, F. Robicheaux and A.R.P. Rau, Phys. Rev. A, **37**, 3655 (1988).

Chapter 12

PSEUDOMOLECULAR ELECTRON CORRELATION IN ATOMS

12.1 The Three-Body Coulomb Problem in Molecular Coordinates

Jan-Michael Rost and John S. Briggs
Fakultät für Physik
Albert-Ludwigs-Universität
Hermann-Herder-Str. 3
D-7800 Freiburg, Germany

Abstract

The D-dimensional scaling properties of the three-body Coulomb problem in atomic and molecular physics have been discussed in part 1 of this volume. Here we examine some new features of the adiabatic molecular problem illustrating the importance of motion near the saddle point of the two-centre potential. Properties of resonant states of the atomic three-body system are shown to be a direct consequence of

D. R. Herschbach et al. (eds.), Dimensional Scaling in Chemical Physics, 471–484.
© 1993 *Kluwer Academic Publishers. Printed in the Netherlands.*

*this saddle-point motion. Its dimensional scaling properties are dis-
cussed briefly.*

Introduction

The dimensional scaling of the three-body Coulomb problem has been
much studied not only for its fundamental importance in atomic and
molecular physics but also because of interdimensional degeneracies
occurring in both the atomic (e.g. helium atom [1-3] and molecular (H_2^+ molecule ion [4]) cases. In the latter case the motion of an
electron in the field of two fixed nuclei separated by the distance R
has been considered. Here we will adopt a unified adiabatic molec-
ular approach by taking the distance $R = |\mathbf{r}_1 - \mathbf{r}_2|$ which joins the
two particles of *like* charge as an *adiabatic* parameter. The motion
of the third particle in the coordinate $\mathbf{r} = (\mathbf{r}_1 + \mathbf{r}_2)/2$ relative to the
centre of mass of the particles of like charge is then solved for fixed
R. For a two-electron atom R represents the interelectronic axis. It
has been amply demonstrated [5,6] that this "molecular" adiabatic
approach can not only explain the nodal characteristics of doubly-
excited atomic resonant states [7] but also give fairly accurate energies
[8]. With R fixed, in either the atomic or molecular case, we discuss
first some new aspects of the correlation diagram (CD) of molecular
 orbital (MO) energies. We show that the three MO quantum numbers arising from the separability of the MO problem contain the five
approximate quantum numbers proposed by Herrick in his classification of doubly-excited states [9]. This equivalence arises from the
motion around the saddle of the two-centre potential of the adiabatic
problem and the nature of this motion is shown to lead to propensity
rules for transitions between resonant states. Finally the dimensional
dependence of the saddle motion is discussed briefly.

MO symmetries and the correlation diagram

In a series of papers (for a review see Herrick [9]) a scheme of classification for doubly-excited atomic resonances was proposed. In this
scheme states can be assigned some (or all) of five quantum numbers.
Three of these, K, T and A arose from a reduction of the $O_4 \times O_4$ prod-

uct group of two non-interacting electrons and subsequently Lin [10] was able to correlate hyperspherical potential curves with these three quantum numbers. In alternative group theory reductions, Kellman and Herrick [11] identified sequences of levels grouped according to rovibrational structure. This led to many papers suggesting approximate rotation-vibration Hamiltonians as models for a two-electron atom.

All of this work suggested, but did not identify, some approximate separability of the two-electron problem in terms of set of collective variables. This separability and the corresponding collective co-ordinates have since been identified by Feagin and Briggs [5]. The collective co-ordinates are those of the relative vector $\mathbf{R}$ of the two electrons and the vector $\mathbf{r}$ of the electronic centre-of-mass (ECM) with respect of the nucleus. (or all) of five quantum numbers. Three of these, K, T and A arose from a reduction of the $O_4 \times O_4$ product group of two non-interacting electrons and subsequently Lin [10] was able to correlate hyperspherical potential curves with these three quantum numbers. In alternative group theory reductions, Kellman and Herrick [11] identified sequences of levels grouped according to rovibrational structure. This led to many papers suggesting approximate rotation-vibration Hamiltonians as models for a two-electron atom.

As is well-documented [12] the two-centre problem is separable in the λ, μ, φ co-ordinates and correspondingly molecular orbital states have quantum numbers n_λ, n_μ, m. These MO assignments are in one-to-one correspondence with parabolic separated atom (SA) $(n_1 n_2 m)$ and spherical united atom (UA) (nlm) quantum numbers as $R \to \infty$ and $R \to 0$ respectively (A full description of the coordinates is provided by the contribution of S.M.Sung to this volume). In the SA limit one-electron parabolic states are produced and the additional designation $A = (-1)^{n_\mu}$ is necessary to specify the linear combination (LCAO), $\Phi_1 + A\Phi_2$, which forms an MO with correct nodal structure. One notes in the correlation diagram (fig. 1) that some MO are depressed smoothly in energy as R decreases from infinity, whereas others are first depressed and then rise in energy so that the MO energy possesses a minimum. This is the phenomenon of promotion (increase of principal quantum number between SA and UA) known to be of fundamental importance for the formation of molecules and in ion-atom collisions. This feature also plays a major role in the classification of MO for the two-electron atomic problem as described below.

The evenness or oddness of n_μ expressed by $A = (-1)^{n_\mu}$ is fundamental to the behaviour of MO energies, although this does not appear to have been pointed out hitherto. For this reason we present, in Fig. 1a and b the separate correlation diagrams for $A = +1$ and $A = -1$ respectively. As discussed extensively in [6], $A = -1$ states have a node on the saddle ($r_1 \approx r_2$) of the two-centre potential, whereas $A = +1$ MO have an anti-node there. Accordingly $A = +1$ MO have more "molecular" character and this is seen qualitatively in Fig. 1 by the fact that the bundle of $A = +1$ MO emanating from a given SA hydrogenic N level are more strongly split in energy at finite R than their $A = -1$ counterparts. The latter are always promoted i.e. $n > N$ while only $A = +1$ MO can correlate SA principal shells N with UA principal shells n = N.

The qualitative appearance of the CD, that of lower levels of a given bundle crossing the remaining members (for $A = +1$ several higher bundles may be crossed) as they are promoted is readily understood [12] and is crucial for the formation both of molecular states and resonant electronic states of atoms. For large R, before molecule formation takes place, the shift of atomic levels is due to the Stark effect. Levels with the minimum value of the "electric" quantum number $k = n_1 - n_2$ for a given m are depressed most. These MO are those in which the two particles of like charge are roughly at 180^0 with the particle of unlike charge between them. Hence the two electric dipoles are maximally oppositely directed to give minimum energy. By contrast states with maximum k have particles of like charge maximally on the same side of the third particle, leading to maximum energy configuration of interacting dipoles. The promotion condition is

$$n - N = [\tfrac{1}{2}(l - m + 1)] \equiv [\tfrac{1}{2}n_\mu] \equiv \begin{cases} n_2, & A = +1 & \text{(1a)} \\ n_2 + 1, & A = -1 & \text{(1b)} \end{cases}$$

where $[x]$ is the integer value of x. From (1) it is readily seen that MO maximally depressed by the Stark effect at large R have n_2 maximum and $n_1 = 0$ to give maximum promotion at small R. It is also clear from (1) that all $A = -1$ MO are promoted and additionally

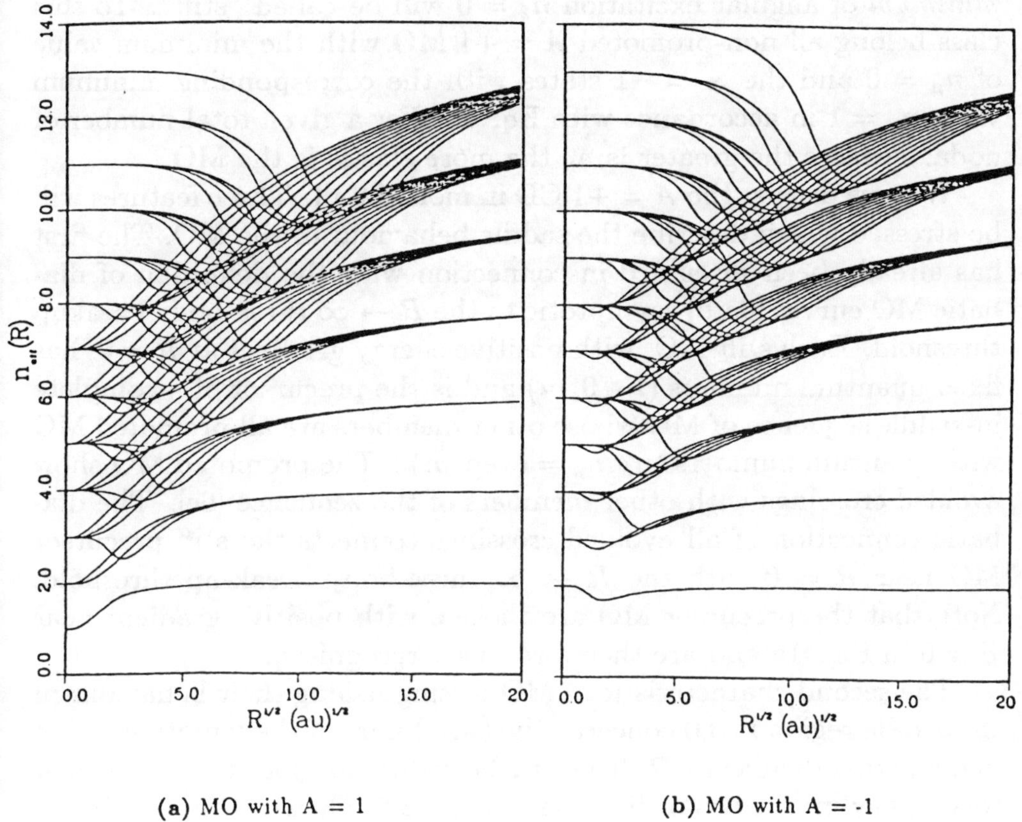

(a) MO with A = 1 (b) MO with A = -1

Figure 1: The correlation diagram of the two-centre Coulomb problem for MO with $m \leq 2$ up to the manifold $N = 7$ of the separated atom. The energy is given in terms of the effective quantum number $n_{\mathrm{eff}} = [2/\mathcal{E}_{im}(R^{1/2})]^{1/2}$.

that promotion is directly proportional to the number of "angular" nodal surfaces n_μ. One also notes that promotion involves a minimum change in binding energy between UA and SA limits but a maximal change in the character of the MO. Such MO will be called "elastic" and are uniquely described by $n_2 \neq 0$. By contrast all MO with a *minimum* of angular excitation $n_2 = 0$ will be called "stiff". To this class belong all non-promoted $A = +1$ MO with the minimum value of $n_\mu = 0$ and the $A = -1$ states with the corresponding minimum value $n_\mu = 1$ in accordance with Eq. 1b. For a given total number of nodal surfaces the greater is n_μ the more elastic is the MO.

We now discuss the $A = +1$ CD in more detail. Three features will be stressed which describe the saddle behaviour of the MO. The first has already been discussed in connection with the definition of diabatic MO curves [6a,b] asymptotic to the $R \to \infty$ three-body breakup threshold. Each stiff MO, with positive energy gradient at $R = 0$, has fixed quantum numbers $(n_\lambda, 0, m)$ and is the precursor of a complete in-saddle sequence of MO whose other members are all promoted MO with quantum numbers $(n_\lambda, n_\mu = \text{even}, m)$. The promoted MO show avoided crossings with other members of the sequence [6c]. The diabatic connection of all avoided crossings connects the stiff precursor MO near $R = 0$ with the $R \to \infty$ three-body break-up threshold. Note that the precursor MO are those e with positive gradient near $R = 0$ in Fig. 1a and are therefore easily recognised.

The second characteristic of MO distinguishing their behaviour in the saddle region $\mathbf{r} \approx 0$ concerns the (λ, φ) part of the $\mathbf{r}$ motion in the plane perpendicular to $\hat{R}$. To describe saddle motion it is convenient to adopt cylindrical co-ordinates $\mathbf{r} = (\rho, z, \varphi)$. Then if we approximate in the region of the saddle point $\lambda \approx 1, \mu \approx 0$ we have

$$\rho = \tfrac{R}{2}[(\lambda^2 - 1)(1 - \mu^2)]^{\frac{1}{2}} \quad \to R((\lambda - 1)/2)^{\frac{1}{2}} \tag{2a}$$

$$z = \tfrac{R}{2}\lambda\mu \quad \to \tfrac{R}{2}\mu \tag{2b}$$

$$\varphi = \varphi. \tag{2c}$$

In contrast to the motion parallel to $\hat{R}$ (described by μ or z), the motion in the plane perpendicular to the $\hat{R}$ axis (described by (λ, φ) or (ρ, φ)) is bound near $\mathbf{r} = 0$. As may be readily shown the motion is that of an isotropic 2-dimensional harmonic oscillator with Hamiltonian

$$H_\perp = -\frac{1}{2}\left(\frac{1}{\rho}\frac{d}{d\rho}\rho\frac{d}{d\rho} + \frac{1}{\rho^2}\frac{d^2}{d\varphi^2}\right) + \frac{1}{2}\omega^2\rho^2 \tag{3}$$

where $\omega = 4R^{-3/2}$. The energies are given by $\hbar\omega(v_2 + 1)$ with

$$v_2 = 2n_\rho + m. \tag{4}$$

According to (2) the saddle nodes (n_ρ, m) go over to MO nodes (n_λ, m) and hence the degeneracy of the two-dimensional isotropic oscillator is expressed as the MO saddle degeneracy condition

$$v_2 = 2n_\lambda + m \tag{5}$$

for various combinations of n_λ and m.

MO interpretation of Herricks' quantum numbers

As discussed [5], the total wavefunction limits the states of total LSt symmetry that can be built upon a given MO internal state with fixed $(n_\lambda n_\mu m)$ quantum numbers and $t = n_\mu + m = l$. For example all σ_g MO ($m = 0, t = 0$) give total states with $(-1)^{S+L} = +1$. Therefore the series $^1S^e, ^3P^o, ^1D^e, \ldots$ is possible for increasing total angular momentum L. Similarly for σ_u MO ($m = 0, t = 1$) we have the complementary spin series $^3S^e, ^1P^o, ^3D^e, \ldots$. For MO with $m \neq 0$ the situation is more complicated. Each member of a series consists of a pair of states with the same angular momentum $L > m$ but with opposite parity and spin. For example, every π_g MO ($m = 1, t = 0$) gives series $(^1P^e, ^3P^o), (^1D^e, ^1D^o) \ldots$. Similarly each π_u MO gives rise to series of pairs of states $(^3P^e, ^1P^o), (^3D^e, ^3D^o) \ldots$. Hence one sees already how the MO picture automatically leads to rotor series whose existence was postulated empirically by Herrick.

One aspect of the MO ordering of two-electron states should be mentioned explicitly. On the basis of ordering states according to group-theoretical considerations Herrick noticed that in zeroth-order, pairs of states e.g. $(^1P^e, ^3P^o)$ or $(^1D^e, ^1D^o)$ are degenerate and he explained this in analogy to the familiar Λ-doubling found in diatomic molecules (Λ corresponds to our m quantum number). Since our MO treatment of 2-electron atoms is identical to that of diatomic molecules, the occurrence of degenerate pairs ($^1P^e, ^3P^o$) or ($^1D^e, ^1D^o$) is *exactly* the Λ-doubling phenomenon. As in molecules the degeneracy is lifted when rotational coupling to other MO of different body-fixed m is taken into account.

In a single-channel adiabatic approximation each state ($LS\pi$) is associated with a potential curve in which vibrational states corresponding to increasing degrees of excitation in R may be defined. However, the vibrational structure of two-electron states introduced by Herrick on the basis of a triatomic molecular model involves three normal modes and only the "breathing" v_1, longitudinal mode (symmetric stretch) corresponds to R. The other two modes appear when the exact λ, μ internal motion is approximated in the region of the saddle point where $\lambda \sim 1, \mu \sim 0$. The corresponding asymmetric stretch (v_3, MO coordinate μ) and bending (v_2, MO coordinate λ) modes are illustrated in fig. 6 of ref. 6a.

Herrick introduced five quantum numbers K, T, A (called by him ν) on the basis of SO_4 group theory and v_2 and d on the basis of the triatomic molecular model of two-electron atoms. All of these five quantum numbers are contained in the three quantum numbers $(n_\lambda n_\mu m)$ unique to each MO.

The MO explanation of the origin of Herricks' quantum numbers is simple. It also explains a feature that Herrick repeatedly found puzzling, namely, why are quantum numbers based on SA wavefunctions and symmetries ($Z \to 0$ limit) valid for classifying resonant states involving intermediate values of R The answer lies in the fact that the MO quantum numbers $n_\lambda n_\mu m$ are valid for all R. In particular for $R \to \infty$ they can be expressed uniquely in terms of the parabolic quantum numbers ($n_1\ n_2\ m$) or ($N\ k\ m$) of the SA state to which each MO is asymptotic. The "extra" quantum number $A = (-1)^{n_\mu}$ is necessary to specify the correlation (this is the ν quantum number of Herrick). It also specifies the LCAO form of the MO in terms of SA states in which electron 1 or electron 2 is bound i.e.

$$|n_\lambda n_\mu m > \approx |n_1 n_2 m >_1 + A|n_1 n_2 m >_2 \quad \text{for } R \to \infty \qquad (6)$$

Herricks' other two labels, d and I ($= L - m$), were derived on the basis of an ad hoc model of a two-electron atom as a triatomic molecule. In particular he was able to organize states into I supermultiplets of states where those corresponding to a given value of v_2 are almost degenerate. We have established that the v_2 quantum number, corresponding to a bending vibration [11] is nothing more than the quantum number $2n_\lambda + m$ describing degenerate saddle-point bending vibrations. Then the degeneracies for fixed v_2 in Herricks' supermultiplet classification have their origin in the underlying degeneracy of MO for fixed $2n_\lambda + m$. In addition Herricks' classification according to $I = L - T$ with $L - T \geq 0$ arises naturally since $T = m$ and since $\mathbf{L} \cdot \hat{R} = m$ then $L - m \geq 0$. In the MO picture $I = L - m$ describes the difference of total angular momentum L and MO angular momentum m. Hence, in an $I = 0$ multiplet, for $m = 2$ the saddle degeneracy first involves D states, for example the ${}^1D^e, {}^1S^e$ pair built on the saddle-degenerate MO $(5g\delta, 4g\sigma)$ with $v_2 = 2$.

There remains the quantum number d, particularly favoured by Herrick. Indeed he was also able to give an interpretation, if not a justification of this number group-theoretically. It will emerge that classification according to d, for fixed N is *exactly* the same as classification according to n_λ denoting the number of elliptical MO nodes. Already in section 2 we have noted that n_λ denotes the "stiffness" of MO and that there is little mixing of MO with different n_λ. The quantum number d arises out of the reduction $SO_4 \times SO_4 = SU_2 \times SU_2 \times SU_2 \times SU_2$ according to irreducible representations in which the operator $\mathbf{J}_>^2$ is invariant. Here $\mathbf{J}_>^2$ is the angular momentum operator with the eigenvalue $d = \max(J_+, J_-)$ where $\frac{1}{2}(\mathbf{L} + \mathbf{B})^2 = J_+(J_+ + 1)$ and $\frac{1}{2}(\mathbf{L} - \mathbf{B})^2 = J_-(J_- - 1)$. In this representation $d = \frac{1}{2}(K + N + T - 1)$. The MO interpretation follows from $K = n_2 - n_1$, $T = m$, to give

$$d = n_2 + m. \tag{7}$$

The real meaning for intra-shell resonances however relies on the fact that for fixed N, d gives the number of λ nodes i.e. $n_\lambda = n_1 = N - n_2 - m - 1 = N - d - 1$.

To summarize we have established that all of Herricks' *five* quantum numbers, arrived at by various physical arguments supported by the SO_4 group classification express different aspects of the MO character through the *three* unique MO numbers $n_\lambda n_\mu m$. The connection is

$$\begin{aligned}
T &= m \\
A &= (-1)^{n_\mu} \\
K &= n_2 - n_1 = [n_\mu/2] - n_\lambda \\
v_2 &= 2n_1 + m = 2n_\lambda + m \\
d &= n_2 + m = [n_\mu/2] + m.
\end{aligned} \qquad (8)$$

Propensity rules

Based on a consideration of the underlying MO nodal structure of two-electron resonant states which has been confirmed recently [7], propensity rules for radiative and non-radiative decay of resonant states (i.e. for their widths) have been given [6b]. Here they will be briefly re-iterated to provide more details of their derivation.

The propensity rules are based on the observation that, to the extent that states of given total symmetry are built upon a single MO, transitions between these states are governed by *selection rules* for transitions between the corresponding MO. In this way the propensity rules arise naturally out of the theory and are not dependent upon empirical deductions. Their simplicity is partly due to the structure of the operators responsible for non-radiative (auto-ionization) a nd radiative decay.

The process of autoionization is the direct analogue of dissociation $H^+ + H(n_1 n_2 m)$ in H_2^+. It proceeds in the MO picture via the non-adiabatic couplings involving the operator ∇_R. These terms separate into the radial and rotational coupling types well-known in molecular and ion-atom collision physics. Non-radiative transitions between MO are induced by off-diagonal matrix elements and are governed by the following selection rules

$$\Delta t = 0 \qquad (9a)$$

$$\text{Radial} \quad < \Phi_f | \partial/\partial R | \Phi_i >; \quad \Delta m = 0 \qquad (9b)$$

$$\text{Rotational} \quad < \Phi_f | l_\pm | \Phi_i > /R^2; \quad \Delta m = \pm 1. \qquad (9c)$$

We note that the combination of (9a) and (9b) implies $\Delta A = 0$ for radial coupling. Similarly (9a) and (9c) imply $\Delta A \neq 0$ for rotational coupling.

These selection rules alone are not sufficiently restrictive to classify states according to the magnitude of their widths. For that a consideration of the facility with which internal saddle $n_\lambda n_\mu$ excitation can couple to vibrational R motion is necessary. Decay occurs preferentially [6b] by radial coupling within saddle sequences corresponding to

$$\Delta n_\mu = 2, \quad \Delta n_\lambda = 0, \quad \Delta m = 0. \tag{10}$$

Within the saddle sequence Eq. 9 implies that a state of the N-manifold decays preferentially to a state of the $(N-1)$-manifold which then dissociates to $e^- + H(N-1)$ or $e^- + He^+(N-1)$ if this is energetically possible. This is also consistent with the Hellman-Feynman theorem for radial coupling

$$< \Phi_f|\partial/\partial R|\Phi_i > = (\mathcal{E}_f - \mathcal{E}_i)^{-1} < \Phi_f|\partial h/\partial R|\Phi_i > \tag{11}$$

showing that decay occurs preferentially to states closest in energy and is particularly effective between states which exhibit an avoided crossing. This also explains why decay rates of in-saddle ($A = +1$) sequences of states are much faster than those of side-saddle ($A = -1$) states. The radial coupling is strongest at the avoided crossings between N and $(N-1)$ manifolds and since the $A = +1$ series have much sharper avoided crossings (see fig.1) the radial couplings are correspondingly greater according to Eq. (11).

All saddle sequences end in a precursor MO having $n_\mu = 0$ or $n_\mu = 1$. For states built on these MO decay is forbidden according to the propensity rule $\Delta n_\mu = 2$. These states must decay by the weaker rotational transition

$$\Delta n_\mu = 1, \quad \Delta n_\lambda = 0, \quad \Delta m = \pm - 1 \tag{12}$$

within a UA n-manifold. These transitions occur with $\Delta A \neq 0$ and therefore imply a change from an anti-node at the saddle, or vice versa. Finally, the least probable transitions are those with

$$\Delta n_\lambda \neq 0 \tag{13}$$

occurring *between* different saddle sequences. That these transitions do not lead to large dissociation widths is plausible when one remembers that excitation in the λ mode corresponds to bending vibrations where the ECM executes stable vibrations perpendicular to $\hat{R}$. These vibrations couple weakly to symmetric stretch R vibrations leading to decay. By contrast $\Delta n_\mu = 2$ transitions correspond to reduction in the asymmetric stretch ECM vibration along $\hat{R}$ which couples directly to the symmetric stretch R vibration. Application of the non-radiative propensity rules are given in ref. 6b.

The key observation leading to propensity rules for radiative transitions is that the dipole operator $\mathbf{D} = \mathbf{r}_1 + \mathbf{r}_2 = 2\mathbf{r}$ i.e. the dipole transition involves only the ECM motion. This gives the selection rule $\Delta t = \pm 1$ directly. The linearized ECM dynamics around the saddle point as the motion of an isotropic harmonic oscillator (3) has already explained the quasidegeneracies in the two-electron spectrum. It can also be shown how a linearization of the ECM motion around the saddle leads to the additional selection rule

$$\Delta v_2 = 0, \pm 1 \tag{14}$$

which can be expressed in terms of the change in MO quantum numbers by the identification $v_2 = 2n_\lambda + m$. This rule incorporates different versions of the radiative propensity rule given by ourselves [6b], by Sadeghpour and Greene [13] and by Gou et al.[14].

Dimensional scaling of saddle-point motion

The $D-$dimensional scaling of the saddle point motion is particularly simple since the two-centre Hamiltonian in cylindrical coordinates has already been discussed by Frantz and Herschbach [4]. The two-centre Hamiltonian for H_2^+ in D dimensions is

$$H_D = \quad -\frac{1}{2}\left[\frac{1}{\rho^{D-2}}\frac{\partial}{\partial\rho}\rho^{D-2}\frac{\partial}{\partial\rho} + \frac{\partial^2}{\partial z^2} - \frac{\mathbf{L}_{D-2}^2}{\rho^2}\right]$$
$$-[\rho^2 + (z - \frac{R}{2})^2]^{-1/2} - [\rho^2 + (z + \frac{R}{2})^2]^{-1/2} \tag{15}$$

Expanding the potential terms about the saddle point $(\rho/R) \to 0$, $(z/R) \to 0$ and keeping only leading terms gives a separable Hamiltonian in which the part perpendicular to $\hat{z}$ can be written

$$H_\perp = -\frac{1}{2}\left[\frac{1}{\rho^{D-2}}\frac{\partial}{\partial\rho}\rho^{D-2}\frac{\partial}{\partial\rho} + \frac{\partial^2}{\partial z^2} - \frac{\mathbf{L}_{D-2}^2}{\rho^2}\right] + \frac{1}{2}\omega^2\rho^2 \tag{16}$$

where, as in (3), $\omega = 4R^{-3/2}$. In (15) and (16), $\mathbf{L}^2_{D-2}$ is the squared orbital angular momentum operator associated with the $(D-1)$-dimensional spherical subspace projected on $\hat{R}$. This operator has eigenvalues
$|m|(|m| + D - 3)$. For $D = 3$, $\mathbf{L}^2_{D-2} = -\partial^2/\partial\phi^2$ with eigenvalues m^2 and one sees that (16) reduces then to (3).

As discussed by Frantz and Herschbach it is more convenient to consider not the eigenfunctions ψ of (16) but the functions $\Psi = \rho^{(D-1)/2}\psi$ whose modulus squared gives the probability density. The functions Ψ have a radial part which are eigenfunctions of the Hamiltonian

$$H'_\perp = -\frac{1}{2}\left[\frac{\partial^2}{\partial\rho^2} - \frac{\lambda_D(\lambda_D + 1)}{\rho^2}\right] + \frac{1}{2}\omega^2\rho^2 \tag{17}$$

where $\lambda_D = m + \frac{1}{2}(D - 4)$. The $(D-1)$-dimensional Hamiltonian (17) is identical in form to that of the three-dimensional oscillator with angular momentum $\ell = \lambda_D$. This Hamiltonian has the eigenvalues $(N + 3/2)\hbar\omega$ where $N = 2n_\rho + \ell$. Therefore the saddle motion, represented by (17), has eigenvalues $(N + 3/2)\hbar\omega$ with

$$N = 2n_\rho + \lambda_D = 2n_\rho + m + \frac{1}{2}(D - 4) \tag{18}$$

This is more generally written as saddle eigenvalues $[v_{D-1} + (D - 1)/2]\hbar\omega$ with

$$v_{D-1} = 2n_\rho + m \tag{19}$$

exactly as in (4) and independent of dimension. Thus we see that for given dimension the saddle degeneracies are preserved. For different dimensions however (18) shows not only the usual degeneracy, that increasing D by 2 is equivalent to increasing m by 1, but a new dimensional degeneracy that increasing D by four is equivalent to increasing $n_\rho(\equiv n_\lambda)$ by one.

Acknowledgement

This work was supported by the DFG within the Sonderforschungsbereich 276 at the University of Freiburg (Germany). One us (JMR) would like to thank D. Herschbach and his group for a pleasant stay in spring 1991 at Harvard University in Cambridge.

References

1. D. E. Herrick, F. H. Stillinger, *Phys.Rev.* **A11**, 42 (1975)
2. D. Z. Goodson, D. K. Watson, J. G. Loeser, D. R. Herschbach, *Phys.Rev.* **A44**, 97 (1991)
3. D. R. Herschbach, *J.Chem.Phys.* **84**, 838 (1986)
4. D.D. Frantz, D. R. Herschbach, *J.Chem.Phys.* **92**, 6668 (1990)
5. (a) J. M. Feagin, J. S. Briggs, *Phys.Rev.Lett.* **57**, 984 (1986)
   (b) J. M. Feagin, J. S. Briggs, *Phys.Rev.* **A37**, 4599 (1988)
6. (a) J. M. Rost, J. S. Briggs, *J.Phys.* **B22**, 3587 (1989)
   (b) J. M. Rost, J. S. Briggs, *J.Phys.* **B23**, L339 (1990)
   (c) J. M. Rost, J. S. Briggs, *Chem.Phys.Lett.* **177**, 321 (1991)
7. (a) J. M. Rost, R. Gersbacher, K. Richter, J. S . Briggs, D. Wintgen *J.Phys.B* **24**, 2455 (1991)
   (b) J. M. Rost, J. S. Briggs, J. M. Feagin *Phys.Rev.Lett.* **66** , 1642 (1991)
8. J. M. Rost, D. Wintgen, to be submitted (1991)
9. D. E. Herrick *Adv.Chem.Phys.* **52**, 1 (1983)
10. C. D. Lin *Phys.Rev.Lett.* **51**, 1348 (1983)
11. M. E. Kellman, D. R. Herrick *Phys.Rev.* **A22**, 1536 (1980)
12. J. D. Power *Proc.Roy.Soc.* **A 274**, 663 (1973)
13. H. R. Sadeghpour, C. H. Greene *Phys.Rev.Lett.* **65**, 313 (1990)
14. B. C. Gou, Z. Chen, C. D. Lin *Phys.Rev.* **A43** , 3260 (1991)

12.2 Valence Electrons in Atoms: Collective or Independent-Particle-Like?

R. Stephen Berry, Sandra C. Ceraulo and Joseph Batka
Department of Chemistry and The James Franck Institute
The University of Chicago
Chicago, Illinois 60637

Abstract

By comparing experimental or accurate theoretical results with others based on approximate models, it is possible to determine which among those models offers the best approximate constants of the motion and quantum numbers to describe particular states. This approach is used to evaluate and compare the extent of validity of independent-particle, Hartree-Fock and collective, molecule-like descriptions of atoms with two valence electrons. The comparisons are made on the basis of overlaps, oscillator strengths, momentum correlation and quadrupole moments. The criterion for each evaluation is the extent of agreement with results obtained from well-converged Sturmian Configuration Interaction wave functions.

Introduction

The question addressed in this paper is, in its most general terms: " To what extent is it appropriate to describe a few-body system in terms of one quantization scheme or another?" Slightly more precisely, we ask " Which simple model, among the choices we can invent, is the best starting approximation for any designated state of a given system?" As our vehicle to study this problem, we use the system of the two valence electrons in the helium and alkaline earth atoms. The models we consider and compare are the Hartree-Fock, independent-particle model and the collective model like that for a linear triatomic molecule.

The history of its relative success is the strongest basis for considering the independent-particle model in which each electron moves in

485

D. R. Herschbach et al. (eds.), Dimensional Scaling in Chemical Physics, 485–498.

the mean field of all the others in the atom. The stimulus for considering a collective, linear triatomic model was the success of a model very
like that [1-4] in describing intrashell states (i.e. same principal quantum numbers, $n_1 = n_2$) of doubly-excited helium, He**. Justification
for pursuing the idea further came from a) examinations of the spatial correlations of the electrons in He** and of the valence electrons
of the alkaline earths; b) overlaps of collective-model wave functions
with well-converged wave functions and c) comparisons of intensities
calculated with model wave functions and with well converged wave
functions.

Here we present a preliminary report of a more extensive set of
comparisons. Some of these, notably overlaps and oscillator strengths,
put the comparisons on a level playing field, so to say, by juxta-
posing collective and independent-particle results which are based on
the same model potentials. Some of these, including the oscillator
strengths, quadrupole moments and momentum correlations, are observables which have at best been measured in a limited number of
cases (oscillator strengths particularly) and in most cases remain un-
measured. However we do in all cases have available the results based
on well-converged wave functions; in the cases in which these can be
compared with experimental counterparts, the results of the comparisons are quite reassuring. We hope that the results presented here
will help stimulate experimental studies of the alkaline earth atoms.

We shall find that one cannot pick one zero-order model as " the
best" . Some properties are better approximated by one model, some
by the other. Some states are described well by both models; some,
by neither. Anticipating our conclusions, we would say, on balance,
that the collective, molecular model is probably a better choice for
describing observables. However this will only emerge from a full
discussion of the calculations and the comparison of all the results.

Review of Methods

This subject, particularly from the viewpoint emphasized here, has
been reviewed at several levels in recent years [5-7], so only a bare
outline of the earlier work will be given here. The systems we have
studied have been the helium atom and the alkaline earth atoms, sys-

tems with two valence electrons that can be distinguished and treated separately from any core. The ground and low-lying states of the helium atom have of course been treated very accurately; we carried out some approximate calculations meant to be efficient mimics of those calculations, by using moderately extensive Hylleraas-Kinoshita basis sets, to determine some of the low-lying states as well as the doubly-excited states that are of greatest interest for their correlation. However most of the " accurate" calculations we have done, and all the calculations used for the properties and comparisons reported here, were carried out with extensive sets of Sturmians [8,9]. The functions we shall hereinafter call our CI functions are composed of 50 to 120 products of pairs of Sturmian functions, which appear to give well-converged wave functions, at least with respect to the properties discussed here. Naturally, since these functions do not contain explicit electron-electron distances, they cannot be expected to reproduce the cusps in the wave functions where the electron-electron distance is zero. Otherwise, they are quite successful at reproducing those observables that have been measured.

Hartree-Fock (HF) functions were computed with single-configuration products of Sturmian functions, with the exponents varied as well as the coefficients in the orbitals [10]. The rotor-vibrator (RV) functions were computed in two forms: one with harmonic oscillator functions for the (assumed normal) stretching vibrations and the other with Morse local stretching functions, symmetrized and antisymmetrized. The symmetric and antisymmetric combinations of local Morse functions gave significantly better representations of the accurate functions than did the harmonic, normal-mode functions. The degenerate bending modes were treated as harmonic, and rigid rotor angular functions were used to represent the rotational motion [11]. The RV functions were determined by maximizing their overlap with the CI functions.

All the wave functions for the alkaline earth atoms were computed with effective core potentials [12]. This includes the CI, HF and RV functions. Some previous comparisons were made of oscillator strengths determined with our CI and RV functions and with HF calculations done by others, using a variety of methods. The results reported here put all three calculations on the same basis.

The calculations include, as said previously, overlaps, conditional probability distributions of the electron probability densities, and these observables: oscillator strengths, quadrupole moments (for states with total angular momentum quantum numbers of 1 or more) and expectation values $\langle p_1 \cdot p_2 \rangle / (|p_1||p_2|)$. (Distributions of this last quantity have also been computed, in preparation for two-electron ionization experiments by electron impact, but are not reported here.) We can proceed to summarize these indicators and then examine them and ask how well each model performs.

Overlaps

A set of squared overlaps is given in Table 1 for states of the alkaline earth elements and for two states of doubly-excited helium. This Table is taken from Batka and Berry [10], who in turn took the values for the CI-RV overlaps from Hunter and Berry [11]. The Table also includes the squares of the RV-HF overlaps so that one can see the similarity of the two zero-order functions.

Apart from a few isolated examples such as the $3s3p\ ^3P$ state of Mg and the $5s6s\ ^1S$ state of Sr, the S and P RV functions have large squared overlaps with their well-converged CI counterparts. The ^1D functions are not nearly as well represented. Their squared overlaps are all below 0.9. The ground states are very well represented by RV functions, with all the squared overlaps of RV and CI ground state functions greater than 0.98.

Table 1. Squares of overlaps of CI functions with HF and RV functions, and of HF with RV functions. The squared overlaps between the CI, RV and HF wavefunctions; $|\langle \Psi_{CI} | \Psi_{RV} \rangle|^2$ are from [11]. All HF states are calculated by varying the the exponential parameter ξ until the energy derivative with respect to ξ, $\frac{\partial E}{\partial \xi_n}$ was less than 5×10^{-4} except for states labelled with *; these were taken to 5×10^{-3}, and states marked **, to 6×10^{-3}, with energy in Hartrees.

Atom	Config.	Term	$\|\langle\Psi_{CI}\|\Psi_{RV}\rangle\|^2$	$\|\langle\Psi_{CI}\|\Psi_{HF}\rangle\|^2$	$\|\langle\Psi_{HF}\|\Psi_{RV}\rangle\|^2$
He	2p2p	$^3P^e$	0.9880	0.9870	0.9856
He	2p2p	$^1D^e$	0.8159	0.9204	0.6759
Be	2s2s	$^1S^e$	0.9966	0.8948	0.8957
Be	2s2p	$^3P^o$	0.9836	0.9869	0.9793
Be	2s2p	$^1P^o$	0.9108	0.9173	0.9105
Be	2s3s	$^3S^e$	0.9587	0.9719	0.9331
Be	2s3s	$^1S^e$	0.9102	0.9545	0.9213
Be	2p2p	$^1D^e$	0.8542	0.7157	0.6434
Be	2p2p	$^3P^e$	0.9856	0.9872	0.9858
Mg	3s3s	$^1S^e$	0.9973	0.9255	0.9275
Mg	3s3p	$^3P^o$	0.9505	0.9840	0.9385
Mg	3s3p	$^1P^o$	0.7970	0.9288	0.7952
Mg	3s4s	$^3S^e$	0.9648	0.9782	0.9385
Mg	3s4s	$^1S^e$	0.9241	0.9625	0.9210*
Mg	3s3d	$^1D^e$	0.6457	0.6902	0.1573**
Mg	3p3p	$^3P^e$	0.9928	0.9791	0.9783
Ca	4s4s	$^1S^e$	0.9963	0.9177	0.9216
Ca	4s4p	$^3P^o$	0.9551	0.9651	0.9221
Ca	4s3d	$^1D^e$	0.5236	0.8541	0.2624
Ca	4s4p	$^1P^o$	0.8649	0.8458	0.7198
Ca	4s5s	$^3S^e$	0.9578	0.9594	0.9232
Ca	4s5s	$^1S^e$	0.9233	0.9247	0.8919*
Ca	4p4p	$^3P^e$	0.9783	0.8784	0.8915
Sr	5s5s	$^1S^e$	0.9944	0.9244	0.9309
Sr	5s5p	$^3P^o$	0.9629	0.9267	0.8874
Sr	5s4d	$^1D^e$	0.7063	0.8324	0.3634*
Sr	5s5p	$^1P^o$	0.9163	0.6837	0.6120
Sr	5s6s	$^3S^e$	0.9684	0.7153	0.7271
Sr	5p5p	$^3P^e$	0.9119	0.5732	0.6257
Sr	5s6s	$^1S^e$	0.7668	0.5879	0.7349
Ba	6s6s	$^1S^e$	0.9896	0.9200	0.9333
Ba	6s5d	$^1D^e$	0.5365	0.8813	0.3320*
Ba	6s6p	$^3P^o$	0.9411	0.8877	0.8541
Ba	6s6p	$^1P^o$	0.8329	0.5319	0.5094
Ba	5d5d	$^3P^e$	0.7109	0.8339	0.4157
Ba	5d5d	$^1S^e$		0.7552	
Ba	6s7s	$^3S^e$	0.9727	0.9232	0.9056
Ba	6s7s	$^1S^e$	0.9362	0.6678	0.7101

The HF functions do less well in representing the ground states. The HF-CI squared overlaps of these functions are about 0.92 except for Be, which is a bit lower. Of the rotor states, the $nsnp\ {}^{3}P^{o}$ and $npnp\ {}^{1}D^{e}$ states, the ${}^{3}P$'s are about equally well represented by one as by the other, and the ${}^{1}D$'s are somewhat better represented by the Hartree-Fock functions although none of the ${}^{1}D$ functions are well fit by either approximate set, the RV or the HF. Some of the states, such as the $npnp\ {}^{3}P$ states of Be and Mg are well-fit by both models but the counterparts of these in Ca, Sr and Ba are significantly better fit by the RV model. Overall, more than half the values given in Table 1 are greater than 0.9. The score for the HF model is higher for light atoms than for heavier ones, presumably because the mean kinetic energies of the valence electrons are lower in the heavier elements, a fact reflected in the ionization potentials. Hence the valence electrons scatter each other and spoil each other's angular momenta more effectively in the heavier atoms.

Spatial Distributions of Conditional Probability

It has proven very useful to examine the distributions in configuration space of the conditional probability that if electron 1 is (momentarily) " at" distance r_1 from the nucleus, then electron 2 is at distance r_2 from the nucleus and that the vectors from the nucleus to electron 1 and to electron 2 form the angle θ_{12} between them. That is, we can compute and plot the conditional probability distribution $p(r_2, \theta_{12}; r_1 = \alpha)$. In fact it is possible to plot the full probability distribution with its three independent variables $p(r_1, r_2, \theta_{12})$ if one uses time as a surrogate for one of the three position coordinates, in an animation. Furthermore one need not use the variables r_1, r_2 and θ_{12}; it is sometimes enlightening to use

$$R = \{r_1^2 + r_2^2\}^{1/2} \tag{1}$$

and the hyperspherical angle $\arctan(r_1/r_2)$ as well as θ_{12} c.f. [11,13,14]. However the choice of r_1, r_2 and θ_{12} best suits our purposes here.

In any graphic representation, in any case, one begins with the absolute square of the full wave function $|\Psi|^2$, moves to the center-of-mass or the Radau origin to eliminate three of the nine independent coordinates and then integrates out the Euler angles to reduce the probability distribution to a function of only three independent variables.

A set of examples of conditional probability distributions for the states of the Magnesium atom are given in Figure 1. A comparable set for Be has been given previously [10] and still earlier for He [11]. It is immediately apparent that the RV distributions are far more like the CI distributions than the HF distributions are. This is particularly intriguing because even in cases of the $3s3p\ ^1P$ and $3s4s\ ^1S$ states for which the HF-CI overlaps are larger, the RV distributions are more like the CI's.

This apparent paradox is not difficult to understand when we examine how even just a little configuration mixing introduces a relatively large angular asymmetry. If only 2% of a ^1S state is composed of $npnp$ and the other 98% is $nsns$, (and we assume that the same radial functions accompany both angular components) then the angular maximum at any given values of r_1 and r_2 is over 21% higher than the corresponding minimum, 180 degrees away. If the wave function contains 10% of the $npnp$ configuration, then the maximum and minimum values at $\theta_{12} = 0$ and 180 degrees differ by over 41%. (Which is maximum and which is minimum depend of course on the relative phase of the $nsns$ and $npnp$ parts.) In Robert Mulliken's words, " A little configuration interaction goes a long way" [R. S. Mulliken, personal communication].

Oscillator Strengths

Oscillator strengths, computed in the dipole length gauge, for five kinds of optically allowed transitions; the resonance $^1S^e - ^1P^o$, the $^1P^o - ^1D^e$, the $^1P^o - ^1S^e$, the $^3P^o - ^3P^e$ and $^3P^o - ^3S^e$, are shown in Figure 2. Values are given for the RV, HF and CI representations, and for those cases determined thus far by experiment. In all cases, the CI and experimental values agree well enough to allow us to say that differences could be due to imperfections in either the the experiments or the theory. In four of the five examples, the RV values agree with the CI values better than the HF values do; only in the last set, the $^3P^o - ^3S^e$ transitions, are the HF values generally closer to the CI and experimental values. The RV and CI values were computed by Hunter and Berry [15] and the HF by Batka and Berry [10]

Quadrupole Moments

Table 2 contains three sets of quadrupole moments, determined from the RV, HF and CI functions for four states of the alkaline earths and doubly-excited He: the $nsnp$ ^3P, the $npnp$ ^1D, the $nsnp$ ^1P and the $npnp$ ^3P states. These are taken from Ceraulo and Berry [16]. The only basis for confidence in the results is the rather good agreement of the quadrupole moment of the He 1s4p 3P state computed (-210 ea_o^2) by Ceraulo and Berry [16] and measured (-289 ea_o^2) by Miller and Freund [17], and the moderately good agreement for the $1s5p$ ^3P state (-473 and -705, computed and measured, respectively) obtained by the same authors [18]. Among all the properties of atoms that reflect effects of correlation, the quadrupoles are probably the most readily amenable to measurement. We hope that the presentation of these values will stimulate some of those measurements.

Angular Distributions

The final properties we present here are obtainable in principle from either $(\gamma, 2e)$ or $(e, 3e)$ experiments in which the angular correlation of the momenta of the outgoing electrons are measured. The full distribution function for the angle between the two momenta can be derived and will be presented elsewhere. Here we give a set of values of the mean deviation of that angle, $\Delta\theta_{12}^{mom}$ from its mean value $\langle\theta_{12}^{mom}\rangle$. A set of these deviations is shown in Fig. 3 for the alkaline earth atoms. The Figure shows only the CI results.

Conclusions

It is apparent, on the basis of the properties presented here, that neither the rotor-vibrator nor the Hartree-Fock representation is overwhelmingly better than the other in all respects for all states, even for the states discussed here in which both electrons are in the same or adjacent shells. State by state, property by property, sometimes

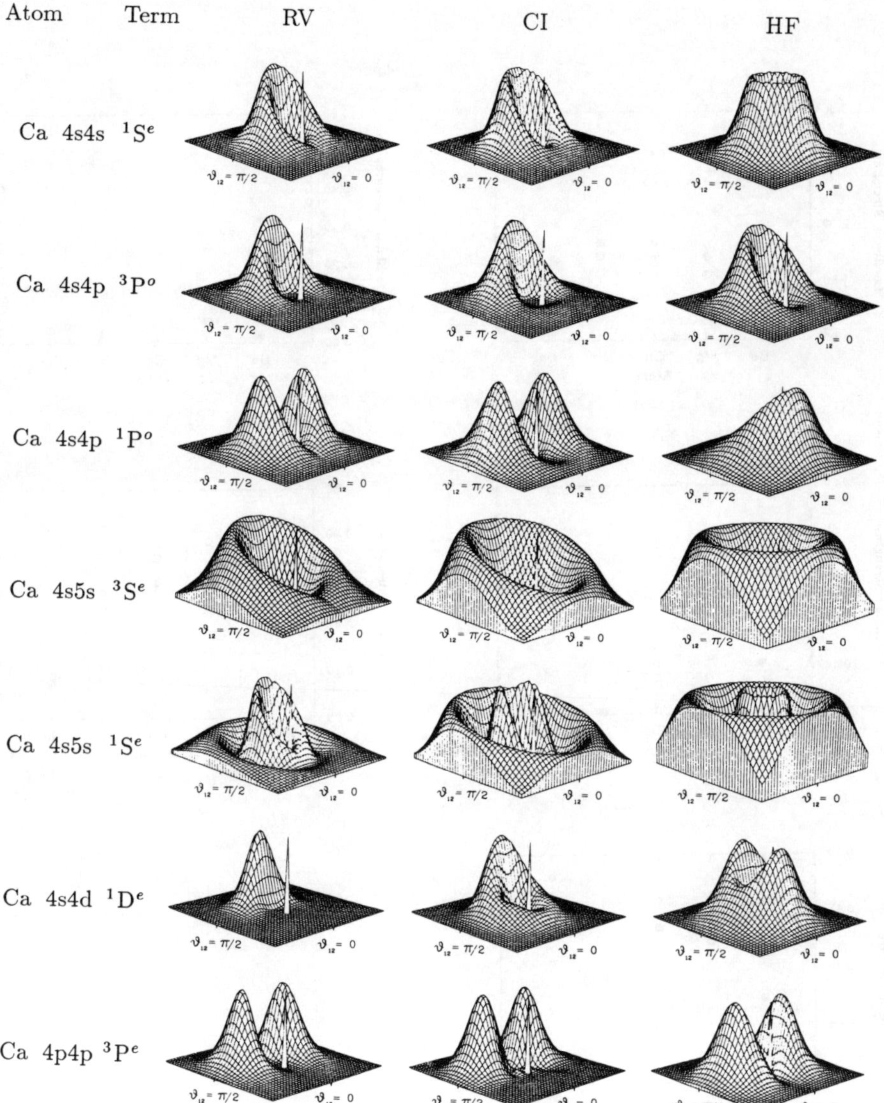

Figure 1: Plots in cylindrical coordinates of the conditional electron probability distribution for various states of Ca. All the peak heights are normalized to one. The distributions were calculated by fixing $r_1 = 2$ Bohr (the large spike present in many of the graphs), $\phi_1 = \phi_2 = 0$, and varying r_2 from 0 to 10 Bohr and θ_{12} from 0 to 2π.

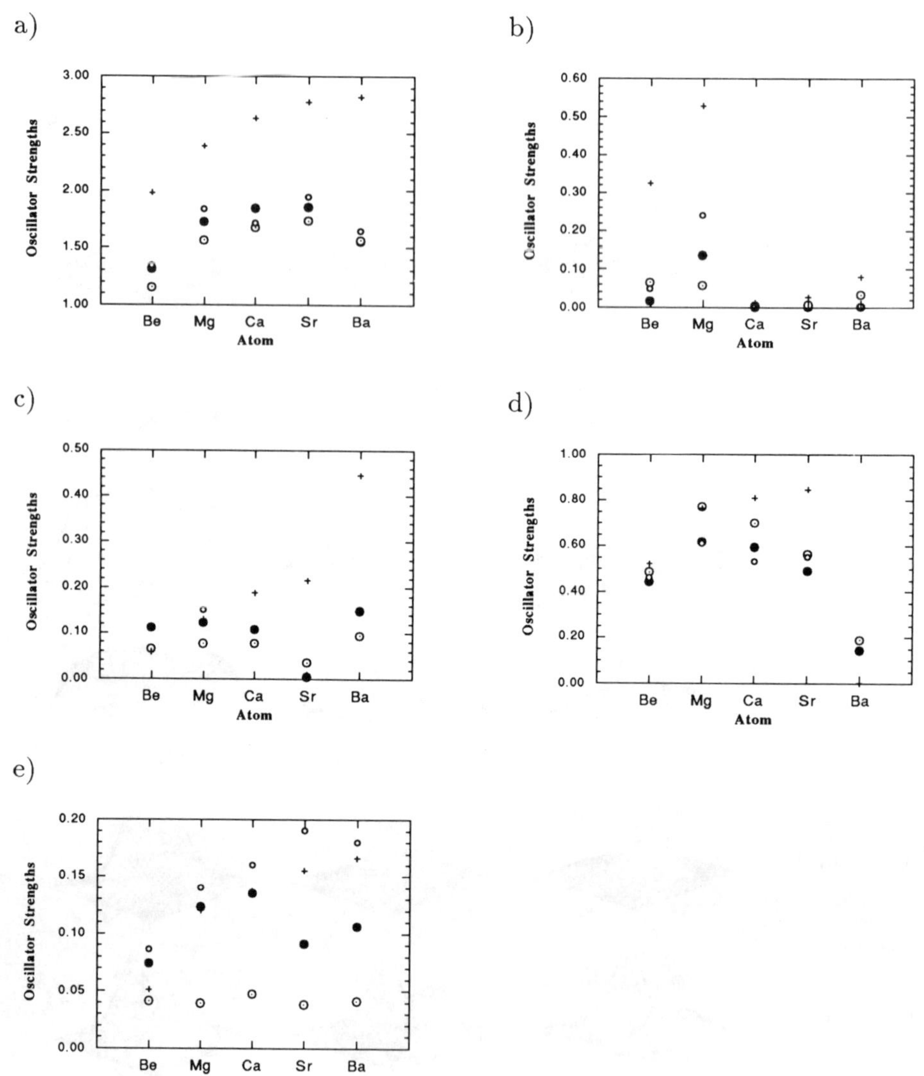

Figure 2. Graphical representation of oscillator strengths listed by transition; a) is $^1S^e \rightarrow\, ^1P^0$ transition, b) is $^1P^0 \rightarrow\, ^1D^e$ transition, c) is $^1P^0 \rightarrow\, ^1S^e$ transition, d) is $^3P^0 \rightarrow\, ^3P^e$ transition, and e) is $^3P^0 \rightarrow\, ^3S^e$ transition. A filled circle represents f_{CI}, circles with dots are for f_{RV}, plusses are f_{HF}, and open circles are f_{EXPT}.

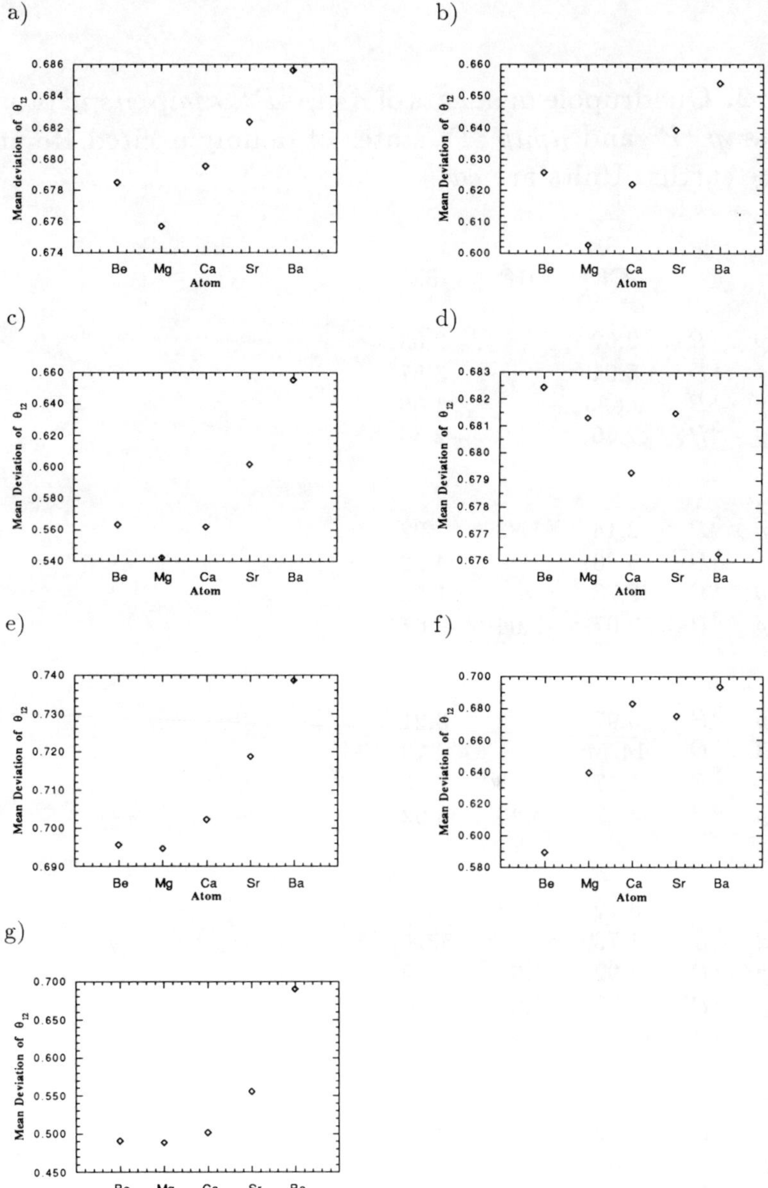

Figure 3. Graphical representations of the calculated deviations of θ_{12}^{mom} in momentum space. They are listed by state; a) is the $^1S^e$ ground, b) the $^3P^0$ rotor, c) the $1P^0$ bend, d) the $^3S^e$ antisymmetric stretch, e) the $^1S^e$ symmetric stretch, f) the $^1D^e$ rotor, g) the $^3P^e$ bend.

Table 2. Quadrupole moments of $nsnp\ ^3P^o$, $npnp$, $nsnd$ or $ns(n-1)d$ $^1D^e$, $nsnp\ ^1P^o$ and $npnp\ ^3P^e$ states of doubly excited He and of the alkaline earths. Units are ea_o^2.

		CI	HF	RV
He				
$2s2p$	^3P	2.62		3.03
$2p2p$	^1D	5.01		3.87
$2s2p$	^1P	0.693		1.66
$2p2p$	^3P	-2.36		-2.36
Be				
$2s2p$	^3P	2.14	1.83	1.92
$2p2p$	^1D	5.80		4.22
$2s2p$	^1P	2.78		1.53
$2p2p$	^3P	-2.07	-2.04	-2.07
Mg				
$3s3p$	^3P	3.97		3.21
$3s3d$	^1D	14.74		9.19
$3s3p$	^1P	5.17		2.20
$3p3p$	^3P	-3.62	-3.43	-3.62
Ca				
$4s4p$	^3P	6.12	6.63	5.32
$4s3d$	^1D	1.73		5.68
$4s4p$	^1P	4.92	6.63	1.99
$4p4p$	^3P	-5.21	-5.64	-5.24
Sr				
$5s5p$	^3P	6.42	6.08	6.01
$5s4d$	^1D	4.62		8.75
$5s5p$	^1P	2.87		1.37
$5p5p$	^3P	-4.74		-4.88
Ba				
$6s6p$	^3P	8.79		8.64
$6s5d$	^1D	2.01		9.54
$6s6p$	^1P	2.86		0.463
$5d5d$	^3P	-4.14		-4.33

one representation is better and sometimes the other. However the RV picture does, on balance, give a more reliable set of predictions–not by any means as good as those of a CI function–and surely represents the spatial distributions of angular correlations through the conditional probability distributions.

What can be the outcome of this inference? If the RV representation is indeed somewhat better, one might guess that expansion of the exact function in a (truncated) basis set of rotor-vibrator functions would converge faster than a conventional expansion in products of one-electron, independent-particle functions. An investigation of this question is in progress now.

Furthermore, one can ask whether the evaluation of approximate representations or of basis sets could be extended. Is it possible to construct a useful quantitative measure of the extent of validity of an approximate model? Can one quantify, for example, how good a pretty good quantum number is? This question has a positive answer for very simple systems [19] and is being investigated for the comparison of HF and RV representations of alkaline earth atoms. Ultimately, the goals of the pursuit of these questions are better physical insight and, as a result, more efficient paths and methods of computation and even possibly the recognition of new physical phenomena that we had previously overlooked.

Acknowledgment

This research was supported by a Grant from the National Science Foundation.

References

1. M. E. Kellman and D. Herrick, *J. Phys. B* **11**, L755 (1978).
2. M. E. Kellman and D. Herrick, *Phys. Rev. A* **22**, 1536 (1980).
3. D. Herrick and M. E. Kellman, *Phys. Rev. A* **21**, 418 (1980).
4. D. Herrick, M. E. Kellman and Poliak, *Phys. Rev. A* **22**, 1517 (1980).
5. R. S. Berry, *The Lesson of Quantum Theory*, (J. deBoer, E. Dal and O. Ulfbeck, eds., Elsevier, New York 1986), p. 241.

6. R. S. Berry and J. Krause, *Adv. Chem. Phys. LXX*, Part 1, 35 (1988).

7. R. S. Berry *Contemp. Phys.* **30**, 1 (1989).

8. J. L. Krause and R. S. Berry, *Phys. Rev. A* **31**, 3502 (1985).

9. J. L. Krause and R. S. Berry, *J. Chem. Phys.* **83**, 5153 (1985).

10. J. Batka and R. S. Berry, (1991) (submitted).

11. J. E. Hunter III and R. S. Berry, *Phys. Rev. A* **36**, 3042 (1987).

12. G. B. Bachelet, D. R. Hamann and M. Schlüter, *Phys. Rev. B* **26**, 4199 (1982).

13. C. D. Lin, *J. Phys. B* **16**, 723 (1983).

14. U. Fano, *Rep. Prog. Phys.* **46**, 97 (1983).

15. J. E. Hunter III and R. S. Berry, *Phys. Rev. Let.* **59**, 2959 (1987).

16. S. Ceraulo and R. S. Berry, *Phys. Rev. A* (1991).

17. T. E. Miller and R. S. Freund, *Phys. Rev. A* **4**, 81 (1971).

18. T. E. Miller and R. S. Freund, *Phys. Rev. A* **5**, 5188 (1972).

19. P. Braier and R. S. Berry, (submitted) (1991).

Epilogue

Having thanked our authors in the Preface, we now wish to thank our readers! Among you we hope soon to have colleagues exploring applications to other problems and developing D-scaling in new ways.

In this book, the role of D is simply that of a robust, workday parameter. However, the concept of dimension has many ramifications in physics as well as mathematics. For interested readers, we cite here a few intriguing discussions [1-5], linked to a vast literature. The question why the physical world should have $D = 3$ has been considered by Aristotle, Ptolemy, Gilbert, Kepler, Kant and, in our own century, Poincaré, Ehrenfest, and Weyl among many others. Thereby the unique properties of three-dimensional space have been shown to be requisite for many phenomena, particularly the stability of matter and the existence of stable, periodic planetary orbits, capable of permitting life to evolve over millions of years.

Yet, currently there is much interest in field theories in many dimensions, which picture our physical world as resulting from "compactification" of a higher dimensional space [5]. Also, atoms and molecules can be reduced to quasi-one-dimensional structures by the huge magnetic field of a neutron star or pulsar [6]. Without resort to such exotic realms, good approximations to $D \geq 3$ atoms exist. The $D = 5$ case, and others of higher odd-D, actually occur as doubly-excited states of two-electron atoms, as described in several portions of this book. The $D = 2$ case is exemplifed, *e.g.*, in impurity states in quantum well systems [7], of practical interest for electronic devices.

–The Editors

Horatio: O day and night, but this is wondrous strange!

Hamlet: And therefore as a stranger give it welcome.
There are more things in heaven and earth, Horatio,
Than are dreamt of in your philosophy.

(Act I, Scene v)

D. R. Herschbach et al. (eds.), Dimensional Scaling in Chemical Physics, 499–500.
© 1993 *Kluwer Academic Publishers. Printed in the Netherlands.*

References for the Epilogue

1. B.B. Mandelbrot, *The Fractal Geometry of Nature* (W.H. Freeman, San Francisco, 1982).
2. F.R. Tangherlini, Nuovo Cimento **27**, 636 (1963); **91B**, 209 (1985); **94B**, 204 (1986).
3. W. Büchel and I.M. Freeman, Am. J. Phys. **37**, 1222 (1969).
4. J.D. Barrow, Phil. Trans. Royal Soc. (London) A **310**, 337 (1983).
5. A.R.P. Rau, Am. J. Phys. **53**, 1183 (1985).
6. J.R. Irvine, *Neutron Stars* (Clarendon Press, Oxford, 1978), pp. 66-80.
7. T. Ando, A.B. Fowler, and F. Stern, Rev. Mod. Phys. **54**, 437 (1982).

Index